Practical ESD Protection Design

Practical ESD Protection Design

Albert Wang
University of California

IEEE PRESS
WILEY

Published by John Wiley & Sons, Inc., Hoboken, New Jersey.
Published simultaneously in Canada.

Library of Congress Cataloging-in-Publication Data is applied for
Hardback: 9781119850403

Cover Design: Wiley
Cover Image: © Albert Wang

MIX
Paper from responsible sources
FSC
www.fsc.org
FSC® C013604

Contents

Author Biography

Professor Albert Wang
Dept. of Electrical and Computer Engineering
University of California, Riverside, USA

After working in the Silicon Valley as an IC design engineer, Wang joined the academia and, currently, is a professor of Electrical and Computer Engineering at the University of California, Riverside, California, USA. His research covers electron devices, analog/mixed-signal (AMX) and RF ICs, design-for-reliability, 3D heterogeneous integration, emerging nanodevices and circuits, and visible light communication systems. His research records include two books, about 300+ peer-reviewed papers, and 16 US patents. His editorial board services include *IEEE Journal of Electron Devices Society*, IEEE *Transactions on Circuits and Systems I*, *IEEE Electron Device Letters*, *IEEE Transactions on Circuits and Systems II*, *IEEE Transactions on Electron Devices*, *IEEE Journal of Solid-State Circuits*, *IEEE Transactions on Device and Materials Reliability*, and *Journal of Engineering*. He has been *IEEE Distinguished Lecturer for IEEE Electron Devices Society*, *IEEE Circuits and Systems Society*, and *IEEE Solid-State Circuits Society*. He was president of *IEEE Electron Devices Society*. He served as a Program Director at the National Science Foundation, USA. He is a Fellow of National Academy of Inventors, USA, and an IEEE Fellow.

Preface

The birth of Germanium (Ge) transistor in 1947, followed by the inventions of integrated circuits (ICs) in Ge and Silicon (Si) in 1958 and 1959, respectively, and particularly the advent of Si complementary metal-oxide-semiconductor (CMOS) IC technology in 1963, led to the prosperity of semiconductor microelectronics, which have forever changed the human society. Semiconductors and ICs laid the foundation of the information technology (IT) era where everything relies on ICs, from internet to internet of everything (IoET), from smartphones to tablets, and from autonomous driving to artificial intelligence (AI). ICs are truly everyday commodity now. Like other products, Performance and Reliability are the two cornerstone attributes of ICs. Electrostatic discharge (ESD) is a daily phenomenon that can cause damages to ICs. ESD failure is a major IC reliability problem ever since IC was born. On-chip ESD protection is therefore required for all ICs, as well as all electronics products. For the microelectronics industry, ICs are sold for monies due to the performance, however, ICs could not be sold without adequate on-chip ESD protection. For consumers, IC performance makes your smartphone powerful and enjoyable, however, one would kill a touch phone instantly if no ESD protection provided for IC chips. Generally, as IC technologies continue to advance into the nanometer domain, while IC performance and complexity increase rapidly, on-chip ESD protection design is becoming extremely more challenging today. How to conduct practical ESD protection designs for ICs is the scope of this book. Developed as a professional reference for IC design engineers and a textbook for upper-level students majoring in microelectronics, this book teaches both fundamentals and practical skills of on-chip ESD protection designs. All design examples discussed in this book are outcomes of our research. The author is therefore very grateful to his graduate students who contributed to the relevant research.

University of California
June 2021

Albert Wang

1

Why ESD?

1.1 A Historical Perspective

Sure, this book means to discuss about electrostatic discharge (ESD) protection designs for integrated circuits (ICs). Yet nothing would prohibit our free minds from wandering around a bit before the show starts.

Imagine a world without electricity? The answer is no-brainer – impossible and scary!

To solve any problem, it is important to know where does it come from. Human curiosity in electricity dates back to our ancient ancestor era. Let us take a flash cyber-trip traveling through the time tunnel back to the ancient past. Around 600 BC, the Greek scientific philosopher, Thales of Miletus, discovered that a piece of amber rubbed with fur can magically attract light things, such as leaves, ash, or dust, because the amber was considered to have gained static *charges*, or, *electrostatic charge* as we call it these days. This *amber effect* is what is called *static electricity* today. Indeed, the loveliness of amber goes well beyond its natural beauty as shown in Figure 1.1. In fact, the English word *electron* came from the Greek word *ēlektron* for amber. Later, William Gilbert conducted serious studies on the attraction associated with rubbing materials, such as amber, and named it *electric* attraction, which led to the publication of DeMagnete in 1600 [1]. The word *electricity* was derived subsequently. The two types of electricity, i.e., vitreous (glass) and resinous (amber) was documented by Charles François de Cisternay DuFay in 1733 [2]. In 1751, Benjamin Franklin gave the terms of "positive" and "negative" for the two types of electrostatic charges in his publication of "Experiments and Observations on Electricity" [3], albeit a reverse definition might have made the life of college freshman students a little easier in understanding the current flowing direction versus the electron flowing direction. Obviously, when Franklin enjoyed his leisure time flying a kite, as shown in Figure 1.2, his brain never stopped roaming in the scientific wonderland. During 1800s, it became evident that electric charge may not be further divided, and Johnstone Stoney gave it the name "electron." Later, Joseph John Thomson's experiments led to the conclusion of the existence of light particles carrying negative charge, and the word "electron" was used for it in 1897. The hard-working Thomson (Figure 1.3) was rewarded with the Nobel Prize in Physics in 1906 "in recognition of the great merits of his theoretical and experimental investigations on the conduction of electricity by gases."

The magic charges have brought in unlimited amount of amazing applications, which have dramatically changed our life today. In a conductor, electric charges can move freely to form the electric current, which fire up an electric lightbulb and light up our life, mostly thanks to Humphry Davy who invented the first electric light in 1809 and Thomas Edison who demonstrated the first carbon filament bulb in 1897, as shown in Figure 1.4. Equally amazing and important are the electric charges at rest, i.e., static charges or electrostatic charges, which when discharge through an air

Practical ESD Protection Design, First Edition. Albert Wang.

Figure 1.1 Elektron in Greek word means Amber that is beautiful both aesthetically and scientifically. (Courtesy of Custom Crystal.)

Figure 1.2 Benjamin Franklin's thought about electricity: (1) the lightning is electricity; (2) the two types of electrical charge are positive (vitreous) and negative (resinous). (Courtesy of Chris Wang.)

Figure 1.3 J. J. Thomson discovered the electron in experiment to study "cathode rays" in 1897. He found that cathode rays consist of charged particles (electrons) that can conduct electricity. (Courtesy of Cavendish Laboratory, University of Cambridge.)

gap, being one version of the ESD phenomena, will produce electromagnetic waves that eventually led to radio-frequency (RF) wireless communications. The ESD-based radio waves were first observed by Heinrich Rudolf Hertz in experiments conducted during 1886–1889 using amazingly simple spark-gap radio transmitter as shown in Figure 1.5 [4]. At the time, Hertz did not realize the importance of his spark-gap radio wave experiments. In his own words, Hertz said *it's of no use whatsoever* and describes his work was *just an experiment that proves Maestro Maxwell was right*. As Hertz stated, *we just have these mysterious electromagnetic waves that we cannot see with the*

Figure 1.4 Thomas Edison debuted the first incandescent lightbulb on 21 October 1897, which burned for about 13 hours. (Courtesy of ThoughtCo.)

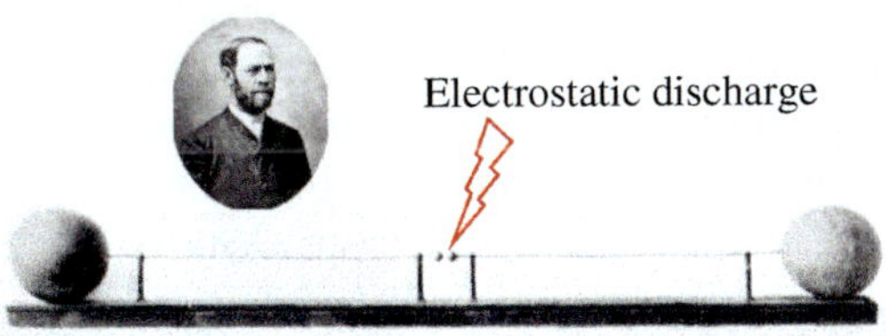

Figure 1.5 Hertz's first spark-gap radio transmitter, which can be called an ESD discharging device, is a capacitor-type dipole resonator comprising a pair of 1-m copper wires separated by a spark gap of about 7.5 mm. When a high voltage is applied through an inductor coil, the spark gap fires up the air and generates standing waves of roughly 50 MHz. (Heinrich Rudolf Hertz/Wikipedia Commons/Public domain.)

naked eye, but they are there, his experiments proved that the airborne electromagnetic waves, initially called Hertzian waves and later named radio waves, exist as predicted by Maxwell.

1.2 ESD and the Dangers

While Hertz's spark-gap transmitter may be regarded as an original ESD discharging device that is *useful*, the ESD phenomena that we are concerned about today are more *harmful* than *useful*, which is the topic of this book.

Electric charge is a fundamental physical property of a matter, which makes it feel a force in an electromagnetic field. There are two types of electric charges. Per Franklin's convention, the electric charges gained by a glass rod rubbing a silk cloth are positive charges (vitreous), while the electric charges obtained by a piece of amber rubbed by a piece of fur are negative charges (resinous). Charge is quantized, meaning a charge carrier can only contain integer number of elementary charges. The SI unit for charge is Coulomb. An elementary charge (denoted as e) is indivisible. Elementary charge is a fundamental physical constant given as $e = 1.602\,176\,634 \times 10^{-19}$ C exactly [5]. One electron has one charge of $-e$.

Electric charge experiences a *force* through an electrostatic field. In modern physics, there exist four fundamental forces: weak force, strong force, electromagnetic force, and gravitational force. Weak force and strong force apply in "short" distance (microscale), while electromagnetic and gravitational forces act in "long" distance (macroscale). These fundamental forces, also referred to as fundamental interactions, can be mathematically described as a field. Electromagnetic force has two components: the electrostatic force that applies to electrically charged particles at rest, and the combined electric and magnetic forces that act on charged particles in motion. Due to the electrostatic force, *like* charges repel each other, while *unlike* charges attract each other. The electrostatic force acting on charges follows the Coulomb's Law as depicted below

$$F = \kappa \frac{Q_1 Q_2}{r} \tag{1.1}$$

where F is the Coulomb's Force, Q_1 and Q_2 are the amount of electrostatic charges contained by the two charge-carrying objects involved, r is the distance in between, and κ is a constant. Therefore, it requires certain amount of energy, defined as *work*, to move a charge in an electric field, which is characterized by the electrostatic potential at related points in the electric field.

Electrostatic charges are created when two objects, at least one of them has to be an electrical insulator or of high electrical resistivity in nature, are brought into contact and then separate from each other. Static electricity reflects an imbalance of electric charges (net positive or negative) inside a charged object. In physics, an isolated system follows the Law of Conservation of Charge, which states that the net charges, the difference of positive and negative charges, are preserved in the universe. Charge conservation does not prohibit static charges from being generated or destroyed. But any gain in charges somewhere at a time will accompany with the loss of the same amount of charges somewhere else. Static charges are associated with electric *charging* and *discharging* for an object. Electric charging puts static charges into an object, while electric discharging removes static charges from an object. Electric charging and discharging are two opposite phenomena associated with static charges, which involve separation and neutralization of positive and negative charges of materials. Electric charging leads to static electricity. Static charge generation requires separation of positive and negative charges through electric charging procedure. Normally, materials are electrically neutral because the atoms have same number of positive and negative charges. When two objects are in contact, electrons may move in between, which causes imbalance of positive or negative charges within each object. Then, when the two objects are separated thereafter, they may retain the charge imbalance, i.e., containing net positive or negative charges. This completes a charging procedure. Therefore, the static electricity phenomena involve contact and separation of materials. There are many electric charging mechanisms. The most common electrostatic charging phenomena observed in our daily life is the *triboelectricity* phenomena, which follow a contact-induced charge separation mechanism. When two objects with different electrical resistivity are in contact, electrons will exchange in between due to different binding force. Upon separation of the two objects, each object will be electrically charged containing either net positive or negative charges. Many magic and fun static electricity phenomena observed in our life follow the contact-induced charge separation procedure. For example, amber rubbed by fur can attract leaves; a plastic comb combing through hair can attract paper scraps; or you feel your hair raising when taking off a hat in a dry day. The triboelectric effect is considered to be related to the materials adhesion phenomenon and dominated by the atomic-scale electron transfer mechanism [6]. Triboelectric effect is generally unpredictable and depends heavily on many factors, such as materials, surfaces, temperature, pressure, and humidity. The Triboelectric series, as given in Table 1.1, is a reference for the tendency of contact-induced electrostatic charge generation based upon the materials properties. Different materials are friendly to either positive or negative charges at varying levels. The farther apart the two involved materials in the Triboelectric series table, the stronger the triboelectric effect, i.e., the easier the two materials will exchange charges. Materials very close to each other may not exchange charges, i.e., triboelectrification may not occur. Figure 1.6 presents a quantified Triboelectric series [7].

Charge-induced charge separation is another electrostatic charging phenomenon commonly observed also referred to as *electrostatic induction*. In an electrically neutral materials, it has equal

Table 1.1 Triboelectric series.

Materials	Most Positive (+)
Air	
Human hands. skin	
Asbestos	
Rabbit fur	
Glass	
Human hair	
Mica	
Nylon	
Wool	
Lead	
Cat fur	
Silk	
Aluminum	
Paper	
Cotton.	
Steel	0
Wood	
Lucite	
Sealing wax	
Amber	
Rubber balloon	
Hard rubber	
Mylar	
Nickel	
Copper	
Silver	
UV resist	
Brass	
Synthetic rubber	
Gold, platinum	
Sulfur	
Acetate, rayon	
Polyester	
Celluloid	
Polystyrene	
Orion, acrylic	
Cellophane tape	
Polyvinylidene chloride (Saran)	
Polyurethane	
Polyethylene	
Polypropylene	
Polyvinylchloride (Vinyl)	
Kel-F (PCTFE)	
Silicon	
Teflon	
Silicone rubber	Most negative (−)

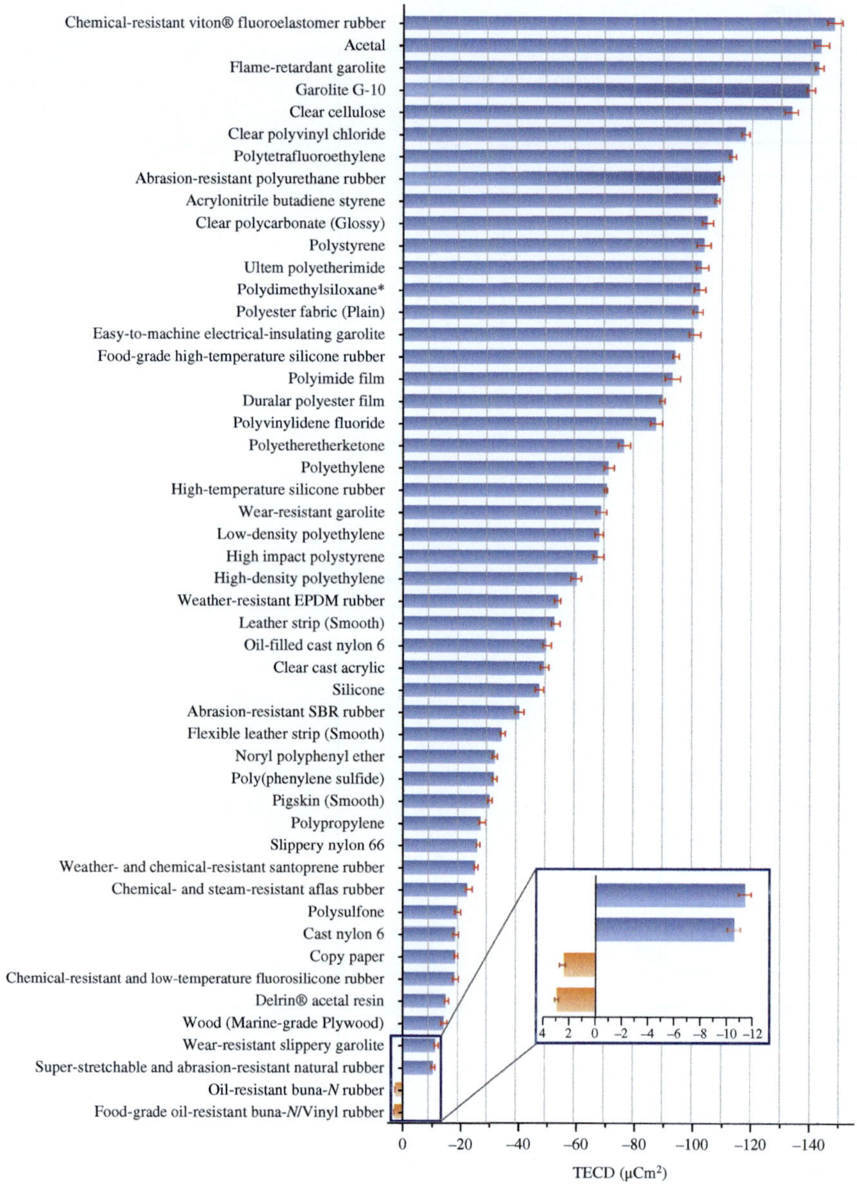

Figure 1.6 The quantified triboelectric series. The error bar indicates the range within a standard deviation. (Zou et al. [7]. Licensed under CC BY 4.0.)

amount of positive and negative charges in it with the electrons carry the negative charges and atomic nuclei hold the positive charges. Since the positive and negative charges are very close to each other in the microatomic scale, the neutral materials do not have locally lumped net charges, therefore, people do not get electric shock when touching an uncharged object in our daily life. In a conductor, like metal, electrons are "free" to move around, while in an insulator, such as amber, electrons are "bound" locally, so they cannot move freely. This makes electrostatic charging very different for conductors and nonconductors. In the scenario involving conductive materials, such as metal rods, if a charged object A, assuming positive, is placed near an electrically neutral conductor B, for example a copper rod without net charge, the electric field generated by the positive charges of A will act on the free electrons in B per the Coulomb's Law. Some free electrons in object B will be attracted to the end closer to object A, effectively leaving positive charges on the far end of the copper rod. Hence, redistribution of electric charges inside the object B occurs. If the copper rod is then grounded, the positive charges on the far end of the copper rod will flow into the ground, leaving the object B negatively charged with net electrons. In the scenario involving nonconductive materials, i.e., dielectric objects such as glass, amber, or rubber, the charge separation procedure is different. Assume a positive-charged object A is placed close to an electrically neutral glass rod B, the electric field produced by the positive charges of A will be experienced by the charges in object B. However, since the electrons inside the dielectric object are not free and bound to atoms or molecules locally, the electrons cannot move to the end close to the object A. Instead, in microscale of a molecule, electrons are attracted closer to the object A, while the positive nuclei are repelled to farther from the object A. This is called *polarization*, which results in electrical *dipoles* at atomic or molecular level. Hence, charge separation or dielectric polarization occurs for the glass rod as governed by the Coulomb's Law. Many everyday electrostatic magics can be explained by the dielectric polarization effect. For example, moving a silk-rubbed glass bar to close to small paper scraps, polarization occurs for the paper scraps where tons of molecular dipoles are created within the paper scraps with the bound electrons being closer to the glass bar. Collectively, the electrostatic force from the positive-charged glass bar will be able to attract these small lightweight paper scraps per the Coulomb's Law. Other induction-based charge separation mechanisms are pressure induction and heat induction. In pressure-induced charge separation, the mechanical stress applied to certain materials will separate positive and negative charges of the materials, which is referred to as piezoelectric effect or *piezoelectricity*. Piezoelectric effect exists in certain crystals and ceramics, such as quartz and zinc oxide (ZnO). Heat-induced charge separation is also called pyroelectric effect or *pyroelectricity*. For certain materials, temperature variation, being heated or cooled, will cause polarization of charges in microscale. The widely used pyroelectric materials are gallium nitride (GaN).

The opposite of static charge generation is the removal (or, *neutralization*) of static charge. Correspondingly, there are electrostatic charging (or static charging) and electrostatic discharging (or static discharging) procedures. With typical electrostatic charging phenomena, through charge separation, explained, let us understand the opposite phenomena – electrostatic discharging, or broadly called ESD. *ESD is a charge neutralization procedure.* In principle, when two objects with different electrostatic potentials are brought into close proximity, either in direct contact or having a small gap in between, transfer of electrostatic charges between the two objects occurs. This process is broadly called *ESD*. In general, ESD phenomenon is a sudden flow of electric current with a short duration between two differently charged objects. Triggering of ESD discharging may be direct contact of the two objects or dielectric breakdown in between in case of a small gap (air or other dielectrics) separating the two objects. The most common and dangerous everyday ESD discharging phenomenon is *lightning*. In natural atmosphere, electrostatic charges can accumulate in thunderstorm clouds, which may be neutralized between two clouds of different electrostatic potentials, or, from a cloud to the ground, resulting in a sudden transfer of static charges with an instantaneous release of energy of gigajoule scale. The air breakdown voltage causing lightning is around 10 000 V/cm (10 kV/cm) [8]. The lightning will zap-heat the air leading to light emission through incandescence, producing a flash, and shock waves of radiation, generating thunders. More commonly and often the fun side, an ESD phenomenon is accompanied by a spectacular spark, which is triggered by abrupt air breakdown at an electric field density exceeding around 4–30 kV/cm, making the air electrical conducting in a sudden. However, most ESD phenomena are invisible to human eyes. While everyday ESD phenomena may typically cause discomfort to human only, such as feeling an electric shock when opening a car door, ESD discharging can pose serious dangers to the industry, causing life hazards in certain environments. For example, movement of fine powders (e.g., granulated grain in a grain silo) and dust clouds in manufacturing plants or flowing of flammable liquids (e.g., gasoline and crude oils) in pipelines can accumulate electrostatic charges, which may be ignited by a tiny ESD spark, causing explosion.

On the other hand, ESD phenomena are extremely harmful to electronic components, particularly for semiconductor ICs [9], which is the concern of this book. The ESD failure problem became a real concern since World War II when highly insulating polymeric materials started to find widespread usage. Electrostatic charges can be easily generated and accumulated in those insulating materials and ESD discharge may shut down machinery in manufacturing plants, causing ESD damages and losses. Yet it was the birth and prosperity of the modern semiconductor microelectronics industry that made people realize the seriousness of ESD failures to the microelectronics industry, the economy, and the society. The invisible ESD phenomena, with a transient level lower than 1000 V, or even as low as 10 V, can cause ESD failures to modern electronics. In the past 70+ years, the microelectronics industry has gained unprecedented successes, which has entirely changed the human life. Germanium (Ge) transistor was invented at the Bell Lab in 1947 [10], which broke the dawn of the microelectronics era. ICs in Germanium and Silicon (Si) were invented in 1958 and 1959, respectively [11, 12]. Si complementary metal-oxide-semiconductor (CMOS) IC technology was invented in 1963 [13]. For more than seven decades, semiconductor IC technologies have been continuously advancing at the pace of the Moore's Law [14]. *Performance* and *Reliability* are the two key aspects for ICs. ESD failure is a main part of IC reliability problems. Every piece of IC may be subjected to ESD failure during its life cycle. As the IC technology reaches to 5 nm node for mass production today, ESD failure is becoming much more serious to the micro/nanoelectronics industry. This is due to the fact that the natural ESD phenomena remain the same as time flies, regardless worse global warming, better ESD awareness, and stricter ESD control measures; while CMOS technology scaling makes tiny ICs more susceptible to ESD transients of the same scale or even lower level. In other word, the ESD phenomena never shrank as the IC technologies have been going through for decades. In fact, as consumer electronics become dominant today, such as smartphones, tablets, and wearable devices, the ESD risks are getting much higher simply because people are touching tiny electronic devices every second. It is obvious that the ESD failure is becoming much bigger an IC reliability concern and challenge today than ever before. Industrial data suggest that up to 30% of all IC field failures are somewhat associated with ESD events of all kinds, which costs the microelectronics industry billions of US dollars in revenue losses annually. ESD failures can be either catastrophic (*hard* ESD failure) or latent (*soft* ESD failure) in nature. Hard ESD failures cause outright damages to ICs, showing immediate IC malfunction or burn out. Soft ESD failures may cause undetectable deterioration of IC performance and future malfunction within the life cycle, i.e., a lifetime issue. In principle, ESD failures are attributed to either high energy associated with large ESD current transients, which causes thermal damages to semiconductors and metal interconnects, or high electric field density associated with large ESD voltage transients, which may breakdown the IC materials, such as a CMOS gate oxide. Therefore, regardless of the origins and phenomena of ESD failures, effective protective measures are needed to protect ICs against ESD failures, which is the topic of this book.

1.3 ESD Protection: The Principles

Obviously, the ESD dangers to semiconductors and ICs are anywhere anytime during the entire lifespan of any microelectronics product, including manufacturing, packaging, shipping, installment to field applications. The ESD risks are unavoidable to ICs because, first of all, electrostatic charge generation occurs any moment during the processes of handling ICs. For example, in typical microelectronics manufacturing settings, as an engineer walks across a floor, triboelectric generation can readily produce static electricity reaching to anywhere from 800 V on a rubber floor to 35 000 V on a carpet. Therefore, it is a no-brainer that one can never decorate an electronic workshop with any fancy carpet as in your bedroom. Sliding electronic devices out of a container can generate triboelectricity easily from 2000 V (plastic tube) to 14 500 V (foam). Relative humidity has strong impact on triboelectricity generation. Lower relative humidity increases triboelectricity. Higher humidity reduces static charging because a thin moisture film on a surface helps to dissipate static charges. Humidity control is therefore important for ESD prevention. Table 1.2 summarizes typical triboelectric charge generation in common workplaces.

Practically, it is impossible to imagine a life without any static electricity. The ESD experience, good or bad, is really an integral part of our everyday life. Therefore, the ESD consequences should never be overlooked. In the industry, electronic devices that are sensitive to ESD failure are called *electrostatic discharge sensitive* (*ESDS*) *device*, often being referred to as *ESDS* devices. The survivability of ESDS devices against ESD failures is a property referred to as the *ESD sensitivity* in the field. Since ESD phenomena always exist, which can damage electronic products including ICs, proper ESD protective measures must be used for electronics, called *ESD protection*. In the real world, the counter-ESD measures can be roughly classified as ESD *Prevention* and ESD *Protection*. The rationale for ESD prevention is to eliminate, more practically, to minimize, as much as possible, electrostatic charge generation, i.e., preventing static electricity from occurring in the first place. ESD prevention includes several measures. First, different materials have different sensitivity to the charge separation and generation processes. Certain materials are extremely sensitive to the triboelectric effect. Hence, it is a good idea to select the materials that are insensitive to static electricity, e.g., per the Triboelectric Series, to make the products "immune" to static electricity. Such a principle is fundamental because it would prevent generation of electrostatic charges, hence, no further ESD issues to worry about. However, this is practically not much of a solution because choice of ESD-insensitive materials for given products and applications are not much at all in the first place. Materials selection may be easier for certain applications, for example better using anti-static floor instead of carpet in a workplace. However, for the semiconductor industry, silicon is the main materials to use and we have to deal with it, anyway, in terms of static charge generation. Second, proper ESD control measures should be established in workplaces and at user ends to eliminate, or more practically, to reduce static charge generation and accumulation in products. A good *ESD Control* (*Static Control*) program has many principles and measures to follow. *Grounding* and *Neutralization* are important concepts for ESD Control. Grounding means to connect all concerned items, including devices, equipment, and personnel, to the same electrical potential level within a working or usage space. Equalizing electrical potential prevents static charge generation and accumulation. Neutralization is a process where excess positive or negative static charges on an object

Table 1.2 Common triboelectric generation at varying humidity related to ICs.

Occurrences	Electrostatic potential at relative humidity (R.H.)	
	10%	55%
Walking across rubber floor	800 V	200 V
Removing DIP devices from plastics	2 000 V	400 V
Moving on bench	6 000 V	400 V
Removing DIP devices from vinyl tray	11 500 V	2 000 V
Walking across vinyl floor	12 000 V	3 000 V
Removing DIP devices from polystyrene foam	14 500 V	3 500 V
Removing PWB parts from bubble wrap	26 000 V	5 500 V
Walking across carpet	35 000 V	7 500 V

DIP, dual-in-line package; PWB, printed wire board.

will be removed through cancellation with the same amount of static charges of opposite polarity. Many specific ESD control means have been widely used in the industry. Wrist and/or ankle straps are commonly used personnel grounding tools, which remove excess static charges from a human body before touching an ESDS item, hence avoiding ESD failures to ESDS devices. ESD protective flooring is extremely critical to preventing static-charging electronic products, such as in IC foundry cleanroom facilities, testing lines, and packaging and assembly plants. In workplaces, an *ESD protected area* (EPA) must be established, which can be a small workstation or workspace, or a large manufacturing floor, where ESD grounding is properly set. EPA area uses static-insensitive materials to prevent static charge generation, and grounded conductive and dissipative materials to avoid static charge accumulation. Figure 1.7 depicts a typical EPA-enabled workstation where everything is connected to the *ESD Common Point Ground*, which is connected to the system/universal ground [15, 16]. Another important ESD control measure is to promote ESD awareness by widely using *ESD Awareness Symbols*. Figure 1.8 shows three commonly used ESD awareness symbols suggested by the EOS/ESD association [16, 17]. The ESD common point ground symbol (Figure 1.8a) indicates the grounding point to ground everything in an EPA area to ensure equal electrical potential, hence, preventing static charge generation. The ESD susceptibility symbol (Figure 1.8b) is used to clearly identify an ESDS item, which must be handled with ESD caution. This symbol basically says "ESD sensitive, do not touch!" The ESD protective symbol (Figure 1.8c) is used to indicate that proper ESD prevention and protection measures are provided to workplaces, tools, and devices. The basic philosophy for ESD prevention may be understood as follows: No Charges – No Discharge. Nevertheless, while various ESD prevention methods can significantly reduce the potential of ESD problems, ESD prevention itself cannot completely eliminate all ESD dangers. ESD protection is therefore required for all electronics. Broadly, ESD protection for anything, not limited to ICs, has a long history and many formats. Back to the fifteenth century, military entities in Europe started to use various ESD protection methods to safely handle munitions. The lightning rod invented by

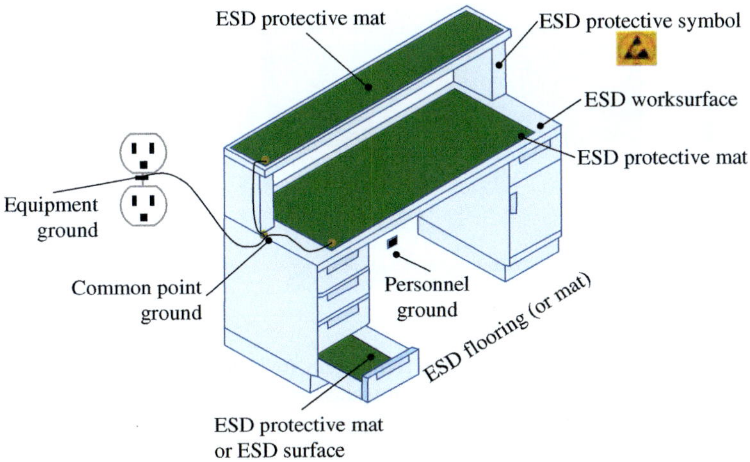

Figure 1.7 Illustration of a typical ESD workstation providing an EPA area. (Reprinted with permission from EOS/ESD Association, Inc.; www.esda.org.)

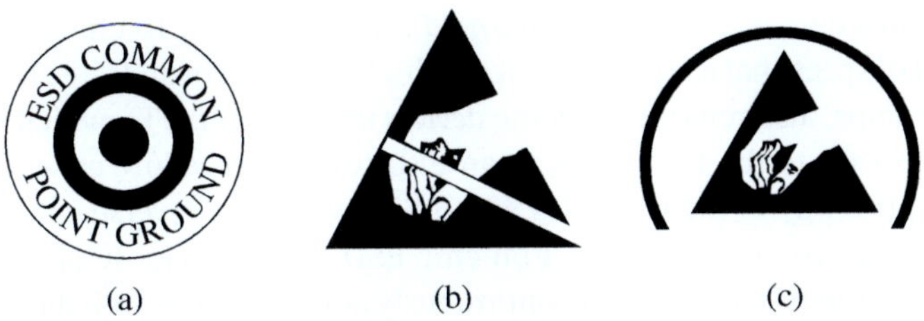

Figure 1.8 Commonly used ESD awareness symbols: (a) ESD common point ground symbol, (b) ESD susceptibility symbol, and (c) ESD protective symbol. (Reprinted with permission from EOS/ESD Association, Inc.; www.esda.org.)

Franklin is still in use for all buildings and high-rises today (Figure 1.9). Precipitation static dischargers are installed on the trailing edges of aircraft to safely discharge the accumulated static electricity into surrounding air as an aircraft flies through rain or snow. Dryer sheets are commonly used in a cloth dryer to prevent and dissipate static charges. For electronics, ESD protection can be at materials, IC, packaging, printed circuit board (PCB), and system levels often in a combination to guarantee ESD safety as much as possible. For example, ESD dissipative materials, high-resistive conducting materials, may be applied to IC packaging to slowly and safely dissipate any static charges accumulated without causing charged device model (CDM) type ESD failures. Standalone ESD protection devices are widely used at system board level for electronic products, e.g., smartphones, which are called *transient voltage suppressors* (TVS). Nevertheless, on-chip ESD protection is always required for ICs, as long as not prohibited by extreme IC performance specifications (Specs). Obviously, one cannot add a lightning rod on an IC chip for ESD protection. ESD phenomena relevant to semiconductors and ICs are unique in that the ESD transients are extremely

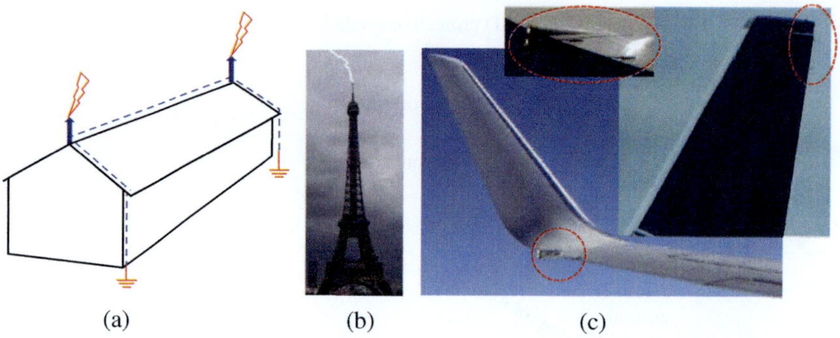

Figure 1.9 ESD protection in real life: (a) a house has a lightning rod system, (b) Eiffel Tower has a lightning rod on the top, and (c) static dischargers (static wicks) are installed at trailing edges of commercial aircraft to protect onboard electronics.

short, typically about 150 ns or less, yet they can easily generate transient current and voltage surges up to a few tens of amperes (A) and kilovolts (kV). Therefore, on-chip ESD protection designs for ICs are unique and challenging. The ESD protection performance, also called *ESD robustness*, for ESD-protected chips is typically evaluated by the *ESD failure threshold voltage* of ICs, normally in units of volts or kilovolts (kV) and referred to as *ESD failure threshold voltage* (*ESDV*). To characterize ICs for ESD robustness, i.e., ESDV specs that are given in the product datasheet of an IC, ESD measurements are conducted for the chips, also referred to as the *devices under test* (DUT), which involves applying ESD transient waveforms to the DUT devices, an ESD testing procedure called *ESD zapping*. From the early days all the way to today, diodes, in both forward and reverse conducting fashions (i.e., Zener diodes), have been widely used for on-chip ESD protection for ICs [9]. Over the past seven decades, as semiconductor technologies continuously advance and IC reliability requirements constantly increase, various ESD protection structures have been developed, such as bipolar junction transistor (BJT) ESD protection devices, metal-oxide-semiconductor field-effect transistor (MOSFET) ESD protection devices and silicon controlled rectifier (SCR) ESD protection structures, and their derivatives of all kinds. Various ESD sub-circuits have also been developed for advanced ESD protection for mixed-signal ICs and system-on-a-chip (SoC) chips. As IC technologies rapidly advance from micrometer nodes to nanometer nodes, novel ESD protection structures have been developed to address the emerging design challenges, such as very large-scale integration (VLSI) chips, high-speed and high-throughput mixed-signal ICs, and high-frequency and broadband RF ICs. Notable design examples include dual-directional and multiple-mode SCR ESD protection structures, diode-triggered silicon controlled rectifier (DTSCR) ESD protection structures, and ultralow parasitic ESD protection structures [18–21]. Particularly, the complex interactions between ESD protection structures and the core circuits under ESD protection have been intensively studied that led to novel ESD-IC codesign techniques [22]. Parallelly, high-triggering-voltage and latch-up-immune ESD protection structures have been developed for high-voltage, high-power ICs. Critically, ESD protection design philosophy has gradually shifted from device-centric ESD protection designs to full-chip-oriented ESD protection designs [22, 23]. Over decades since 1960s, the field of on-chip ESD protection has never been settled, and today, ESD protection research is becoming more and more active.

1.4 ESD Protection: More or Less?

From the previous discussions, it is clear that ESD is everyday phenomenon that nobody can avoid. On-chip ESD protection is therefore required for all ICs. Practically, the strategy for developing ESD protection solutions has two aspects to consider: Science and Marketing. First, there are obviously scientific and technical needs of ESD protection for electronics. On the one hand, it is straightforward that engineers must understand the scientific fundamentals of ESD phenomena and ESD protection mechanisms including the WHYs and HOWs. ESD events involve fast and large current and voltage transients, when applied to electronics, causing unbearable internal heating and electric field stressing that result in thermal failures and/or dielectric breakdown. Therefore, on-chip ESD protection is required to provide a low-resistance (low-R) conducting path to discharge the electrostatic charges without overheating and/or electric field over-stressing to ICs. For a given ESD transient, the ESD energy is there, which must be discharged safely into the ground without overheating ICs. Similarly, the substantial electric field induced by an ESD surge must be defused safely through low-resistive conduction. Therefore, there are clearly hard-core science and technical concerns for on-chip ESD protection. On the other hand, any product designs must consider economic impacts, marketing competition, and consumer psychology. Electronic products are made to sell. If any key function does not "look" good or makes customers "feel" better and "worth" the dollars compared to competitive alternatives on the market, your products cannot be sold. Also, time-to-market is an important consideration for new IC product development today. At the time when averaging costs reached to one hundred million U.S. dollars and beyond in developing a new smartphone SoC chip at 10 nm node, a possible ESD failure, quite often in IC designs, will require lengthy and painful debug and redesign efforts, hence, seriously delaying new product release and missing the narrow market window. It would not only result in significant revenue losses but also put the fate of a company in jeopardy. The above discussions naturally lead to typical engineering questions: for ESD protection designs, how high is too high, how low is too low, and what is adequate? In deciding on *More* or *Less* for ESD protection, the following factors should be considered on a big picture as a good ESD protection design strategy. First, one has to meet the basic ESD protection target, for example 2 kV human body model (HBM) ESD protection is commonly accepted as a basis bar for most ICs. Without providing adequate ESD protection, your ICs may not be sold well, or, even if being sold, you may expect substantial field returns from the customers later. Second, more ESD protection is better only IF it would not adversely affect the IC specs. There are many negative impacts of ESD protection on ICs: Almost all ESD protection are based on PN junctions, which inherently introduce parasitic capacitance, leakage, and noises. These ESD-induced parasitic effects can seriously affect performance of core IC under ESD protection. ESD protection structures also take substantial die area on a chip. ESD protection device layout is often irregular. Hence, overall, more ESD protection using a given ESD structure type means more negative impacts on IC performance and chip density. Particularly, the ESD-induced parasitic effects may have deadly impacts on high-speed, high-through, high-frequency, and broadband ICs, for example, >10 Gbps or >10 GHz. Consequently, you often see much lower ESD protection for high-data-rate I/O pads of high-speed datalink chips or high-frequency pins of RF ICs. Third, one must balance the needs for core IC performance and ESD robustness by excising thorough ESD-IC co-design in order to simultaneously achieve both IC specs and ESD protection. Careful IC-ESD design trade-off is critical in practical designs. Fourth, IC designers should always keep the customer's needs in mind and consider the marketing. Given similar or little better IC specs, higher ESD protection would offer your IC products a critical advantage over competitors on the market. The common user psychology is to buy better things using same dollars, and ESD protection is a considering factor when consumers shop for ICs. Keep in mind that what you think about ESD protection as a product designer is much less important compared to what the customers think they would want for ESD protection. To this end, there is often no rocket science for ESD protection. System vendors and end users want a peace in mind in terms of reliability. In recent years, as IC technologies advanced into sub-28 nm nodes, and IC performance and chip complexity continuously increased, ESD protection for advanced ICs becomes extremely challenging. One key reason is that ESD protection never followed the Moore's Law to shrink in sizes. The negative factors for ESD protection, including ESD-induced parasitic effects, ESD layout, and size issues (collectively called *ESD-induced design overhead*), make it extremely difficult to maintain the usual ESD protection targets for advanced CMOS ICs. Accordingly, an industrial ESD workgroup of ESD engineers, "the Industry Council on ESD Target Levels," was established in 2006 with a mission "to review the ESD robustness requirements of modern IC products for allowing safe handling and mounting in an *ESD protected area*." The Industrial Council recognized that the current industry ESD qualification target levels are "unsupportable" and released several White Papers to recommend "suitable" (lower) ESD target levels [24–27]. The White Paper believes that the common ESD targets are unnecessary on manufacturing sites with basic ESD control management, for example the common 2 kV ESD target in HBM ESD protection is too high because 500 V would be ESD safe. Hence, it recommends significant reduction of ESD qualification target level to 1 kV HBM (down from the common target of 2 kV) [24] and 250 V CDM (down from commonly 500 V) for modern ICs [25]. The key rationale of the White Papers is that ESD "awareness" is common now and ESD "control" is comprehensive in manufacturing sites today, hence, the ESD danger is much "lower." The White Papers further promote ESD protection at system board level as an alternative solution to the "unsupportable" on-chip ESD protection requirements [26, 27]. However, one must never forget that any ICs are to be used by the customers, being system vendors to make electronic products including smartphones and touch pads, and the consumers who use smartphones or wear smart watches. For the everyday consumers in a real world, the ESD phenomena stay the *same* today compared to ten years ago, hence, the ESD dangers never shrank today. Further, consumer electronics are more touch-based today that essentially increases the likelihood and level of ESD dangers, while the advanced IC technologies are much more vulnerable at 7 nm node compared to at 180 nm node. Also, since the new product development costs increase exponentially for today's microelectronics products, e.g., iPhones and Tesla cars, hence, the system vendors actually require extra ESD protection at both IC and board levels to minimize return losses from customers. Asking the system vendors to increase system-level ESD protection while dramatically reducing on-chip ESD protection will not fly. Therefore, regardless of what perfect and strict ESD control program are in place on manufacturing sites, never imagine to ask the consumers (e.g., grandparents) in the street to think about ESD control before using a smartphone. This is a classic example to argue that IC designers must keep the end-users in minds in terms of ESD protection designs when designing IC products. Product designers and manufacturers must take care of the product reliability and not try to shift the reliability burden to users. It is indeed that ESD protection design is becoming more and more challenging for advanced IC technologies. It is important to keep up the research and development efforts to continuously explore novel and transformative ESD protection solutions for advanced IC technologies.

1.5 ESD Protection: Evolution to Revolution

The previous discussions cover a long and interesting history of ESD research and development spanning around seven decades and the preceding section clearly states the needs for novel ESD protection solutions for advanced IC technologies. Since 1970s, significant R&D efforts have been devoted to continuously improve ESD protection for ICs, from simple PN junction diodes to more sophisticated gated and shallow trench isolation (STI) diode ESD protection structures in advanced technologies; from ESD diodes to BJT and MOSFET ESD protection structures; from single diode to diode-string ESD protection structures for less ESD-induced parasitic capacitance; from grounded-gate MOSFET (ggMOS) to gate-coupled MOS (gcMOS) for lower triggering voltage; from single-device to subcircuits for ESD triggering assistance; from single-finger ESD device to multiple-finger ESD protection structures for improved ESD discharging uniformity; from MOSFET ESD to SCR ESD structures for high-voltage ICs; from regular SCR to DTSCR for lower ESD-induced parasitic capacitance and lower ESD triggering voltage; from single-stage ESD structure to two-stage ESD protection for improved CDM ESD protection; and so on. ESD protection design principles gradually shifted from device-centric individual/standalone ESD protection structure designs to circuit-oriented full-chip ESD protection designs. ESD design practices have been evolving from experience-based trial-and-error design, to CAD-based ESD protection design for design optimization and predication, and to full-chip ESD protection circuit design verification by CAD. ESD protection design goal also changes from focusing on ESD protection level to balancing both ESD protection and core IC performance for advanced ICs. On the other hand, almost all-traditional ESD protection structures rely on PN junction-based structures for ESD discharging, except for a few inductor-based ESD protection subcircuits that are inherently narrow band in nature. As IC technologies march into sub-10 nm regime, the traditional PN junction-based ESD structures have fundamental disadvantages. For example, the ESD-induced parasitic effects, including ESD-induced parasitic capacitance (C_{ESD}), leakages (I_{leak}), and noises, are quickly becoming relatively more significant and unbearable to high-performance ICs. In the meantime, complete whole-chip ESD protection requires too many ESD protection structures that are consuming relatively too much Si area for complex ICs with hundreds to thousands of pads. Further, simply imagine future chips comprising non-CMOS non-Si devices, such as various nano devices, micro/nano-electromechanical systems (MEMS/NEMS) devices, bioinspired devices, etc., it is naturally to expect that the traditional inside-Si PN-junction-based ESD structures will not be acceptable for future chips. Hence, it is important to think abnormally about ESD protection and imperative to explore truly disruptive ESD protection methods, from ESD protection mechanisms to structures. For example, a backend-based above-IC graphene-based mechanical switch structure concept and graphene-based ESD interconnects were reported, which are fundamentally different from any traditional in-Si PN-based ESD protection solutions [28, 29]. In other words, ESD protection designs have been advancing since 1970s, yet the progresses have been incremental and evolutionary so far. Revolution in on-chip ESD protection is needed for future chips.

References

1 (a) Gilbert, W. (1600). *De Magnete, Magnetisque Corporoibus, et de Magno Magnete Tellure: Physiologia noua, Plurimis & Argumentis, & Experimentis Demonstrata*, (in Latin). London: Peter Short. (b) (1893). *On the Loadstone and Magnetic Bodies, and on That Great Magnet the Earth: A New Physiology, Demonstrated with Many Arguments and Experiments* (trans. P.F. Mottelay).

New York: Wiley. (c) (1900). *On the Magnet, Magnetic Bodies Also, and on the Great Magnet the Earth: A new Physiology, Demonstrated by Many Arguments & Experiments* (transl. S.P. Thompson). London: Chiswick Press.

2 de Cisternay du Fay, C.F. (1733). Premier mémoire sur l'électricité, Histoire de l'électricité. *Philos. Trans. R. Soc.* 38: 1734.

3 (a) Franklin, B. *Experiments and Observations on Electricity*, English Edition 1.1 printed April 1751, 1–88. sold by E. Cave at St. John's Gate. (b) Letter from Benjamin Franklin to Peter Collinson (28 March 1747). *Founders Online, National Historical Publications and Records Commission (NHPRC), 2016*. Retrieved March 14, 2017.

4 Hertz, H.R. (1887). On electromagnetic effects produced by electrical disturbances in insulators. *Sitzungsber. d. Berl. Akad. d. Wiss.*; in *Wiedemann's Ann.* (34), 273, 1888; reprinted in: *Hertz*, pp. 95–106, 1962.

5 2018 CODATA Value: elementary charge. *The NIST Reference on Constants, Units, and Uncertainty*, NIST, 20 May 2019. Retrieved 2019-05-20.

6 Xu, C., Wang, A.C., Zou, H. et al. (2018). Raising the working temperature of a triboelectric nanogenerator by quenching down electron thermionic emission in contact-electrification. *Adv. Mater.* 30 (38): e1803968. https://doi.org/10.1002/adma.201803968.

7 Zou, H., Zhang, Y., Guo, L. et al. (2019). Quantifying the triboelectric series. *Nat. Commun., 1427* 10 (1): 10.1038/s41467-019-09461-x.

8 Lowke, J.J. (1992). Theory of electrical breakdown in air. *J. Phys. D: Appl. Phys.* 25 (2): 202–210. https://doi.org/10.1088/0022-3727/25/2/012.

9 Wang, A. (2002 Edition). *On-Chip ESD Protection for Integrated Circuits: An IC Design Perspective*. Springer ISBN-13: 978-0792376477.

10 Brattain, W.H. (entry of 15 December 1947). Laboratory notebook, case 38139-7, Bell Laboratories archives.

11 Kilby, J.S. (1964). Miniaturized electronic circuits. US Patent No. 3, 138, 743 (filed in 1959).

12 Noyce, R.N. (1961). Semiconductor device-and-lead structure. US Patent No. 2, 981, 877A (filed in 1959).

13 Wanlass, F.M. (1967). Low stand-by power complementary field effect circuitry. US Patent No. 3, 356, 858.

14 Moore, G.E. (1965). Cramming more components onto integrated circuits. *Electronics*: 114–117.

15 EOS/ESD Association (2014). *Fundamentals of Electrostatic Discharge, Part 3 – Basic ESD Control Procedures and Materials*. EOS/ESD Association.

16 ESD Association Standard ANSI/ESD S8.1.

17 EOS/ESD Association, Inc., Rome, NY.

18 Wang, A.Z., Tsay, C.H., and Deane, P. (2002). Dual-direction over-voltage and over-current IC protection device and its cell structure. US Patent No. 6, 365, 924.

19 Wang, A.Z., Tsay, C.H., and Deane, P. (2001). Method for manufacturing a dual-direction over-voltage and over-current IC protection device and its cell structure. US Patent No. 6, 258, 634.

20 Wang, A.Z. (2003). Single structure all-direction *ESD* protection for integrated circuits. US Patent No. 6, 512, 662.

21 Wang, A.Z. (2003). Bonding pad-oriented all-mode ESD protection structure. US Patent No. 6, 635, 931.

22 Wang, A., Feng, H., Zhan, R. et al. (2005). A review on RF ESD protection design. *IEEE Trans. Electron Devices* 52 (7): 1304–1311. https://doi.org/10.1109/TED.2005.850652.

23 Lin, L., Wang, X., Tang, H. et al. (2009). Whole-chip ESD protection design verification by CAD. *Proceedings of IEEE EOS/ESD Symposium*, pp. 28–37.

24 Industry Council on ESD Target Levels (2010). White Paper 1: A Case for Lowering Component Level HBM/MM ESD Specifications and Requirements, Rev. 2.0, Industry Council on ESD Target Levels, October 2010.

25 Industry Council on ESD Target Levels (2009). White Paper 2: A Case for Lowering Component Level CDM ESD Specifications and Requirements, Rev. 1.0, Industry Council on ESD Target Levels, March 2009.

26 Industry Council on ESD Target Levels (2010). White Paper 3: System Level ESD. Part I: Common Misconceptions and Recommended Basic Approaches, Rev. 1.0, Industry Council on ESD Target Levels, December 2010.

27 Industry Council on ESD Target Levels (2012). White Paper 3: System Level ESD. Part II: Implementation of Effective ESD Robust Designs, Rev. 2.0, Industry Council on ESD Target Levels, September 2012.

28 Ma, R., Chen, Q., Zhang, W. et al. (2016). A dual-polarity graphene NEMS switch ESD protection structure. *IEEE Electron Device Lett.* 37 (5): 674–676. https://doi.org/10.1109/LED.2016.2544343.

29 Chen, Q., Ma, R., Zhang, W. et al. (2016). Systematic characterization of graphene ESD interconnects for on-chip ESD protection. *IEEE Trans. Electron Devices* 63 (8): 3205–3212. https://doi.org/10.1109/TED.2016.2582140.

2

ESD Failure Analysis

2.1 ESD Failure Analysis

Though common electrostatic discharge (ESD) failure mechanisms, such as human body model (HBM), machine model (MM) and charged device model (CDM), have been well discussed, ESD *failure analysis* (FA) remains to be a highly debatable topic in ESD protection designs, particularly when dealing with results obtained using less stable CDM ESD test models and for less obvious latent ESD damages, as well as in correlating the ESD failure threshold values measured with the ESD failure signatures observed. The contributing factors to the complexity of ESD failure analysis include ESD test models, ESD testing equipment, and ESD failure mechanisms. Nevertheless, it is very beneficial for integrated circuit (IC) and ESD designers to understand ESD failure analysis in general, including ESD failure mechanisms, ESD FA techniques, ESD failure signatures, and ESD design debugging, which will be discussed in this chapter.

A common question from an IC designer on ESD failure analysis is often *Why would I bother considering ESD failure analysis when things seem to be uncertain anyway?* Indeed, this is a valid and legitimate question to ask since ESD protection design cannot be guaranteed by performing ESD failure analysis only. It is common that papers and technical reports are typically case-based, i.e., describing *specific* ESD FA results that are only valid for a *specific* device made in a *specific* process technology and characterized using a *specific* tester per a *specific* ESD testing model. Vague and even controversial results are often seen in publications on ESD failures, showing substantial variations across different ESD test models (e.g., HBM, MM, CDM, International Electrotechnical Commission [IEC], transmission-line-pulsing [TLP], very fast TLP [VFTLP]) for different process technologies (e.g., from 180 nm, 65 nm, and 45 nm nodes, to 28 nm, 14 nm, and 10 nm nodes, and to 7 nm, 5 nm, and 3 nm nodes). Still, ESD FA work can be very helpful in debugging ESD protection circuit design failures and in ensuring successes of future ESD protection designs. What it takes is for IC designers to think about ESD FA analysis in a fuzzy way instead of using the traditional digital mindset when understanding ESD FA information. Today, ESD FA remains a powerful and informative technique in practical ESD protection designs. ESD FA examples will be discussed in the following sections.

2.1.1 ESD Failure Criteria

What are typical ESD failure criteria commonly used to evaluate on-chip ESD protection designs for ICs? In general, the fundamental ESD failure criteria are to examine any changes in circuit function and performance "Specs" of IC chips under ESD protection [1]. Typically, the specs of an IC are measured *before* and *after* ESD stressing against the IC design targets. If either malfunction

Practical ESD Protection Design, First Edition. Albert Wang.

or unacceptable deterioration in circuit specs is observed for an IC after ESD stressing, ESD failure occurs. Therefore, the ESD failure criteria are determined by the deterioration in IC specs, which, theoretically include all function parameters listed on the datasheet of an IC product. However, the common practice in checking ESD failure thresholds is to focus on a few critical specs parameters for an IC depending upon its applications. For example, for a power amplifier (PA), the gain and power efficiency are the key specs; for a wireless transceiver IC, the center frequency and streaming data rate are the key specs; and for an oscillator, the clock rate is certainly the key parameter. In fact, it is up to the IC designer to determine what IC specs parameters will be examined as suitable ESD failure criteria for given IC products. On the other hand, it is the common practice for the industry that leakage current is always used to evaluate the integrity of an IC chip against ESD failures.

2.1.2 Hard and Soft ESD Failures

Broadly speaking, there are two ESD failure categories: *hard* and *soft* ESD failures. An ESD failure is defined as hard ESD failure, also called *catastrophic* ESD failure, if, during ESD testing, clear malfunction is observed for an IC part stressed by an ESD event of certain level. Hard ESD failure is destructive and irreversible. Hard ESD failure is typically used to characterize the ESD protection level for ICs, or referred to as ESD protection robustness. Often, ESD hard failure may be observed as complete destruction of ICs, e.g., no gain and no function for a PA die. Yet, most of times, ESD-induced malfunction appears as substantial and unacceptable deterioration of key IC specs. For example, the power gain and efficiency of a PA tested decrease by 20% after 2 kV HBM ESD zapping, then the PA fails at 2 kV HBM ESD test. The vendor of an IC product may define the scale of ESD-induced specs deterioration as "hard" failure for a specific IC, e.g., 10% or 20% drop in gain, which is normally selected according to system applications or per customer requirements.

ESD soft failure, also called *latent* ESD failure, refers to ESD failure phenomena where no obvious IC malfunction is observed after certain level of ESD stressing; however, some subtle changes in IC specs may be detected. For example, the leakage current, originally very low, may increase from 1 to 5 nA, but not yet changing by orders of magnitude in measurement, or the power gain of an PA chip decreases slightly, e.g., by 5%. Typically, an IC chip suffering from ESD soft failure is not destructive and can still be used in the system. However, uncertain and unexpected long-term effects may occur during the lifetime of an IC. While ESD hard failure can be readily observed during ESD testing, identification of ESD soft failure is extremely difficult in typical accelerated tests. Similarly, many ESD hard failures can be well understood, which helps to improve ESD protection designs; however, most ESD soft failures remain mysterious in practical ESD protection designs. In summary, hard failure is destructive and irreversible that causes immediate IC malfunction, while soft failure is nondestructive and reversible that may not affect IC performance for the time being. Hard failure can be readily observed in ESD testing and is commonly used to evaluate ESD protection robustness of ICs, while soft failure is often unnoticeable that affects the lifetime of IC products.

2.2 ESD FA Techniques

A variety of techniques and instruments can be used for ESD FA analysis, both electrically and physically. ESD qualification for ICs, i.e., ESD protection level that is specified in a product datasheet, is evaluated by performing ESD zapping tests, which will be discussed in details in Chapter 3. ESD protection capability of ICs is characterized against the predefined ESD failure criteria by measuring electrical characteristics, i.e., IC specs, using automatic test equipment (ATE) in typical industrial environments. While electrical characterization can detect ESD failures of IC chips, details in ESD damages are typically inspected by visual and physical examinations. The results obtained during ESD FA procedures will be analyzed to identify ESD failure signatures to understand ESD failure mechanisms and to correlate between electrical testing and ESD damages. ESD failure analysis is critical to debugging ESD protection design problems and achieving future ESD protection design successes.

Typical ESD FA checking procedures consist of visual inspection, electrical measurement, and physical examination. Visual inspection for ESD damages is done by simply looking at a wafer, after ESD stressing, under a microscope that may visually reveal severe ESD hard failures, e.g., blowout in dielectric or metal interconnects of a dead IC die. When no clear visual sight can be seen for an IC after ESD stressing, electrical characterization must be conducted for all electrical parameters or a set of key specs preset, which will be compared against the predefined ESD electrical failure criteria, such as gain, bandwidth, and data rate, etc. The margin for ESD failure criteria is determined by IC designers per company rules and customer requirements. After ESD stressing, hard failure occurs to an IC if deterioration in electrical specs exceeds the electrical failure criteria, which often does not show any easy-to-see visual ESD damages under a microscope. In many cases, even if no ESD hard failure is identified in electrical characterization, ESD soft failure may occur to an IC suffering from ESD stressing. Typically, in soft failure cases, the IC specs are still within the datasheet range. Increase in leakage current can be detected, which however does not show a big jump by orders of magnitude as typically observed in ESD hard failure cases. Some lifetime characterization methods, such as accelerated stressing or aging techniques, may be used and modified for ESD soft failure examination. If ESD hard or soft failures occurs, ESD FA physical examination can be performed to study details of ESD failures. Physical examination of ESD damages requires proper preparation of IC samples, typically involving decapping and deprocessing an IC die by removing the thick passivation dielectric layers to open up the metal interconnects and Si area for inspection by using sophisticated inspecting instruments to reveal multiple-dimensional morphological and/or atomic information with high resolution at both surface and subsurface levels where ESD-induced damages can be observed.

A variety of visual and physical inspecting instruments are available for ESD FA inspection. The simplest, but most efficient, tool is an optical microscope that allows easy identification of hard damages at the surface or subsurface level after delayering. The main limitation of optical microscopy is its limited magnification, normally below 1500× with resolution to 1 μm or so. Much more powerful and sophisticated microscopy tools that offer very high resolution include scanning electron microscope (SEM), transmission electron microscope (TEM), atomic force microscope (AFM), scanning force microscope (SFM), liquid crystal analysis, light emission microscope, etc. [2]. Another useful contactless electrical test technique is the electron beam test system (E-beam).

SEM microscopy offers very high magnification (~150 000×) and resolution. IC dies are deprocessed layer by layer and inspected by SEM, which can readily reveal visible defects caused by ESD stressing. A SEM system basically does surface analysis; however, limited three-dimensional image and atomic information can be obtained by using sample tilting skill and back-scattered electron technique, respectively. TEM microscopy features very high magnification that allows a resolution better than two angstroms in three-dimensional microstructural analysis. The disadvantage is that TEM requires sample preparation that is both complicated and destructive. AFM utilizes one surface of a stylus to scan over another surface, i.e., the IC sample, and to detect the atomic forces between the two surfaces. The mapping of atomic force variations while scanning the sample is recorded for topographical and sectional analysis. Since ESD failures are usually accompanied by heat generation or photon emission, the liquid crystal and light-emission analysis techniques can be very useful in ESD FA studies. In liquid crystal analysis, the wafer is placed on a controlled heat chuck and covered by selected liquid crystals. While the chip under powered operations, a thermally colored image is taken where hot spots, i.e., ESD failure defects, will appear as dark-colored spots. A light emission microscopy technique is also referred as photon emission microscopy (EMMI) technique, which generates a spectral image by detecting photons emitted from a powered IC die induced by current, avalanche electron–hole generation, thermal radiation, tunneling current through dielectric, and inter-band transitions [3, 4]. EMMI is a very powerful ESD FA analysis tool because ESD failures often result in high current, heat generation, junction breakdown, and dielectric breakdown, which generate large amounts of photons during ESD discharging and failures. EMMI technique can pin-down an ESD defect at a few *nanometer* scale. More attractively, EMMI is an in-operando inspection technique that can monitor ESD discharge processes while an IC is in normal operation mode, hence, possibly to reveal ESD failure evolution procedures under ESD stressing, which is critically informative and useful for ESD protection design optimization. The downsides of such sophisticated techniques include high cost-of-ownership, complexity in sample preparation, and requirements for high operation skills. Indeed, there is no free lunch in a real world.

2.3 ESD Failure Signatures

As discussed previously, ESD failures in ICs fall into two categories: hard failures and soft failures. ESD hard failures result in immediate IC malfunction including either unacceptable specs degradation or a dead chip. ESD hard failures are generally associated with either thermal damages or dielectric breakdown. ESD thermal damages occur in metal interconnects, contacts, and vias, and silicon or any semiconductors, which are typically caused by joule heating generated by large transient ESD discharging currents. Since transient ESD discharging currents are both fast and large, while IC materials typically have poor thermal conductivity, the ESD-induced heating is inherently localized, leading to tiny *hot spots* inside an IC die. When the local heating raises the hot spot temperature to the given melting temperature threshold, e.g., 1414 °C (or 1687 K) in Si, 660 °C (or 933 K) in aluminum (Al), or 1085 °C (or 1358 K) in copper (Cu), ESD thermal failure will occur in Si substrate and/or metal interconnects. On the other hand, dielectric breakdown is associated with the high-localized electric field density induced by large ESD transients, which, when exceeding a specific dielectric breakdown electric field density threshold, e.g., $\sim 10^6$ V/m in SiO_2 in complementary metal-oxide-semiconductor (CMOS), ESD-induced voltage breakdown damage will occur in IC dielectrics. While ESD failure mechanisms are well understood for hard failures, ESD soft failure phenomena are still rather mysterious, which require further research. In general, ESD soft failures are not clearly noticeable during electrical measurements and physical inspection. Soft failures may cause a slight change in electrical specs, leading to time-dependent performance degradation and reduction in IC lifetime. Typical ESD soft failures may induce materials integrity problems, resulting in *time-dependent dielectric breakdown* (TDDB) of gate oxide layers and minor-to-moderate increase in leakage current. Many ESD hard failures can be categorized into unique *ESD failure signatures*, which provide useful information for ESD failure debugging and ESD design optimization for ICs. In this section, typical ESD failure signatures are discussed to show the two opposite aspects of ESD FA: being very informative or being rather confusing and even completely useless.

The first ESD FA example to discuss is a classic two-stage primary–secondary ESD protection circuit depicted in Figure 2.1, which consists of a primary ESD protection device (ESD_p) that is a thick oxide gate (FOX) NMOS device (Thick- or ThickG-NMOS), a secondary ESD protection

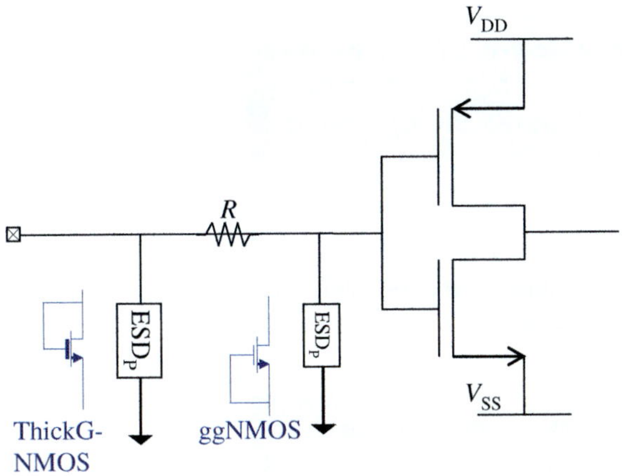

(a)

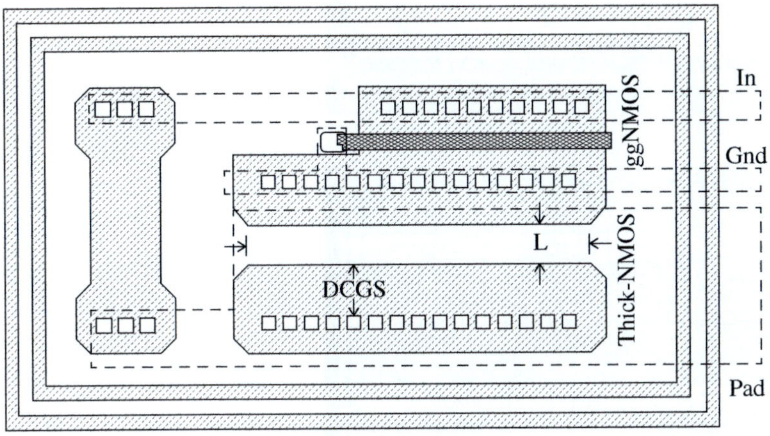

(b)

Figure 2.1 ESD failure analysis for a classic two-stage ESD protection circuit block containing primary and secondary ESD protection structures: (a) schematic and (b) layout. (Clark et al. [5].)

(ESD_s) that is a grounded-gate NMOS device (ggNMOS), and a diffusion resistor, *R*, between the ggNMOS and Thick-NMOS ESD devices [5]. The design concept is that the ggNMOS can be triggered by an ESD pulse at a low triggering voltage, which then helps to turn on the Thick-NMOS that is designed to handle large ESD transient current. The ESD failure phenomena of this primary–secondary ESD protection circuit under ESD zapping per HBM, MM, and CDM testing methods were studied. Figure 2.2 shows the ESD damage images for this two-stage ESD protection structure. Figure 2.2a shows that under HBM ESD stressing, ESD damages are observed at the two ends of the thick-NMOS device, featuring thermal filament from the contacts at the drain end to the gate region and spreading into the source end. This is typically referred as a *D–S silicon filament* defect. Figure 2.2b shows similar D–S filament damages in one IC sample after MM ESD zapping, where in addition to the D–S filament damages at the two ends, extra D–S filament defects across the drain region were observed, which may be associated with the oscillatory MM waveforms. Since these MM pulse waves are strong and last long (~30 ns), each

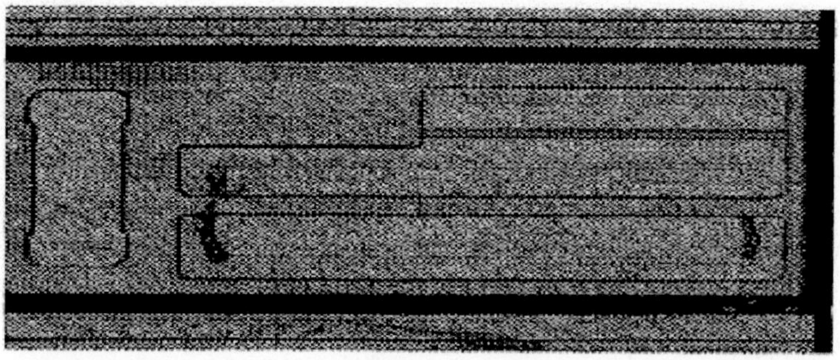

(a)

(b)

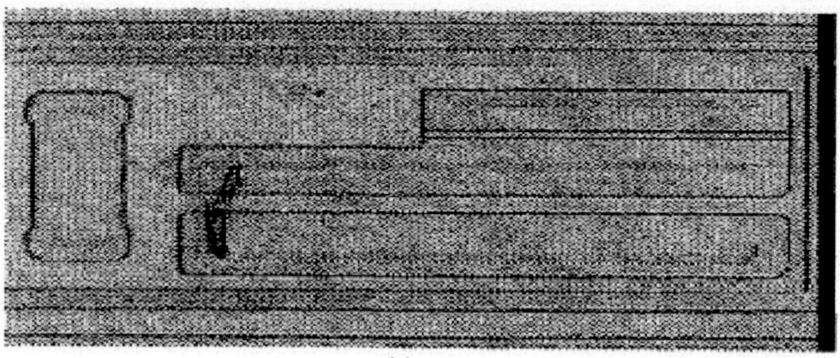

(c)

Figure 2.2 Images for ESD damages under ESD zapping for the primary–secondary ESD protection structure shown in Figure 2.1 under (a) HBM stressing, (b) MM stressing, and (c) CDM stressing. (Clark et al. [5])

oscillatory MM waveform peaks can cause similar D–S filament damage, but likely at different locations. Figure 2.2c depicts the same type of D–S filament damage in one IC sample after CDM ESD zapping, which shows less severe ESD damage than that under HBM zapping. It seems that the two finger ends of the primary–secondary ESD protection structure are most vulnerable to ESD stresses of any type (i.e., HBM, MM, or CDM), likely due to nonuniform ESD discharging current flow that may cause large current crowding at the finger ends due to layout discontinuity (Figure 2.1b) and large thermal gradient at the boundary of heated shallow drain diffusion regions and the external portion. This observation suggests that the end contacts seem to be the weakest points under ESD stressing. Accordingly, two layout design improvements, as shown in Figure 2.3, were used to resolve the end-damage problem per careful ESD failure analysis.

Obviously, if on-chip ESD protection is adequate, an IC chip shall be immune to ESD damages. However, practically, ESD protection structures may not work the way they are designed.

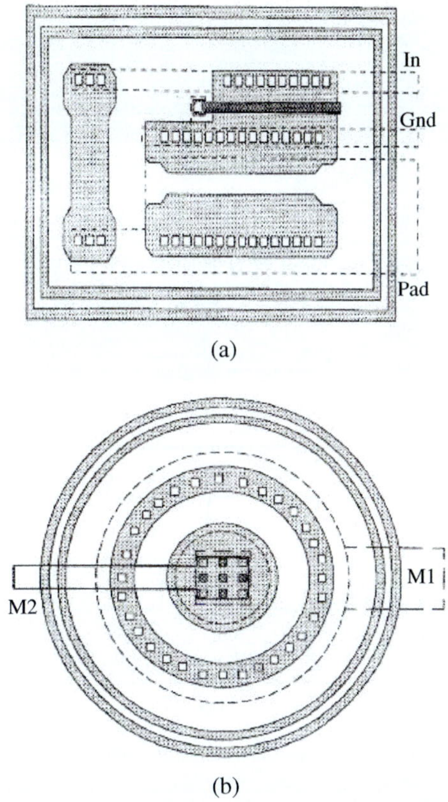

Figure 2.3 Improved layout design for the two-stage ESD protection circuit shown in Figure 2.1 where the Thick-NMOS was re-designed to avoid the end-ESD-damage problem: (a) round-corner layout and (b) closed-loop layout. (Clark et al. [5].)

Consequently, ESD damages to internal circuit may occur during an ESD event. Very often, a parasitic device inside an IC core circuit under ESD protection may be unintentionally turned on during an ESD event, forming a parasitic ESD discharging path that competes against the intentionally designed ESD protection structure. Apparently, any internal parasitic structure cannot be optimized for high current-handling capability; hence, ESD damages may occur inside an IC core circuit being protected. Figure 2.4 presents an example for such an internal circuit ESD failure case, where an HBM ESD-induced D–S filament in Si was observed in the NMOS transistor of an input buffer circuit fabricated in a 0.35 μm retrograde *n*-well salicided CMOS technology [6]. Such D–S Si filament ESD damages are also common failure signatures in metal-oxide-semiconductor field-effect transistors (MOSFET) ESD protection devices that are regularly observed during HBM and CDM ESD zapping tests [7, 8]. The D–S filament ESD failure mechanism is associated with the large ESD current transient discharging through a triggered lateral NPN bipolar transistor in ICs. Another classic ESD failure signature is the source or drain contact damage. Figure 2.5 illustrates two such exemplar NMOS ESD damages due to CDM zapping in a gate-coupled NMOS (gcNMOS) ESD protection structure implemented in a 0.35 μm CMOS technology [8]. The terminal-to-terminal Si filament ESD failure signature is also common in other ESD protection structures, for example in ggNMOS-triggered silicon-controlled rectifier (SCR) ESD protection structures. Figure 2.6 shows examples of such ESD damages in a low-triggering SCR ESD protection structure made in 0.35 μm CMOS process [8], where the difference is that a D–S Si filament occurs under positive CDM ESD zapping because the ggNMOS device conducts currents (Figure 2.6b), while an anode–cathode Si filament happens in

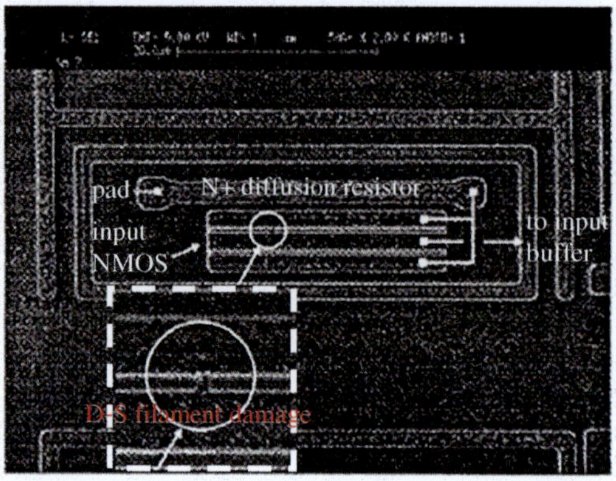

Figure 2.4 Example for ESD damage in internal IC core circuit: D–S Si filament ESD failure observed in an input buffer NMOS transistor. The inset is a close-up image. (Smith [6].)

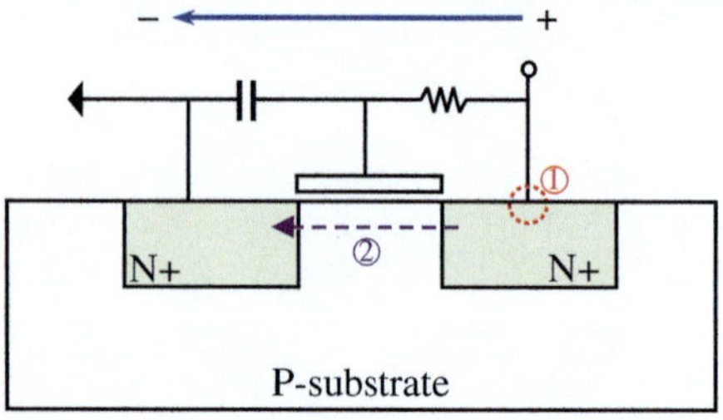

(a)

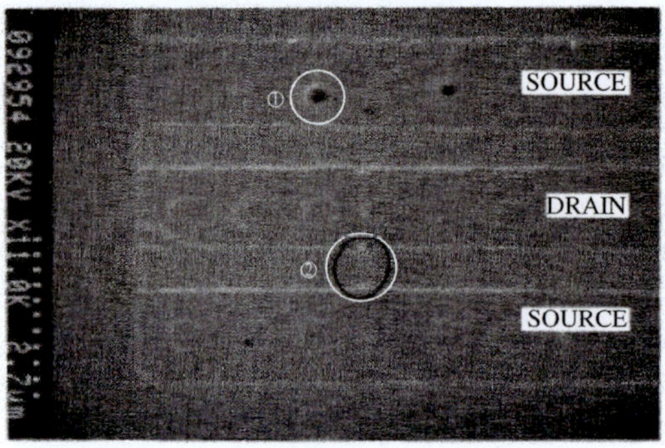

(b)

Figure 2.5 D/S contact (type ①) and D–S Si filament (type ②) CDM ESD damages in a gcNMOS ESD protection structure: (a) cross-section view, and (b) FA image. (Duvvury and Amerasekera [8].)

negative ESD stressing due to SCR operation (Figure 2.6c). The cross-sectional view in Figure 2.6a explains the ESD filament failure mechanisms. Figure 2.7 shows another contact-spike ESD failure signature by SEM for ESD-induced Al/Si melt-through in a CMOS gate array circuit [9].

ESD damages to gate oxide layers are considered another common ESD failure signature in CMOS ICs, which can occur during any ESD stresses mode, e.g., HBM, MM, and CDM ESD zapping. ESD-induced oxide failures can appear at different locations in CMOS ICs and in various styles even for the same IC chip depending upon actual ESD zapping methods. Figure 2.8 presents an example for ESD-induced gate oxide damages in data buses of a data communication IC circuit made in 1.5 μm CMOS technology under different ESD stressing tests. It is readily observed that similar gate oxide ESD failure signatures appear in the IC dies under HBM, MM, and CDM ESD zapping tests. However, it is also clear that different ESD zapping methods result in gate oxide ESD failures in different locations on the IC chip. Figure 2.8a shows ESD defect occurring at an internal

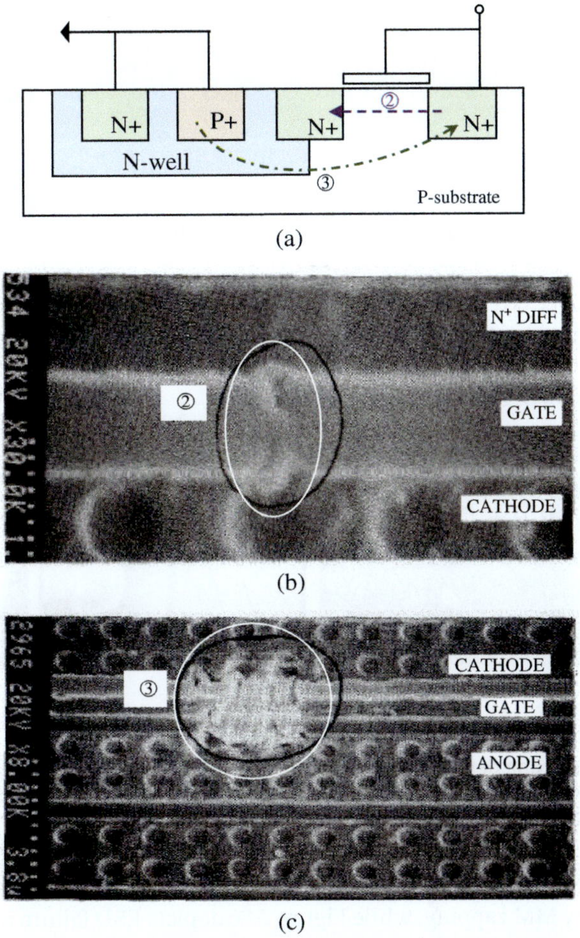

Figure 2.6 Typical terminal-to-terminal Si filament ESD damages in a ggNMOS-triggered SCR ESD protection structure: (a) cross-section view, (b) D–S filament (type ②) in ggNMOS under positive CDM zapping, and (c) anode–cathode filament (type ③) in SCR under negative CDM stressing. (Duvvury and Amerasekera [8].)

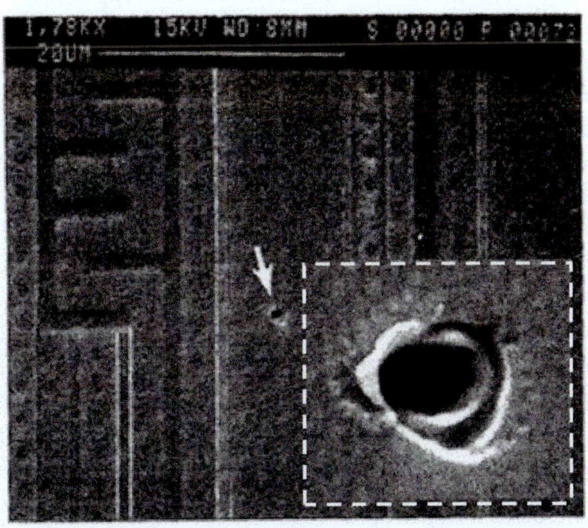

Figure 2.7 A contact spiking ESD failure signature in a CMOS gate array circuit. (Kiefer et al. [9].)

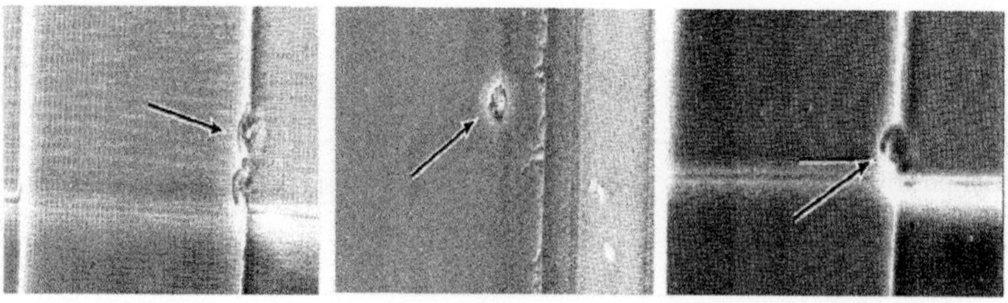

(a) (b) (c)

Figure 2.8 Similar ESD failure signatures were observed in gate oxide of a data communication chip under HBM, MM and CDM ESD zapping, however, occurring at different locations: (a) in internal NMOS for HBM, (b) in internal PMOS for MM, and (c) in internal NMOS for CDM. (Kelly et al. [10].)

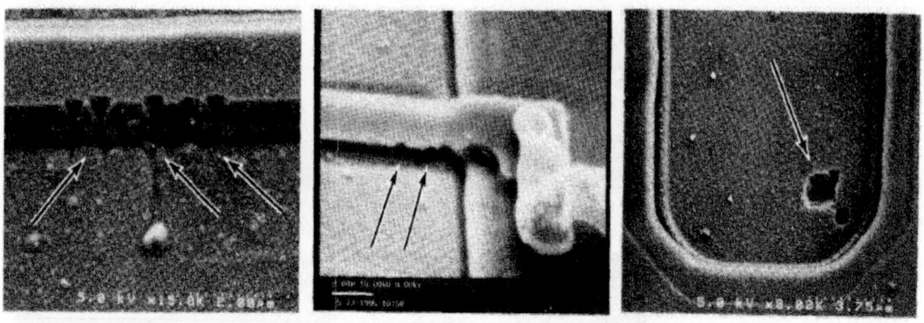

(a) (b) (c)

Figure 2.9 Different ESD zapping tests resulted in different ESD failure signatures in an audio IC chip: (a) oxide defect in a NMOS ESD protection device under MM zapping, (b) oxide damage in an internal NMOS by CDM zapping, and (c) contact spiking damage in an ESD protection resistor during HBM zapping. (Kelly et al. [10].)

NMOS device under HBM zapping, Figure 2.8b reveals ESD damage in an internal PMOS device by MM zapping, while Figure 2.8c depicts ESD failure in an internal NMOS device caused by CDM stressing [10]. Figure 2.9 shows another ESD FA example where gate oxide defects occur either in NMOS ESD protection device under MM zapping (a) or in an internal NMOS device during CDM ESD zapping (b); however, contact-spiking damage (c) appears in an ESD protection resistor under HBM zapping for the same audio IC chip made in a 1.5 μm CMOS process [10]. These two examples clearly demonstrate that ESD failure mechanisms may vary for the same IC chip depending upon ESD zapping methods. Figure 2.10 shows a light emission image for ESD damages in a two-finger ggNMOS ESD protection structure made in a 0.35 μm CMOS technology [11]. It is observed that the ESD-induced defects (hot spots marked by arrows) are rather evenly distributed along the layout fingers between the drain contacts and gates, suggesting uniform triggering of the multiple-finger ESD protection structure upon ESD zapping, a desired design feature attributed to using an lightly doped drain (LDD) blocking mask to prevent LDD implantation in the ESD protection devices, which is commonly used to improve ESD protection capability of MOSFET ESD protection structures in CMOS by reducing the electric field density at drain junction corners. In comparison, hot spots were observed in one finger only for the same IC chip without using LDD-blocking technique.

Another common ESD failure signature is ESD damages in metal interconnects due to joule heating generated by large transient ESD discharging currents, occurring in both Al and Cu interconnects layers in ICs. Figure 2.11 shows Al extrusion-like ESD damages occurring in Ti/Al/Ti interconnects of an IC made in a 0.25 μm CMOS technology, where the ESD zapping results in local

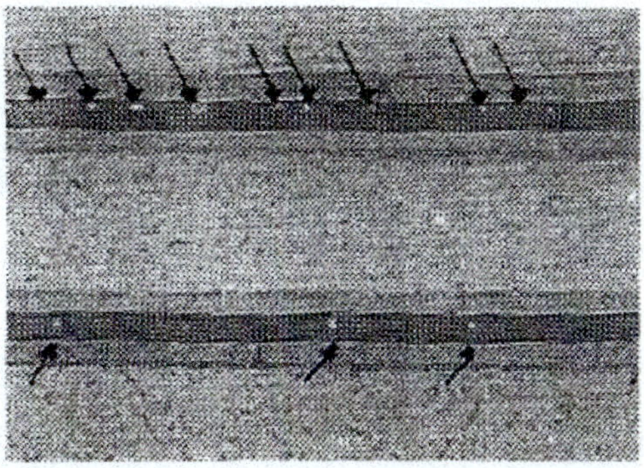

Figure 2.10 Evenly-distributed ESD-induced hot spots appear in the drain contact to gate regions in both fingers of the two-finger ggNMOS ESD protection structure using LDD-blocking mask, indicating uniform ESD turn-on, an improvement over non-uniform ESD triggering in its counterpart with LDD implantation. (Richier et al. [11].)

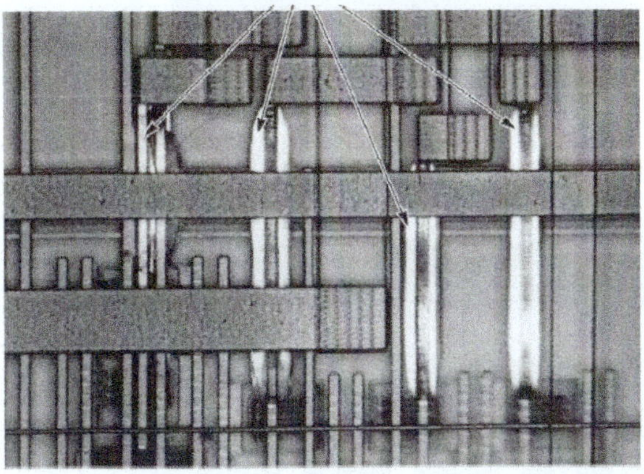

Figure 2.11 ESD-induced metal extrusion defects in Al interconnects under HBM ESD stressing. (Voldman [12].)

dielectric cracks that were filled up with melted Al due to overheating [12]. In another FA example of ICs made in 0.18 μm CMOS with Cu interconnects, Figure 2.12 depicts different ESD-induced failures in cladded-Cu interconnects under HBM stressing, including dielectric cracking, extrusion, blistering, and displacement [12]. From thermal failure view point, obviously, Cu interconnects is more ESD-robust than Al interconnects, which is a major benefit for ICs because narrower Cu interconnects metal lines can be used for the same level of ESD protection compared with using Al, hence substantially reducing the ESD-metal-induced parasitic capacitance that becomes increasingly unacceptable to high-performance ICs at advanced technology nodes. ESD damages to metal interconnects are also sensitive to the metal thickness. In one study [13], metal power supply buses of 5 μm wide with different metal layer thicknesses, i.e., thickness of 0.5 μm and scaled thickness of 0.45 μm, were used in N+/*n*-Well diode ESD protection structures and HBM ESD zapping was applied to the IC. It was observed that no ESD damage to the thicker metal bus appearing with ESD zapping up to 10 kV; however, ESD defects in forms of metal vaporization and electro-thermal migration were discovered in the thinner metal lines as shown in Figure 2.13. Proper metal interconnects design for ESD protection is a subtle design task that involves careful design balance to address the needs for achieving adequate ESD robustness and reducing ESD-metal-induced parasitic capacitances simultaneously.

Careful ESD FA analysis helps to understand the ESD protection structure triggering procedures and the ESD failure mechanisms. In [14], light emission microscopy technique was employed to investigate the ESD triggering procedures of ggNMOS ESD protection structures in a 0.5 μm silicided LDD CMOS technology. Both silicide-blocking and LDD-blocking techniques were used to

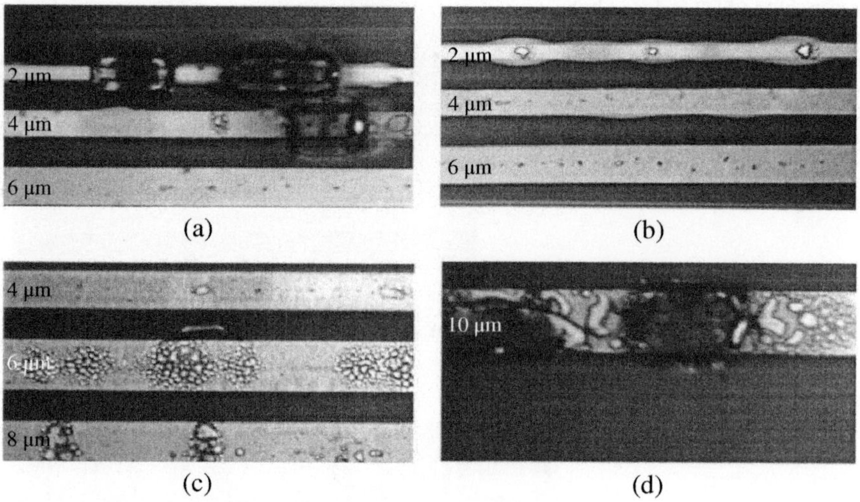

Figure 2.12 ESD-induced various defects in Cu interconnects shows ESD failure evolution under HBM ESD stressing (a) dielectric cracking, (b) extrusion, (c) blistering, and (d) displacement. (Voldman [12].)

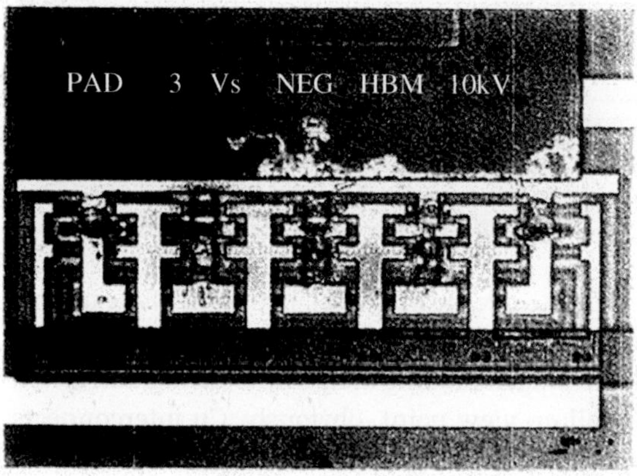

Figure 2.13 HBM ESD-induced metal vaporization and electro-thermal migration defects in the scaled metals in a diode ESD protection structure. (Voldman and Gross [13].)

improve ESD protection performance. Two ggNMOS ESD devices with and without LDD regions were studied for their ESD triggering behaviors by applying both DC and TLP stressing. Figure 2.14 depicts the live EMMI images for the ggNMOS ESD devices under DC stressing to show ESD triggering sequences for non-LDD (a) and LDD (b) ggNMOS devices, respectively. Figure 2.14a depicts the trigging procedures for the non-LDD ggNMOS device (using LDD-blocking mask), which shows that lighting spots started at the corners of the channel in image A at the bias of 10.2 V and 20 μA. As the bias increases continuously, the observed light emission region changes from lighting spots to continuous lighting lines, i.e., from image B (450 μA) to image C to image D (10.9 V and 2 mA). At stage D, the avalanche current was high enough to trigger the whole ggNMOS device and driving it into the snapback-discharging region. After the snapback threshold, very bright and localized light emission mega spots appear at the ends of the channel and the lighting spot started to hop between the two ends of the channel as the avalanche current continuously increases, corresponding to images E to H (60 mA). Therefore, the triggering voltage of this ggNMOS ESD device is about $V_{t1} \sim 10.9$ V. A continuous light emission line before triggering indicates a uniform turn-on mechanism for the non-LDD ggNMOS ESD device, which is highly desirable. Initiation of lighting spots,

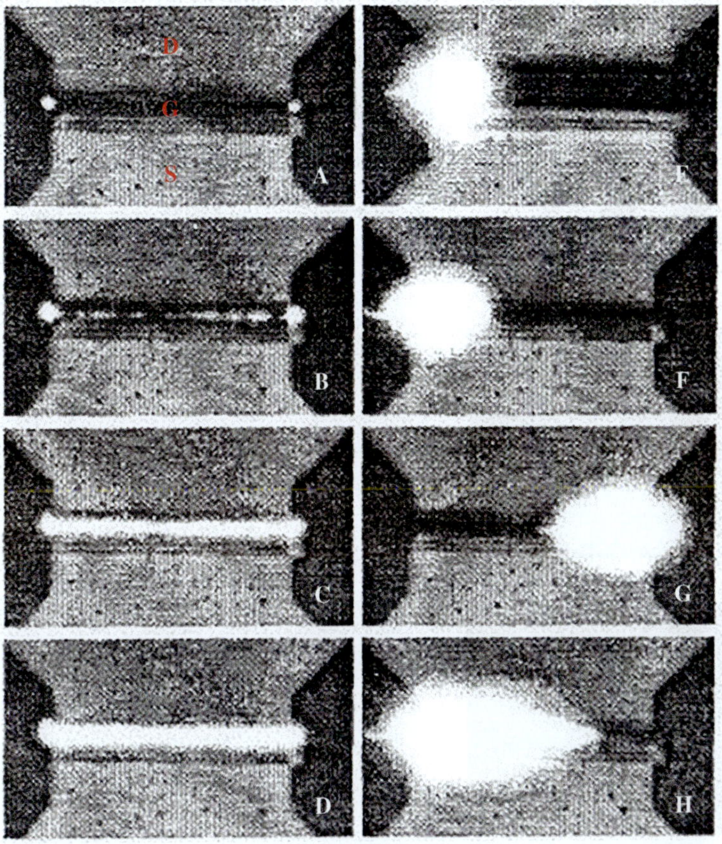

(a)

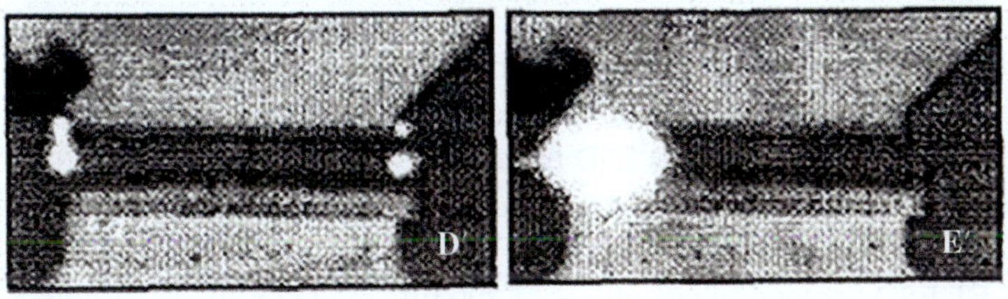

(b)

Figure 2.14 Real-time EMMI imaging under DC stressing for (a) non-LDD ggNMOS and (b) an LDD ggNMOS ESD protection devices shows uniform ESD triggering for the non-LDD ggNMOS ESD protection device, reflected by a continuous light emission line in image D, but non-uniform ESD triggering for the LDD ggNMOS ESD protection device as shown in image D′. (Russ et al. [14])

i.e., avalanche currents, at the corners of the channel is attributed to the strong electric field density formed at the sharp junction corners. The hopping of lighting spots after snapback is attributed to the thermal nature that caused the triggered regions switching. In comparison, Figure 2.14b shows that, for the LDD ggNMOS ESD structure (not using LDD-blocking mask), while light emission was also initiated at the junction corners, the lighting spots remained in the corner regions all the way to the triggering threshold (image D′), leading to snapback conduction. After that, strong and localized light emission (image E′) was observed at the end of the channel, which clearly suggests nonuniform triggering for the LDD ggNMOS ESD devices as oppose to the uniform turn-on

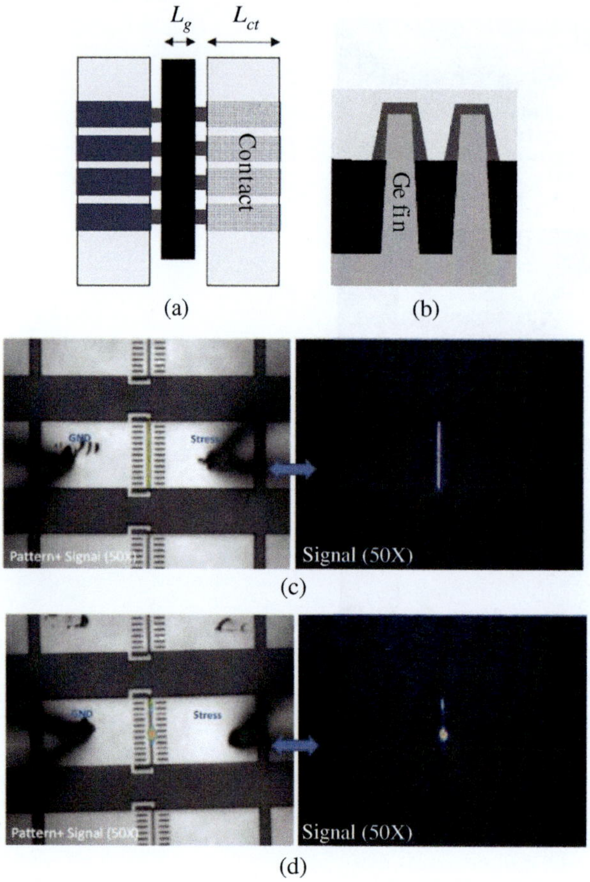

Figure 2.15 FA analysis for 300-fin ggNFET ESD protection structures (a/b) made in 20 nm Ge-fin FinFET stressed by TLP and monitored by EMMI shows (c) uniform ESD discharge across fins due to large S/D ballasting resistance of Ge, and (d) hot spots due to un-uniform ESD conduction due to reduced Ge-S/D resistance using SiGe:P/Si:P replacing Ge. (Boschke et al. [15].)

observed in the non-LDD ggNMOS ESD devices. This difference in ESD triggering sequences is due to doping profile variation at the drain junction corner in the channel. This observation supports the common understanding that LDD doping degrades ESD protection performance in MOSFETs and an LDD-blocking mask should be used to improve ESD robustness by eliminating LDD doping in MOSFET ESD devices. Figure 2.15 depicts FA analysis using EMMI to monitor transient ESD discharging procedures in ggNFET ESD protection structures of 300 fins made in 20 nm undoped Ge-fin FinFET CMOS when stressed by TLP [15]. The large S/D resistance of Ge fin serves as a ballasting resistance that ensures uniform ESD discharging across the Ge-fin array, as shown in Figure 2.15c. For the ggNFET using Si:P/SiGe:P on top of the Ge-fin to dramatically reduce the large Ge-fin S/D resistance, the reduced ballasting-*R* resulted in un-uniform ESD discharging across the fin array; hence, severely degraded ESD protection, leading to local hot spots as shown Figure 2.15d.

AFM is a powerful tool for ESD FA analysis where its quantitative image helps to reveal weak and subtle ESD defects due to both hard and soft failures. Figure 2.16 presents AFM FA images for an NMOS ESD structure fabricated in a 0.5 μm salicided LDD CMOS technology, revealing its soft-to-hard ESD failure evolution under TLP stressing [16]. Images by optical microscope show soft failure (Figure 2.16a) starting to appear at the drain under a 200 ns TLP stressing, which evolves into D–S filament hard failure (Figure 2.16b) after stressed by TLP pulse of 500 ns. This soft-to-hard ESD failure evolution procedure was confirmed by quantitative AFM analysis as given in Figure 2.16c,d. In Figure 2.17, AFM images reveal ESD-induced soft defect, a very tiny defect hole (~0.19 μm × 0.14 μm), in oxide sidewall within a CMOS gate circuit under CDM ESD zapping, where the corn-shaped defect hole is clearly depicted by high-magnification AFM image [17].

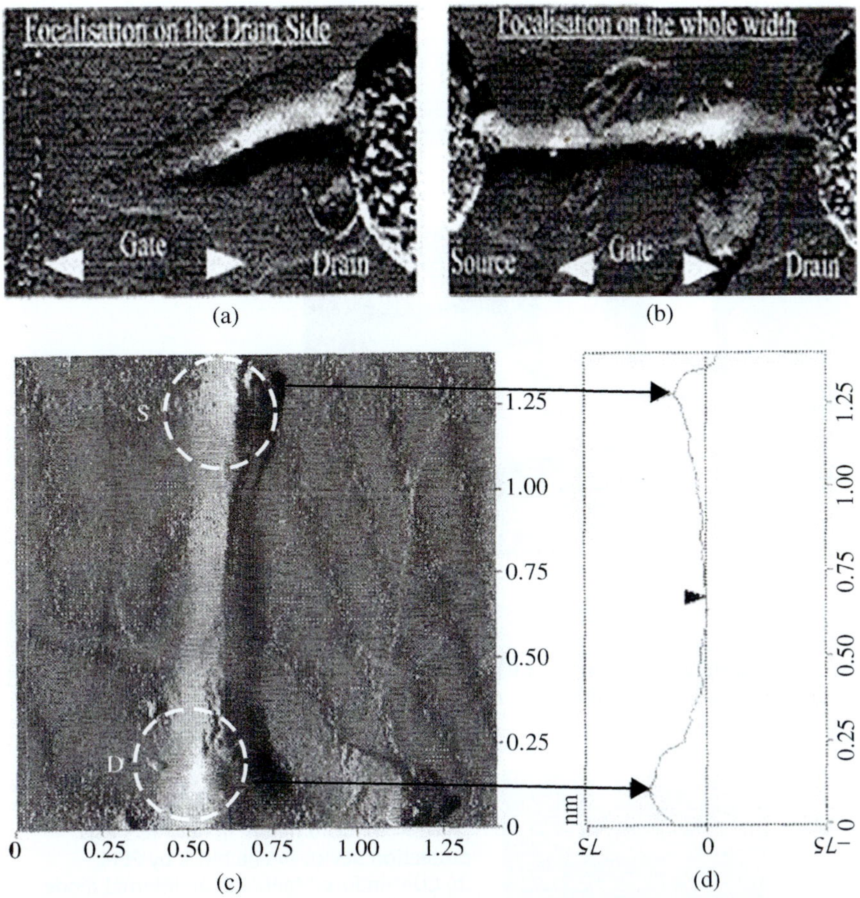

Figure 2.16 Optical microscopy and AFM images reveal the soft to hard ESD failure evolution procedures in a MOS device after TLP stressing: (a) a soft defect starts at Drain stressed by a 200 ns TLP pulse, (b) the soft defect develops into a D–S filament hard failure after 500 ns TLP stressing, and (c) and (d) confirmation by quantitative AFM imaging. (Salome et al. [16].)

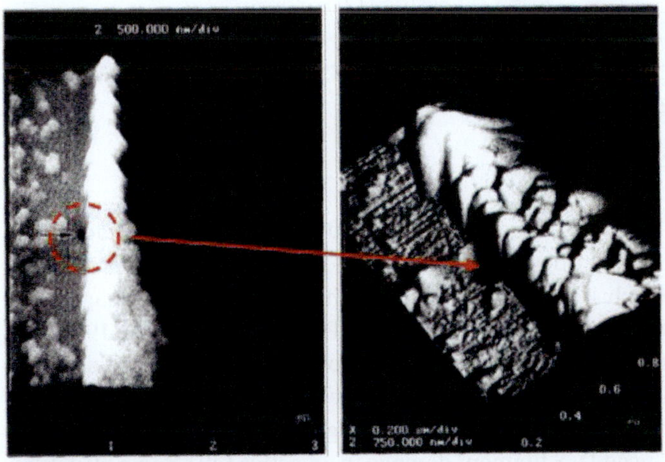

Figure 2.17 AFM images reveal ESD-induced soft defect in oxide sidewall in a CMOS input gate circuit after CDM ESD zapping. (Colvin [17].)

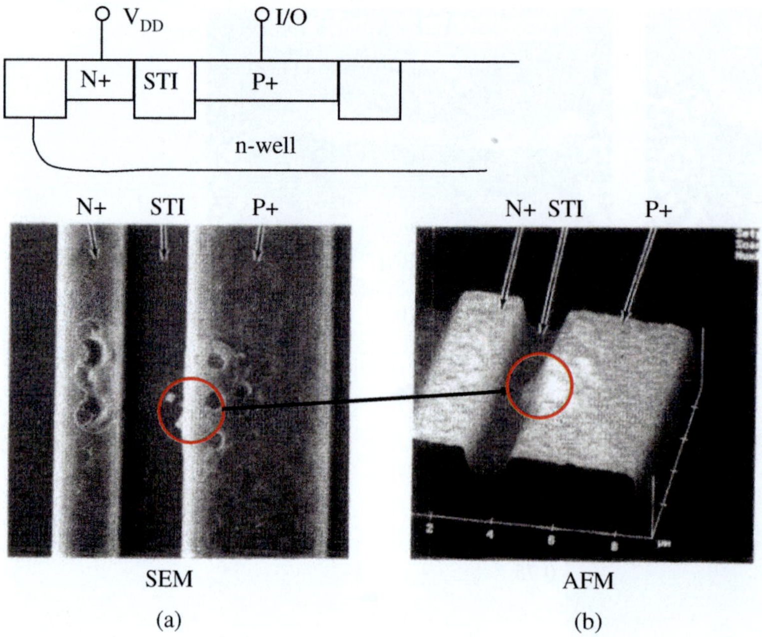

Figure 2.18 SEM shows under-surface ESD defect in a P+/n-Well ESD protection diode within a CMOS DRAM chip (a), which is confirmed by the high-resolution AFM image (b). (Never and Voldman [18].)

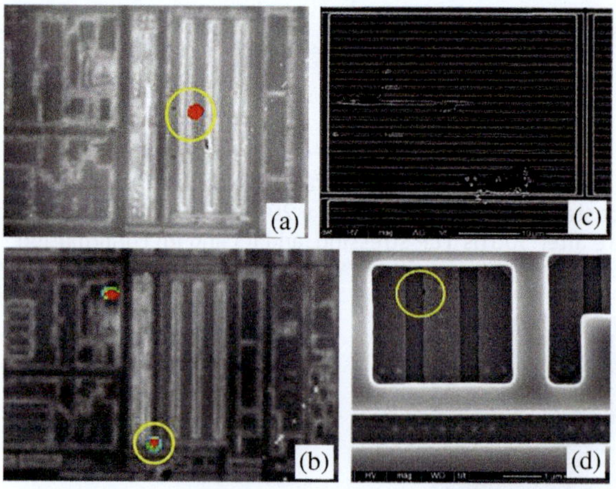

Figure 2.19 ESD damages are observed for the 800-pin mixed-signal SoC in 45 nm CMOS: (a) HBM-induced failure in ggNMOS ESD protection device in one block by PEM, (b) CDM-induced failure in an internal diode next to the ggNMOS ESD protection device by PEM, (c) HBM-induced D–S filament failure in ESD clamp in a different block by SEM, and (d) CDM-induced oxide damage in internal MOSFET in a different block by SEM. (Smedes [19].)

In another example, SEM image in Figure 2.18 shows an under-surface ESD defect underneath the shallow trench isolation (STI) plug, after deprocessing, in a P+/n-Well diode ESD protection structure in a CMOS DRAM chip, which is confirmed by the high-resolution AFM image [18]. Photoelectron microscopy (PEM) is another powerful tool for ESD FA analysis. Figure 2.19 shows an FA example for large 800-pin mixed-signal system-on-a-chip (SoC) chip fabricated in a 45 nm CMOS, where PEM was used to easily locate HBM-induced ESD failure in ggNMOS ESD device (a) and CDM-induced failure in an internal diode parallel to the ggNMOS ESD device (b), and SEM was also used to reveal HBM-induced D–S filament damage in an active ESD clamp structure (c) and CDM-induced oxide defect in an internal MOSFET (d) in different blocks [19].

ESD failures in compound semiconductors can be different from that in Si ICs. Figure 2.20 depicts an FA example for AlGaN/GaN high electron mobility transistor (HEMT) structures by

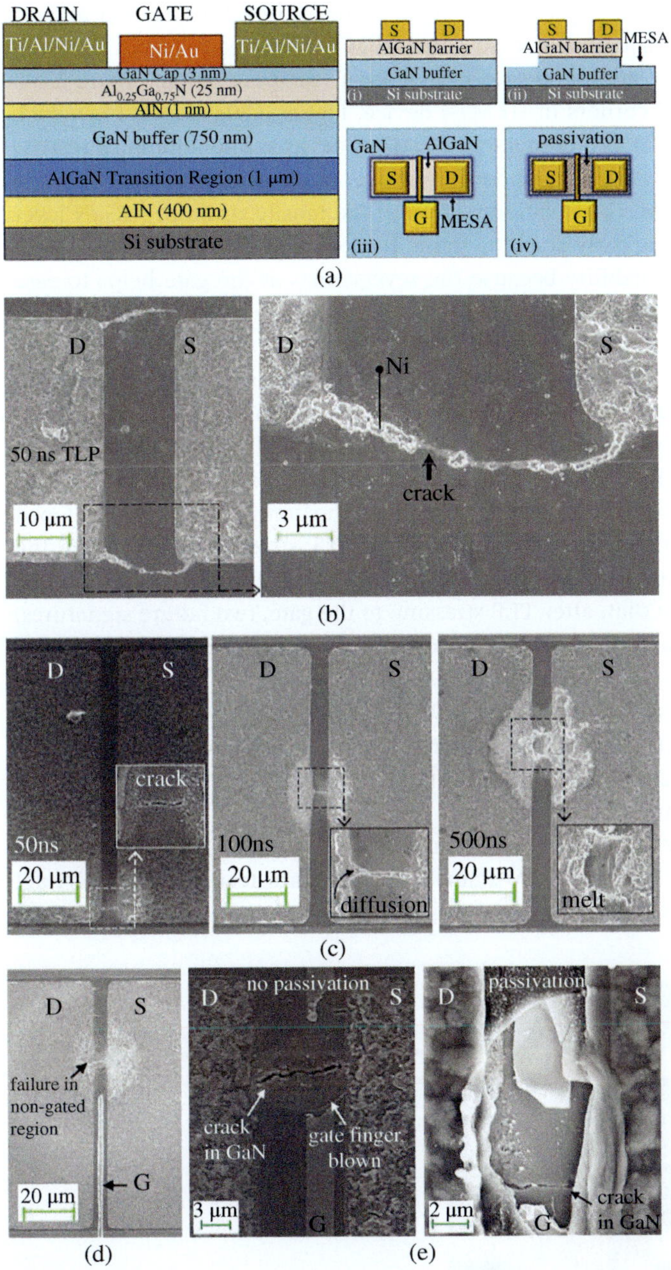

Figure 2.20 FA example for AlGaN/GaN HEMT devices by TLP stressing: (a) device splits of (i) without MESA isolation and gate, (ii) MESA-isolated without gate, (iii) MESA-isolated with gate and (iv) MESA-isolated with gate and surface passivation, (b) cracks at D/S corners in HEMT-i (50 ns), (c) damages within channels, not at D/S corners, in HEMT-ii, (d) failure occurring in channel in HEMT-iii (partially gated), and (e) channel crack and gate blown-off in HEMT-iv. (Modified from Shankar et al. [20]).

TLP stressing [20]. To study the structural impacts on ESD protection, a set of HEMTs was studied including four device splits as shown in Figure 2.20a: HEMT-i has no MESA isolation and no Schottky gate; HEMT-ii has MESA isolation, but no gate; HEMT-iii has both MESA isolation and gate; and HEMT-iv has MESA isolation and gate as well as surface passivation. Figure 2.20b shows ESD damage occurring at D/S corners in HEMT-i device. Figure 2.20c reveals ESD crack failures within the channels, but not at D/S corners, in HEMT-ii devices. The difference in ESD failures may be explained that HEMT-i device suffers from overall stress crowding at the S/D corners, while HEMT-ii devices use MESA isolation to release the tensile strain in AlGaN layer near the drain corners, hence, avoiding early cracking there. It was found that the gate in HEMT (HEMT-iii) serves to improve ESD reliability because the reverse-bias at the gate helps to ease the electrical field density at the drain end when stressing the drain by TLP pulses, which is confirmed in Figure 2.20d, where a partially gated HEMT-iii device shows channel crack damage in the nongated region only. Figure 2.20e depicts similar ESD failures, i.e., both channel crack and gate blown-off, occurring in the devices without passivation (HEMT-iii) and with SiN passivation (HEMT-iv), suggesting that the passivation may not affect ESD reliability of HEMT devices. Further, real-time imaging was used to study ESD failure evolution, as depicted in Figure 2.21 for HEMT-ii devices that clearly shows how ESD failure originated at the drain corner and developed into a crack across the channel from the drain to the source as TLP stressing increases. In the next example, a wide bandgap SiC nMESFET with MESA isolation was evaluated by HBM and TLP stressing [21]. Figure 2.22 shows that, after TLP stressing to the gate, two failure signatures, damage to the unmetallized part of the gate and a gate-to-source shot occurred to the SiC MESFET device.

ESD FA analysis for multiple-wall carbon nanotubes (MWCNTs) was discussed in [22] including both suspended carbon nanotube (CNT) and collapsed CNT structures as shown in Figure 2.23. After TLP stressing, ESD failures were observed in both types of MWCNT devices, which were mainly attributed to self-heating during ESD discharging.

An ESD FA analysis example for GaN-based light-emitting diodes (LEDs) by ESD zapping was discussed in [23]. Figure 2.24a depicts the V-shape LED structure featuring InGaN/GaN multiple-quantum well (MQW) layers and a p-cap layers grown at varying temperatures: 900,

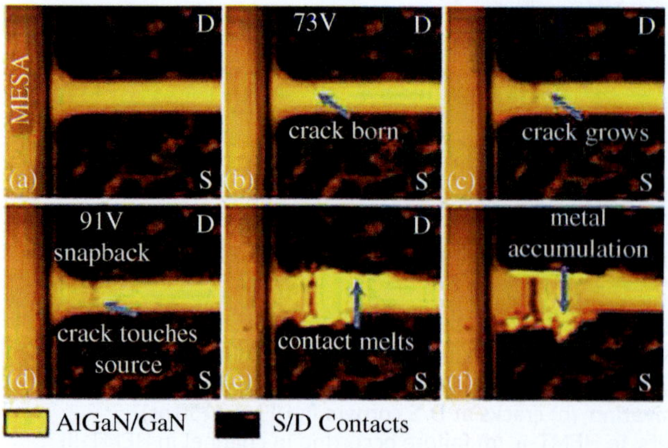

Figure 2.21 FA for HEMT-ii by on-the-fly monitoring reveals failure evolution of the ESD-induced crack in the channel from the drain across the channel to the source under 50 ns TLP stressing (a) structure without gate and with MESA isolation, (b) mechanical crack starts from drain, (c) crack propagates, (d) crack reaches to source, (e) contact metals melt at drain, and (f) metal melting diffuses to source. (Shankar et al. [20]).

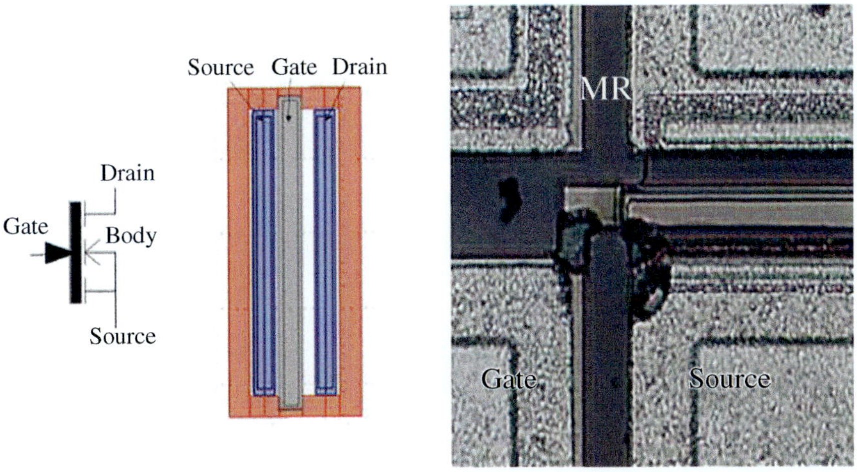

Figure 2.22 ESD FA analysis example for a SiC *n*MESFET with MESA isolation by HBM and TLP stressing shows damages: failure at the unmetallized part of the gate and gate-to-source shot after stressing the gate with the drain floating. (Modified from Phulpin et al. [21].)

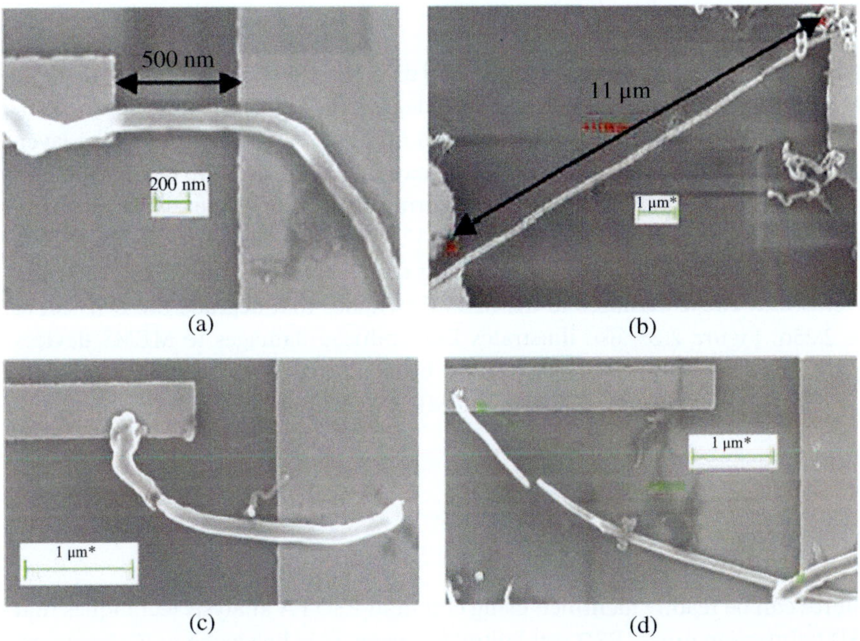

Figure 2.23 ESD FA example for multiple-wall carbon nanotubes (MWCNTs) were characterized by TLP stressing, showing self-heating related failures for both suspended CNTs (a and c) and collapsed CNTs (b and d). (Mishra and Shrivastava [22].)

1000, and 1040 °C. It was found that increase in the p-cap growth temperature can enhance the LED ESD capability under negative ESD zapping from 1100 to 3500 V. Figure 2.24b shows a dead spot, associated with the V-shape defects, occurring randomly in the 900 °C-grown p-cap LEDs. Figure 2.24c shows a large dead area failure near bonding pad, featuring highest electric field, in 1040 °C-grown p-cap LEDs.

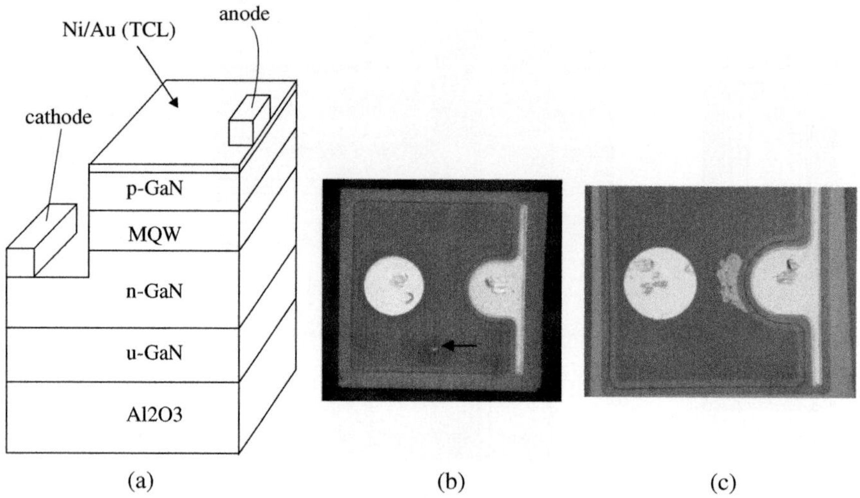

Figure 2.24 ESD FA analysis example for GaN-based LED by ESD zapping tests: (a) device structure, (b) random dead spot defects observed in LED with 900 °C-grown p-cap, and (c) large dead area failure defects near bonding pad observed in LED with 1040 °C-grown p-cap. (Su et al. [23].)

ESD stresses may affect MEMS devices too. Reference [24] discusses an ESD FA analysis example for an SiGe-based MEMS RF switch structure. Figure 2.25 depicts the SiGe MEMS structure comprising a 300 nm thick poly-SiGe beam suspended over an air gap above a SiC protective layer. Voltage-controlled actuation will cause mechanical deformation (D) of the SiGe beam to trigger the RF switch function, which is characterized by measuring the out-of-plane displacement of the SiGe beam as a function of the control voltage, i.e., *D–V* curve as designed. The SiGe MEMS structures have two splits, each having an air gap spacing of G1 = 600 nm and G2 = 1200 nm. Unexpected ESD events may cause damages to the MEMS switches that degrade the *D–V* curve as shown in Figure 2.25d. Figure 2.25 also illustrates ESD-induced damages to MEMS devices by HBM ESD zapping, showing a minor defect in SiGe beam in MEMS-G1 device after 300 V HBM stressing (b) and a major damage to SiGe beam in MEMS-G2 after 600 V HBM zapping (c), respectively.

2.4 ESD Soft Failures

While ESD hard failures can be readily identified using common ESD FA analysis techniques that often show clear ESD failure signatures, ESD soft failure has been a big headache to IC designers. Unlike hard failures that result in immediate IC malfunction, soft failures do not affect IC functions immediately. Soft failures lead to time-dependent ESD damages that may affect IC specs, which generally fluctuate within the datasheet specs range. Soft ESD failures typically result in increased leakage currents at junctions and CMOS gates, and a drift in the threshold voltage of MOSFETs. Soft ESD failures can be characterized by evaluating the TDDB, oxide noises and lifetime, and the results are typically analyzed statistically [25–30]. In one soft ESD FA example, soft defects in 190 Å gate oxide layers in NMOS devices after DC and HBM stressing were investigated [25]. Two types

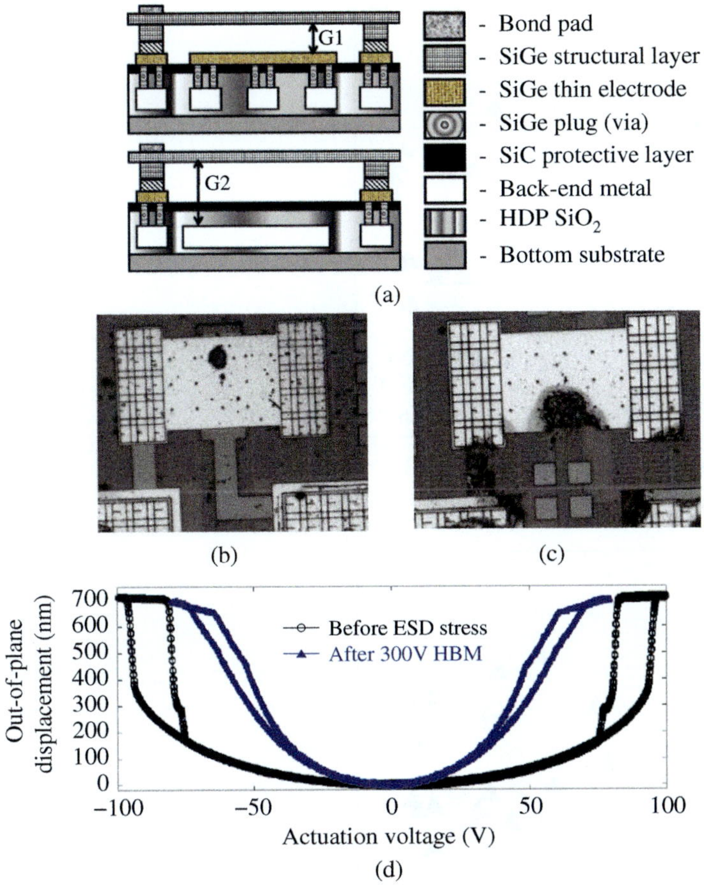

Figure 2.25 ESD FA example for poly-SiGe beam based RF MEMS structures by HBM zapping test: (a) two different MEMS device structures featuring gaps G1 and G2, (b) MEMS-G1 suffering minor ESD damage after 300 V HBM zapping, (c) MEMS-G2 showing a major ESD damage after 600 V HBM zapping, and (d) MEMS-G2 showing degradation in *D*–*V* curve after 300 V HBM zapping. (Sangameswaran et al. [24].)

of soft failures in the CMOS gate oxide films were discovered: increase in gate leakage current and drift of threshold voltage due to the trapped charges in the gate (Type I), and soft defects in the drain junction caused by the drain leakage current (Type II), both induced by DC and HBM stressing. The soft failures in the gate oxide and drain junction were examined by checking the TDDB lifetime. The leakage currents measured range from a few pico Ampere (pA) to several hundreds of milli Ampere (mA). Figure 2.26 shows that no meaningful lifetime reduction due to soft failures in the gate oxide, stressed to a gate leakage level up to several hundreds of nA, was observed. Thermal annealing of 200 °C can substantially recover the gate leakage degradation, which however does not affect the lifetime. On the other hand, it is found that soft failures in the drain junction led to significant reduction in the lifetime even at a lower current stress level. Both drain leakage and lifetime induced by soft failures in the drain cannot be recovered by thermal annealing. In another soft FA example studying ultrathin oxide films of 2.2–4.7 nm stressed by DC and TLP pulses, the soft oxide defects are attributed to TDDB electron trap generation and charge trapping in oxides,

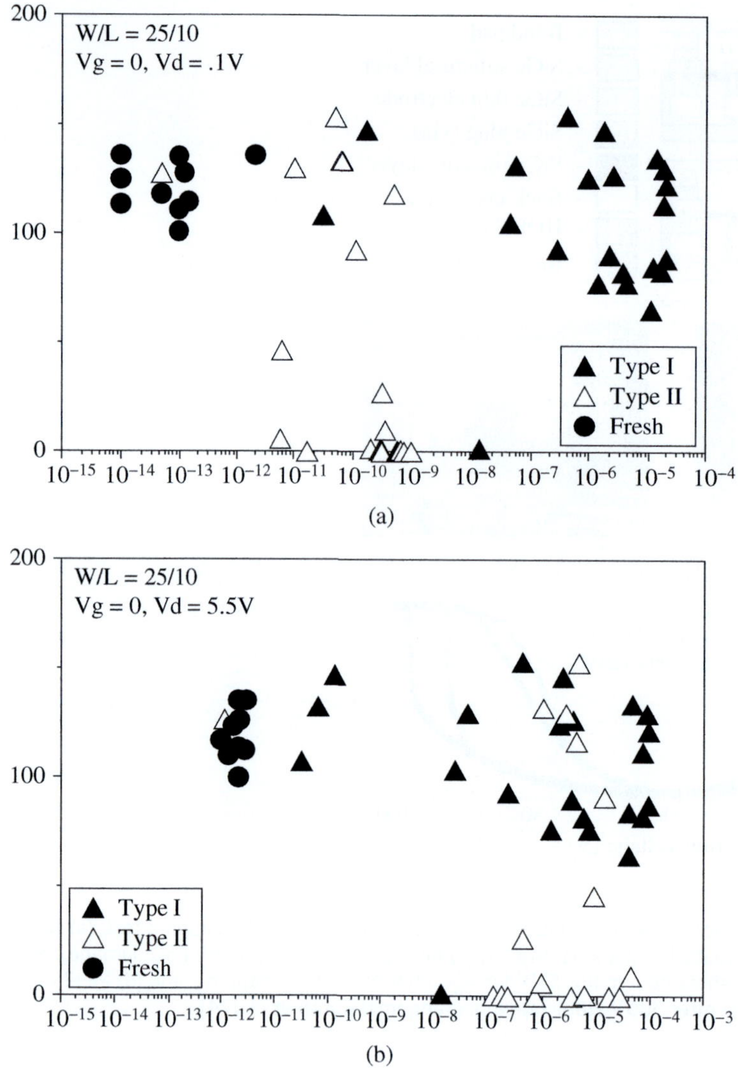

Figure 2.26 Soft failures were studied by measuring TDDB lifetime ~ I_d curves for NMOS devices in fresh and stressed conditions, showing two soft defect types (I and II): (a) measured at V_d = 0.1 V, and (b) tested at V_d = 5.5 V. (Song et al. [25].)

and quantitative analysis was conducted by correlating the stress-induced leakage current (SILC) with the stressing pulse duration as shown in Figure 2.27 [26]. SILC is defined as the difference between the after-stress leakage and before-stress leakage currents, i.e., SILC = $I_{AS} - I_{BS}$. The trap generation rate is proportional to SILC/I_{BS}, which is strongly related to the stressing pulse width as shown in Figure 2.27, indicating that DC stressing results cannot predict soft ESD failures in oxide films correctly [26]. The relationship is modeled as following:

$$\frac{\text{SILC}}{I_{BS}} \propto t^n \tag{2.1}$$

where t is the stressing pulse width and the fitting factor n is extracted as $n \approx 0.33$ for the oxide films studied.

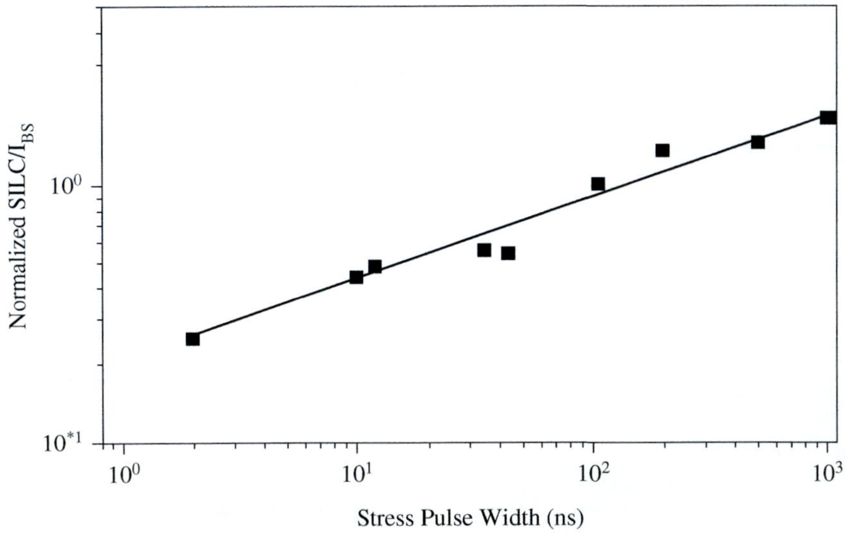

Figure 2.27 ESD soft failure in ultrathin CMOS gate oxide films was studied by correlating the normalized SILC/I_{BS} with the stressing pulse duration (t). (Wu et al. [26].)

2.5 ESD Failure Correlation

There exist many different ESD test models and industrial standards for ESD stressing tests, including HBM, MM, CDM, TLP, and IEC models. A large variety of different ESD testers, homemade or commercially developed, have been used for ESD zapping tests. The question that is often asked by IC designers and customers is how to correlate the ESD testing results, i.e., the ESD protection capability typically stated in the ESD failure voltage threshold level (ESDV) in kilo-voltage (kV), for the same IC tested using different ESD zapping methods or the same ESD zapping test using ESD testers from different vendors. Unfortunately, accurate or even only reasonable correlation between different ESD testing methods and/or testers remains a rather challenging task, mainly because some uncertainties with existing ESD testing mechanisms and ESD testing settings. For example, CDM ESD testing is notoriously unstable and extremely sensitive to the unavoidable parasitic effects associated with both IC dies and packaging, as well as specific CDM testers, testing environments, and even tester operators. Even worse, it is reported that the existing CDM zapping method developed for zapping the traditional pad-based CDM ESD protection circuits may be fundamental faulty, or, at least oversimplified, which may also contribute to the un-reproducibility and instability of CDM testing that will be discussed in details in Chapter 16. In this section, a few examples are presented to show how successful and unsuccessful it can be when trying to correlate ESD test results obtained using different ESD testing models and testers.

In the first example, a set of different IC chips was characterized using different ESD zapping models, i.e., HBM, MM, and CDM testers, at different vendor/user locations, and the ESD testing results are correlated unsuccessfully [10]. The IC devices tested include ASIC disk driver ICs, audio ICs, data communication interface ICs, and automotive control ICs, designed and fabricated in 0.9 μm, 1.2 μm, and 1.5 μm CMOS technologies, respectively. Figure 2.28 presents the ESD failure images for the same automotive control IC made in 1.2 μm CMOS (Device-1) zapped by HBM, MM, and CDM testers. ESD FA analysis shows that, for the same IC chips, different HBM-induced ESD damages occurred when zapped using two different HBM testers, i.e., poly-silicon extrusion

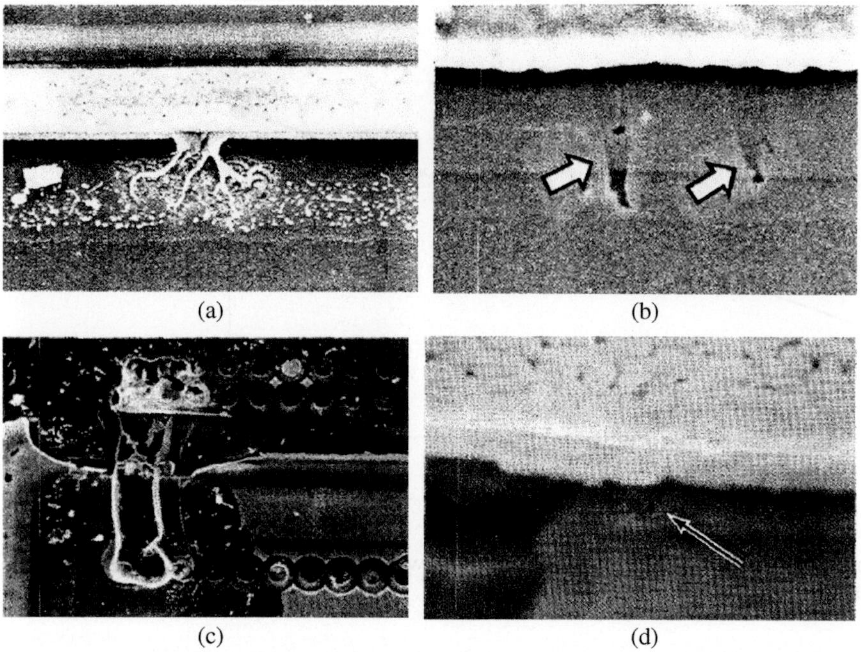

(a) (b) (c) (d)

Figure 2.28 Failure signatures in an automotive IC (Device-1) by different ESD stressing methods: (a) poly-Si extrusion in NMOS ESD protection device by HBM tester 1, (b) oxide damage in internal NMOS by HBM tester 2, (c) metal failure in NMOS ESD protection by MM zapping, and (d) gate oxide damage and poly-Si filament in internal PMOS by CDM zapping. (Kelly et al. [10].)

appeared in an NMOS ESD protection device by HBM Tester-1 (Figure 2.28a) and gate oxide damage in an internal NMOS device by HBM Tester-2 (Figure 2.28b). The same IC chips zapped by MM ESD stressing method resulted in metal melting damage in NMOS ESD protection device (Figure 2.28c). When stressed using a CDM zapping tester, the same IC chips suffered both oxide and poly-silicon filament ESD damages in an internal PMOS device (Figure 2.28d). Next, ESD FA analysis of a communication IC chip (Device-2) shows similar oxide defect types under HBM, MM, and CDM stressing, however, occurring at different spots on the chip, as depicted in Figure 2.8 where the observed oxide damages include oxide defect in an internal NMOS under HBM zapping (a), oxide failure in an internal PMOS by MM stressing (b) and oxide defect in NMOS by CDM zapping (c), respectively. Lastly, ESD FA analysis for an audio IC chip (Device-3) shows different ESD failures due to different ESD zapping methods as depicted in Figure 2.9, where MM stressing led to damage in the protection resistor and oxide failure in NMOS ESD protection structure (a), CDM zapping resulted in oxide defect in an internal NMOS device (b) and HBM stressing caused contact spiking damage in the ESD protection resistor (c), respectively. The multiple-chip multiple-tester study clearly shows that accurate and straightforward correlation of ESD zapping results using different ESD testing models (i.e., HBM, MM, and CDM) or even different ESD testers of same ESD testing model may not readily exist in terms of ESD failure signatures and/or ESD failure thresholds.

In the second FA study, reasonable HBM–TLP correlation was observed for a set of NMOS devices fabricated in 0.5 μm and 0.35 μm CMOS technologies [31]. It shows good matching between the HBM and TLP stressing results for the 0.5 μm NMOS devices and the SEM image in Figure 2.29 illustrates the same kind of contact spiking damages. However, poor correlation was observed

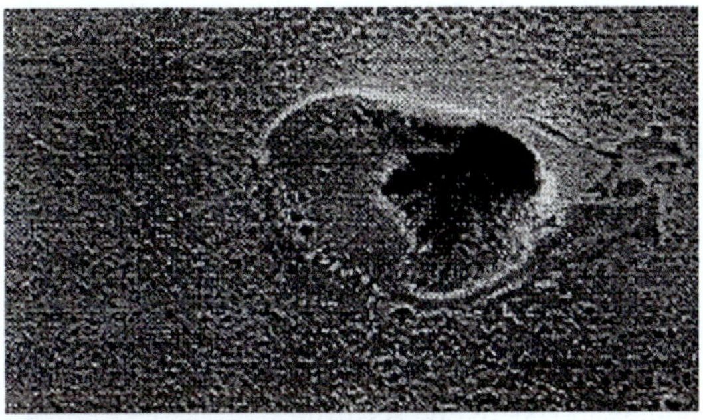

Figure 2.29 Same contact spiking defects observed in NMOS devices in 0.5 and 0.35 μm processes characterized by HBM and TLP stressing, indicating a good HBM–TLP correlation in failure signature type. (Stadler et al. [31].)

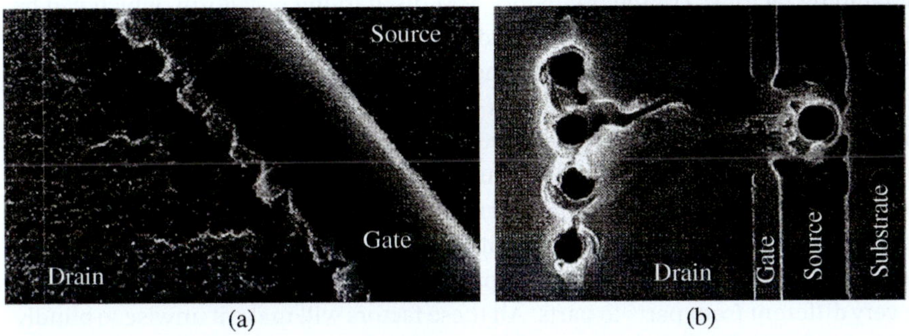

Figure 2.30 Different ESD failure signatures are observed in a 0.35 μm NMOS characterized by different ESD stressing methods: (a) 1.3 A HBM, and (b) 1.5 A TLP, showing poor HBM–TLP correlation. (Stadler et al. [31].)

between the HBM and TLP results for the 0.35 μm NMOS devices. The FA analysis also reveals different ESD failure signatures under HBM and TLP stressing, i.e., poly-Si filament and weak Si melting in drain region due to HBM zapping (Figure 2.30a), and a D–S Si filament damage by TLP stressing (Figure 2.30b), respectively. This study further suggests that while TLP testing is nondestructive and provides rich details of transient ESD discharging characteristics (e.g., I–V curves and leakage currents) that is extremely useful for ESD protection design simulation and optimization, the TLP testing technique may not accurately reflect or correlate with HBM ESD stressing events in a real world. Therefore, using TLP testing to characterize HBM zapping events is still a research topic that requires careful consideration in practical ESD protection designs. In one study of ggNMOS ESD protection structures made in 0.25 μm CMOS at both wafer and package levels, it found that the simplified TLP–HBM correlation formula, Eq. (2.2), has poor accuracy, while including a series resistance in the discharging channel (R_S) may improve the TLP–HBM correlation as given by Eq. (2.3) [32].

$$V_{t2\text{-HBM}} = I_{t2\text{-TLP}} \times R_{\text{HBM}} \tag{2.2}$$

and

$$V_{t2\text{-HBM}} = I_{t2\text{-TLP}} \times \left(R_S + R_{\text{HBM}}\right) \tag{2.3}$$

where $R_{\text{HBM}} = 1.5\ \text{k}\Omega$ is the human body resistance used in the HBM ESD model, $I_{t2\text{-TLP}}$ is the second breakdown current obtained by TLP testing, and $V_{t2\text{-HBM}}$ is the equivalent HBM ESD protection level. R_S can be extracted by either the least square fitting method [33] or the lumped element model method [34]. In one effort to quantitatively correlate HBM and MM ESD protection capability [35], under the simplified hypothesis of ESD thermal failure mechanism being associated with the ESD pulse energy, the HBM and MM ESD failure voltage levels were estimated by simply equating the incident energies of HBM and MM pulse waveforms. In the nutshell, the ESD pulse waveforms are modeled mathematically and the equivalent circuits for HBM and MM ESD zapping set-ups are analyzed. The ratio of HBM to MM failure voltages was approximated as

$$\frac{V_{HBM}}{V_{MM}} \approx \sqrt{\frac{C_{MM} \times R_{HBM}}{C_{HBM} \times R_{MM}}} \tag{2.4}$$

where V_{HBM} and V_{MM} are the ESD failure voltage thresholds, C_{HBM} and C_{MM} are the charge storage capacitances, and R_{HBM} and R_{MM} are total discharging path resistances for HBM and MM test models, respectively. Assume $C_{HBM} \approx 150\ pF$, $R_{HBM} \approx 1500\ \Omega$, and $C_{MM} \approx 235\ pF$, and R_{MM} can be estimated from damping factor (ζ) of the oscillatory MM discharging waveform, which can be measured. Using $\zeta = 4.2$–3.2 for the given MM tester and IC parts stressed, it gives $R_{MM} \approx 17$–$22\ \Omega$. Hence, a ratio of $V_{HBM}/V_{MM} \approx 10$–$12$ was received for the specific case analyzed. This analysis suggests that the MM ESD failure voltage may be 10 times lower than that of HBM ESD failure for a given IC chip. While the ballpark ratio seems to be lovely for IC designers, one must understand that many uncertain factors will seriously affect the accuracy and usefulness of such a quantitative correlation. For example, ESD failures may not be thermal failure and may be caused by electric field-induced dielectric breakdown, and the component values for the ESD equivalent circuit are quite different among various ESD test standards, and particularly, the parasitic parameters of the IC products can be very different from parts to parts. All these factors will make it unwise to blindly use such oversimplified ESD failure rating ratios in the real world.

In summary, indeed, it would be wonderful if accurate correlation formulas exist for ESD protection ratings among various ESD testing models, including HBM, MM, CDM, IEC, TLP, and VFTLP, etc., which may make it easier for IC designers and users to understand the complex ESD failure problems and compare ESD robustness of IC products. It might be that such quantitative correlation does exist scientifically. Unfortunately, it is not here yet. The above examples clearly tell that ESD failure correlation between the ESD protection ratings using different ESD test models or even just different ESD testers of the same ESD testing model can be uncertain, confusing, and very challenging in practical ESD protection design and evaluation.

2.6 ESD Failure Models

While ESD FA can be very informative and useful, particularly for debugging ESD design failures, the FA techniques are generally qualitative that does not provide any numeric guidelines to guide ESD protection designs for IC engineers. Further, ESD FA work can be too complicated and luxury for IC designers to understand and to use in practical designs. Much like SPICE device models for IC simulation, accurate ESD models are essential for on-chip ESD protection circuit design by simulation, which allows both design prediction and optimization before Si tape-out. While the old saying was "analog design is an art" and, similarly, ESD protection design has been largely considered experience determined, such a design mindset is no longer suitable for modern IC designs at advanced technology nodes including ESD protection designs. The key reasons are that advanced ICs are too sophisticated and fabrication costs are so high that any design iteration becomes unaffordable. Today, it is unimaginable that any IC tape-out will be allowed without full-chip IC design verification by simulation. This is particularly important for on-chip ESD protection designs. Unfortunately, experience-based trail-and-error ESD design approaches still sort of dominate in the industry, mainly due to lack of accurate ESD device models and sufficient ESD simulation tools. The complexity for ESD modeling is well beyond that of normal IC device modeling, such as SPICE models. Several key factors make ESD modeling extremely difficult. First, ESD discharging behaviors involve *multiple coupling effects*, i.e., transient-electro-thermal-materials-process-device-circuit-layout level coupling, which must be addressed simultaneously in ESD simulation. Second, the ultrafast nature (ns–ps) of ESD pulsing and discharging results in extremely localized overheating in ESD protection devices, appearing as ESD hot spots in ICs, that makes setting the thermal boundary conditions, such as thermal resistance and capacitance, in IC ESD simulation almost impossible. Third, the ESD-induced hot spots in ICs during different ESD stressing events can be unpredictable, making it very difficult to solve the thermal dissipation equations because the ESD-induced heating sources are generally random. This said, constant research efforts have been devoted to developing useful ESD device models, including analytical, behavioral, and physics-based modeling, to enable ESD protection design simulation at circuit level. Simple ESD failure models are discussed in the following.

In a sense, ESD failure modeling can be straightforward: An ESD transient results in large and fast ESD discharging current, which generates heat that leads to localized hot spots. The increase in temperature during an ESD event seriously affects IC electrical parameters, most of them are temperature-dependent, which generally worsens the ESD overheating. When the temperature increases to certain melting threshold point in semiconductors (e.g., 1687 K in Si) or metals (e.g., 933 K in Al and 1358 K in Cu), ESD-induced thermal failure occurs, leading to IC malfunction and damage. Figure 2.31 depicts a simplified ESD device-modeling concept where an ESD heating source, i.e., a parallelepiped-shaped hot spot featuring dimensions of *a*, *b*, and *c*, is assumed for an MOSFET ESD protection device. Therefore, ESD heat dissipation behaviors can be described by solving the heat conduction equation under suitable boundary conditions for the ESD protection device, which predicts the ESD-induced temperature inside the ESD protection structure in time domain and, hence, predicts the ESD thermal failure threshold [36]. Because the thermal diffusion length for Si is no longer than a few μm for an ESD pulse of ~150 ns, much smaller than the physical dimensions of Si wafers and IC dies, the heat equation will be solved for an infinite medium. This condition is not valid in vertical direction because IC devices reside in the active layer very close to the Si surface and thermally insulated by the dielectrics on top. Within the physical boundaries

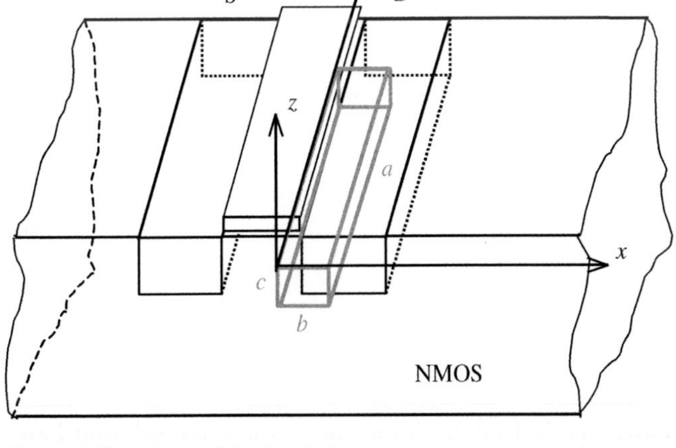

Figure 2.31 A parallelepiped-shaped heat source model assumed for an NMOS device. (Courtesy of IEEE.)

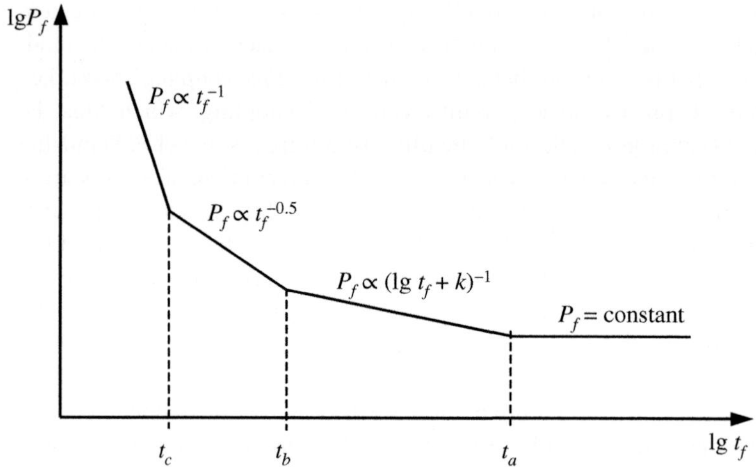

Figure 2.32 Illustration for a four-segment ESD failure model. (Courtesy of IEEE.)

of $-\infty < x < \infty$, $-\infty < y < \infty$ and $-\infty < z \leq 0$, the ESD-induced heat equation can be solved as follows:

$$P_f = \left(\frac{A}{t_f} + \frac{B}{\sqrt{t_f}} + \frac{C}{\log t} + D \right) (T_C - T_0) \tag{2.5}$$

where T_C is the critical temperature corresponding to the ESD failure threshold, T_0 is the ambient temperature, P_f is defined as the *power-to-failure* for the power required to cause the ESD failure under an ESD pulse of a fixed duration, t_f is defined as the time-to-failure for the time needed to cause the ESD failure under a specific ESD stressing power, and A, B, and C are related coefficients. The thermal diffusion time constants associated with the three dimensions (A, B, C) are given as follows:

$$t_a = \frac{a^2}{4\pi D}, t_b = \frac{b^2}{4\pi D}, t_c = \frac{c^2}{4\pi D} \tag{2.6}$$

where D is the thermal diffusion constant of silicon. Figure 2.32 depicts a common four-segment ESD failure model. Using the Green's function method, the ESD-induced heat dissipation equation can be more rigorously solved as follows:

$$P_f = \frac{A'abc\left(T_C - T_0\right)}{t_f}, 0 < t_f < t_c \tag{2.7}$$

$$P_f = \frac{B'ab\left(T_C - T_0\right)}{\sqrt{t_f} - \frac{\sqrt{t_c}}{2}}, \quad t_c < t_f < t_b \tag{2.8}$$

$$P_f = \frac{C'a\left(T_C - T_0\right)}{\log\left(\frac{t_f}{t_b}\right) - 2 - \frac{c}{b}}, t_b < t_f < t_a \tag{2.9}$$

$$P_f = \frac{D'a\left(T_C - T_0\right)}{\log\left(\frac{a}{b}\right) - 2 - \frac{c}{2b} - \sqrt{\frac{t_a}{t_f}}}, t_a < t_f \tag{2.10}$$

where A', B', C', and D' are related coefficients. Indeed, both the concept and the solution look quite nice mathematically. Unfortunately, the use of too many hypotheses and assumptions in such oversimplified ESD modeling makes it practically useless, though conceptually and mathematically beautiful. The key defects for such ESD modeling methods come from many uncertainties, such as where is the real ESD hot spot? What is the real shape of an ESD hot spot? What are the true dimensions for the ESD hot spot? What are the real electrical and thermal boundary conditions for the specific ESD hot spot? How to accurately extract the thermal resistance and capacitance at the boundary of any ESD hot spots? Without a clear and quantitative answer to such critical questions, any ESD modeling method may be useless for real-world ESD protection designs. What makes things more difficult is that ESD thermal failure is just one type of ESD failure mechanism. Other ESD failure mechanisms, such as dielectric breakdown and soft ESD failures, must also be accurately modeled in practical ESD protection designs and simulation.

References

1 Wang, A., Feng, H., Zhan, R. et al. (2005). A review on RF ESD protection design. *IEEE Trans. Electron Devices* 52 (7): 1304–1311. https://doi.org/10.1109/TED.2005.850652.

2 Richards, B. and Footner, P. (1992). *The Role of Microscopy in Semiconductor Failure Analysis, Oxford Microscopy Handbook*, vol. 5. Oxford University Press.

3 Deboy, G. and Kolzer, J. (1993). Fundamentals of light emission from silicon devices. *Semicond. Sci. Technol.* 9: 1017–1032.

4 Kolzer, J., Boit, C., Dallmann, A. et al. (1992). Quantitative emission microscopy. *J. Appl. Phys.* 71 (11): R23–R41.

5 Clark, N., Parat, K., Maloney, T., and Kim, Y. (1995). Melt filaments in n^+pn^+ lateral bipolar ESD protection devices. *Proceedings of EOS/ESD Symposium*, pp. 295–303.

6 Smith, J. (1998). A substrate triggered lateral bipolar circuit for high voltage tolerant ESD protection applications. *Proceedings of EOS/ESD Symposium*, pp. 63–71.

7 Amerasekera, A., van den Abeelen, W., van Roozendaal, L. et al. (1992). ESD failure modes: characteristics, mechanisms, and process influences. *IEEE Trans. Electron Devices* 39 (2): 430–436.

8 Duvvury, C. and Amerasekera, A. (1995). Advanced CMOS protection device trigger mechanisms during CDM. *Proceedings of EOS/ESD Symposium*, pp. 162–174.

9 Kiefer, S., Milburn, R., and Rackley, K. (1993). EOS induced polysilicon migration in VLSI gate array. *Proceedings of EOS/ESD Symposium*, pp. 123–127.

10 Kelly, M., Diep, T., Twerefour, S. et al. (1995). A comparison of electrostatic discharge models and failure signatures for CMOS integrated circuit devices. *Proceedings of EOS/ESD Symposium*, pp. 175–185.

11 Richier, C., Maene, N., Mabboux, G., and Bellens, R. (1997). Study of the ESD behaviour of different clamp configurations in a 0.35 um CMOS technology. *Proceedings of EOS/ESD Symposium*, pp. 240–245.

12 Voldman, S. (1998). The impact of technology scaling on ESD robustness of aluminum and copper interconnects in advanced semiconductor technologies. *IEEE Trans. Compon. Packag. Manuf. Technol. – Part C* 21 (4): 265–277.

13 Voldman, S. and Gross, V. (1993). Scaling, optimization and design considerations of electrostatic discharge protection circuits in CMOS technology. *Proceedings of EOS/ESD Symposium*, pp. 251–260.

14 Russ, C., Bock, K., Rasras, M. et al. (1998). Non-uniform triggering of ggnMOS investigated by combined emission microscopy and transmission line pulsing. *Proceedings of EOS/ESD Symposium,* pp. 177–186.

15 Boschke, R., S.-H. Chen, M. Scholz et al. (2017). ESD ballasting of Ge FinFET ggNMOS devices. *Proceedings of IEEE IRPS*, pp. 3F-3.1–3F-3.6.

16 Salome, P., Leroux, C., Mariolle, D. et al. (1997). An attempt to explain thermally induced soft failures during low level ESD stresses: study of the differences between soft and hard NMOS failures. *Proceedings of EOS/ESD Symposium*, pp. 337–345.

17 Colvin, J. (1993). The identification and analysis of latent ESD damage on CMOS input gates. *Proceedings of EOS/ESD Symposium*, pp. 109–116.

18 Never, J. and Voldman, S. (1995). Failure analysis of shallow trench isolation ESD structures. *Proceedings of EOS/ESD Symposium,* pp. 273–288.

19 Smedes, T. (2012). Analysis of ESD fails in a 45 nm mixed signal SoC. *Proceedings of EOS/ESD Symposium*, pp. 1–9.

20 Shankar, B., Raghavan, S., and Shrivastava, M. (2020). Distinct failure modes of AlGaN/GaN HEMTs under ESD. *IEEE Trans. Electron Devices* 67 (4): 1567–1574. https://doi.org/10.1109/TED.2020.2974508.

21 Phulpin, T., Isoird, K., Tremouilles, D. et al. (2018). Contribution to silicon-carbide-MESFET ESD robustness analysis. *IEEE Trans. Device Mater. Reliab.* 18 (2): 214–223.

22 Mishra, A. and Shrivastava, M. (2016). New insights on the ESD behavior and failure mechanism of multi wall CNTs. *Proceedings of IEEE IRPS*, pp. EL-8-1–EL-8-5.

23 Su, Y.K., Chang, S.J., Wei, S.C. et al. (2005). ESD engineering of nitride-based LEDs. *IEEE Trans. Device Mater. Reliab.* 5 (2): 277–281.

24 Sangameswaran, S., Cherman, V., De Coster, J. et al. (2012). Design and fabrication of SiGe MEMS structures with high intrinsic ESD robustness. *Proceedings of IEEE IRPS*, pp. 3E.4.1–3E.4.6.

25 Song, M., Eng, D., and MacWilliams, K. (1995). Quantifying ESD/EOS latent damage and integrated circuit leakage currents. *Proceedings of EOS/ESD Symposium*, pp. 304–310.

26 Wu, J., Juliano, P., and Rosenbaum, E. (2000). Breakdown and latent damage of ultra-thin gate oxides under ESD stress conditions. *Proceedings of EOS/ESD Symposium*, pp. 287–295.

27 Greason, W., Kucerovsky, Z., and Chum, W. (1993). Latent effects due to ESD in CMOS integrated circuits: review and experiments. *IEEE Trans. Ind. Appl.* 29: 88.

28 Woodhouse, J. and Lowe, K. (1988). ESD latency: a failure analysis investigation. *Proceedings of EOS/ESD Symposium*, p. 47.

29 Tunnicliffe, M., Dwyer, V., and Campbell, D. (1993). Latent damage and parametric drift in electrostatically damaged MOS Transistors. *J. Electrostat.* 31: 91.

30 DiMaria, D. (1999). Electron energy dependence of metal-oxide-semiconductor degradation. *Appl. Phys. Lett.* 75 (16): 2287–2295.

31 Stadler, W., Guggenmous, X., Egger, P. et al. (1997). Does the ESD failure current obtained by transmission-line pulsing always correlate to human body model tests? *Proceedings of EOS/ESD Symposium*, pp. 366–372.

32 Lee, M., Liu, C., Lin, C. et al. (2000). Comparison and correlation of ESD HBM obtained between TLPG, wafer-level, and package-level tests. *Proceedings of EOS/ESD Symposium*, pp. 105–110.

33 Giordano, A. and Hsu, F. (1985). *Least Square Estimation with Applications to Digital Signal Processing*, Chapter 2. New York: Wiley.

34 Verhaeg, K., Roussel, P., Groeseneken, G. et al. (1993). Analysis of HBM ESD testers and specifications using a 4th order lumped element model. *Proceedings of EOS/ESD Symposium*, pp. 129–137.

35 Kuznetsov, V. (2018). HBM, MM, and CBM ESD ratings correlation hypothesis. *IEEE Trans. Electromagn. Compat.* 60 (1): 107–114.

36 Dwyer, V., Franklin, A., and Campbell, D. (1990). Thermal failure in semiconductor devices. *Solid State Electron.* 33: 553–560.

3

ESD Test Models and Standards

3.1 ESD Origins

Electrostatic charging and discharging are two opposite phenomena commonly seen in everyday life. Static electricity is associated with separation and neutralization of positive and negative charges in different materials. "*Charging*" puts static charges into an object, while "*discharging*" removes static charges from a charged object. There are different "charging" mechanisms for static charge generation, such as tribo-charging, charge induction, ion beam charging, and photoelectric charging. Electrostatic charges are created through charge separation procedures, i.e., when two objects of different resistivities, of which at least one object is an insulator or featuring very high resistivity, are brought together, either in direct or close proximity, electrons can transfer in between, resulting in an imbalance of electric charges (net positive or negative) in the objects, respectively. In the common contact-induced tribo-charging event, electrons exchange between two objects of different resistivities and net static charges are retained in each object when separated. This is how does the magic play in a fun show where a plastic comb is used to brush hairs and then it can magically pick up paper scraps. Table 1.1 summaries the tendency of static charge generation of different materials, where the further apart the two materials involved, the easier the triboelectricity "charging" can be with the quantified triboelectricity generation illustrated in Figure 1.6. In a typical electrostatic induction event involving electrical conductors, if an electrically neutral conductor B is placed inside the electric field of a charged object A, some free electrons in B will be attracted to the end closer to A per the Coulomb's Law, causing charge redistribution inside B. If B is then grounded, the free charges on the far side will flow into the ground, leaving net charges inside B, hence "charging" the conductor B. In case of "charging" dielectric materials per the electrical charge induction due to no free electrons inside an insulator, electrical polarization caused by the Coulomb's force will create microdipoles, which shows the macroscale charge separation effect, e.g., attracting paper scraps by a silk-rubbed glass rod. Similar induction-type "charging" procedures are observed through pressure induction and heat induction for specific materials, such as piezoelectric or pyroelectric materials.

"Discharging" is the opposite procedure of "charging," which is a charge neutralization procedure typically referred to as electrostatic charge (ESD). Generally, when two objects with different potentials are brought together, either in direct contact or in close proximity, sudden static charge exchange occurs in between, which results in fast and large ESD transients, or called *ESD pulses*. Natural lightning, gap sparks, and shocking when touching a car door are common ESD events in our everyday life. Unfortunately and unavoidably, ESD discharging can cause severe damages to electronics, including semiconductors, IC chips, as well as any consumer products, such as smartphones and tablets. The invisible ESD transients, featuring current pulses as much

Practical ESD Protection Design, First Edition. Albert Wang.

Table 3.1 Triboelectricity generation in workplace.

Events	Electrostatic voltage at relative humidity		
	10% RH	40% RH	55% RH
Walking across carpet	35 000 V	15 000 V	7500 V
Walking across vinyl tile	12 000 V	5000 V	3000 V
Worker at bench	6000 V	800 V	400 V
Removing DIPs from plastic tubes	2000 V	700 V	400 V
Removing DIPs from vinyl trays	11 500 V	4000 V	2000 V
Removing DIPs from Sryrofoam	14 500 V	5000 V	3500 V
Packing PCBs in foam container	21 000 V	11 000 V	5500 V

as tens of Amperes and voltage surges as high as tens of kilo Volts in short duration ranging from a few nS to a few 100s of nS, can readily kill electronic devices and ICs, and your iPhone. The ESD dangers to electronics can be readily comprehended by looking into the ESD surge levels listed in Table 3.1, which are commonly observed in typical life and working settings. The electronic components (a.k.a., *devices*, loosely defined as items including resistors, diodes, transistors, ICs, etc.) susceptible to ESD failure are referred to as electrostatic charge sensitive (ESDS) devices. The ESD susceptibility property of electronics is referred to as *ESD sensitivity*. ESD protection is therefore required for all electronics, from devices, to IC chips, to system products, in a real world. ESD protection capability, often referred to as ESD robustness, is commonly characterized in terms of the ESD failure threshold voltage (a.k.a., *ESD withstand voltage*, noted as ESDV) of electronics in units of volts or kilovolts (kV), and recently, evaluated for the *ESD failure threshold current* (ESDI) for charged device model (CDM) ESD events. It is hence important to properly and accurately characterize ESD vulnerability of any electronics and robustness of any ESD protection solutions, which is accomplished by stressing the *devices under test* (DUT) using certified ESD tester, which is a testing procedure called *ESD zapping*. Accordingly, to ensure meaningful, accurate, and comparable ESD testing, various industrial ESD testing models and standards have been developed over the past decades for practical ESD protection evaluation. This chapter discusses the commonly used ESD testing models some of which are well established and others are still being actively studied.

3.2 HBM Model

Different ESD phenomena have different ESD mechanisms per their origins, which are described by different ESD test models according to which different industrial ESD testing standards have been developed. Several governmental, industrial, and professional organizations proposed their own ESD test standards tailored to their specific product needs, including that from Department of Defense (DoD), the EOS/ESD Association (ESDA), IEEE, Joint Electron Device Engineering Council (JEDEC), the American National Standard Institute (ANSI), the Automotive Electronics Council (AEC), and the International Electrotechnical Commission (IEC), etc. The most commonly experienced ESD event in everyday life is associated with charging and discharging a human body, which is characterized by a *human body model* (HBM). An HBM ESD event describes a human body-induced ESD discharge procedure where static charges accumulated inside a human body will be discharged into an electronic component (i.e., a device) when touching the device. The HBM ESD-induced electrostatic transient (a.k.a., a pulse, surge or waveform) will stress the device being touched, possibly causing ESD failure to the device. With a global population of around seven billion, countless actual HBM ESD discharge waveforms will be produced every minute by different human bodies when touching countless different electronic devices in a real world. HBM ESD test models are developed to generalize the real-world ESD phenomena caused by human bodies for ESD characterization purpose. The original, well-understood, reliable, and popular HBM model was initially proposed as *MIL-STD-883E Method 3015.7* by DoD for military electronics in 1980s [1]. Figure 3.1a depicts the simplified equivalent circuit for HBM ESD testing (a.k.a., ESD zapping) per Method 3015.7, where a large lumped capacitor (C_{ESD}) representing a human body that is charged to a voltage level of V_{ESD} and the electrostatic charges accumulate inside the human body. When touching a device as triggered by a switch (S1), ESD discharging starts where the stored charges are discharged into the DUT through a discharging resistor (R_{ESD}), resulting in an ESD current pulse stressing the DUT as depicted in Figure 3.1b. Method 3015.7 defines the components of the HBM model as listed in Table 3.2, where the typical values are $C_{ESD} = 100\,\text{pF}$ and $R_{ESD} = 1500\,\Omega$, which represent a "typical" human body – NO question will be taken here please. To guarantee the absolute "Human Right" of each human body among the global population of ~7 billion when using a finger to touch a DUT device, Method 3015.7 mandates that any HBM ESD discharge waveforms produced by any "fingers" must be the same as depicted in Figure 3.1b with the HBM waveform parameters defined in Table 3.3 at given $V_{ESD} = 4000\,\text{V}$, well, this is how any actual HBM ESD tester (a.k.a., *ESD simulator*) on the market must deliver regardless of any tricks used to build up a zapping instrument by a tester vendor. Using an HBM ESD tester complying with Method 3015.7, the ESD robustness (i.e., ESD protection level) of an DUT (e.g., an IC chip) can be characterized per

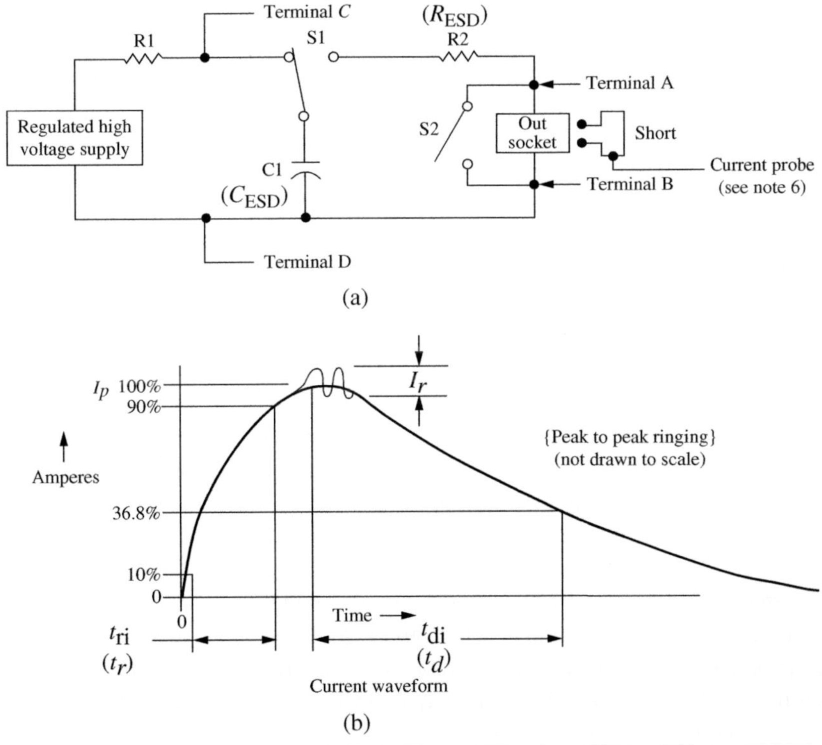

Figure 3.1 Original HBM test model per Method 3015.7: (a) equivalent circuit where C_{ESD} (C1) is pre-charged and then discharges into DUT via R_{ESD} (R2), and (b) standard HBM discharge waveform featuring t_r (t_{ri}) and t_d (t_{dI}). ([1]. Courtesy MIL-STD-883E.)

Table 3.2 Method 3015.7 equivalent circuit Specs.

Elements	Specs
R1	10^6 to 10^7 Ω
R2 (R_{ESD})	1500 Ω ± 1%
C1 (C_{ESD})	100 pF ± 10%
S1	HV relay
S2	Open during ESD zapping

Table 3.3 Method 3015.7 waveform Specs.

Parameters	Values
t_r (t_{ri})	<10 ns
t_d (t_{dl})	150 ± 20 ns
I_{osc} (I_r)	<15% I_p (No ringing after 100 ns)

the following procedures. First, an ESD tester must be validated before conducting any zapping test by shorting the DUT and check the ESD waveforms produced under ±4000 V zapping. The measured ESD waveforms must be the same as given in Figure 3.1b with the waveform parameters defined in Table 3.3, including *rise time* (t_{ri}, more commonly, t_r), *delay time* (t_{dl}, or, more commonly, *decay time* or loosely called *duration*, t_d), *peak current* (I_p), and *ringing* (I_r, the oscillation around I_p). Next, DUT devices can be tested. Each new device must be tested using three positive and three negative ESD pulses with a minimum pulse interval of one second. The recommended pin combination when zapping an IC DUT follows: In Step-1, each pin is zapped with respect to (i.e., grounding) the DUT Ground (GND) pin(s), while all other pins not being tested are open, or floating (i.e., against-GND zapping). In Step 2, each pin is zapped with respect to each combination set of all like-named power supply pins (e.g., V_{DD}, V_{SS}, V_{CC}, V_{EE}), while all other pins not being tested remain open (i.e., against-supply zapping). In Step 3, each input and output pin (i.e., signal pin) is zapped with respect to a combination of all other input/output pins not under testing (i.e., signal-pin to signal-pin zapping), while keeping all other pins floating. After each HBM zapping, the DUT IC will be measured for its performance (i.e., Specs). If no damage or specs degradation beyond the ESD failure criteria is preset, then the DUT passes this zapping level. Otherwise, the DUT IC fails at the given ESD zapping level, and the ESDV level will be printed in the IC datasheet. In the early days, ESD protection capability for ICs was often categorized into several ESD classes as listed in Table 3.4 for easy comparison, though showing no scientific meaning.

The HBM ESD model equivalent circuit given in Figure 3.1a is obviously oversimplified, which cannot produce the standard HBM ESD discharge waveform depicted in Figure 3.1b as required by Method 3015.7. Mathematically, since the HBM model circuit of Figure 3.1a does not have any inductive component (i.e., $L_{ESD} = 0$ H); hence, when an HBM ESD event occurs, the ESD current pulse waveform produced will peak immediately at $t = 0$ second, shown as the dashed line in Figure 3.2, which means that there would be no way to protect any DUT ICs because any ESD protection structure will require a finite response time (i.e., $t_1 > 0$ second) to turn on an ESD discharging path. Additionally, an ESD pulse waveform produced by the HBM model circuit in Figure 3.1a certainly does not comply with the required HBM waveform defined in Figure 3.1b. To address the problem, a modified second-order HBM ESD model equivalent circuit is proposed as shown in Figure 3.3 that contains a parasitic inductor ($L_{ESD} \sim 6$–$10\,\mu$H) associated with the

Table 3.4 Method 3015.7 HBM ESDV classification.

Classes	ESDV levels
Class 1	<2 kV
Class 2	2–4 kV
Class 3	>4 kV

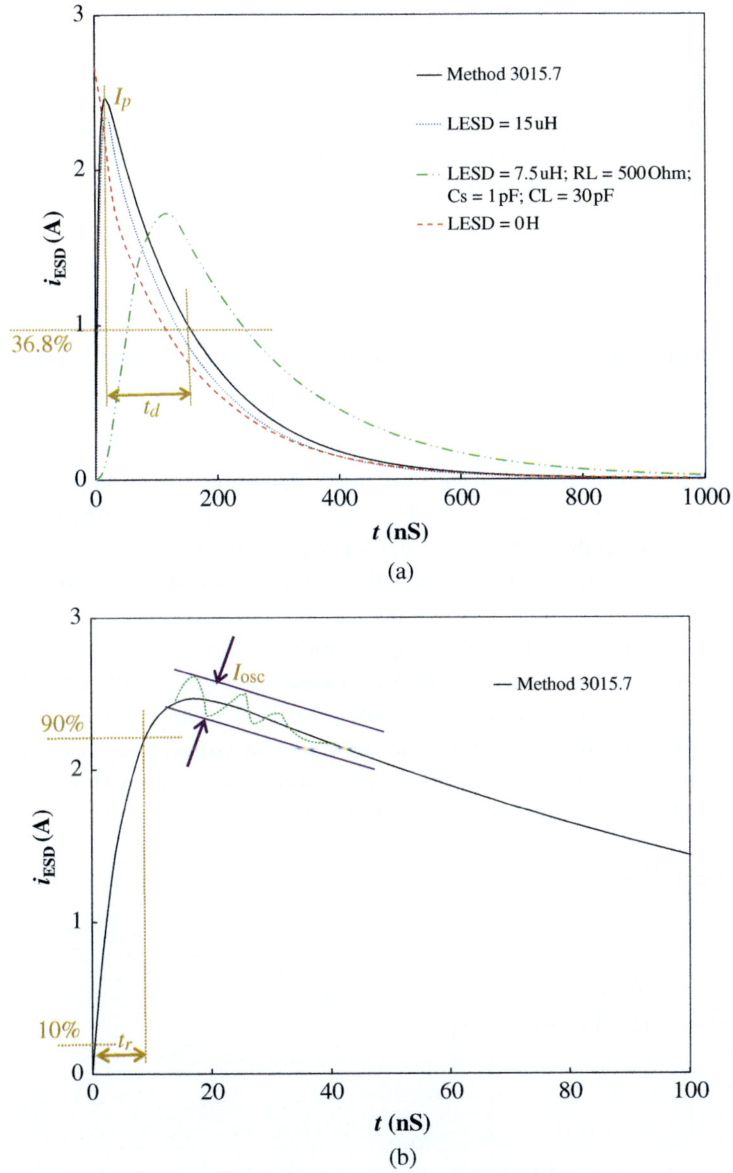

Figure 3.2 Illustration for HBM ESD waveforms: (a) HBM waveform variation per model circuit component values shows impacts of HBM tester parameters, and (b) zoom-in shows a standard HBM waveform complying with Method 3015.7. ([1]. Courtesy MIL-STD-883E.).

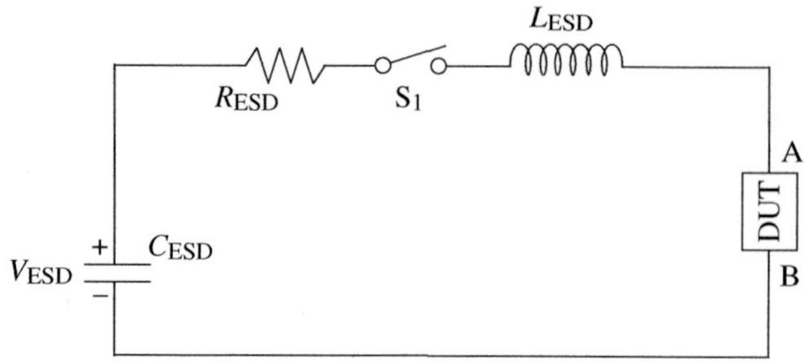

Figure 3.3 A second-order HBM ESD model equivalent circuit includes a parasitic L_{ESD} in the ESD discharging path.

inductive effects of the ESD discharge set-up in a real world. Theoretically, the L_{ESD} will delay an ESD discharge event, which allows a time, i.e., a response time t_1, for an ESD protection structure to be turned on to conduct the ESD transient, hence, protect ICs against possible ESD failures in field. Accordingly, the HBM ESD discharge current waveform can be derived by solving the second-order Kirchhoff current equation for this circuit, resulting in

$$i_{ESD}(t) = \frac{V_{ESD}}{2L_{ESD}\sqrt{\alpha^2 - \omega_0^2}}(e^{\beta_1 t} - e^{\beta_2 t}) \tag{3.1}$$

where i_{ESD} is the transient HBM discharging current, V_{ESD} is the ESD voltage caused by the electrostatic charges stored in a human body (C_{ESD}), $\alpha \equiv \frac{R_{ESD}+R_L}{2L_{ESD}}$ is the waveform damping factor, $\omega_0^2 \equiv \frac{1}{\sqrt{L_{ESD}C_{ESD}}}$ is the natural frequency of the waveform, and $\beta_{1,2} \equiv -\alpha \pm \sqrt{\alpha^2 - \omega_0^2}$. Since HBM ESD discharge is an overdamped discharge procedure, $\alpha > \omega_0$ holds, meaning $L_{ESD} < 56.26\,\mu H$. Typically, $L_{ESD} = 1.5$–$15\,\mu H$ for $t_r = 2$–20 ns is expected. For the HBM waveform defined by Method 3015.7, the required $t_r = 10$ ns needs an $L_{ESD} \approx 7.5\,\mu H$ for a common HBM ESD tester. Obviously, the component values of the HBM model equivalent circuit can substantially affect the ESD discharge waveforms as shown in Figure 3.2, which means that designing and making a good HBM ESD zapping tester is not trivial, which is especially true for making CDM ESD testers. Theoretically, more bells and whistles can be included to make a wonderful HBM model circuit that can always produce whatever ESD discharge waveform required for ESD zapping. Figure 3.4 illustrates an exemplar fourth-order HBM model circuit containing a stray capacitance (C_S) in a human body and a C_{DUT} accounting for the parasitic capacitances associated with both a test board and a DUT device [2, 3], which leads to a formula for the HBM discharge current as [4],

$$i_{ESD}(t) = V_{ESD}C_{ESD}\frac{\omega_0^2}{\sqrt{\alpha^2 - \omega^2}}e^{-\alpha t}\sinh\left(\sqrt{\alpha^2 - \omega_0^2 t}\right) \tag{3.2}$$

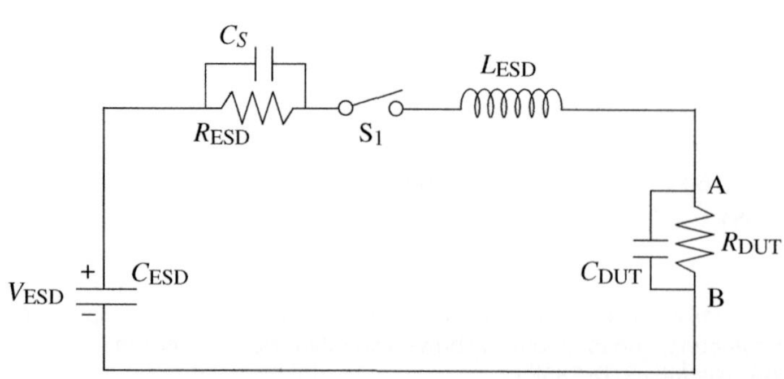

Figure 3.4 A fourth-order HBM model equivalent circuit includes more parasitic elements in the ESD discharge set-up.

These extra components, i.e., parasitic effects in an ESD zapping set-up, make it extremely challenging to design reliable ESD testers, because the unexpected parasitic parameters in both the ESD testers and DUTs can make the ESD zapping waveforms incompliant with the standard ESD stressing waveforms required by ESD test models, as shown in Figure 3.2. This is an extremely important thing to consider in ESD zapping tests in order to avoid the junk-in-junk-out problem. In part to address these uncertainties, many newer HBM ESD test models have been developed based on Method 3015.7 to better regulate the complex ESD zapping practices, which practically makes life not easy for ESD characterization [5–8].

One of the latest HBM test model for component-level ESD characterization is *ANSI/ESDA/JEDEC JS-001-2017*, which was developed jointly by ESD Association and JEDEC [5]. With the core originating from Method 3015.7, ANSI/ESDA/JEDEC JS-001-2017 provides tons of details to, at least, justify its existence. Figure 3.5 shows the standard HBM waveform by ANSI/ESDA/JEDEC JS-001-2017, which is similar to that by Method 3015.7: rise time $t_r \approx t_{ri}$, decay time $t_d \approx t_{dl}$, maximum peak current at shorting $I_{ps\text{-}max} \approx I_p$, and maximum ringing current $I_R \approx I_r$. Considering

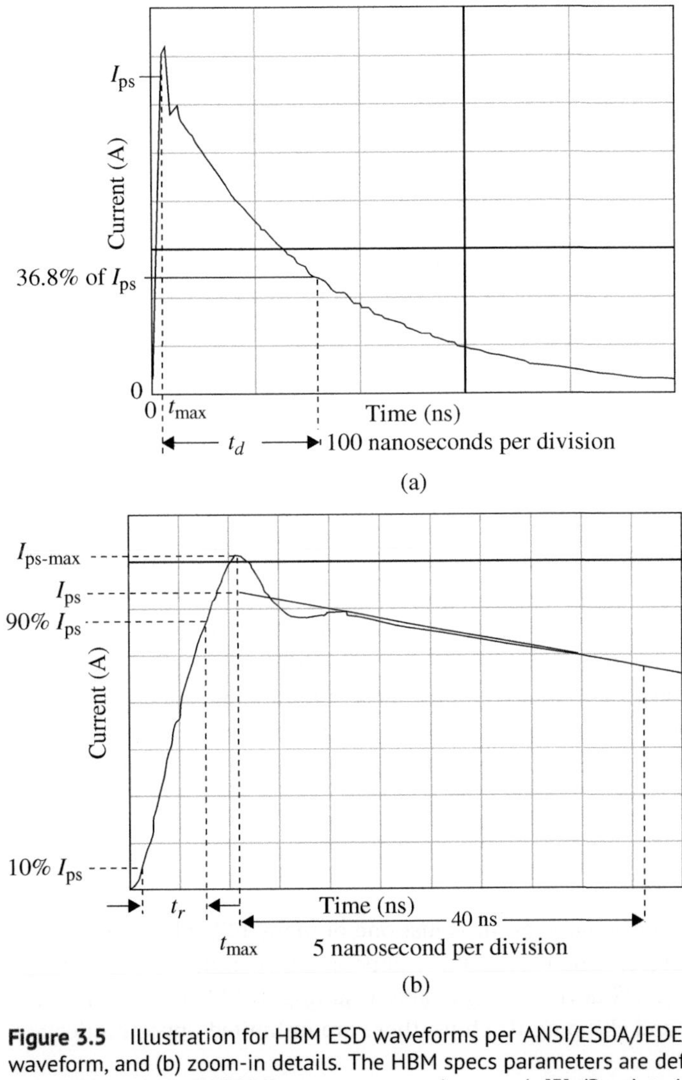

Figure 3.5 Illustration for HBM ESD waveforms per ANSI/ESDA/JEDEC JS-001-2017: (a) full HBM waveform, and (b) zoom-in details. The HBM specs parameters are defined similarly, but slightly different from that in Method 3015.7: $t_r \approx t_{ri}$, $t_d \approx t_{dl}$, $I_{ps\text{-}max} \approx I_p$ [5]. (Reprinted with permission from EOS/ESD Association, Inc.; www.esda.org.)

Table 3.5 Waveform Specs by ANSI/ESDA/JEDEC JS-001-2017.

V level	I_{ps}/short (A)	t_r/short (ns)	t_d/short (ns)	I_R (A)
50 (optional)	0.027–0.040	2.0–10	130–170	15% I_{ps}
125 (optional)	0.075–0.092	2.0–10	130–170	15% I_{ps}
250	0.15–0.18	2.0–10	130–170	15% I_{ps}
500	0.30–0.37	2.0–10	130–170	15% I_{ps}
1000	0.60–0.73	2.0–10	130–170	15% I_{ps}
2000	1.20–1.47	2.0–10	130–170	15% I_{ps}
4000	2.40–2.93	2.0–10	130–170	15% I_{ps}
8000 (optional)	4.80–5.87	2.0–10	130–170	15% I_{ps}

Table 3.6 ESD classification by ANSI/ESDA/JEDEC JS-001-2017.

ESD classes	ESDV (V)
0Z	<50
0A	50–125
0B	125–250
1A	250–500
1B	500–1000
1C	1000–2000
2	2000–4000
3A	4000–8000
3B	≥8000

the ringing effect, a new nominal peak current at short-circuit (I_{ps}) is defined by back-extrapolating following the decaying HBM ringing waveform from t_{max} to t_{max+40} (i.e., 40 ns beyond), where t_{max} corresponds to $I_{ps\text{-}max}$. The HBM tester waveform verification routines require five positive and five negative pulses at each nonoptional zapping voltage levels as listed in Table 3.5. For ESD zapping test, a sample of three devices should be stressed using at least one positive and one negative pulse with a minimum time interval of 100 ms, and the recommended ESD zapping voltage levels are given in Table 3.5. The suggested pin combinations for ESD zapping are more complicated than that per Method 3015.7. The recommended ESD classification for DUT is given in Table 3.6.

3.3 MM Model

Obviously, human body-induced ESD discharge event is just one of many ESD phenomena in a real world. Generally, any charged object, when touching or approaching closely to an electronic component, will discharge into the ESDS device, and possibly resulting in ESD failures. During manufacturing as well as handling and shipping IC chips, there are unlimited opportunities for an IC part getting zapped by metallic machinery or tools, which are precharged just like a human body. Such machinery pieces include everything you can imagine on a manufacturing floor, for example, a robot carrying a Si wafer, a bonding machine, an automatic test equipment (ATE). A charged machine is similar to a charged human body, but having different electronic parameters. One can easily image that a machine may have a larger charge storage reservoir (i.e., C_{ESD}), smaller discharging inductance (i.e., L_{ESD}), and much smaller discharging resistance (i.e., R_{ESD}) compared to a human body, which, in turn, will produce a very different ESD discharge waveform. Such machine-induced ESD events are characterized by a *Machine Model* (MM), which was initially proposed by Japan Electronics and Information Technology Industries Association (JEITA) in 1981 as a very fast worst-case HBM model, which was followed by many other MM ESD test standards developed by the industry over years, including JEDEC, AEC, and ESDA [9–12]. Figure 3.6 depicts a simplified equivalent circuit for MM ESD model, for which the early MM test standards typically define the parameters as $C_{ESD} = 200$ pF, $L_{ESD} \sim 0$ H, and $R_{ESD} \sim 0\,\Omega$. Similar to HBM test standards, the details for an equivalent circuit model and an MM tester can be varying, however, a qualified MM zapping simulator must deliver an MM waveform defined by an MM test standard, for example, the one shown in Figure 3.7 and Table 3.7 per the AEC-Q100-003-Rev-E standard [12]. Uniquely, an MM waveform is oscillatory and faster than a typical HBM waveform. Also, the peak current in MM zapping is much higher than that in HBM stressing. Together, MM-induced ESD failures are likely different from that caused by HBM ESD events. Due to the negligible L_{ESD} and

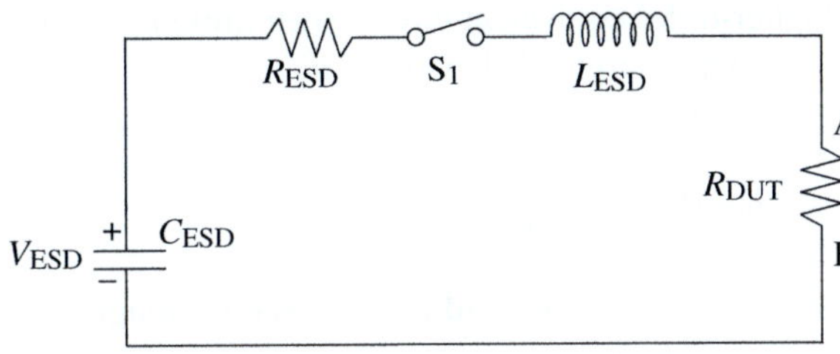

Figure 3.6 A general equivalent circuit for MM ESD model.

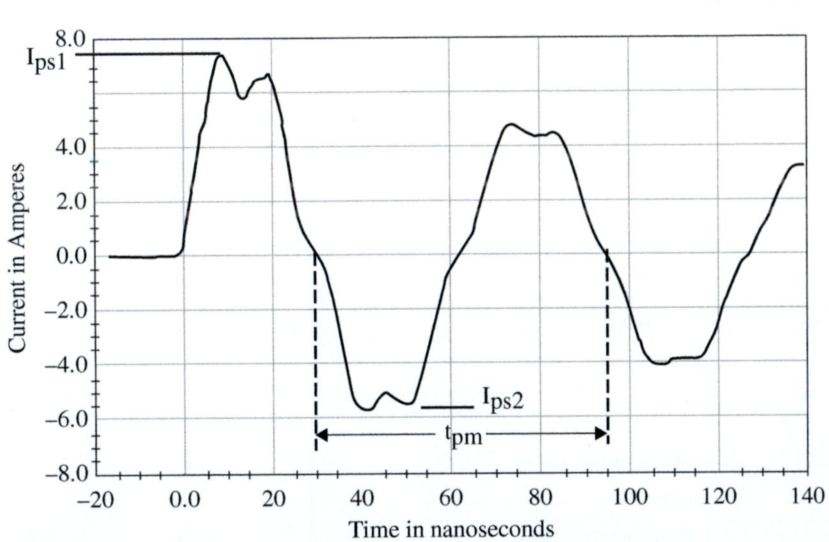

Figure 3.7 A standard MM discharge waveform, under 400 V zapping at Short Wire condition, is oscillatory [12]. The key waveform parameters are the first positive peak current (I_{ps1}), the second positive peak current, (I_{ps2}) and the major pulse period (t_{pm}). (Courtesy by AEC-Q100-003-Rev-E. Courtesy AEC.)

Table 3.7 MM waveform Specs (Short/400 V) [12].

Zapping (V)	First positive current peak, I_{ps1} (A)	Second positive current peak, I_{ps2} (A)	Major pulse period, t_{pm} (ns)
100	1.5–2.0	67–90% I_{ps1}	66–90
200	3.0–4.0	67–90% I_{ps1}	66–90
400	6.0–8.1	67–90% I_{ps1}	66–90
800	11.9–16.1	67–90% I_{ps1}	66–90

Courtesy by AEC-Q100-003-Rev-E.

R_{ESD} specified in the industrial MM models, it is obviously that an MM tester may be very sensitive to any parasitic inductance and resistance in the MM discharging path, both from the tester and the DUT IC itself, which leads to a major MM testing uncertainty problem – a grand challenge in real-world MM zapping test practices. Figure 3.8 illustrates this MM testing uncertainty problem by comparing a standard MM waveform per AEC-Q100-003-Rev-E, and those with slightly different L_{ESD} and R_{ESD}. Clearly, even small, yet unavoidable, parasitic effects from an IC package may substantially change the MM waveforms, which will result in a junk-in-junk-out MM ESD testing problem. Obviously, using the oversimplified equivalent circuit with $L_{ESD} \sim 0$ H and $R_{ESD} \sim 0\,\Omega$ will not be able to produce the required oscillatory MM waveform as defined in any MM test standards. A higher-order equivalent circuit with finite L_{ESD} and R_{ESD} must be used to model the real-world MM ESD discharging behaviors, which is shown in Figure 3.6, where the total discharge resistance is given as $R = R_{ESD} + R_{DUT}$. An analytical model can then be derived as

$$i_{ESD}(t) = \frac{V_{ESD}}{L_{ESD}\omega_d} e^{\frac{-R}{2L_{ESD}}t} \sin(\omega_d t) \tag{3.3}$$

where $\omega_d \equiv \sqrt{\omega_0^2 - \alpha^2}$ and $\omega_0 > \alpha$. With enough wisdom and complexity, a suitable equivalent circuit can always be constructed to generate the desired ESD discharge waveform exactly same

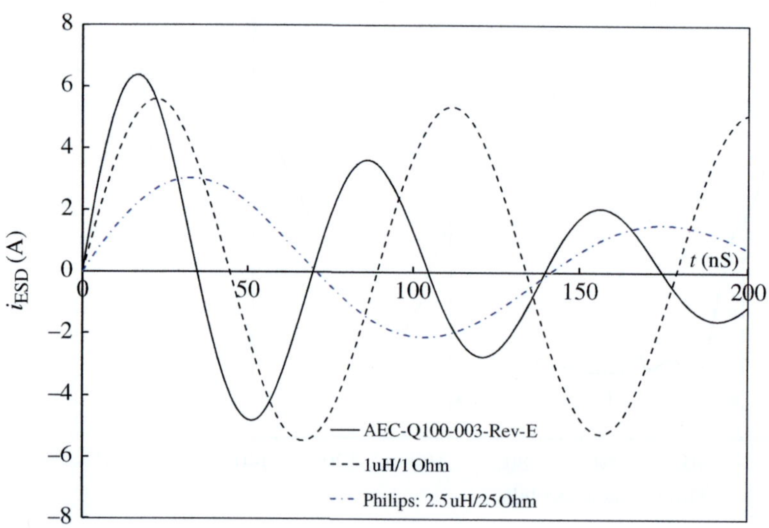

Figure 3.8 Equivalent circuit component values can significantly affect the MM ESD waveforms produced, causing testing uncertainties.

as that defined in an MM test standard. Nevertheless, practically, making a reliable MM zapping tester that can repeatedly produce the required MM waveform has never been an easy task. In fact, unable to reliably reproduce the standard MM waveforms in ESD testing has so far been a major problem, which is attributed to uncertainty in both MM ESD protection designs and zapping tests in a real world. One of the root causes to the MM test uncertainty is certainly associated with the negligible L_{ESD} and R_{ESD} suggested in MM models because any inevitable parasitic effects in the MM discharging path and inside an DUT can substantially alter the MM equivalent circuit, resulting in substantial variations in the MM discharge waveforms, which lead to testing uncertainty often observed in MM zapping practices. Interestingly, it was recently suggested that the MM test model is no longer needed in evaluating ESD susceptibility of ICs, which can be covered by use of HBM and CDM ESD test models in combination. Consequently, JEDEC and ESDA announced discontinuation of their MM testing standards, claiming that MM evaluation is not necessary for ESD qualification of ICs [13]. The main reasons cited include MM is redundant to HBM at device level since it produces the same failure mechanisms; MM test has less repeatability then HBM test due to greater sensitivity to parasitic effects; there are no significant engineering studies supporting MM testing, and "machine model" was misnamed for "low-voltage HBM" events of fast "metal-to-metal" contact discharge nature that can be covered by CDM test model. However, engineers should be cautious about any suggestion of abandoning the MM test model simply for the sake of engineering difficulties (e.g., MM uncertainties) and economic burdens (e.g., testing workload and delay in time-to-market). In a real world, unique ESD phenomena that were loosely defined as MM events do exist, which cannot be completely covered by HBM and CDM models. For example, MM waveforms are low-damped oscillation containing many significant discharging current peaks that are very different from typical fast-attenuated CDM waveforms. MM waveforms are also much faster and feature significantly higher peak currents than HBM waveforms. Therefore, while imagining a simple correlation factor (i.e., a voltage ratio range) between ESDV measured per HBM and MM may be too optimistic and simple-minded [14], simply abandoning the MM model due to the "reasons" mentioned earlier may not be a wise idea either. In fact, more research efforts must be given to thoroughly investigate the scientific properties of MM phenomena and develop engineering know-hows for MM ESD protection designs and MM testing. Indeed, though we are accustomed to treating things digitally today, the real world is analogue that requires a mindset of being able to think things in a somewhat fuzzy way. To oversimplify an engineering problem due to lack of a reliable solution is not a solution to any real-world design problem.

3.4 CDM Model

Rapid advances in silicon CMOS technologies come along with aggressive dimensional scaling, which is another ESD danger emerging as a grand, and somewhat hopeless, IC reliability challenge that often causes CMOS gate oxide breakdown. A CDM was developed to describe this new ESD phenomenon. In the real world, an ESDS device, such as ICs, can be charged possibly due to tribo-charging or field induction throughout its whole lifetime: from wafer fabrication, to measurements, to packaging and assembly, to handling, to shipping, all the way to the users' hands for everyday applications. The electrostatic charges generated are stored inside the ESDS device, throughout the inner body in random and mysterious ways. When the ESDS device gets a chance to touch a ground, e.g., soldering an IC to a printed circuit board (PCB), the electrostatic charges stored within the ESDS device will discharge into the ground, which produces an ultrafast (in picosecond scale)

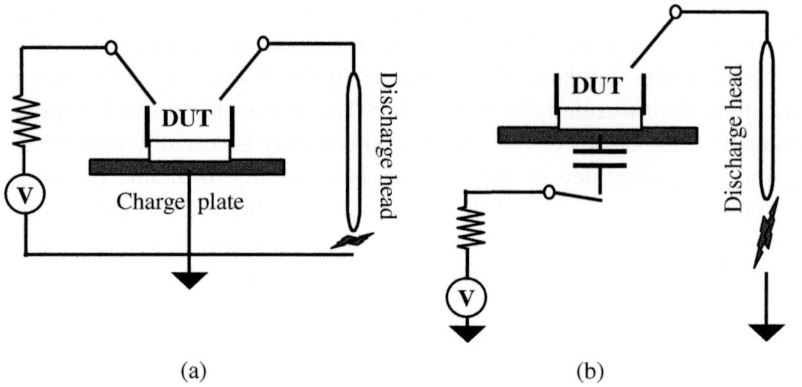

Figure 3.9 CDM charging has two formats: (a) direct charging, e.g., contacting or triboelectricity charging, and (b) non-contact field induction via a high-voltage source.

and very strong (up to tens of Amperes) ESD pulse that can cause internal ESD damage to the IC. This type of ESD event is categorized as CDM ESD event, which is entirely different from the HBM ESD event. A CDM ESD event consists of both charging and discharging procedures. *CDM charging* is done by either direct charging or field induction charging as depicted in Figure 3.9. In direct charging, the ESDS device gains static charges by direct contact with a charge generator (often through metal-to-metal contact) as shown in Figure 3.9a. CDM indirect charging is realized by electric field induction through dielectric isolation without conductive contact as shown in Figure 3.9b. When grounding, *CDM discharging* occurs through a grounded pin of the IC part in a real world, which is simulated in CDM ESD testing by using a grounded CDM discharge probe head (i.e., a pogo pin) to touch an IC pin.

There are many CDM ESD test models and standards of different flavors proposed and developed by various entities over years [15–18]. One early day CDM model was the JEDEC JESD22-C101-A, which schematically depicts the suggested CDM tester set-up and the required CDM discharge waveform as shown in Figure 3.10 [16]. It is clear that the CDM discharge waveform is very unique, featuring an ultrafast oscillatory and rapidly damped waveform shape with several current peaks as shown in Figure 3.10. The CDM discharge waveform specs are summarized in Table 3.8, showing an ultrashort pulse rise time of $t_r < 400$ ps. Figure 3.11 depicts a simplified second-order CDM testing model circuit that can produce ultrafast multiple-peak oscillatory CDM waveforms. Different

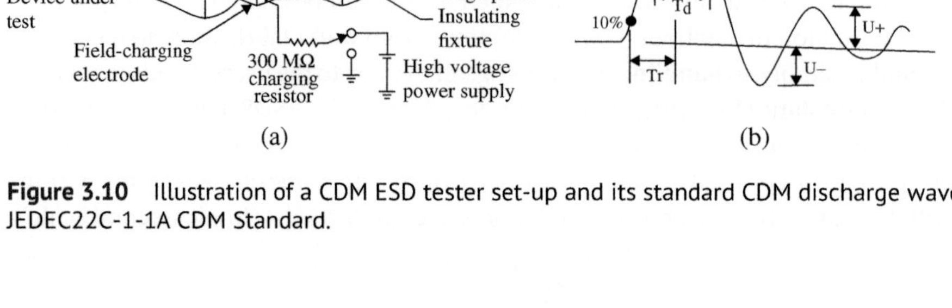

Figure 3.10 Illustration of a CDM ESD tester set-up and its standard CDM discharge waveform defined in JEDEC22C-1-1A CDM Standard.

Table 3.8 Short-circuit CDM ESD waveform specs per JESD22-C101-A (short-wired and C_{ESD} = 6.8 pF).

V_{ESD} (V)	Peak current I_p (A)	Undershoot I_{p2} (A)	Overshoot I_{p3} (A)	Full width at half height (FWHH) (ns)	t_r (ns)
500	$5.75 \pm 15\%$	$<50\% \ I_p$	$<25\% \ I_p$	1.0 ± 0.5	<0.4

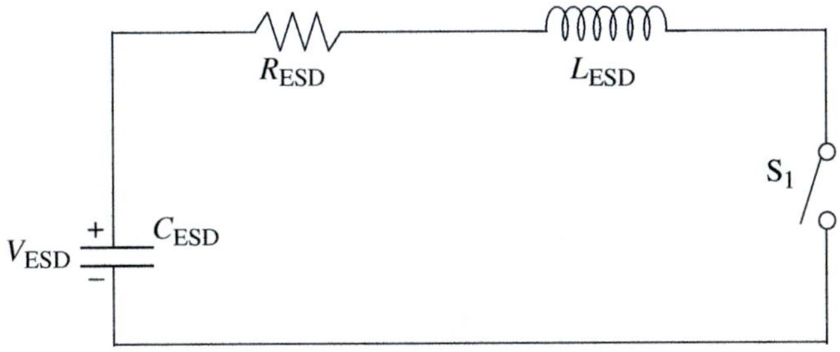

Figure 3.11 A second-order CDM ESD discharge model circuit where the ESDS DUT is represented by C_{ESD}.

from the *external-oriented* HBM and MM ESD models where the charge reservoir (C_{ESD}) is with an external object and the DUT is subject to stresses from the incoming ESD pulses (a "foreigner"), CDM models an *internal-oriented* ESD event though the C_{ESD} and the DUT in the equivalent circuit appearing to be the same. It is generally considered that the CDM test set-up features negligible discharging L_{ESD} (~nH) and R_{ESD} (~Ω), which means that the CDM discharge waveform is extremely sensitive to the parasitic effects in the CDM zapping set-up including the CDM tester and the DUT. Figure 3.12 illustrates the sensitivity of CDM waveforms to parasitic L_{ESD} and R_{ESD}, showing that a slight change in the parasitics will make the ESD discharge waveform very different from

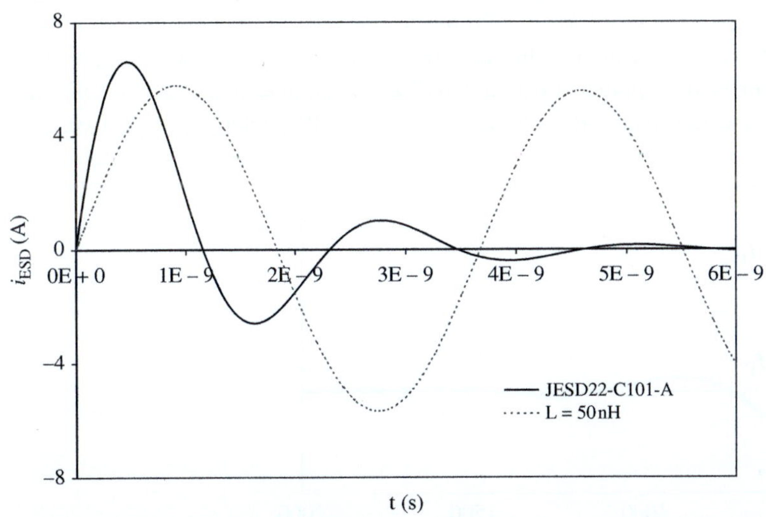

Figure 3.12 ESD discharge waveforms produced by a CDM zapping tester (C_{ESD} = 6.8 pF and V_{ESD} = 500 V) is very sensitive to the parasitic effects associated with the discharging channel and the DUT itself. The solid line is from JESD22-C101-A and the dashed line corresponding to L = 50 nH as an example for parasitic effect.

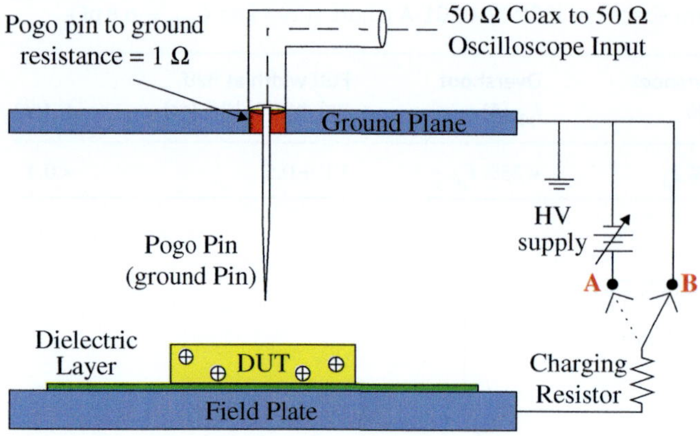

Figure 3.13 A tester schematic for field-induced CDM (FICDM) per ANSI/ESDA/JEDEC JS-002-2018.

that defined in an industrial CDM standard, e.g., JESD22-C101-A. Such sensitivity makes it very challenging to make high-quality CDM ESD testers, designed following whatever industrial CDM test standards, which can reliably reproduce CDM zapping waveforms. To address this hard challenge, modified CDM test models have constantly been proposed over years. One of the recent-day CDM test model is the ANSI/ESDA/JEDEC JS-002-2018 [18], which describes the CDM test set-up and CDM discharge procedures in great details. Figure 3.13 illustrates the proposed field-induction charged device model (FICDM) testing hardware schematic, showing that an ESDS DUT is placed on top of a conductive *field plate* separated by an insulating layer. FICDM testing comprises *charging* and *discharging* steps. At the charging step, a high-voltage (HV) supply is applied to the field plate to raise its potential to a given voltage level, which generates a strong electric field that will induce static charges inside the DUT IC. At discharging step, the conducting pogo-pin will lower down to contact the selected pin-under-test (PUT) of the DUT, which initiates discharge of the stored static charges into the ground, generating an ultrafast CDM stressing waveform as shown in Figure 3.14 and, possibly, resulting in CDM ESD failure to the DUT IC. Obviously, the stressing CDM waveforms must be verified before conducting any CDM zapping tests, which should be done by using two types of verification plates (a.k.a., metal discs): a small size of $C_{ESD} \sim 6.8\,pF$ and a large size of $C_{ESD} \sim 55\,pF$, each corresponding to small and large IC packages, respectively, to

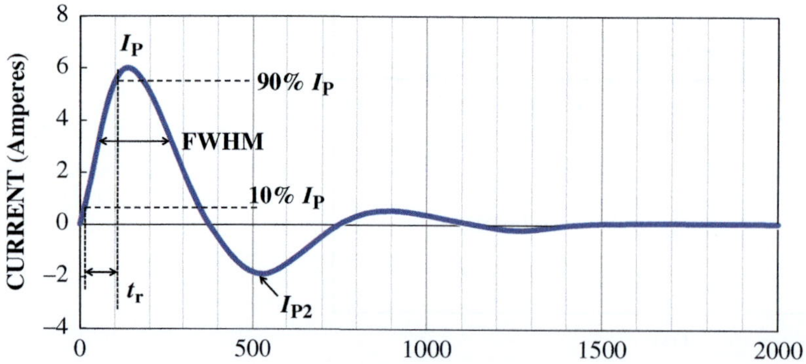

Figure 3.14 A typical CDM tester discharge waveform per ANSI/ESDA/JEDEC JS-002-2018. (Reprinted with permission from EOS/ESD Association, Inc.; www.esda.org.)

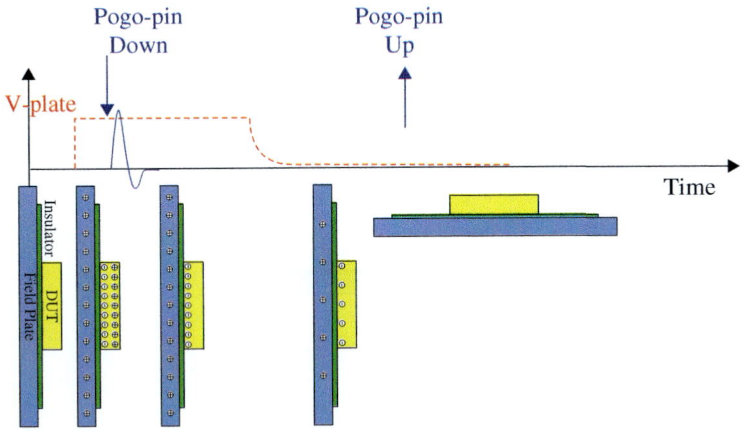

Figure 3.15 Illustration of single CDM discharge procedures per ANSI/ESDA/JEDEC JS-002-2018.

simulate the charge storage reservoir of the DUT IC. The suggested CDM zapping test procedures include single and dual discharge methods. Figure 3.15 depicts the details of the *single* discharge procedures where single positive or single negative CDM ESD pulse is applied to DUT for individual CDM discharge with the following sequences: an uncharged fresh DUT is placed on the field plate; the field plate is raised to a given voltage level per CDM zapping levels, which induces static charge pairs inside the DUT IC; the pogo-pin is lowered down to touch the PUT of the IC resulting the first CDM discharge originated internally from the DUT IC (i.e., DUT *discharging*: DUT is stressed by first/one/only DUT discharge event of a positive CDM pulse produced by discharging the positive charges inside the DUT into GND); charges of opposite polarity will stay inside the DUT (i.e., DUT *charging*: resulting in net electrons in the DUT); the voltage of the field plate is resistively and slowly reduced to ground zero (i.e., through resistive current leaking), which simultaneously causes the net charges stored in the DUT to slowly drain into the GND through the pogo-pin (second DUT discharge) that is still in contact with the PUT pin; and finally, the pogo-in is lifted up and separated from the PUT, returning the DUT IC back to its net-zero uncharged state and ready for the next CDM zapping routine. It is worth note that the single discharge procedure involves only one CDM discharge to stress the DUT device. Figure 3.16 depicts the details of the *dual* discharge procedures where single positive and single negative CDM ESD pulses are applied to produce one pair of alternating polarity CDM discharges to zap the DUT with the following testing sequence: the uncharged fresh DUT is placed on the field plate; the field plate is raised to a given voltage level (e.g., positive) per CDM zapping levels, which induces static charge pairs inside the DUT IC; the pogo-pin is lowered down to touch the PUT of the IC resulting in the first CDM discharge originated internally from the DUT IC (i.e., *first* DUT *discharging*: DUT is stressed by the first DUT discharge event of a positive CDM pulse generated through discharging the positive charges inside the DUT into GND); charges of opposite polarity will stay inside the DUT (e.g., net negative charges); the pogo-in is lifted up to separate from the PUT, leaving net negative charges in the DUT IC (i.e., DUT *charging*); the voltage of the field plate is resistively and slowly reduced to ground zero (i.e., through resistive current leaking) to remove all charges (e.g., positive charges) in the field plate; the pogo-pin is lowered down a second time to touch the PUT pin again to quickly drain the net negative charges (electrons) left in the DUT IC, resulting in a second CDM discharge that will stress the DUT a second time (i.e., *second* DUT *discharging*: negative). The dual discharge procedure is then completed, and the DUT device is ready for the next CDM zapping routine. It is worth to note

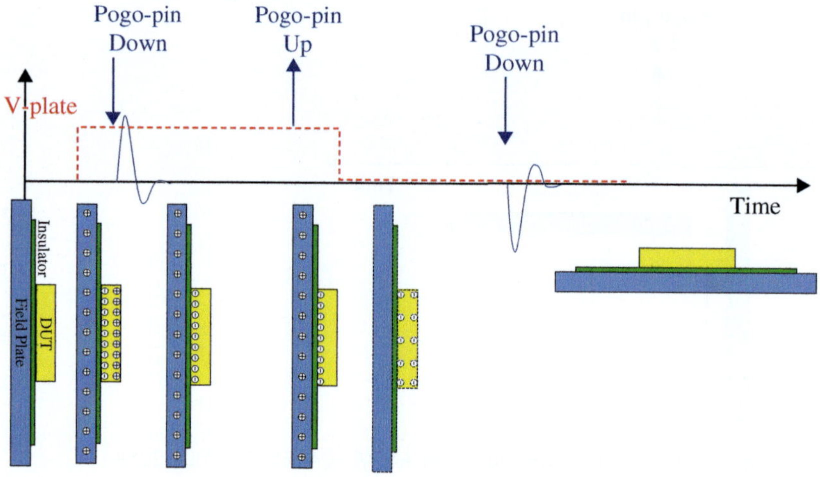

Figure 3.16 Illustration of dual CDM discharge procedures per ANSI/ESDA/JEDEC JS-002-2018.

Table 3.9 Short-circuit CDM ESD waveform specs per ANSI/ESDA/JEDEC JS-002-2018 (High-bandwidth ≥ 6 GHz, short-wired and C_{ESD} = 6.8 pF).

	Peak current I_p (A)		Undershoot I_{p2} (A)		Full width at half maximum (FWHM) (ns)		t_r (ps)	
V_{ESD} (V)	Small	Large	Small	Large	Small	Large	Small	Large
125	1.4–2.3	2.3–3.8	<70% I_p	<50% I_p	250–600	450–900	<250	<350
250	2.9–4.3	4.8–7.3	<70% I_p	<50% I_p	250–600	450–900	<250	<350
500	6.1–8.3	10.3–13.9	<70% I_p	<50% I_p	250–600	450–900	<250	<350
750	9.2–12.4	15.5–20.9	<70% I_p	<50% I_p	250–600	450–900	<250	<350
1000	12.2–16.5	20.6–27.9	<70% I_p	<50% I_p	250–600	450–900	<250	<350

that the dual discharge procedure involves one pair of CDM discharge, positive and negative that stress the DUT device twice. Table 3.9 summarizes the suggested CDM ESD zapping voltage levels and the corresponding CDM waveform specifications and Table 3.10 lists the recommended CDM stressing classification for easy comparison of IC product susceptibility per ANSI/ESDA/JEDEC JS-002-2018. Very interestingly, while the CDM ESD robustness is still commonly specified in terms of ESD susceptible voltage levels (e.g., 100 V or 500 V in the product datasheet), it is recently suggested that a better way to specify IC CDM ESD protection capability is using CDM stressing current level, instead of CDM voltage level. Kind of confusing? The CDM uncertainty speaks out the reality – there are still plenty of unknowns with CDM ESD phenomena, CDM ESD protection designs, and CDM ESD testing methods. The world of ESD, particularly for CDM ESD, is definitely not black and white. In fact, it is recently discovered that the existing and commonly used CDM ESD test models may likely be fundamentally faulty [19]. This is indeed a huge problem, isn't it? More details about the misconception of existing pad-based CDM ESD protection method and test set-ups will be discussed in Chapter 16. Again, it is extremely important for IC designers to understand the ESD things as much as possible both to achieve good ESD protection designs and to mitigate ESD design frustrations.

Table 3.10 ANSI/ESDA/JEDEC JS-002-2018 CDM ESDV classification.

Classes	ESDV levels (V)
C0a	<125
C0b	125–<250
C1	250–<500
C2a	500–<750
C2b	750–<1000
C3	≥1000

3.5 IEC Model

These days, IEC 61000-4-2 test model has often been used for ESD evaluation of ICs [20]. It is important to understand that IEC 61000-4-2 was initially developed to characterize the ESD survivability of electronic systems, i.e., the equipment under test (EUT), as opposed to the component level (i.e., devices, hence, DUT) ESD test models, such as HBM, MM, and CDM. IEC 61000-4-2 provides guidelines for evaluating EUT subjected to ESD stresses from adjacent charged objects including human body and machine. Figure 3.17 depicts a simplified IEC ESD model equivalent circuit consisting of $C_{ESD} = 150\,\text{pF}$ (charge storage capacity) and $R_{ESD} = 330\,\Omega$ (discharge resistance), which produces an IEC zapping waveform defined in Figure 3.18 with the waveform specifications given in Table 3.11. Similarly, no construction details are provided by IEC standard for making an IEC tester except that the IEC waveforms generated must comply with the specs defined in IEC 61000-4-2. Uniquely, an IEC ESD waveform features ultrafast rise time of $t_r = 700$ ps–1 ns and huge current spike (up to tens of Amperes), quite similar to a CDM ESD pulse, as well as a fairly long stressing duration of ~80 ns, sort of comparable to an HBM ESD pulse. This suggests that IEC ESD risks somewhat represent those from both HBM and CDM ESD events in terms of ESD failures associated with the energy-related thermal failures and dielectric breakdown induced by electric field density from the fast voltage spikes, respectively. IEC testing is quite involving. There are two IEC discharge methods: *contact discharge* and *air discharge*. Contact discharge is an IEC testing method where the electrode of a test generator (i.e., an IEC gun) is held in direct contact with the EUT for ESD discharging. Air discharge is an alternative IEC testing method where the charged electrode of the IEC gun is brought close to the EUT and ESD discharge is actuated by an ESD spark

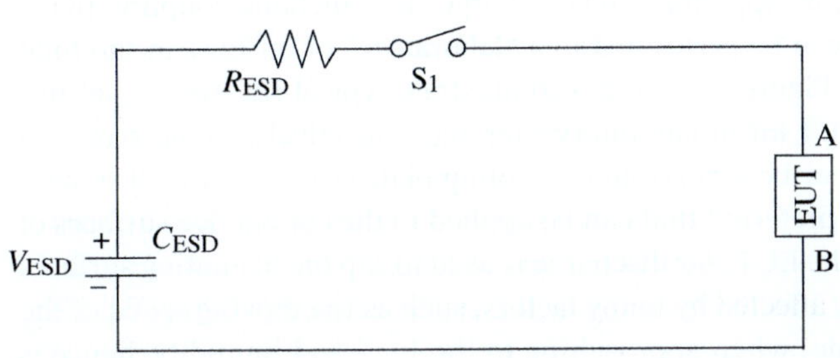

Figure 3.17 A simplified IEC ESD test model equivalent circuit. The parasitic inductive effects from the tester and the EUT/DUT, not shown, can substantially affect the IEC discharge waveforms.

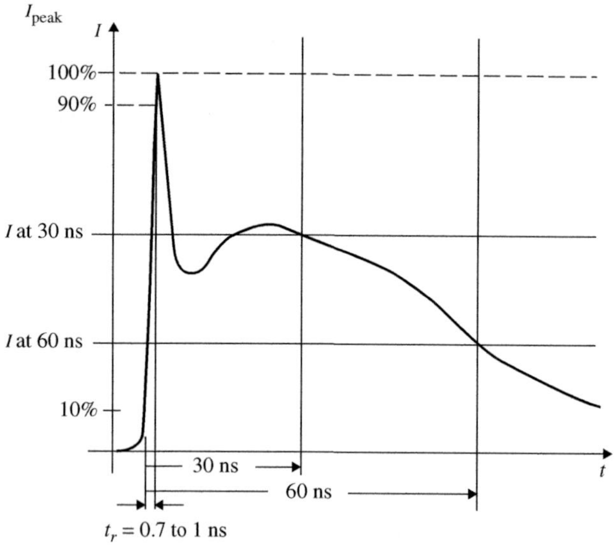

Figure 3.18 Standard IEC ESD discharge waveform per IEC 61000-4-2. (The author thanks the International Electrotechnical Commission (IEC) for permission to reproduce Information from its International Standards. All such extracts are copyright of IEC, Geneva, Switzerland. All rights reserved. Further information on the IEC is available from www.iec.ch. IEC has no responsibility for the placement and context in which the extracts and contents are reproduced by the author, nor is IEC in any way responsible for the other content or accuracy therein.)

Table 3.11 IEC discharge waveform specs (IEC 61000-4-2).

V_{ESD} (V)	First peak current I_p (A) ± 10%	Current at 30 ns (A)	Current at 60 ns (A)	Duration (ns)	t_r (ns)
2000	7.5	4	2	~80	0.7–1.0
4000	15	8	4	~80	0.7–1.0
6000	22.5	12	6	~80	0.7–1.0
8000	30	16	8	~80	0.7–1.0

to the EUT via air. IEC discharges can be applied to EUT in two different ways: *direct* application and *indirect* application. Direct application is done by applying the ESD discharge directly to the EUT. Indirect application is performed by applying ESD discharges to a metallic coupling plane in the vicinity of the EUT. IEC testing can be performed in a "laboratory" or on EUT in the final installation (i.e., post-installation test). Figures 3.19 and 3.20 illustrate typical IEC testing set-ups for table-top EUT and floor-standing EUT for in-laboratory zapping, respectively. In indirect ESD application, the metal coupling plate may be a horizontal coupling plate (HCP) or a vertical coupling plate (VCP). Contact discharge is preferred that can be applied to the conductive surfaces of the EUT or to the coupling plates near the EUT. Air discharge is used to zap the insulating surfaces of the EUT. Air discharge may be easily affected by many factors, such as the moving speed of the IEC gun tip (i.e., the discharge electrode) when approaching to the EUT and humidity, hence is less reliable compared with the direct discharge method. While the IEC model was developed for system-level ESD testing, the trend seems to be that more aggressive semiconductor companies are

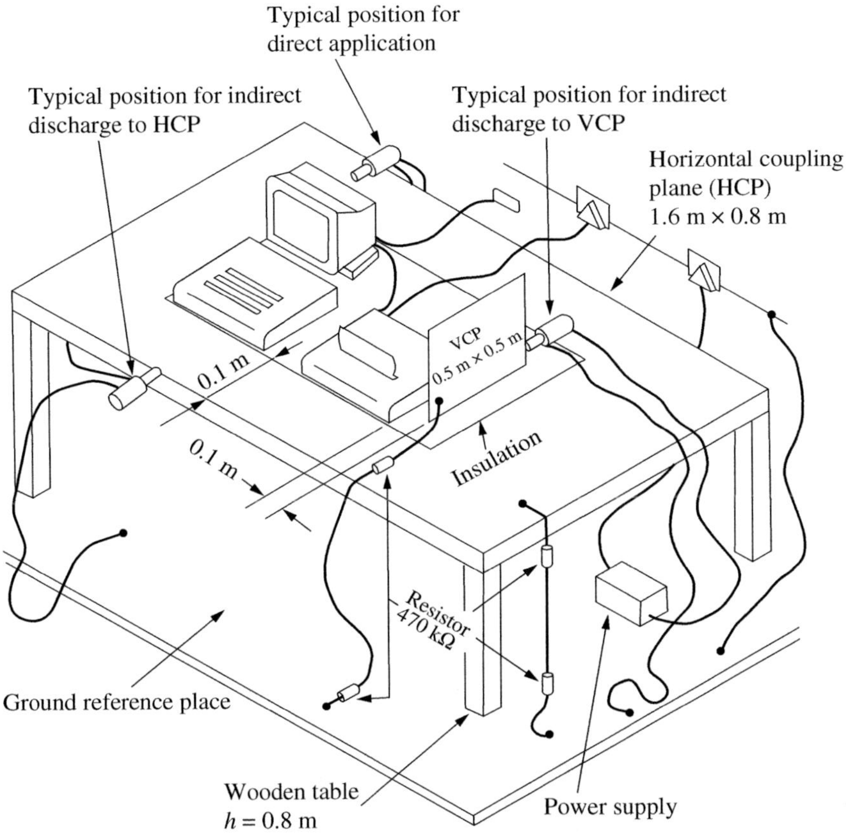

Figure 3.19 Illustration of IEC testing set-up for desk-top EUT in laboratory. (The author thanks the International Electrotechnical Commission (IEC) for permission to reproduce Information from its International Standards. All such extracts are copyright of IEC, Geneva, Switzerland. All rights reserved. Further information on the IEC is available from www.iec.ch. IEC has no responsibility for the placement and context in which the extracts and contents are reproduced by the author, nor is IEC in any way responsible for the other content or accuracy therein.)

applying IEC method to IC ESD characterization. The rationale may be better understood from the product engineering perspective, i.e., ESD protection has both scientific and marketing needs in the real world. It is also important to note that IEC 61000-4-2 does not define the inductive effect, which is commonly considered in the scale of nH that is so small that the IEC ESD discharge waveforms produced by an IEC zapping gun may be readily altered by the inevitable parasitic inductance in the IEC test set-up, including from the IEC tester and from the EUT/DUT objects.

3.6 TLP Model

The various ESD test models discussed previously, including HBM, MM, CDM, and IEC, are generally referred to as ESD zapping test models since the common procedure is to use the relevant real-world ESD pulses to zap a DUT to evaluate its ESD susceptibility. These ESD zapping models follow a straightforward Pass/Fail test sequence, i.e., detecting a DUT pass or failure threshold voltage threshold in terms of the DUT malfunction or performance degradation caused by ESD

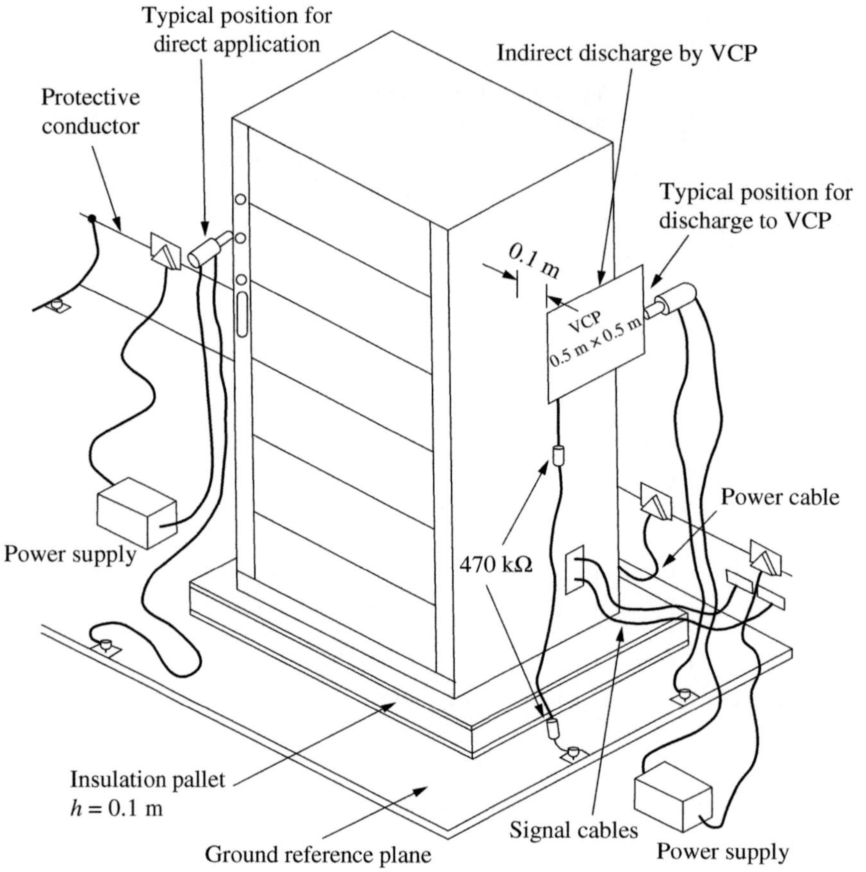

Figure 3.20 Illustration of IEC testing set-up for floor-standing EUT in laboratory. (The author thanks the International Electrotechnical Commission (IEC) for permission to reproduce Information from its International Standards. All such extracts are copyright of IEC, Geneva, Switzerland. All rights reserved. Further information on the IEC is available from www.iec.ch. IEC has no responsibility for the placement and context in which the extracts and contents are reproduced by the author, nor is IEC in any way responsible for the other content or accuracy therein.)

stressing. The outcome of such ESD zapping test is to provide a pass/fail ESDV value that can be printed on the DUT product datasheet to show the ESD robustness of an IC product. Beyond that, such ESD zapping test provides no further information as to why ESD failure occurs and offers nothing to improve ESD protection designs. Such ESD zapping test is also destructive to DUT ICs. For IC designers, it is highly desirable to know the transient ESD discharging behaviors and to understand the possible ESD failure mechanisms in order to optimize and predict ESD protection designs for ICs, which unfortunately cannot be supported by HBM, MM, CDM, and IEC ESD test models. Fortunately, such insightful information on ESD discharges can be offered by a transmission-line-pulsing (TLP) technique [21–23]. The TLP ESD testing technique was developed as a powerful alternative to the commonly used pass/fail-type destructive HBM zapping method. Figure 3.21 depicts the conceptual schematic for a TLP testing set-up and its simplified equivalent circuit. In the TLP testing system, a long transmission line, replacing the C_{ESD} in an HBM zapping tester, is precharged to a given voltage level per ESD testing levels and the static charges stored will then discharge into the DUT via a piece of transmission line, hence stressing the DUT IC. A TLP

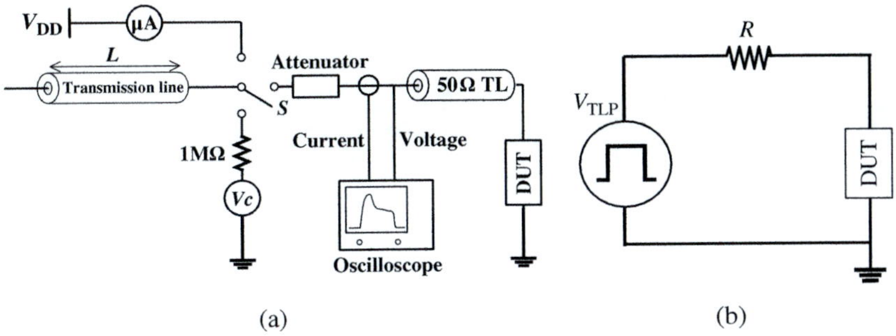

Figure 3.21 A simplified circuit schematic for (a) a TLP testing set-up, and (b) its over-simplified equivalent circuit model.

tester can be considered as an *L–R* network that can produce a well-defined square wave with the rise time (t_r) and duration (t_d) similar to that for an HBM ESD waveform. Therefore, a TLP method is used to emulate real-world HBM zapping waveforms in terms of ESD pulse speed and energy involved to characterize ESD protection designs. With a sophisticated system design, a TLP tester can produce square waveforms to model the HBM ESD waveforms as depicted in Figure 3.22. From Figure 3.22, in order to equate the TLP and HBM waveforms in terms of the transient ESD responses and ESD pulse energy involved, a commercial TLP tester typically sets the square wave as $t_r \sim 10$ ns and $t_d \sim 100$ ns. The TLP $t_r \sim 10$ ns is selected to be the same as that of HBM for the speed concern. The TLP $t_d \sim 100$ ns is set to be shorter than that for HBM ($t_r \sim 150$ ns) in order to ensure similar ESD-induced energy dumped upon the DUT, which is because the HBM waveform tail is weak in energy compared to the flat TLP square waveform. The TLP-generated square pulse will be applied to stress DUT via a short coaxial cable. A broadband oscilloscope is used to monitor the transient voltage and current waveforms of the DUT. A pair of coaxial voltage and current sensors are used to capture the transient voltage and current of the DUT. Therefore, the transient ESD discharging *I–V* curve for the DUT IC during TLP stressing can be captured for further analysis. The TLP method is generally considered to be nondestructive because the TLP pulse shape and duration can be readily and accurately controlled by the transmission line, avoiding destructing the DUT IC. The TLP testing can be understood using Figure 3.23: during TLP testing, a train of TLP-produced square waveforms are applied to stress the DUT one by one, with the TLP pulse height starting from very

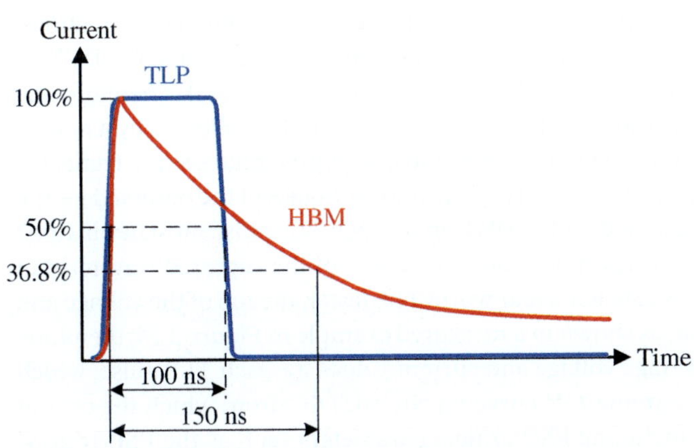

Figure 3.22 A comparison of TLP and HBM pulse waveforms for ESD discharge equivalence.

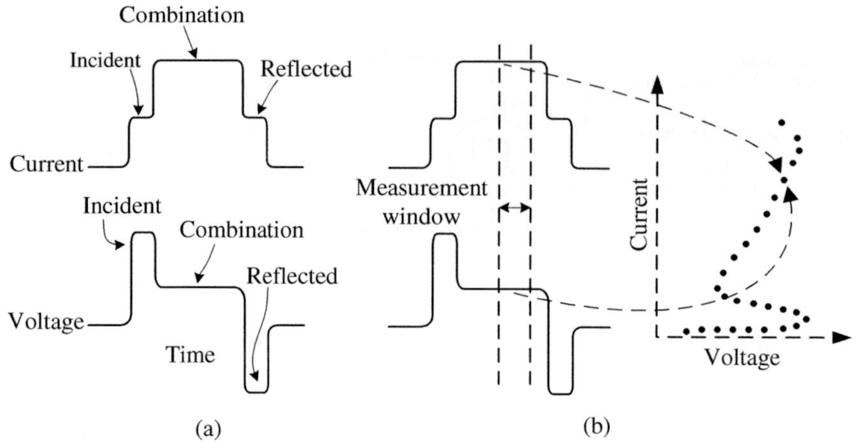

Figure 3.23 Illustration of TLP pulse waveforms across a DUT stressed by a TLP pulse train: (a) incident and reflected TLP pulses over DUT, and (b) obtaining transient ESD I–V curve of the DUT under TLP pulse train stressing.

low level and increasing gradually per the ESD testing plan. Completing the TLP stressing using the TLP pulse train will result in a transient ESD discharge I–V curve, captured by a broadband oscilloscope, for the ESD protection structure designed. After each TLP pulse, the leakage current of DUT will be captured, typically at the normal IC supply voltage. When an TLP pulse reaches to a certain high level, ESD failure will occur, which typically corresponds to an abrupt jump in the DUT leakage current, relevant to the thermal failure of the DUT device measured. According to the transmission line theory, the relationship between incident and reflected components of the TLP waveforms with respect to the DUT can be determined using the following equation:

$$V_{\text{Ref}} = V_{\text{Inc}} \cdot \frac{Z_L - Z_0}{Z_L + Z_0} \tag{3.4}$$

$$I_{\text{Ref}} = -I_{\text{Inc}} \cdot \frac{Z_L - Z_0}{Z_L + Z_0} \tag{3.5}$$

where V_{Inc} and I_{Inc} are the incident voltage and current, V_{Ref} and I_{Ref} are the reflected voltage and current, Z_L and Z_0 are the impedance of the DUT and transmission line, respectively. At a higher pulse level when the DUT is turned on, Z_L of DUT is usually smaller than $Z_0 = 50\,\Omega$ of the transmission line. Therefore, V_{Ref} is negative, while I_{Ref} is positive according to the equations. For the TLP testing set-up operating in the time domain, the incident pulse applied to the DUT IC and the reflected pulse from the DUT are separated by a measurable time interval. Therefore, the waveforms observed will consist of a beginning section (incident pulse only), followed by a middle section of the combined incident and reflected pulses, and finishing with a final section (reflected pulse only), as depicted in Figure 3.23a. Due to the varying waveform shapes of the observed for the DUT, the voltage and current values measured for the DUT under each TLP pulse must be obtained within a small pulse window, which is typically a section of 70–90% of the captured waveform in the time domain as indicated in Figure 3.23b. For a real-world TLP tester, the top of the voltage and current waveforms captured are not flat, as shown in a measured example in Figure 3.24; therefore, integration is required to obtain the average voltage and current values for each TLP pulse, which finally produce the transient ESD discharging I–V curve for the DUT IC, from which the critical transient ESD discharging behaviors, including ESD-critical parameters such as the ESD triggering voltage (V_{t1}), ESD discharging resistance (R_{ON}) and ESD thermal failure current (I_{t2}) can be accurately contained for ESD design analysis. Figure 3.25 provides a measured ESD discharging I–V curve by TLP testing.

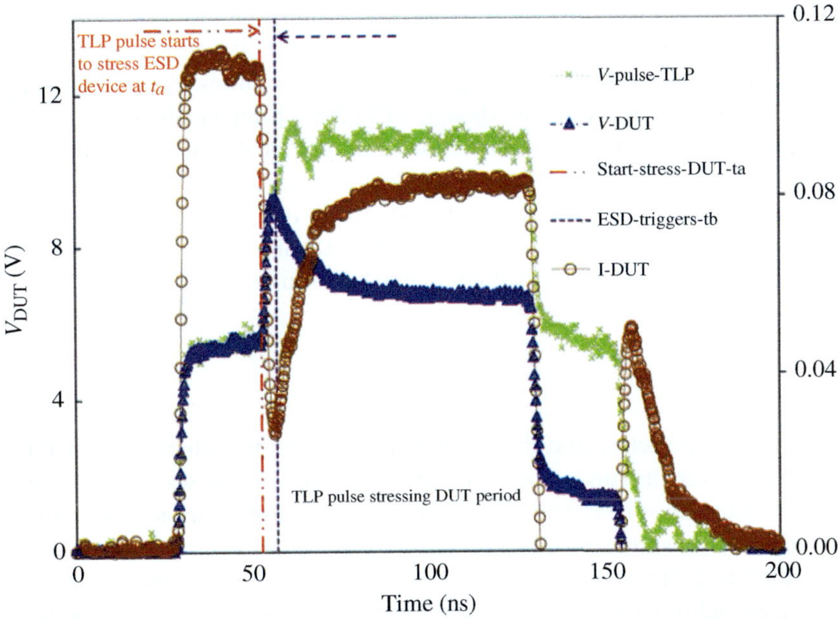

Figure 3.24 Example of measured TLP-produced square waveform (Green) that is applied to stress the DUT and the measured transient voltage (Blue) and current (Brown) waveforms of the DUT IC. It is clear that while the incident TLP pulse is flat, the measured voltage and current waveforms are not flat as ideally expected. Hence integration over a time window (typically 70–90%) is needed to estimate the average V and I values for each TLP pulse.

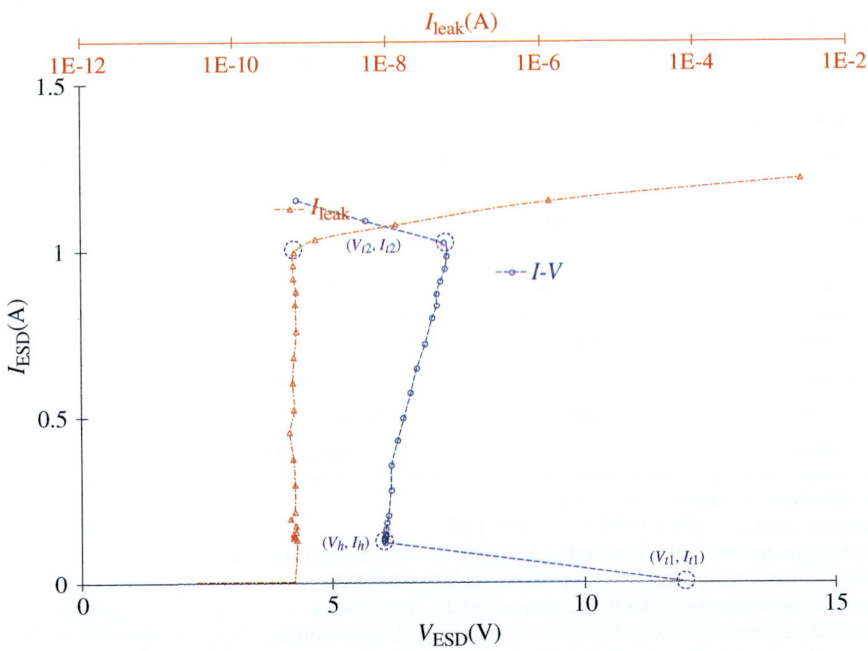

Figure 3.25 Example for TLP-measured transient ESD discharging I–V and I_{leak}–t curves of a DUT IC stressed offers rich details of the ESD discharging behaviors of the DUT device.

Similar to using TLP testing to emulate real-world HBM ESD events, the ultrafast CDM ESD stressing can be modeled by a very fast version of TLP method, i.e., VFTLP ESD testing method [24]. A VFTLP is designed to produce ultrafast and high-spiking square pulses with its t_r and t_d equivalent to that for an CDM waveform. However, though VFTLP has been widely used to characterize ESD protection structures for CDM ESD protection, it only evaluates the ultrafast CDM ESD response property of an ESD protection device, which does not characterize the whole-chip CDM ESD protection functions, as commonly misunderstood, due to its external zapping nature (i.e., zapping IC pads with pad-based ESD protection structures). Details on the problem of the existing VFTLP testing for CDM ESD protection evaluation will be discussed in Chapter 16.

3.7 Summary

Naturally, ESD events occur when electrostatic charges transfer between two objects of different electrical potentials, resulting in very fast and strong transient ESD discharge pulses that can damage electronics. ESD survivability must be characterized for electronics using various ESD testers built per different ESD test models and standards, including HBM, MM, CDM, and IEC models that are developed according to different ESD discharge origins. These ESD test models can be unified into one common second-order RLC equivalent circuit model with different component parameters as summarized in Table 3.12. Various industrial ESD test standards were developed to characterize ESD susceptibility of varying ESD types and tuned for different electronics product needs. Generally, an ESD test standard defined the ESD discharge waveform specifications for stressing DUT devices without providing any details for how to build an ESD zapping tester. HBM test model is by far the most reliable and widely used ESD testing method. On the other hand, MM, CDM, and ICE test models are very sensitive to the parasitic effects of the ESD test set-ups due to the vary small L_{ESD} that can be readily altered by parasitic inductance of the DUT, which is especially true for CDM model. While HBM, MM, and IEC test models simulate the external-oriented ESD discharge events, CDM model addresses the internal-oriented, self-discharging ESD phenomena. HBM, MM, CDM, and IEC test models are considered as pass/fail zapping test methods that offer no information on the transient ESD discharging behaviors and are destructive to ICs. TLP and

Table 3.12 Typical equivalent circuit parameters for different ESD test models.

Models	C_{ESD} (pF)	R_{ESD} (Ω)	L_{ESD} (μH)
HBM	100	1500	7.5[a)]
MM	200	0[b)]	Minimized[c)]
CDM	6.8	~1[d)]	Minimized[e)]
IEC	150	330	Minimized[f)]

a) Not defined in the standards, commonly used value.
b) "Normally zero series-resistance". (ANSI-ESDSTM5.2-1999 [10]).
c) "Low inductance", but not defined in the standards. (ANSI-ESDSTM5.2-1999 [10]).
d) Pogo-pin to GND resistance = 1 ohm. (ANSI-ESDA-JEDEC JS-002-2018 [18]).
e) Not defined in the standards. "the parasitics in the charge and discharge paths should be minimized since the resistance inductance-capacitance (RLC) parasitics in the equipment greatly influence the test results". (ANSI-ESDA-JEDEC JS-002-2018 [18]).
f) Not defined in the standards, should be minimized (~nH).

VFTLP models, on the other hand, were developed to emulate the real-world HBM and CDM ESD events, respectively, allowing capturing the transient ESD discharging *I–V* characteristics, which provide details on transient ESD behaviors and insights for ESD protection design optimization and prediction by CAD simulation. ESD test models are essentially living test models that have constantly been revised according to new understanding of ESD phenomena, especially at advance IC technology nodes down to nanometer scale. For example, a new human metal model (HMM) was proposed recently trying to define how electronic components, e.g., ICs, can be characterized per the IEC 61000-4-2 ESD model that was initially developed for system-level ESD evaluation, but has been recently adopted to characterize IC devices [25]. Indeed, while major uncertainties still exist for ESD test models, mainly for CDM model, thoroughly understanding the existing ESD test models and standards are important for IC designers and ESD design engineers to develop reliable ESD-robust electronic products.

References

1 MIL-STD-883E, Method 3015.7 (1989). *Electrostatic Discharge Sensitivity Classification*. Dept. of Defense, Test Method Standard, Microcircuits.

2 Van Roozendaal, L., Amerasekera, A., Bos, P. et al. (1990). Standard ESD testing of integrated circuits. *Proceedings of EOS/ESD Symposium*, pp. 119–130.

3 Amerasekera, A. and Verweij, J. (1992). ESD in integrated circuits. *Qual. Reliab. Int.* 8: 259–272.

4 Verhaege, K., Roussel, P., Groeseneken, G. et al. (1993). Analysis of HBM ESD testers and specifications using a 4th order lumped element model. *Proceedings of EOS/ESD Symposium*, pp. 129–137.

5 ANSI/ESDA/JEDEC JS-001-2017 (2017). *For Electrostatic Discharge Sensitivity Testing – Human Body Model (HBM) – Component Level*. An American National Standard jointly developed by ESD Association and JEDEC.

6 ESD STM5.1-2007 (2007). *Revised, Electrostatic Discharge Sensitivity Testing – Human Body Model Component Level*. the ESD Association.

7 JEDEC JESD22-A114-F (2007). *Electrostatic Discharge Sensitivity Testing Human Body Model*. the Electronics Industries Alliance.

8 AEC-Q100-002 Rev-C (1998). *Human Body Model Electrostatic Discharge Test*. the Automotive Electronics Council.

9 EIAJ-Standard, IC-121-1988 (1988). *Testing Method 20: Electrostatic Destructive Test*. Electrostatic Discharge.

10 ESD STM5.2-2012 (2012). *Electrostatic Discharge Sensitivity Testing: Machine Model - Component Level*. the ESD Association.

11 EIA/JESD22-A115-A (1997). *Electrostatic Discharge (ESD) Sensitivity Testing Machine Model (MM)*. JEDEC, the Electronics Industries Alliance.

12 AEC – Q100-003 Rev-E (2001). *Machine Model Electrostatic Discharge Test*. the Automotive Electronics Council.

13 JEP172A (2014). *Discontinuing Use of the Machine Model for Device ESD Qualification*. JEDEC and ESDA.

14 Industry Council on ESD Target Levels (2007). White Paper 1: A Case for Lowering Component Level HBM/MM ESD Specifications and Requirements.

15 ESD STM5.3.1-1999 Revised (1999). *Electrostatic Discharge Sensitivity Testing – Charged Device Model Component Level*. the ESD Association.

16 JEDEC JESD22-C101-A (2000). *Field-Induced Charged-Device Model Test Method for Electrostatic Discharge-Withstand Thresholds of Microelectronic Components*. the Electronics Industries Alliance.

17 AEC – Q100-011 Rev-A (2001). *Charged Device Model Electrostatic Discharge Test*. the Automotive Electronics Council.

18 ANSI/ESDA/JEDEC JS-002-2018 (2018). *For Electrostatic Discharge Sensitivity Testing – Charged Device Model (CDM) – Device Level*. An American National Standard jointly developed by ESD Association and JEDEC.

19 Di, M., Wang, H., Zhang, F. et al. (2019). Does CDM ESD protection really work? *Proceedings of IEEE Workshop on Microelectronics and Electron Devices (WMED)*. https://doi.org/10.1109/WMED.2019.8714145.

20 IEC 61000-4-2 (2008). *Electromagnetic Compatibility, Part 4: Testing and Measurement Techniques, Section 2: Electrostatic Discharge Immunity Test*. the International Electrotechnical Commission (IEC).

21 Maloney, T. and Khurana, N. (1985). Transmission line pulsing technique for circuit modelling of ESD phenomena. *Proceedings of EOS/ESD Symposium*, pp. 49–54.

22 Barth, J., Verhaege, K., Henry, L., and Richner, J. (2000). TLP calibration, correlation, standards, and new techniques. *Proceedings of EOS/ESD Symposium*, pp. 85–96.

23 ANSI/ESD STM 5.5.1-2016 (2016). *ESD Association Standard Test Method for Electrostatic Discharge (ESD) Sensitivity Testing – Transmission Line Pulse (TLP) – Component Level*. the ESD Association.

24 ESD TR5.5-03-14 (2014). *ESD Association Technical Report for Electrostatic Discharge (ESD) Sensitivity Testing – Very Fast – Transmission Line Pulse (TLP) - Round Robin Analysis*. ESD Association Standard Practice for Electrostatic Discharge Sensitivity Testing – Human Metal Model (HMM) Component Level.

25 ESD SP5.6-2019 (2019). *ESD Association Standard Practice for Electrostatic Discharge Sensitivity Testing – Human Metal Model (HMM) Component Level*. the ESD Association.

4

ESD Protection Devices

Now that we understood thoroughly the electrostatic discharge (ESD) phenomena, ESD failures, and ESD characterization models and techniques, we are ready to discuss ESD protection designs. This chapter explains basic ESD protection solutions, including ESD protection mechanisms and commonly used ESD protection device structures, which are foundation and building blocks for more advanced and complex ESD protection circuits to be discussed in later chapters. Good ESD and integrated circuit (IC) designers must understand these ESD protection fundamentals in order to achieve successes of ESD protection designs, rationally and creatively, without gambling.

4.1 On-Chip ESD Protection Mechanisms

4.1.1 Switch for ESD Discharge

As discussed previously, thermal failure and voltage breakdown are the two main ESD hard failure mechanisms. ESD thermal failure is associated with joule heating produced by the large transient current induced by an ESD pulse. In general, semiconductors, e.g., silicon, are poor thermal conductors; meanwhile, ESD pulses generate very high and fast transient currents. Further, the ESD discharge current typically routes through IC subsurface, which is covered by dielectric layers that are normally thermal insulators. Consequently, the overall result is localized heating, i.e., hot spots, which leads to thermal damage when the local temperature increases to the melting temperature threshold in either Si or metal interconnects. ESD voltage breakdown failure is associated with the strong and fast electric field density generated by the large voltage transient induced by an ESD pulse. When the localized electric field density reaches to the breakdown threshold of dielectric layers, e.g., ~8–10 MV/cm in typical gate oxide (SiO_2) in complementary metal-oxide-semiconductor (CMOS) [1], or Si PN junctions, voltage breakdown failure will occur. For example, in 28 nm CMOS with gate oxide of ~35 Å, the gate breakdown voltage is as low as $BV_{ox} \sim 2.2$ V for core MOSFET's, easily being destroyed by an ESD pulse.

Obviously, any ESD protection solution must be able to handle the large and fast voltage and current pulses generated during ESD discharge events. Accordingly, there are two basic principles for on-chip ESD protection: first, to safely discharge ESD currents via a low-impedance conduction path without generating too much heat; and second, to clamp bonding pad voltage to a sufficiently low level without causing voltage breakdown. Apparently, a two-terminal "*Switch*" would be the real ESD protection solution in theory. As depicted in Figure 4.1, an ESD switch (i.e., shunt ESD switch) work just like a lighting switch on a wall that can turn a lamp on and off. The two terminals of an ESD switch are named as *anode* (A) and *cathode* (K). Here is how on-chip ESD protection

Practical ESD Protection Design, First Edition. Albert Wang.

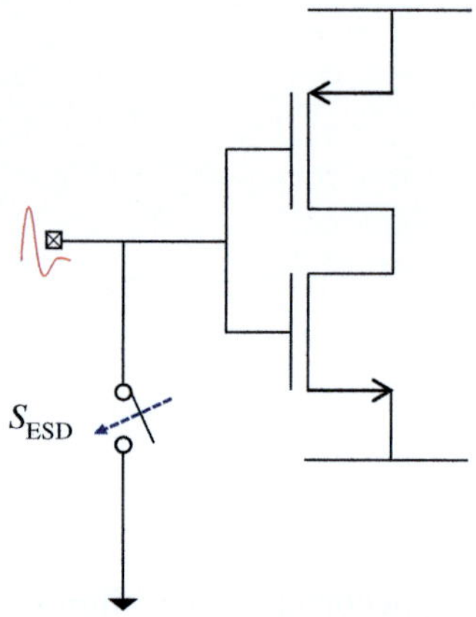

Figure 4.1 Concept of a shunt ESD switch method for on-chip ESD production: the incident ESD pulse can be discharged into GND via a shunting switch to prevent possible ESD thermal and/or voltage breakdown failures.

works in principle: an ESD switch at an IC pad, connected to electrical ground (GND), remains OFF during normal IC operations, hence not interfering with IC functions. When an incident ESD pulse occurs at the pad, the ESD protection switch can be turned ON to form a low-R conduction path to discharge the ESD transient into the ground (GND), therefore prevents the ESD surge from running into the IC core and causing internal ESD failures. In street language, an ESD protection switch serves as a "guard" at the "door" (i.e., a pad) to stop any "external intruders" (i.e., an ESD pulse made of charges) from running into the "room" (i.e., IC core). In concept, there are several key attributes required for any good ESD protection structure. First, unlike a lighting switch, the ESD switch must be able to turns ON and OFF very fast in response to the fast-incoming ESD surges. This is fundamentally why one cannot just use a lamp switch for on-chip ESD protection even if physically possible. Second, in ON state, the ESD switch must have very low resistance (low-R) so as to not generate much heat while discharging an ESD pulse, and certainly ensuring low clamping voltage at the IC pad to avoid voltage breakdown. Third, an ESD switch should have as little parasitic effects as possible, hence, not affecting IC circuit performance while achieving ESD protection. Fourth, an ESD protection switch must be "tiny" and can be integrated into an IC chip to be useful. Accomplishing these desired attributes are not trivial, which has drawn endless R&D efforts since semiconductor transistors and ICs were invented. Figure 4.2 illustrates a second ESD protection scenario using ESD switches (i.e., shunt-series ESD switch) for on-chip ESD protection, which, conceptually, intends to enhance ESD protection from that shown in Figure 4.1. Considering the fact that ESD failure is caused by an external ESD pulse appearing at an IC pad, therefore, in addition to using one ESD switch (i.e., S_{ESD1}) to effectively shunt the incident ESD transient to GND, a

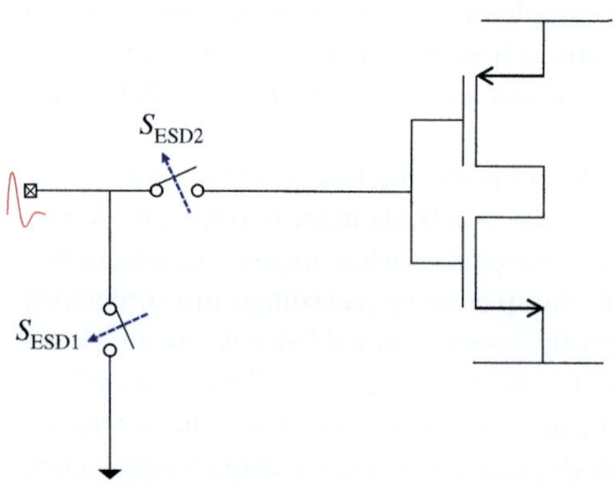

Figure 4.2 Concept of a shunt-series ESD switch method for on-chip ESD production shows dual ESD protection functions: both discharging the transient ESD energy into GND and clamping the pad voltage, and buffer an incident ESD transient from the IC core.

second ESD switch (i.e., S_{ESD2}) may be placed between the pad and the node to the IC core, which, during an ESD event and as the S_{ESD1} is closed to discharge the ESD pulse to GND, will be turned OFF simultaneously as an extra measure to further block the incoming ESD surges from running into the IC core circuitry. Of course, this shunt-series ESD switch method may not be easily implemented as ideally wished in practical IC designs. An easy alternative ESD protection could be series ESD switch method as shown in Figure 4.3, where the ESD switch is connected between a pad and the input node of IC core circuitry. The general idea for using the series ESD switch method is that it remains ON during normal IC operations and will be turned OFF by an incident ESD pulse to block it from getting into the IC core, hence, realizing ESD protection. However, using the series ESD switch method may not be practically possible as wished. Comparing the shunt ESD switch (Figure 4.1) and series ESD switch (Figure 4.3) methods reveals a major difference. Using the shunt ESD switch method, the large and fast energy transient associated with an ESD pulse can be efficiently discharged into the ground through a shunt ESD switch without causing ESD-induced thermal and breakdown damages to ICs. However, using the series ESD switch method only, while turning OFF the series ESD switch can block the incoming ESD pulse, the large transient ESD energy cannot be discharged into the ground and will be "hanging" over at the series ESD switch, which will induce fast and significant electric field density that will likely cause ESD breakdown. An alternative thinking for the series ESD switch is either to slow down the incident ESD pulse, allowing enough time to turn on the shunt ESD switch for ESD discharging, or to act as a buffer between the incoming ESD pulse and an internal IC node, preventing direct zapping the internal node. This concept has been used in multiple-stage charged device model (CDM) ESD protection designs.

In real-world IC designs, ESD protection devices (i.e., ESD switches) of different and suitable types are used to protect every IC pad including input/output (I/O) and power supplies (e.g., V_{DD}, V_{SS}), as illustrated in Figure 4.4. ESD zapping may occur to any IC pads, causing ESD failures. Accordingly, ESD events to an IC can be classified according to the way an incident ESD pulse stresses an IC pad: positively from I/O to GND or a negative supply V_{SS} (*PS mode*), negatively from I/O to GND or V_{SS} (*NS mode*), positively from I/O to a positive supply V_{DD} (*PD mode*), and negatively from I/O to V_{DD} (*ND mode*), as well as positively from V_{DD} to GND or V_{SS} (*DS mode*), and negatively from V_{DD} to GND or V_{SS} (*SD mode*) [2]. All ESD protection devices must be in OFF state during normal IC operations. Under ESD stressing, an ESD protection device must be turned ON to establish a low-R discharging path to shunt the incoming ESD pulse, hence providing full-chip ESD protection. There are two classic ESD protection mechanisms featuring two typical ESD discharge characteristics in terms of the transient ESD discharging current–voltage (I–V) curves as depicted in Figure 4.5. The first ESD discharging characteristics is readily modeled by a *simple*

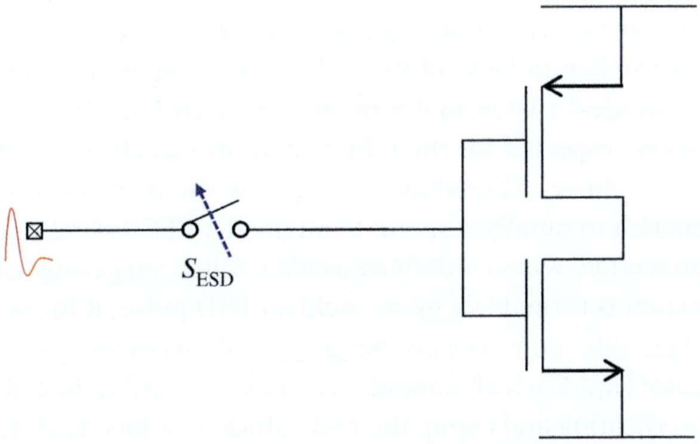

Figure 4.3 Concept of a series ESD switch method for on-chip ESD production: a buffer between an incident ESD pulse and internal IC core, but cannot discharge the ESD transient into GND.

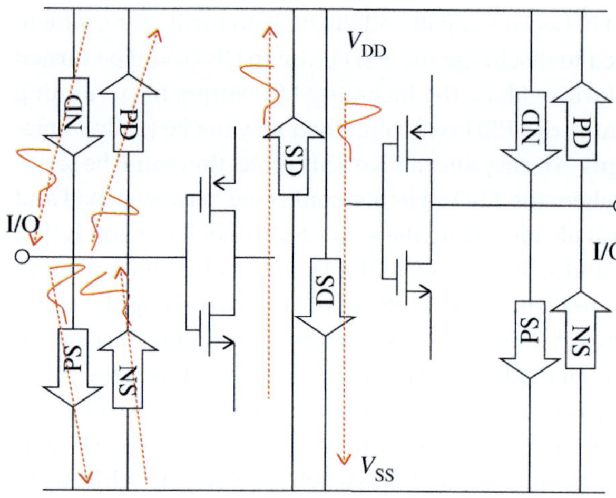

Figure 4.4 Illustration of full-chip ESD protection scheme: multiple ESD protection devices are needed per pad to form a complete on-chip ESD protection network to ensure a low-*R* ESD discharge path between any two pads on the chip.

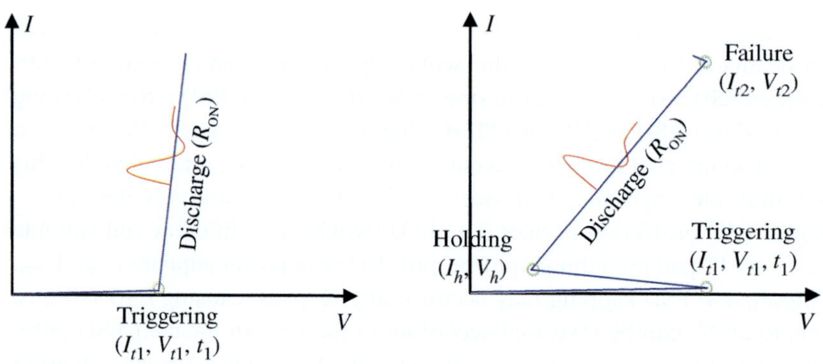

Figure 4.5 Illustration of two typical *I*–*V* characteristics facilitating ESD protection: simple diode turn-on and snapback *I*–*V* curves.

turn-on I–V curve, typically seen in diode conduction that is an easy analogue of a switch. An ESD diode switch stays OFF during normal IC operations and will be turned ON by an incident ESD pulse to shunt the ESD surge to GND. Alternatively and quite popularly, an ESD protection device featuring a *snapback I–V* characteristic can be used as an ESD switch. An ESD switch device is characterized by a set of electrical parameters collectively known as the *ESD-Critical Parameters* [3, 4]. Typical ESD-critical parameters are used to describe the key functionalities of ESD discharge, which are ESD *Triggering Threshold* featured by ESD *triggering voltage* (V_{t1}), *triggering current* (I_{t1}), and *triggering time* (t_1). V_{t1} defines the control voltage required to turn ON an ESD switch to kick-off the ESD discharging procedure. Certainly, no real-world ESD switch may be an ideal switch that remains 100-ly OFF during normal IC operations, and a finite current is always expected for the OFF-state ESD switch that is denoted as the ESD triggering current, I_{t1}, at V_{t1}. Since ESD pulses are very fast, the response time of an ESD switch designed must be fast enough to rapidly respond to an incident ESD transient to swiftly trigger the ESD switch for ESD protection, which is defined as ESD switch triggering time, t_1, or, ESD *response time*. When an ESD switch is turned ON by an incident ESD pulse, it forms a low-*R* conduction path to discharge the alien ESD surge. The discharging capability of an ESD switch is evaluated by the *Discharging Resistance* (R_{ON}), which should be as low as possible in order to handle large ESD transients without overheating and clamp the pad voltage to a low level. Unfortunately, though a diode is simple for

ESD protection, its ESD-critical parameters may not be suitable for many ICs, such as triggering voltage, clamping voltage, and discharging resistance. An ESD switch featuring a snapback *I–V* characteristic offers some unique electrical properties desired for good ESD discharge functions, which requires a few more ESD-critical parameters to depict: ESD *holding* voltage (V_h) and current (I_h), as well as *thermal breakdown* voltage (V_{t2}) and current (I_{t2}). V_h typically determines the pad clamping voltage, while I_{t2} roughly reflects the *ESD current-handling capability* of an ESD switch device that is more meaningfully denoted as a thermal failure current density, J_{t2}. Another commonly used ESD-critical parameter is the ESD *leakage* current (I_{leak}) measured at an on-chip supply voltage (V_{DD}, V_{SS}, V_{CC}, V_{EE}, etc.). Since a snapback *I–V* characteristic is also the signature for the latch-up effect in CMOS, essentially, ESD discharge is the same as CMOS latch-up with one key difference: only controllable (not random) latch-up effect can be used for ESD protection. Therefore, snapback-based ESD protection is essentially a *controlled latch-up* function. Carefully designing the V_h and I_h is important to ensure that a snapback-type ESD switch will work as an ESD protection device, not an uncontrollable latch-up devil to ICs. Interesting, isn't it? The world is never black and white. Analog design is always an art. On-chip ESD protection design is nothing less than an artistic work. While the ESD protection principle is crystal-clear and very simple, a good ESD protection design does burn quite many brain cells to fine-tune a device structure and to manipulate the ESD-critical parameters in order to achieve desirable specs: low R_{ON} and I_{leak}, short t_1, accurate V_{t1}, V_h, and V_{t2}, and high current-handling capability (I_{t2} and J_{t2}).

4.1.2 ESD Protection: Active versus Passive

An ESD protection switch may be realized in countless device structures in real-world designs. Indeed, a designer's brain will be the only limit. Conceptually, all ESD protection switch structures can be classified as either *active* or *passive* devices. Active ESD protection switches are made of active electronic devices that require an electrical bias to turn ON a conduction channel to discharge ESD pulses. Active ESD switches are the most commonly used on-chip ESD protection structures, including diodes, bipolar junction transistors (BJTs), metal-oxide-semiconductor field-effect transistors (MOSFETs), silicon-controlled rectifiers (SCRs), and their derivatives and subcircuits, which will be discussed in details in this book. On the other hand, passive ESD switch largely remains as a concept. Since active ESD protection structures mostly rely on PN junctions for conduction, there exist inevitable disadvantages inherent to PN junctions, including various parasitic effects, e.g., ESD-induced leakage, capacitance and noises, and ESD device size, which are collectively referred to as the ESD design overhead. Details on ESD design overhead will be discussed in later chapters. An ideal passive ESD switch may be a mechanical switch structure that has inherent advantages over electronic devices, i.e., no PN-induced parasitic effects. Conceptual passive ESD switch structures were recently reported, e.g., nano crossbar array and graphene NEMS switch array [5, 6].

Alternatively, passive electronic filters and resonators may be used for ESD protection, typically for RF IC designs, as illustrated in Figure 4.6. The rationale for using filter type ESD protection for RF ICs and wireless systems seems to be straightforward: First, wireless systems use RF ICs operating in varying frequency spectrum, often at high frequencies. Second, ESD pulses are fast transients, meaning very high frequencies. Third, RF ICs, typically for broadband and high data rate applications, are very sensitive to ESD-induced parasitic effects, which is actually a huge design challenge for any RF IC designers. Therefore, commonly used PN-based active ESD switch structures are not ideal for RF ESD protection. On the other hand, if the operation frequencies of RF ICs to be protected are separated from ESD transient frequencies in the spectrum, as shown in Figure 4.7, why

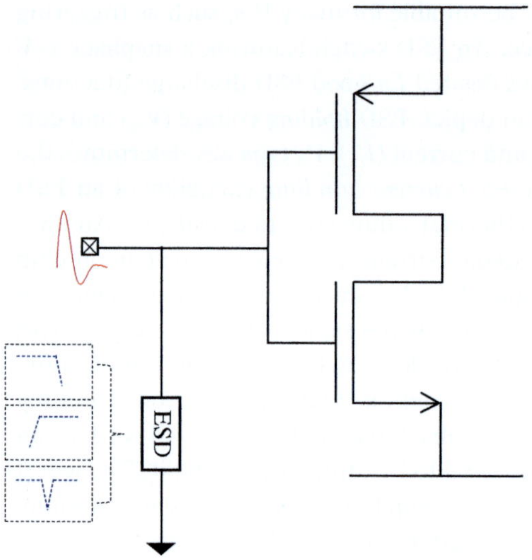

Figure 4.6 Concept of using filters or resonators for ESD protection: high-pass filter, low-pass filter or notch filter.

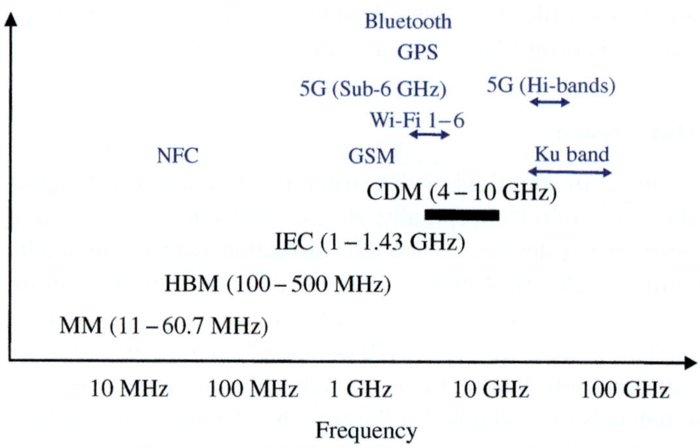

Figure 4.7 Frequency spectrum comparison: ESD protection versus common wireless applications.

not using electronic filters for ESD protection? A filter is a device that can reject signals of certain frequency while pass signals at other frequency. Assume an RF transceiver IC for high-band 5G mobile systems require robust ESD protection (of course, because HBM and CDM ESD stresses are expected every time you touch your smartphone), since the 5G high-band is above 28 GHz that is much higher than any ESD pulse frequency (i.e., <10 GHz), then, a low-pass filter (LPF) may be placed at an IC pad to GND to shunt the incident ESD pulses. Under normal smartphone usage, the 28 GHz + signals (your chat or music) will fly through the I/O ports without being affected by the ESD protection LPF filter. During ESD events, the filter will shunt the incident ESD transients into GND, hence, protect the 5G IC. Therefore, such an ESD filter is considered a high-pass filter (HPF) to its load (i.e., RF signals at the input port of the RF IC protected), while being LPF to the ESD transient itself, i.e., rejecting the 28 GHz + RF signals while passing the under-10 GHz ESD pulses. Though the concept for filter-based ESD protection is simple for RF ICs, it is not practically useful or easy to design in a real world due to several brain-burning reasons: Most RF ICs to be protected are operating in the same frequency spectrum with ESD pulses, hence, impractical to separate the two groups of signals (RF angels and ESD devils) easily and cleanly. Inductor-based ESD filters and resonators are extremely frequency-dependent, making inductive ESD structures very tricky in RF ESD protection designs, though simple in the concept. Inductors used for ESD protection are also too large in sizes. More importantly, unlike an active ESD switch that can be optimized for minimum discharge resistance R_{ON}, the ON state of an ESD filter may still have substantial frequency-dependent on-resistance that is dominated by the series resistance (R_s) of inductors. In addition, the ON/OFF transition of a filter ESD protection structure can be blur, compared to a relatively clear-cut ON/OFF of typical active ESD switches. Figures 4.8 and 4.9 depict exemplar inductor ESD filter and inductor-capacitor (LC) filter ESD protection structures, respectively. The LPF ESD structure in Figure 4.8 is good for RF ICs operating at frequencies well above ESD pulse frequencies. The LC notch filter (band-stop filter) shown in Figure 4.9 can be designed for a given resonance frequency fitting a specific ESD model, e.g., HBM (100–500 MHz) or CDM (4–10 GHz) ESD events, which may be suitable for RF ICs running at varying frequencies. Regardless, the design problem of substantial R_s and not-so-sharp ON/OFF transition will always be the challenge. Now, let us return to the core of this book to discuss the commonly used active ESD protection devices.

Figure 4.8 Concept of using single inductor filter for ESD protection.

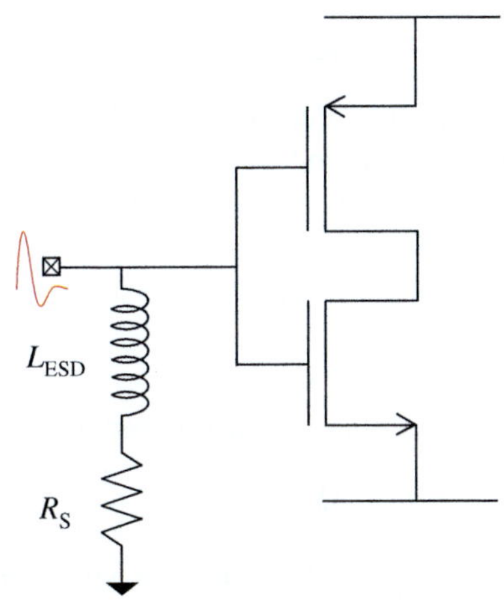

Figure 4.9 Concept of using LC resonator for ESD protection.

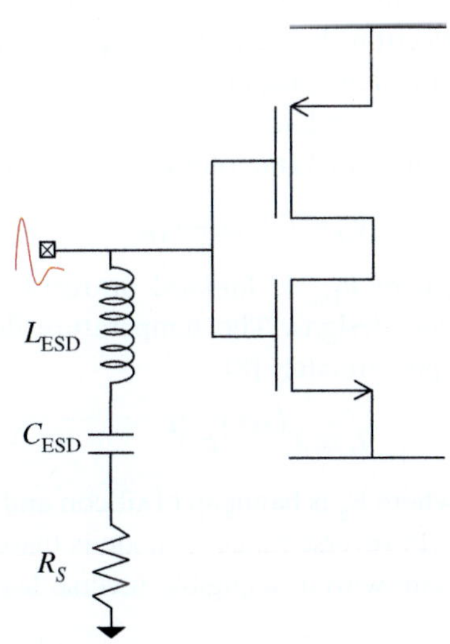

4.2 Diode for ESD Protection

The old wisdom says simplicity is beauty, which translated into the language of IC designs becomes: the simpler, the better; well, if the design specs are met. This is particularly the Bible for analog IC designers. In the world of ESD protection, a diode is the simplest electronic device acting as an active ESD switch. Complex circuitry is used to deliver sophisticated functions. Similarly, complex ESD protection structures and circuits have been developed to be advanced ESD protection for various special reasons. However, design complexity always comes with adverse effects, i.e., ESD design overhead, including more parasitic effects and larger footprint, which will affect core IC performance more. Therefore, a diode remains to be the preferred on-chip ESD protection solution if ever adequate ESD protection can be delivered.

Diodes can be used for ESD protection in many different ways: either in forward or reverse conduction modes, as well as acting as a single-device or diode networks. Further, a diode (mostly a PN junction) is the foundational unit for almost all other active ESD protection structures and circuits, including BJT, MOSFET, and SCR-based ESD protection solutions. Hence, it is important to understand how a diode works as an ESD protection device.

4.2.1 Diode Device Physics

Figure 4.10 depicts the global picture for I–V characteristics of a typical PN junction diode. In forward mode, a PN diode is biased forwardly from P-region to N-region until reaching to the forward turn-on voltage, typically at $V_{ON} \sim 0.65\,\text{V}$ for a silicon diode. Consequently, the diode starts forward conduction with its I–V characteristic modeled by the Shockley equations for an ideal diode [7, 8]. The total diode forward current is given by

$$i_D = I_s\left(e^{\frac{v_D}{nV_T}} - 1\right) \tag{4.1}$$

and its saturation current follows

$$I_s = Aqn_i\left(\frac{D_P}{L_P N_D} - \frac{D_N}{L_N N_A}\right) \tag{4.2}$$

where v_D and i_D are total biasing voltage across and total current through the diode, $n \approx 1$–2 is the ideality factor, $V_T = \frac{kT}{q}$ is the thermal voltage, k is Boltzmann constant, T is temperature, q is single electron charge, A is diode junction size, n_i is intrinsic carrier density, $D_P, D_N, L_P,$ and L_N are carrier diffusion constants and diffusion lengths, respectively, and N_A and N_D are impurity densities for donors and acceptors. A diode can be globally depicted by a piece-wise-linear model as shown in Figure 4.10 and follows an equivalent circuit model given by

$$v_D = i_D R_{DF} + V_{ON} \tag{4.3}$$

where R_{DF} is forward on-resistance of the diode, which is very important for ESD protection designs. The temperature dependence, also a critical factor in ESD operations, follows approximately [8]

$$I_s \approx T^{\left(3+\frac{\gamma}{2}\right)} e^{-\frac{E_g}{kT}} \tag{4.4}$$

where E_g is bandgap of silicon and γ is a constant.

In reverse mode, a diode is biased reversely from the P-region to N-region, hence stays in OFF state with a negligible reverse leakage current inherent to a PN junction, $i_R = -i_D \approx I_S$. As the

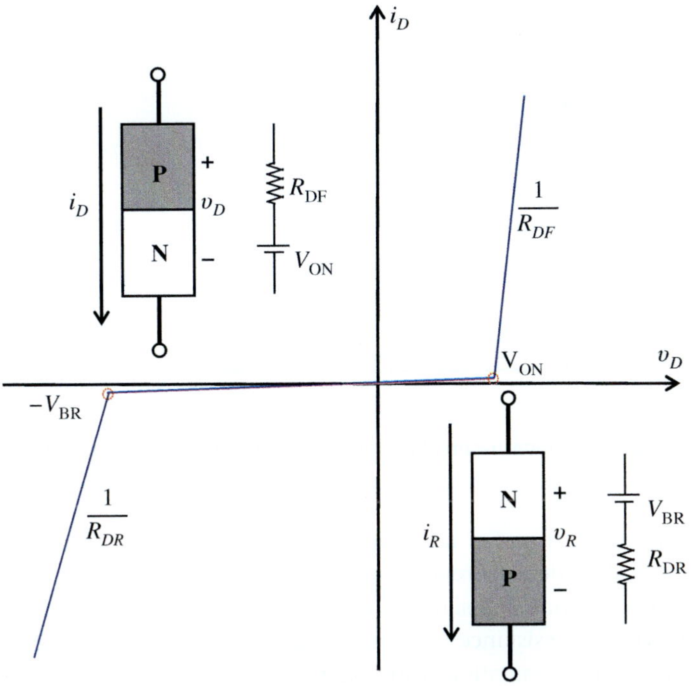

Figure 4.10 Typical *I*–*V* characteristics for a diode can be described by piece-wise-linear models.

negative bias increases to the reserve breakdown voltage, $v_D = -v_R = -V_{BR}$, avalanche breakdown occurs at the PN junction, resulting in a dramatical increase in reserve conduction current. A piece-wise-linear model is shown in Figure 4.10 that follows the equivalent circuit model as

$$v_R = i_R R_{DR} + V_{BR} \tag{4.5}$$

where R_{DR} is equivalent diode on-resistance in reverse mode. Typical diode reverse breakdown is associated with avalanche multiplication, or impact ionization, in the depletion region following [8, 9]

$$V_{BR} = \frac{\varepsilon_s E_{max}^2}{2qN_B} \tag{4.6}$$

and

$$M = \frac{1}{1 - \left(\frac{v_R}{V_{BR}}\right)^n} \tag{4.7}$$

where ε_s is semiconductor permittivity, E_{max} is maximum electric field in the depletion region, N_B is impurity density of the lower-doped region in a single-sided abrupt PN junction, and n is a constant.

4.2.2 Diode in ESD Protection

A diode can operate in both forward and reverse modes for ESD protection. In forward mode, the previously discussed diode device physics generally governs, except that a diode operates

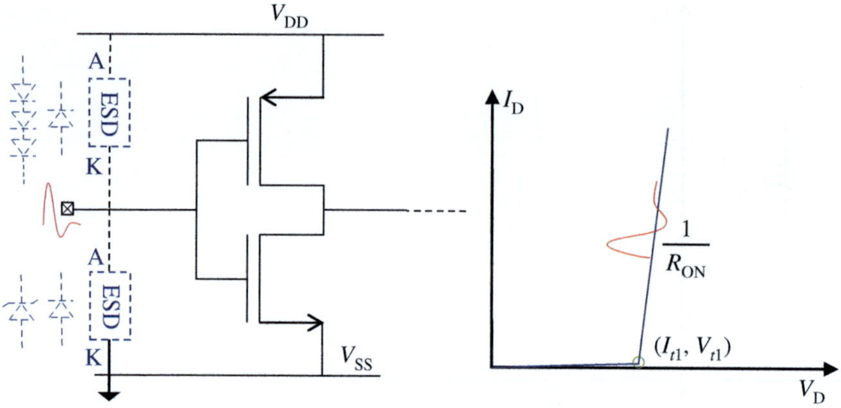

Figure 4.11 Conceptual schematics for exemplar diode-based ESD protection schemes and typical ESD discharging I–V curve.

in high-current mode, which is generally true for all devices as ESD protection devices. In high-current mode, both drift and diffusion current components have to be considered. The voltage drop across the intrinsic PN junction is insignificant compared to the ohmic drop of the diode series resistance (r_s) and dynamic resistance (r_d) under high injection (~10^4 A/cm^2). Generally, the diode I–V characteristics in ESD operation is approximated as [8]

$$i_D \propto e^{\frac{v_D}{2V_T}} \tag{4.8}$$

Figure 4.11 illustrates conceptual schematics for diode-based on-chip ESD protection schemes as examples. Briefly, a diode ESD protection device is placed at an IC pad with respect to GND and/or power supply bus (e.g., V_{DD}). For instance, a reverse-biased diode is connected between I/O pad and GND that stays OFF during normal IC operations. Under ESD stressing, the diode will be triggered at $V_{t1} \approx V_{BR}$, which creates an ESD discharge path from I/O pad to GND to shunt the incident ESD pulse, hence protects the IC. Single diode in forward biasing mode cannot be used for ESD protection for ICs operating with supplies $V_{DD} > 5$ V even if the DC voltage drop across the diode is lower than diode forward turn-on voltage, because a typical fluctuation of 10% in power supply may accidently trigger a forward diode ($V_{ON} \sim 0.65$ V in Si), causing IC malfunction. In comparison, R_{ON} for a forward-biased diode is lower than that for a reverse-biased diode. Consequently, ESD protection using parasitic (not intentionally designed) reverse PN junction is not preferred in practical IC ESD protection designs. One common solution to this problem is to optimize the diode for reverse conduction, which is similar to a Zener diode where $V_{t1} \approx V_Z$, i.e., the Zener diode turn-on voltage. Another advantage of using reverse diodes for ESD protection is that, typically, the PN junction reverse breakdown voltage can be designed with a wide range in a given IC process, nice for on-chip ESD protection of mixed-signal ICs with multiple power domains. On the other hand, a forward diode-string can be used as ESD protection that offers unique features, such as reduced ESD-induced parasitic capacitance (C_{ESD}) due to the series PN junctions, while setting the ESD triggering voltage to the desired level, $V_{t1} \approx nV_{ON}$ for a series of n didoes in forward mode. One major disadvantage of a forward diode-string for ESD protection is that the leakage current (I_{leak}) may be enlarged due to the Darlington effect in the diode-string. In practical ESD protection designs, one ought to burn lots of brain cells to optimize didoes for ESD robustness, for example, to minimize the series r_s and dynamic r_d of a PN junction diode in order to minimize ESD overheating.

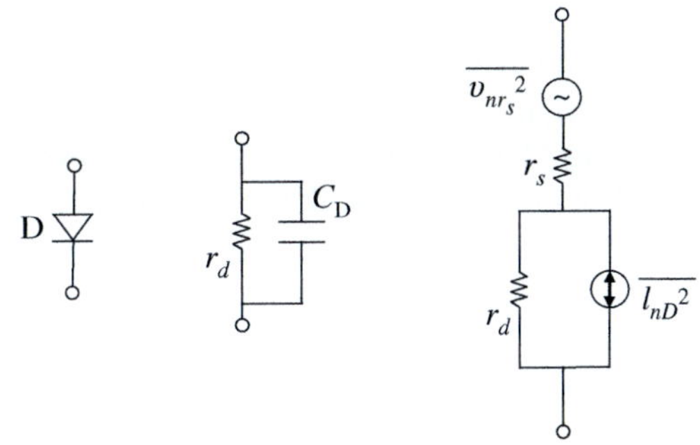

Figure 4.12 Diode equivalent circuit models for ESD-induced parasitic capacitance and self-generated noises.

4.2.3 Diode ESD Parasitic Modeling

There is no free lunch in this world. Everything comes with a cost, which is true for on-chip ESD protection designs too, even if using a single diode for ESD protection. Almost all active ESD switch structures rely on PN junction(s) embedded in Si for ESD discharging, which comes with parasitic effects inherent to PN junctions, including parasitic capacitance (C_{ESD}), leakage (I_{leak}), and noises. These parasitic effects are part of the total ESD design overhead, which can seriously affect IC performance and size, especially for large and complex ICs at advanced technology nodes. This is a major ESD protection design problem that will be discussed in details in later chapters. This section discusses typical ESD-induced parasitic effects for diode ESD devices. First, PN junction introduces capacitances, including depletion capacitance and diffusion capacitance, which, together, form the undesired C_{ESD}. ESD-induced C_{ESD} is a major design problem for high frequency and broadband RF ICs. Further, ESD-induced C_{ESD} leads to noise coupling problem that is another design challenge to analog, mixed-signal (AMX) and RF ICs. Second, leakage current always exists in a PN junction, which is becoming a major problem in advanced IC technologies. Third, an ESD protection device itself will generate extra noises (ESD *self-generated noise*) due to its leakage and series diffusion resistance. Figure 4.12 depicts the common parasitic components of an ESD diode and the small-signal equivalent circuit model for the ESD-induced parasitic capacitance and noises, where $C_{ESD} \approx C_D$ includes both depletion and diffusion capacitances of the PN junction. For ESD noise consideration, r_d and r_s are the dynamic and series resistances for the PN diode, respectively. The ESD-induced self-generated noises are characterized as r_s-induced thermal noise voltage power ($\overline{v_{nr_s}^2}$) and its intrinsic noise current power ($\overline{i_{nD}^2}$) comprising shot and flicker noises. The noise power spectral densities are approximated as [10, 11],

$$\overline{v_{nr_s}^2} = 4kTr_s\Delta f \tag{4.9}$$

and

$$\overline{i_{nD}^2} = 2qi_D\Delta f + K\frac{i_D^a}{f}\Delta f \tag{4.10}$$

where Δf is the noise bandwidth concerned, and K and a are process and device-related coefficients. These small-signal ESD parasitic models can be included in circuit simulation and analysis to evaluate the adverse impacts of ESD protection structures on circuit performance.

4.3 BJT for ESD Protection

Naturally, a bipolar junction transistor (BJT) may be considered as a promising candidate for ESD protection, because the active device function, i.e., electrical amplification of a BJT, shall boost the current conduction capability, hence, enhances the ESD discharging capacity of a BJT ESD protection device. In fact, while a PN junction is a brick for active ESD switches, a BJT is really the underlying functional building cell for almost all active ESD protection structures and circuits, including MOSFET and SCR ESD protection devices, and their derivatives and subcircuits. It is therefore important to understand BJT operations.

4.3.1 BJT Device Physics

Figure 4.13 depicts an exemplar NPN BJT device structure and amplification scheme in common emitter (CE) mode. The BJT operates in three functional modes depending upon the biasing conditions. The BJT stays OFF, i.e., in the cut-off region, when its two PN junctions are both reverse-biased, i.e., $V_{BE} < 0$ and $V_{BC} < 0$. When both PN junctions are forward-biased, i.e., $V_{BE} > 0$ and $V_{BC} > 0$, the two PN junctions are in ON states and the BJT operates in the linear region (a.k.a., saturation region for BJTs), behaving as a resistor. When the emitter junction is forward-biased and the collector junction is reverse-biased, i.e., $V_{BE} > 0$ and $V_{BC} < 0$, the BJT is driven into the active region where a small signal can be substantially amplified. For an NPN BJT transistor working in active amplification mode, electrons (i.e., minority carriers) in the emitter (E) are injected into the base (B) region where a small portion of the electrons will be zeroed out with the holes (i.e., majority carriers) in the base through the recombination process. Since the base width (W_B) of a well-designed BJT is very narrow, the majority of the electrons injected from the emitter into the base will immediately reach to the collector (C) junction and are swept into the collector region by the electrical field. Similarly, holes in the base are injected into the emitter, however, the hole injection volume is much smaller compared with the electron injection flow due to the leveled impurity doping schemes in the emitter, base, and collector, i.e., $N_E > N_B > N_C$, hence, forming a regular $N^{++}P^{+}N$ BJT structure. A battery provides external base current flowing into the base to continuously supply holes needed to maintain the BJT amplification. The terminal I–V characteristics of an NPN BJT transistor are depicted in Figure 4.14 and approximately follow the equations below [8]. The total collector current is approximated by

$$i_C = I_s e^{\frac{v_{BE}}{V_T}} \tag{4.11}$$

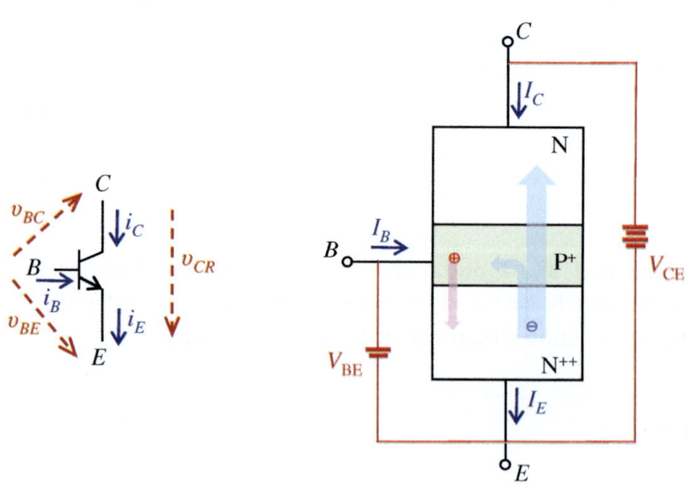

Figure 4.13 Operation of an intrinsic NPN BJT transistor.

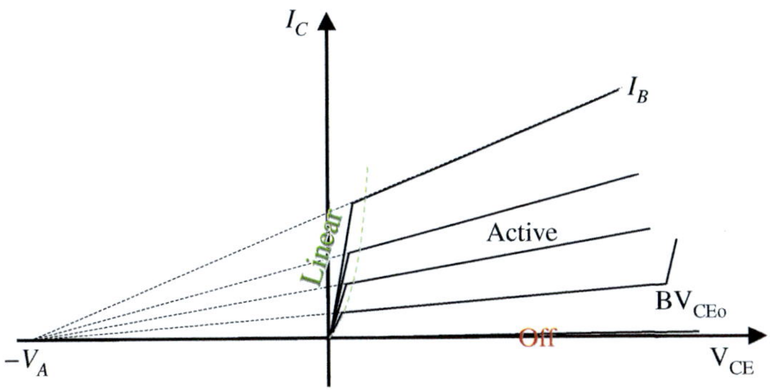

Figure 4.14 Typical common-emitter (CE) BJT $I-V$ characteristics in normal operations.

with

$$I_s = \frac{A_E q D_n n_i^2}{N_A W_B} \tag{4.12}$$

the total base current is given by

$$i_B = I_s \left(\frac{D_p}{D_n} \frac{N_A}{N_D} \frac{W_B}{L_p} + \frac{1}{2} \frac{W_B^2}{D_n \tau_b} \right) e^{\frac{v_{BE}}{V_T}} \tag{4.13}$$

and the total emitter current follows

$$i_E = i_B + i_C \tag{4.14}$$

where i_E, i_B, and i_C are total emitter, base, and collector currents, v_{BE} is total emitter junction bias, A_E is emitter junction size, W_B is base width, N_A and N_D are acceptor and donor impurity densities (i.e., N_E, N_B, N_C in each region, respectively), D_P and D_n are free carrier diffusion coefficients, L_p and L_n are carrier diffusion lengths, and τ_b is minority carrier lifetime in the base. The CE and CB (common-base) mode current gains, i.e., β_F and α_F in forward amplification, are derived roughly as

$$\beta_F \equiv \frac{I_C}{I_B} = \frac{1}{\frac{D_p}{D_n} \frac{N_A}{N_D} \frac{W_B}{L_p} + \frac{1}{2} \frac{W_B^2}{D_n \tau_b}} \tag{4.15}$$

and

$$\alpha_F \equiv \frac{I_C}{I_E} = \frac{\beta_F}{1 + \beta_F} \tag{4.16}$$

where I_E, I_B, and I_C are terminal currents, and footnote F denotes BJT forward amplification operation. Practical BJT transistors are affected by the Early effect, which is characterized by the Early Voltage (V_A). As the reverse biasing voltage across the collector junction continuously increases, reverse voltage breakdown (BV) will eventually occur at the collector junction due to either avalanche breakdown or punch-through breakdown if W_B is narrow enough, following the formulas below,

$$M = \frac{1}{1 - \left(\frac{v_{CB}}{BV_{CBo}} \right)^n} \tag{4.17}$$

and

$$BV_{CEO} = BV_{CBo} (1 - \alpha_F)^{\frac{1}{n}} \tag{4.18}$$

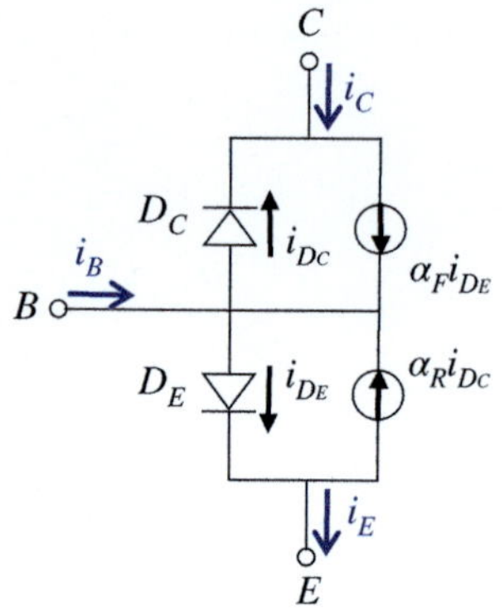

Figure 4.15 Ebers–Moll model equivalent circuit for an NPN BJT.

where BV_{CEO} and BV_{CBO} are open-base and open-emitter breakdown voltages, respectively. This avalanche multiplication effect plays an important role in BJT ESD protection operations. Figure 4.15 presents the equivalent circuit model for an NPN BJT per the Ebers–Moll Model, facilitating universal BJT operations in both directions, i.e., forward (F) and reverse (R) operations, as approximated by

$$i_E = \frac{I_s}{\alpha_F}\left(e^{\frac{v_{BE}}{V_T}} - 1\right) - I_s\left(e^{\frac{v_{BC}}{V_T}} - 1\right) \tag{4.19}$$

and

$$i_C = I_s\left(e^{\frac{v_{BE}}{V_T}} - 1\right) - \frac{I_s}{\alpha_R}\left(e^{\frac{v_{BC}}{V_T}} - 1\right) \tag{4.20}$$

and

$$i_B = \frac{I_s}{\beta_F}\left(e^{\frac{v_{BE}}{V_T}} - 1\right) + \frac{I_s}{\beta_R}\left(e^{\frac{v_{BC}}{V_T}} - 1\right) \tag{4.21}$$

4.3.2 BJT in ESD Protection

Figure 4.16 illustrates several simple ESD protection schemes using BJT transistors. The Case-Q_1 depicts the concept for BJT-based I/O-to-GND ESD protection where the NPN BJT Q_1 is placed at I/O pad and connected to GND. In normal IC operations, since the collector junction is

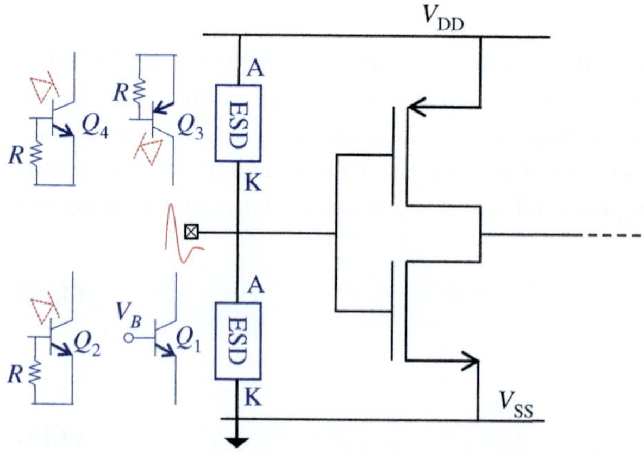

Figure 4.16 Illustration for BJT-based ESD protection schemes with four simple implementation cases.

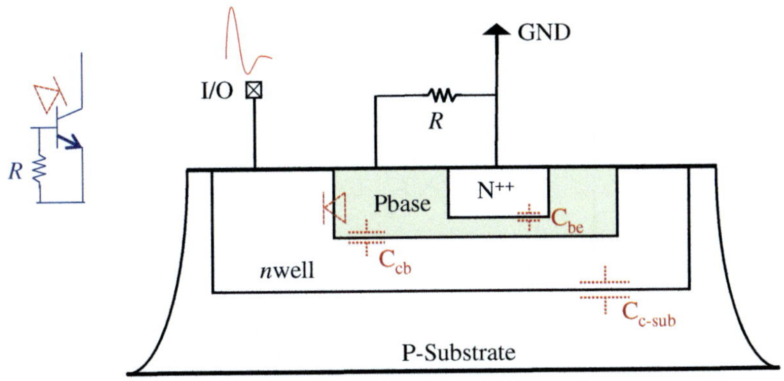

Figure 4.17 Illustration of schematic and cross-section of NPN BJT for I/O-to-GND ESD protection.

reverse-biased, Q_1 stays OFF, hence not affecting IC functions. Under ESD stressing, if a magic hand (i.e., a V_B bias) is applied to the base immediately as the incident ESD pulse occurs at the pad, Q_1 can be triggered to discharge the ESD transient into GND, hence providing ESD protection. Case-Q_2 is a simple implementation of Case-Q_1 in practical designs where a trigger-assisting resistor (R) is connected between the B and E terminals. As shown in Figure 4.17 in its cross-sectional view, a positive ESD pulse at I/O pad (with respect to GND) will reverse-bias the collector junction of NPN Q_2 until reaching to breakdown. The large avalanche breakdown current will flow through the external R, building up a potential to quickly turn-on the emitter of Q_2. Consequently, Q_2 is triggered into forward application mode to discharge the incoming ESD transient into GND and protect the IC core. If a negative ESD pulse occurs at I/O pad with respect to GND, ESD discharge will take place through a parasitic diode in forward mode, i.e., the collector junction of Q_2. For NPN ESD protection structures, ESD discharge typically follows a snapback I–V characteristic as shown in Figure 4.18. In practical designs of BJT-based ESD protection structures, it is important to carefully design the ESD-critical parameters, including the ESD triggering voltage (V_{t1}), holding voltage (V_h), and discharge resistance (R_{ON}). In practices, there are countless ways to design various BJT-based ESD protection structures and subcircuits using both NPN and PNP transistors. For example, Figure 4.16 depicts two exemplar cases for I/O-to-V_{DD} ESD protection including use PNP in Case-Q_3 or NPN in Case-Q_4. The key principle in designing BJT-based ESD protection is to make sure that the BJT ESD devices must stay in OFF state during normal IC operations and would not be mistriggered by supply voltages or any strong signals in absence of ESD events. It is also important to know that BJT ESD protection devices operate in

Figure 4.18 Typical snapback ESD discharge I–V characteristic for a BJT ESD protection device.

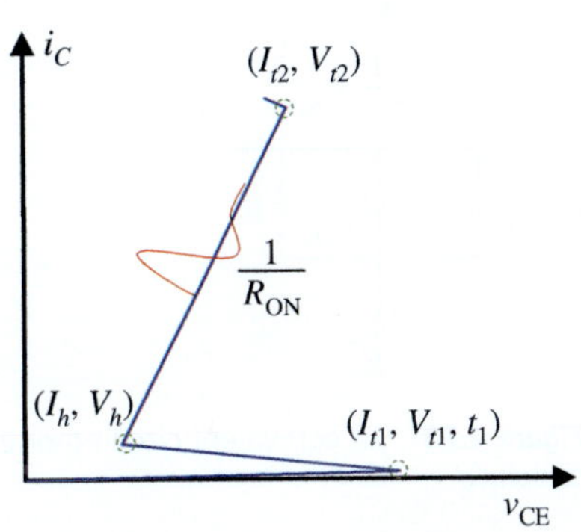

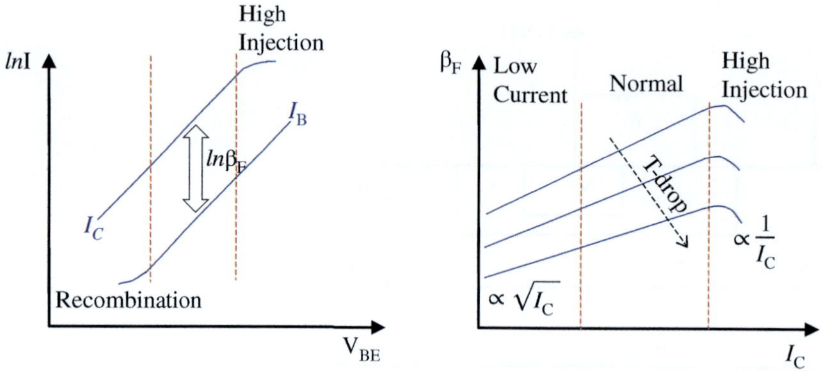

Figure 4.19 BJT gain depends on biasing conditions and reduces at high current injection, which corresponds to ESD protection mode.

high-current mode. As shown in Figure 4.19, high current injection can significantly reduce BJT gain (β) [8, 12]. In ESD discharge mode, the ESD current is actually orders of magnitude higher than a typical "high injection" current in a normal BJT transistor; hence, careful design optimization is needed to make a BJT transistor strong and effective in conducting ESD transient currents.

4.3.3 BJT ESD Parasitic Modeling

ESD-induced design overhead must be carefully considered when using BJT for ESD protection. Particularly, since a BJT has several PN junctions in IC format, much more ESD-induced parasitic effects, such as C_{ESD}, I_{leak}, and noises, are expected that will affect core IC performance more severely, even though the BJT ESD device stays in OFF state during normal IC operations. Compared to an ESD diode, the parasitic network for an ESD BJT is more involving and its complexity depends entirely on the BJT ESD protection topologies used. Take the NPN BJT for I/O-to-GND ESD protection scheme shown in Figure 4.17 as an example, the total ESD-induced C_{ESD} can analyzed using a capacitance network shown in Figure 4.20. The BC junction capacitance (C_{cb}) and the collector-substrate capacitance ($C_{c\text{-}sub}$) dominate because the two junctions are reverse-biased. The BE junction capacitance (C_{be}) is negligible due to its forward-biasing mode. Consequently, the total ESD-induced C_{ESD} of the NPN BJT ESD device operating in Figure 4.17 mainly comes from the two parallel capacitors, C_{cb} and $C_{c\text{-}sub}$. Next, one needs to consider the ESD-induced noise effects in the NPN ESD protection structure, which is depicted by the

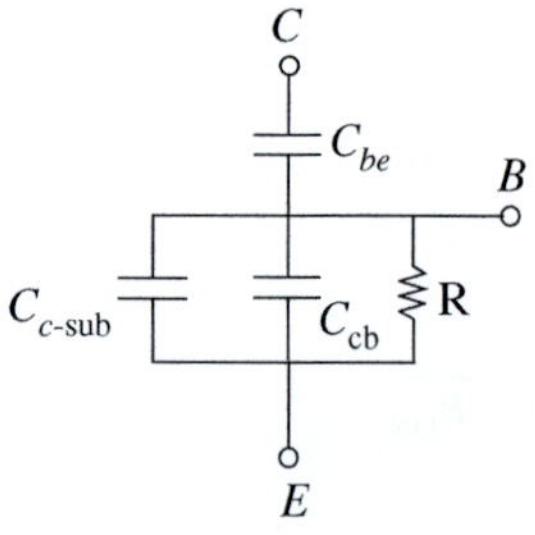

Figure 4.20 An equivalent circuit models ESD-induced parasitic C_{ESD} for a BJT-based ESD protection structure.

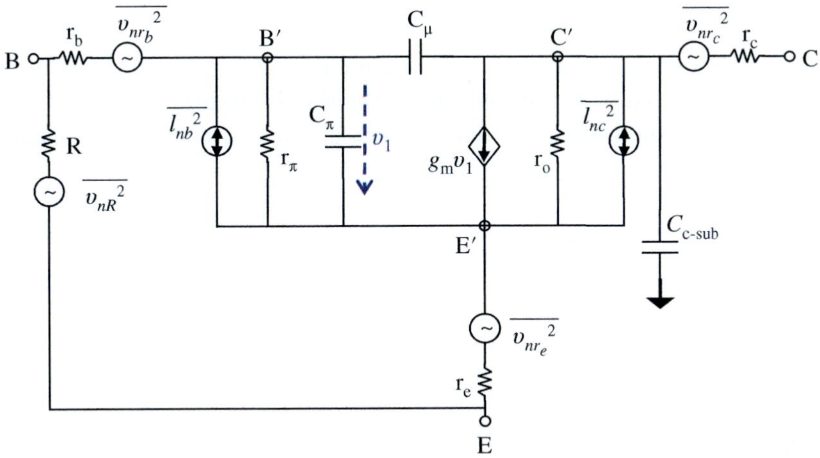

Figure 4.21 An equivalent circuit models ESD-induced noises for a BJT-based ESD protection structure.

equivalent noise circuit model shown in Figure 4.21 with each ESD noise generators approximated by the formulas below [10],

$$\overline{i_{nC}^2} = 2qi_C\Delta f \tag{4.22}$$

$$\overline{i_{nb}^2} = 2qi_B\Delta f + K_1\frac{i_B^a}{f}\Delta f \tag{4.23}$$

$$\overline{v_{nr_b}^2} = 4kTr_b\Delta f \tag{4.24}$$

$$\overline{v_{nr_e}^2} = 4kTr_e\Delta f \tag{4.25}$$

$$\overline{v_{nr_c}^2} = 4kTr_c\Delta f \tag{4.26}$$

and

$$\overline{v_{nR}^2} = 4kTR\Delta f \tag{4.27}$$

where R is external resistance, r_b, r_e, and r_c are series resistances of BJT diffusion regions, K_1 and a are device-specific constants, and Δf is the frequency bandwidth of interest. The noises considered included thermal noises, shot noises, and Flicker noise. Other noise sources may be included if known and for different BJT-based ESD protection structures.

4.4 MOSFET for ESD Protection

MOSFET has been widely used for on-chip ESD protection in CMOS ICs for decades. There are many varieties for MOSFET-based ESD protection structures and subcircuits, including grounded-gate MOSFET (ggNMOS and ggPMOS) and gate-coupled MOSFET (gcNMOS and gcPMOS) ESD protection structures, and their derivatives. In recent years, it is the general understanding that designing MOSFET-based ESD protection structures is becoming very challenging due to the relatively lower ESD discharging efficiency and large ESD design overhead. It is very important to understand the underlying basics in order to enhance MOSFET ESD protection designs.

4.4.1 MOSFET Device Physics

Figure 4.22 depicts the normal operation of a typical enhancement mode N-channel MOSFET (NMOS). Channel conduction of a MOSFET is controlled by its threshold voltage (V_{th}). An NMOS transistor operates in three different modes depending upon its biasing conditions. When $V_{GS} < V_{th}$, NMOSFET is in the cut-off region and stays OFF. When $V_{GS} > V_{th}$ and $V_{GD} = V_{GS}- V_{DS} > V_{th}$, an *n*-channel is induced throughout the NMOS FET, making it acting as a resistor, hence operating in the linear region (a.k.a., triode mode in MOSFET). When $V_{GS} > V_{th}$ and $V_{GD} = V_{GS}-V_{DS} < V_{th}$, the channel is ON at the source end, but OFF at the drain end, making the NMOS FET operating in the active region (a.k.a., saturation for MOSFETs) and being able to amplifying electrical signals. The terminal *I*–*V* characteristics, including second-order effects, for an NMOS FET approximately follow the equations below [8, 12]. The MOSFET threshold voltage is given by

$$V_{th} = V_{tho} + \frac{\sqrt{2qN_A\varepsilon_s}}{C_{ox}}(\sqrt{2\varphi_f + V_{SB}} - \sqrt{2\varphi_f}) \tag{4.28}$$

with the intrinsic threshold voltage defined as follows:

$$V_{tho} = \phi_{ms} + \frac{Q_b}{C_{ox}} - \frac{Q_{SS}}{C_{ox}} + 2\varphi_f \tag{4.29}$$

The drain current in linear region is given as

$$i_D = \mu_n C_{ox}\frac{W}{L_{eff}}\left[(V_{GS} - V_{th})V_{DS} - \frac{1}{2}V_{DS}^2\right] \tag{4.30}$$

and that in active (saturation) region follows the Square Law

$$i_{Dsat} = \frac{1}{2}\mu_n C_{ox}\frac{W}{L_{eff}}(V_{GS} - V_{th})^2(1 + \lambda V_{DS}) \tag{4.31}$$

The subthreshold drain current is given by

$$i_{Dsub} = \mathrm{K}\mu_n C_{ox}\frac{W}{L_{eff}}e^{\frac{V_{GS}-V_{th}}{nV_T}}\left(1 - e^{-\frac{mV_{DS}}{nV_T}}\right) \tag{4.32}$$

The relevant device parameters include the metal-semiconductor work function difference (ϕ_{ms}), the surface depletion region space charges (Q_b), the fixed charges in gate oxide (Q_{SS}), the strong inversion Fermi level ($2\varphi_f$), the unit gate oxide capacitance (C_{ox}), the electron mobility (μ_n), the channel width (W), the effective channel length (L_{eff}), the equivalent Early voltage ($\frac{1}{\lambda}$) associated with channel-length modulation effect, and K, m and n being device-specific coefficients.

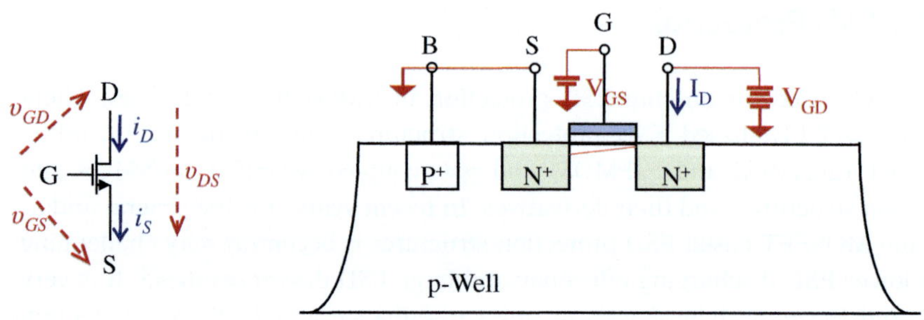

Figure 4.22 Biasing condition and cross-section for an enhancement mode NMOS FET operating in amplification mode.

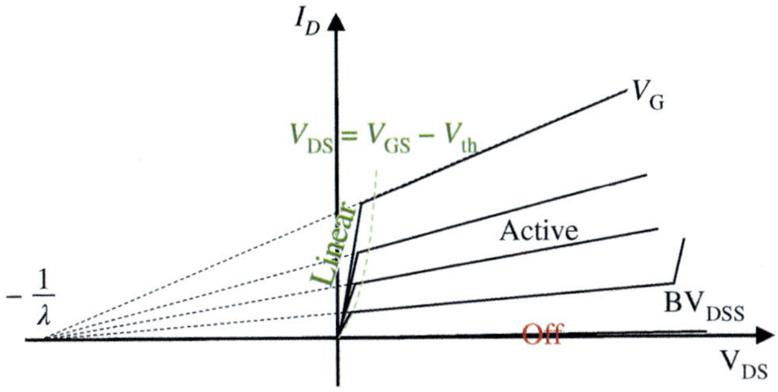

Figure 4.23 Typical common-source (CS) NMOSFET *I*–*V* characteristics in normal operations.

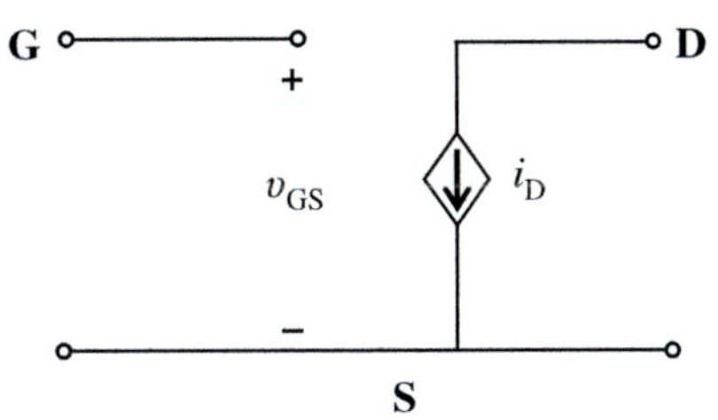

Figure 4.24 A large-signal equivalent circuit model for NMOS transistor.

Figure 4.23 illustrates typical terminal *I*–*V* characteristics for an NMOSFET in normal operations. For ESD protection design, MOSFET breakdown voltage is an important parameter. For long-channel MOSFETs, the drain-to-source breakdown (BV_{DSS}), is caused by the drain-to-substrate junction breakdown (BV_{DB}) following equations (4.6) and (4.7). For short-channel MOSFETs, BV_{DSS} is associated with either parasitic lateral NPN breakdown or punch-through breakdown. Considering the parasitic NPN of drain-substrate-source, BV_{DSS} if approximated by

$$BV_{DSS} = BV_{CEO} \approx \frac{BV_{DB}}{\sqrt[n]{2}} \left(\frac{L_G}{L_{eff}} \right)^{\frac{2}{n}} \tag{4.33}$$

where BV_{CEO} is the open-base CE breakdown of the parasitic NPN and L_G is drawn gate length, equivalent to base width of the lateral NPN. Another breakdown mechanism that is particularly devastating to ESD protection is gate dielectric breakdown, which is a major concern in ESD protection designs for advanced CMOS. Figure 4.24 shows the corresponding large-signal equivalent circuit model for an NMOSFET.

4.4.2 ggMOS in ESD Protection

There are many ways to use MOSFETs for ESD protection. Figure 4.25 illustrates a few examples of using classic grounded-gate MOSFET (ggNMOS or ggPMOS) to protect an I/O pad. Figure 4.26 depicts the mechanism of a ggNMOS ESD protection device, which may be the most popular MOSFET ESD protection solution in CMOS ICs. A ggNMOS ESD protection device has its gate (*G*) grounded by shorting to the source (*S*) and the well body (*B*), and B is typically connected to the most negative potential node (GND or V_{SS}) on a chip for an NMOS transistor. For PS ESD stressing mode, an ggNMOS ESD device serves as an active ESD switch with its anode (A, i.e., D) connected to an I/O pad and its cathode (K, i.e., G = S = B) connected to GND. As a positive ESD transient appears at the I/O pad with respect to GND (or V_{SS}), the drain-well junction (DB) is reverse-biased, quickly reaching to its junction breakdown due to the fast and large ESD pulse.

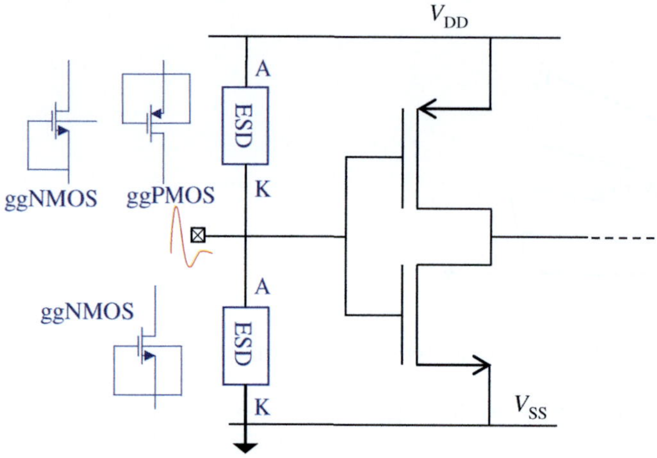

Figure 4.25 Various I/O ESD protection schemes using ggNMOS and ggPMOS ESD protection structures on a chip.

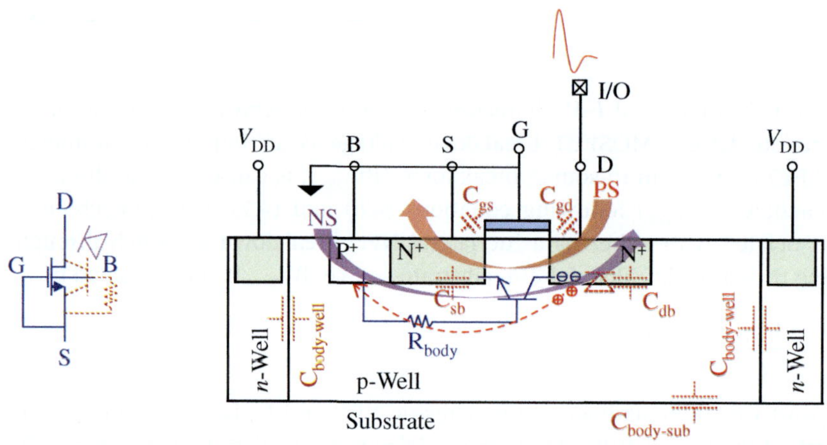

Figure 4.26 Illustration of ggNMOS ESD protection mechanism under PS ESD stressing mode from I/O-to-GND.

Avalanche multiplication takes place and generates sea of electron–hole pairs. The hole current flows into GND via the p-well body region (B grounded) and builds up a potential lateral (V_R) across the lateral parasitic body resistance (R_{body}). Since B and S are electrically shorted, V_R actually appears positively across the BS PN junction. Under a fast and large ESD pulse, the BS junction will be forward turned on quickly and triggers the parasitic lateral NPN transistor Q (DBS). The rest of the story shall be similar to that for an NPN BJT ESD protection device discussed previously. As a result, ggNMOS is turned on at a triggering point (V_{t1}) by an ESD pulse and forms a low-impedance (R_{ON}) conduction path to shunt the incoming ESD pulse into GND. This completes the ggNMOS ESD protection operation. Essentially, the parasitic lateral NPN BJT works for ggNMOS in ESD protection, hence, featuring a snapback I–V characteristic. Apparently, the ggNMOS ESD efficiency is dominated by the parasitic NPN BJT, whose β determines R_{ON} and ESD holding V_h. If a negative ESD pulse comes to the I/O pad w.r.t. GND (i.e., NS ESD mode), a forward-biased parasitic body-drain diode (BD) will be turned on to discharge the NS ESD transient. One main reason for grounding the gate is to ensure "zero" leakage of ggNMOS ESD protection structure under normal operations. A ggNMOS ESD protection structure has many advantages. First, it provides an active ESD discharging path enhanced by NPN amplification, although in one direction only for PS ESD stressing. Second, it is a natural device in CMOS ICs. A ggNMOS ESD protection structure has disadvantages too. For example, it cannot be included into circuit simulation using SPICE due to its snapback *I–V* characteristic. For NS ESD stressing, ggNMOS is weak because the parasitic BD junction is typically not optimized for handling large ESD transient currents. Examples of I/O-to-V_{DD} ESD protection schematics are illustrated in Figures 4.27 and 4.28. Figure 4.27 depicts a classic ggPMOS ESD protection structure connected between I/O pad and V_{DD}, similar to ggNMOS I/O-to-GND ESD protection. Under negative I/O-to-V_{DD} ESD zapping (ND), a parasitic lateral PNP BJT will be triggered to discharge the ESD pulse. During positive I/O-to-V_{DD} ESD stressing (PD), the parasitic Drain-Body diode will discharge the ESD transient. Figure 4.28 shows an incorrect way of using ggNMOS for I/O-to-V_{DD} ESD protection. During fast I/O-to-V_{DD} ESD stressing in PD mode, V_{DD} pad is transiently equivalent to an *ac*-GND, hence, forms a parasitic lateral PNP

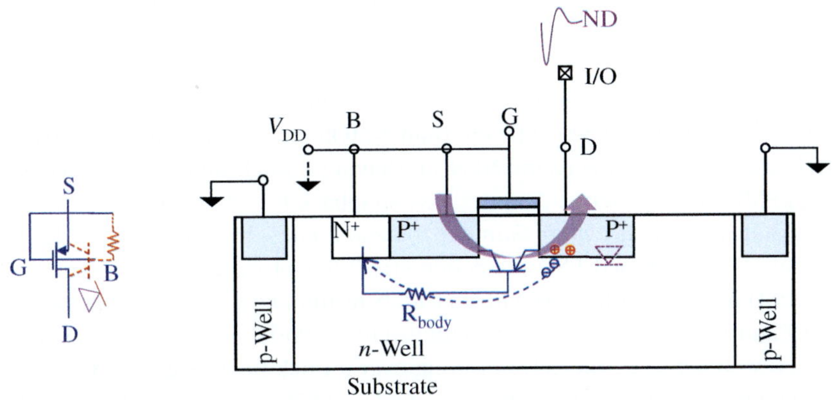

Figure 4.27 Illustration of ggPMOS ESD protection mechanism under ND ESD stressing mode from I/O-to-V_{DD}. At ESD stressing, a V_{DD} supply pad is equivalent to *ac*-grounding, hence making ggPMOS transiently gate-grounded during ESD discharging.

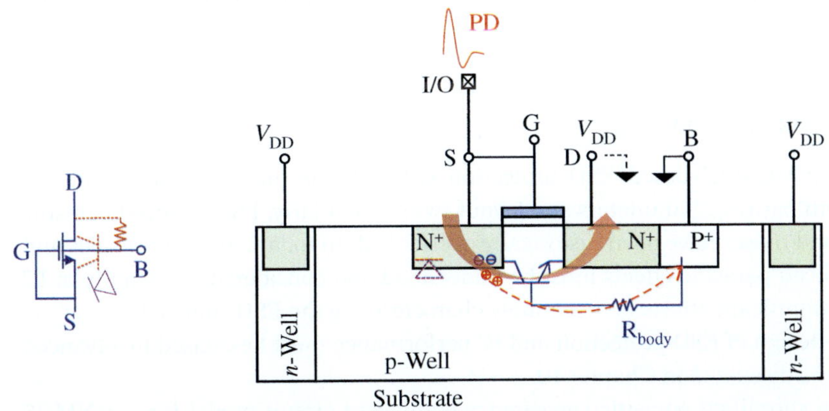

Figure 4.28 Illustration of an *incorrect* way of using ggNMOS ESD protection for I/O-to-V_{DD} ESD stressing. Though a parasitic lateral PNP BJT may discharge the ESD pulse, the gate is directly exposed to and stressed by an ESD pulse, easily causing gate breakdown failure.

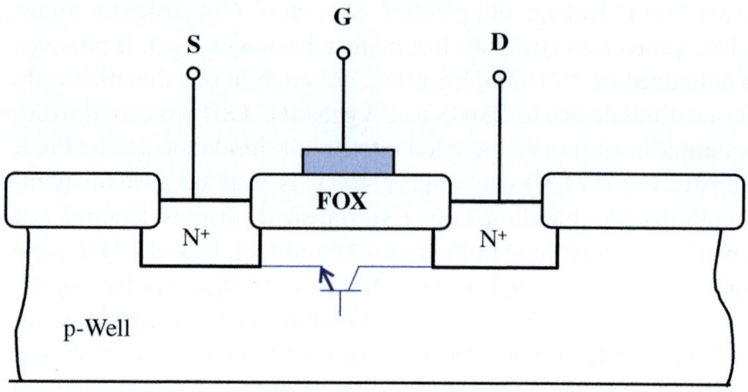

Figure 4.29 A FOX-MOS ESD structure uses thick oxide as gate.

BJT to discharge the ESD pulse. However, since the gate (G) is exposed to an ESD pulse directly, though G = S makes $V_{GS} = 0$, V_{GD} will be finite, possibly causing gate failure under ESD stressing. The risk is actually rather clear: the gate will be zapped directly by an incident ESD pulse from an I/O pad. Indeed, one has to be very cautious in ESD protection designs at chip level while being artistic. It is also well known that PMOS is less efficient than NMOS for ESD discharge because, obviously, the lateral PNP is not as nice as the lateral NPN, i.e., having a much smaller β due to low hole mobility. Typically, ggPMOS ESD structure does not have snapback *I–V* characteristics. Interestingly, as the old saying states – *there are more than one route to Rome* – imagination is indeed the only limit in designing MOSFET-based ESD protection structures of varieties. For example, one alternative MOSFET ESD protection often used in old CMOS technologies is a thick-gate MOS ESD device as shown in Figure 4.29 where a parasitic equivalent MOSFET is formed associated with a piece of metal or poly-silicon (i.e., the gate) running over a thick field oxide (FOX) layer between two N^+ diffusion regions. Under ESD zapping, the incident ESD pulses can be discharged by the parasitic lateral NPN in this FOX-MOSFET, which is obviously much less efficient than its BJT counterpart in a regular thin-gate ggNMOS, mainly because the large equivalent base width W_B limited by a large layout dimension determined by FOX layout. On the other hand, a FOX-MOS ESD device can withstand much higher gate voltage, making it possible to connect *G* to *D* for certain ESD designs.

4.4.3 MOSFET ESD Parasitic Modeling

One major concern for MOSFET-based ESD protection structures is the ESD-induced design overhead including both parasitic parameters and layout size. Particularly, ESD-induced parasitic capacitance, leakage, and noises have been historically overlooked. In today's IC technologies and chip designs, ESD-induced parasitic effects must be understood and considered thoroughly in IC designs. This requires significant efforts for accurately characterizing the ESD-induced C_{ESD}, I_{leak}, and noises. Careful co-design of ESD protection and IC performance must be excised in advanced IC designs, which will be discussed in Chapter 10.

Figure 4.30 depicts a simplified parasitic capacitance equivalent circuit model for a ggNMOS ESD protection structure. The total C_{ESD} comprises a capacitor network including the gate-source and gate-drain overlap capacitances (C_{gs}, C_{gd}), the source-body and drain-body junction capacitances (C_{sb}, C_{db}), the body to guard-ring junction capacitance ($C_{body\text{-}well}$) and the body-substrate

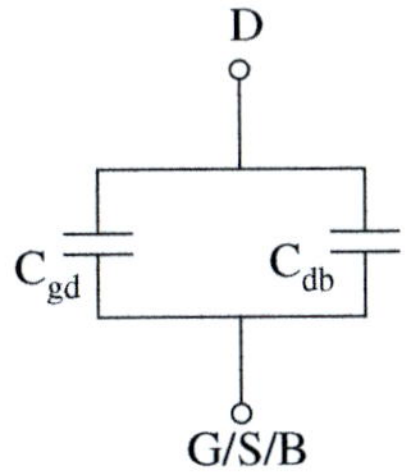

Figure 4.30 Simplified parasitic C_{ESD} equivalent circuit model for a ggNMOS ESD protection structure.

capacitance ($C_{\text{body-sub}}$). Take into account of ggNMOS ESD connection and biasing condition, the total ggNMOS parasitic C_{ESD} is dominated by C_{gd} and C_{db} in series as shown in Figure 4.30.

Figure 4.31 gives a simplified ESD noise equivalent circuit model. The ESD-induced noise sources include thermal noise due to channel conduction ($\overline{i_{nCh}}$), Flicker noise associated with i_D ($\overline{i_{nf}}$), induced-gate noise ($\overline{i_{ng}}$), noise generated by the distributed gate resistance ($\overline{i_{nr_g}}$), as well as thermal noises associated to series diffusion resistors r_d and r_s, i.e., $\overline{i_{nr_d}}$ and $\overline{i_{nr_s}}$. The ESD-induced noise power can be derived as [5],

$$\overline{i_{nD}^2} = \overline{i_{nCh}^2} + \overline{i_{nf}^2} = 4kT\gamma g_m \Delta f + \frac{K i_D^a}{f C_{ox} L_{eff}^2} \Delta f \tag{4.34}$$

$$\overline{i_{ng}^2} = \frac{16}{15} kT\delta\omega^2 C_{gs}^2 \Delta f \tag{4.35}$$

$$\overline{i_{nr_g}^2} = \frac{4kT}{3r_g} \Delta f \tag{4.36}$$

$$\overline{i_{nr_d}^2} = \frac{4kT}{r_d} \Delta f \tag{4.37}$$

and

$$\overline{i_{nr_s}^2} = \frac{4kT}{r_s} \Delta f \tag{4.38}$$

where γ, α, K, and δ are process- and device-specific coefficients, r_g is distributed gate resistance, r_d and r_s are drain and source series resistances, respectively. Since an ESD protection device is normally off, the drain current (i_D) is much smaller than typical subthreshold current.

Together, the ESD-associated small-signal equivalent circuit models for MOSFET ESD protection structures can be included in SPICE circuit simulation to evaluate impacts of ESD parasitic effects on core IC performance, which will be discussed in details in later chapters using various design examples.

4.5 SCR for ESD Protection

Silicon controlled rectifier (SCR) device is widely considered one of the most efficient and robust ESD protection structure due to its deep snapback I–V characteristics, being able to handle large ESD currents without much overheating and to clamp pad voltage to a very low level, both attributed to its strong BJT gain amplification. Nevertheless, designing SCR-based ESD protection structures is also very challenging, mainly due to its inherent latch-up behavior that must be

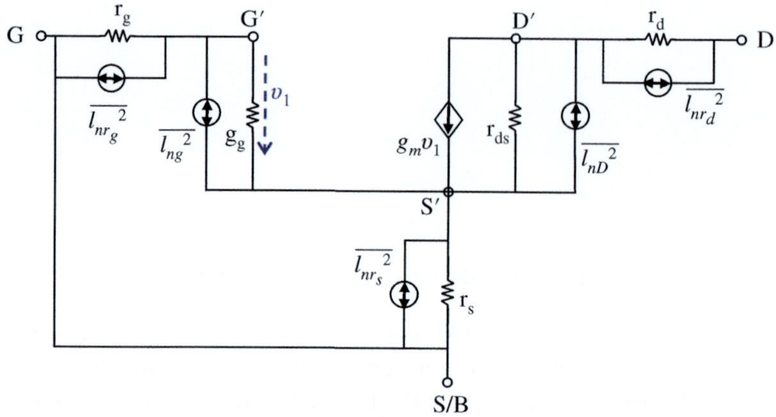

Figure 4.31 ESD-induced noise equivalent circuit model for a ggNMOS ESD protection device.

controlled. It is hence important for IC designers to thoroughly understand SCR device basics and SCR ESD protection mechanisms.

4.5.1 SCR Device Physics

Designing good SCR ESD protection structure is very challenging due to the genetic devil inside an SCR structure – latch-up. As shown in Figure 4.32, in a basic CMOS inverter circuit unit, there exists a parasitic SCR structure formed by a pair of parasitic lateral PNP BJT (Q_1) and vertical NPN BJT (Q_2), leading to a significant gain amplification, i.e., $\beta_1 \times \beta_2$, in terms of common-emitter current gains. An intrinsic SCR structure is a four-layer, $P_1N_2P_3N_4$, device with an anode (A) and a cathode (K) terminals and two control gates (G1 and G2), as shown in Figure 4.33. Typical impurity concentration levels in an SCR device is denoted by $P_1N_2P_3N_4 = P^+N^-PN^+$. A classic SCR device features an asymmetric snapback I–V characteristic as depicted in Figure 4.34. The total SCR I–V curve can be partitioned into five segments: a reverse blocking region ①, a reverse breakdown region ②, a forward blocking region ③, a negative resistance region ④, and a forward conduction region ⑤. An SCR device functions as follows: Starting from region ①, when a negative bias is applied to terminal A w.r.t K, both junctions J1 (P_1N_2) and J3 (P_3N_4) are reverse-biased, while junction J2 (N_2P_3) is positive-biased. There is no current conduction path from terminal A to terminal K, hence, SCR is in reverse blocking state (OFF). As the negative bias V_{AK} continuously increases until reaching to the reverse breakdown point, either by avalanche multiplication in depletion regions

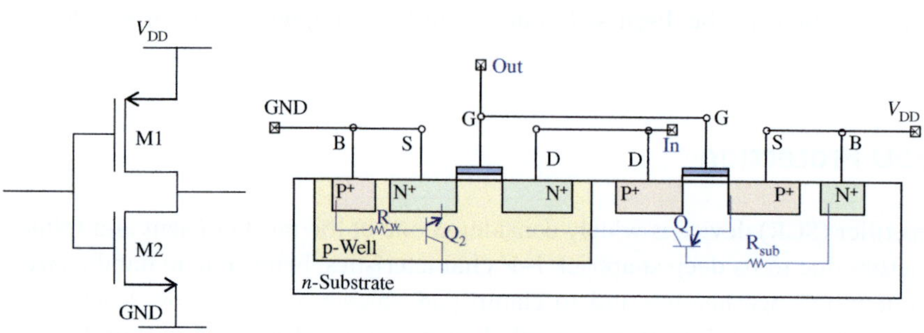

Figure 4.32 A cross-section for a classic CMOS inverter shows a parasitic SCR structure inside.

or punch-through breakdown, the SCR device will be driven into reverse breakdown region ② and starts to conduct current reversely. The total V_{AK} mainly drops across J1 junction and causes avalanche in its depletion region. The reverse breakdown voltage (BV_R) of $P_1N_2P_3$ is approximated by [8]

$$BV_R \approx BV_{DJ_1}(1-\alpha_1)^{\frac{1}{n}} \tag{4.39}$$

where BV_{DJ1} is reverse breakdown of junction J1 and n is a constant. In forward-biasing mode, when a small positive V_{AK} is applied from terminal A to terminal K, junctions J1 and J3 are forward-biased, while J2 is reverse-biased. At this moment, there still does not exist a conduction channel between terminals A and K since J2 is off, and SCR stays in forward-blocking region ③ (OFF). As the V_{AK} continuously increases, breakdown eventually occurs at the collector junction of Q_1, which produces a seeding current to trigger significant current regeneration in SCR. The SCR current regeneration mechanism is depicted by its equivalent circuit model shown in Figure 4.33: the collector current of Q_1 (I_{C1}) supplies the base current (I_{B2}) for Q_2 and pushes Q_2 into active mode; in turn, the collector current of Q_2 (I_{C2}) sources the base current (I_{B1}) for Q_1. As a result, as long as the current gain product is greater than unity, i.e., $\beta_1\beta_2 \geq 1$, the current regeneration sustains and a SCR device can function [8, 13]. The triggering of current regeneration will sweep the SCR through a negative resistance region (④) into a low-impedance, high-current, low-voltage forward ON state (⑤). What determines forward breakdown voltage, BV_{AK}? A math genius may readily point it to the breakover point of $\frac{dV_{AK}}{dI_A} = 0$ on the I–V curve, but it is useless for an IC designer who prefers to understand the electronic meaning to guide circuit designs. In general, there are two different SCR trigger methods, i.e., voltage-triggering by gradually stepping up V_{AK} or current-triggering using a $\frac{dV}{dt}$ transient by injecting a seeding current from the gate (i.e., G1 and/or G2). For example, in V_{AK}-induced avalanche breakdown and assuming no gate

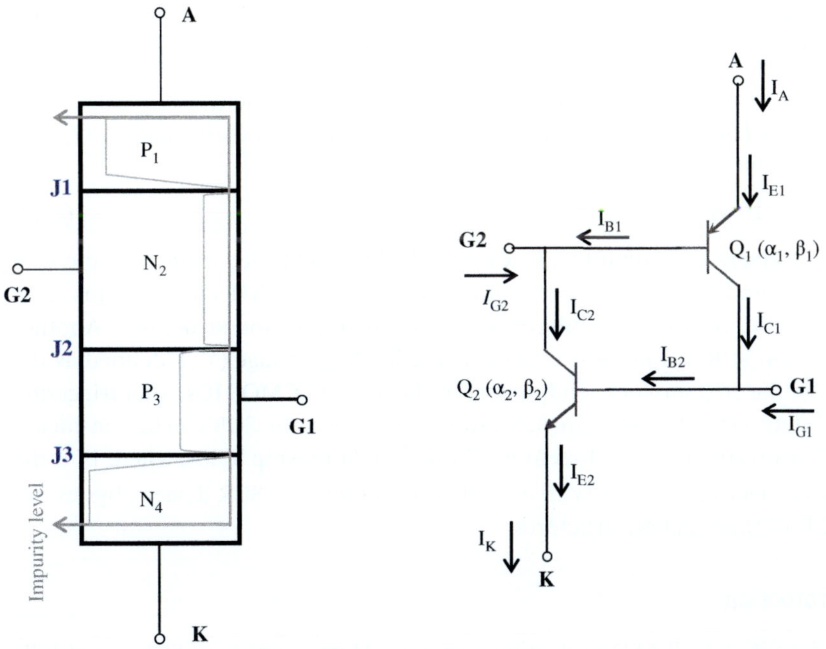

Figure 4.33 A generic SCR device and its equivalent circuit depict SCR operational mechanism.

injection, i.e., $I_{G1} = I_{G2} = 0$, from Ebers–Moll model and Kirchhoff current law, it comes,

$$I_A = I_{E1} = I_{E2} = I_K \tag{4.40}$$

$$I_{B1} = (1 - \alpha_1)I_A - I_{CO1} \tag{4.41}$$

and

$$I_{C2} = \alpha_2 I_K + I_{CO2} \tag{4.42}$$

Since

$$I_{B1} = I_{C2} \tag{4.43}$$

it hence gives

$$I_A = \frac{I_0}{1 - \alpha_1 - \alpha_2} \tag{4.44}$$

where $I_0 = I_{CO1} + I_{CO2} \ll I_A$. These formulas explain SCR current regeneration conditions. Then, what determines BV_{AK}? Assuming avalanche multiplication occurs in the junction J2 depletion region as depicted in Figure 4.34, the main collector currents, I_{C1} and I_{C2}, come from the injected hole and electron currents, i.e., I_p and I_n, which are multiplied by a multiplication factor, M (assuming $M \approx M_p \approx M_n$, same for both holes and electrons). The total terminal current is given by

$$I = I_A = I_K = M_p I_p + M_n I_n = M(\alpha_1 I_A + I_{CO1}) + M(\alpha_2 I_K + I_{CO2}) \tag{4.45}$$

Hence, it comes

$$\frac{1}{M(J2)} = \alpha_1 + \alpha_2 + \frac{I_0}{I_A} \tag{4.46}$$

For avalanche breakdown in the junction J2 depletion region, the multiplication factor can be expressed by

$$M(J2) = \frac{1}{1 - \left(\frac{V_{J2}}{BV_{DJ2}}\right)^2} \tag{4.47}$$

where V_{J2} and BV_{DJ2} are reverse biasing and breakdown voltages for junction J2, respectively. Assuming the condition of $I \gg I_0$ holds, by equating M-factor equations (4.46) and (4.47), the forward triggering voltage can be approximated by

$$V_{t1} = BV_{AK} \approx BV_{DJ2}(1 - \alpha_1 - \alpha_2)^{\frac{1}{n}}. \tag{4.48}$$

Formula (4.48) quantitatively states that the SCR forward triggering is determined by the well breakdown voltage (N_2P_3) and can be greatly enhanced by the parasitic BJT current gains. This forward-breakover point is critical to designing SCR-based ESD protection structures. Another important parameter in the SCR snapback *I–V* curve is the holding voltage (V_h), defined as the current needed to sustain the SCR forward conduction (i.e., latch-up in CMOS ICs) after triggering occurs. In the SCR ON state, since J_1 and J_2 are forward-biased, hence, the SCR can be considered as a P_1-I-N_4 diode in forward conduction. Therefore, a $V_h \propto \frac{W}{\tau_{eff}}$ relationship holds, where W is the I-region width and τ_{eff} is effective lifetime. With a good understanding of SCR device physics, we are ready to discuss SCR ESD protection structures.

4.5.2 SCR in ESD Protection

Figure 4.35 depicts a simplified SCR ESD protection structure. More varieties of SCR-based ESD protection structures will be discussed in Chapter 5. As shown in Figure 4.35, a SCR structure

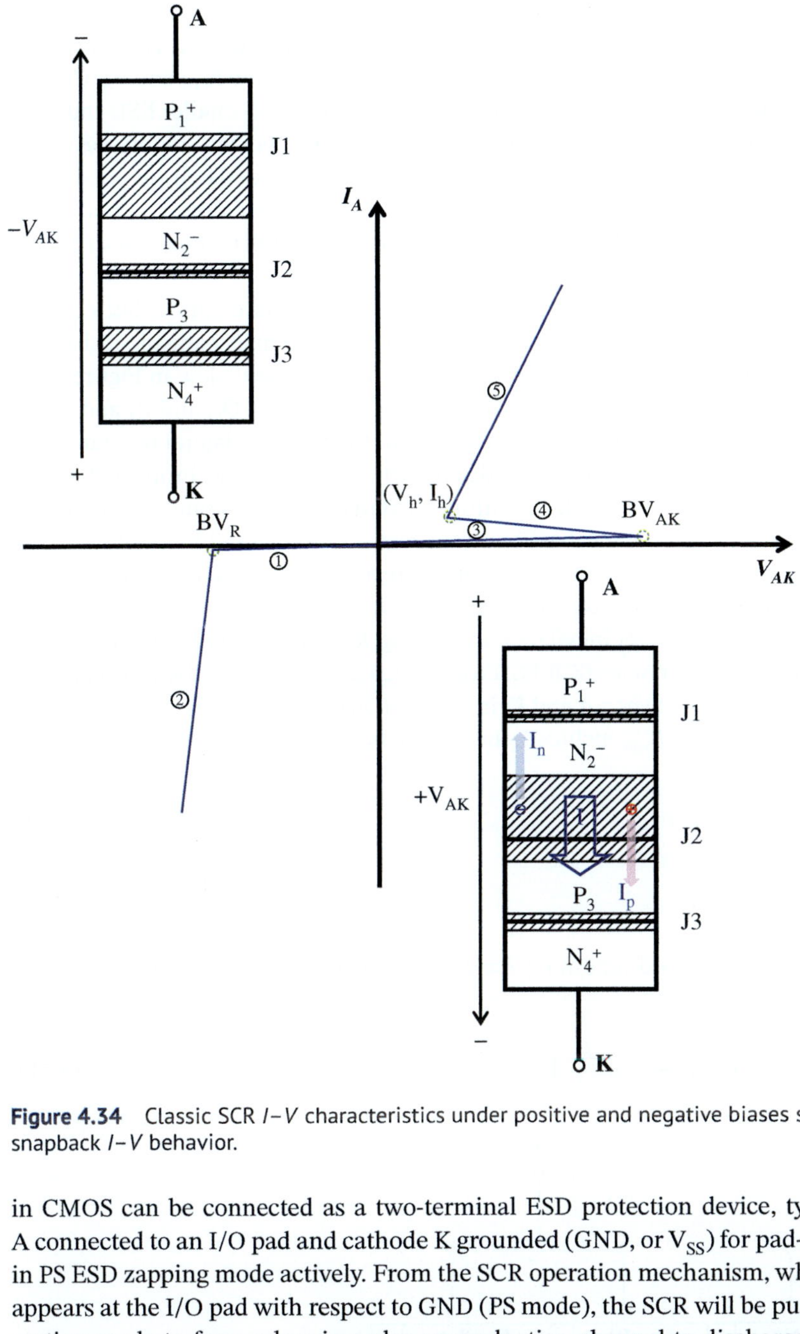

Figure 4.34 Classic SCR $I-V$ characteristics under positive and negative biases shows an asymmetric deep snapback $I-V$ behavior.

in CMOS can be connected as a two-terminal ESD protection device, typically, with the anode A connected to an I/O pad and cathode K grounded (GND, or V_{SS}) for pad-to-GND ESD protection in PS ESD zapping mode actively. From the SCR operation mechanism, when a positive ESD pulse appears at the I/O pad with respect to GND (PS mode), the SCR will be pushed into current regeneration mode to form a low-impedance conduction channel to discharge the ESD current pulse safely. The ESD triggering voltage V_{t1} is determined by the SCR forward breakdown voltage, i.e., $V_{t1} \approx BV_{AK}$, which must be designed to a specific value to fit into the ESD design window. During ESD discharge, as SCR moves into ON state, the I/O pad voltage will be clamped to a low and safe level, i.e., V_h, to prevent ESD-induced voltage breakdown failures. On the other hand, if a negative ESD pulse comes to I/O pad with respect to GND (NS mode), ESD discharge will occurs through the parasitic diode of p-well/n-sub (J2) junction via forward conduction. Therefore, a SCR ESD device provides active low-R_{ON} discharge in one direction and offers an asymmetric ESD discharging *I–V* characteristics. The main advantage for SCR ESD protection device is its large current-handling capacity associated with is PNP-NPN BJT coupling effect, resulting in very low R_{ON} and V_h. Consequently, an SCR ESD protection device is more area efficient, translating into smaller ESD-induced design overhead, including smaller footprint and lower parasitic capacitance – highly desirable for advanced ICs. The disadvantages of SCR ESD protection device include latch-up possibility, unfriendly to SPICE circuit simulation due to its snapback *I–V* curve, and poor reserve-mode ESD protection when using a parasitic PN diode. It is noteworthy that an SCR ESD protection structure is essentially a controlled latch-up device.

Originating from the generic SCR functions shown in Figure 4.33, more subtle, but critical, design factors must be considered for SCR ESD protection structures, such as parasitic resistance impacts, and $\frac{dV}{dt}$ and $\frac{dI}{dt}$ effects under ESD stressing. From the cross-sectional view in Figure 4.35, the SCR ESD parasitic effects are analyzed as following. Since P_1 and N_2 of Q_1, and P_3 and N_4 of Q_2 are shorted, respectively, to form the anode A and cathode K, the parasitic lateral resistances of N-substrate (R_W) and P-substrate (R_{sub}) must be included in SCR circuit analysis. From SCR equivalent circuit model depicted in Figure 4.36, the shunting currents flowing through the parasitic R_W (I_W) and R_{sub} (I_{sub}) will break up the ideal SCR current regeneration closed-loops of $I_{C2} = I_{B1}$ and $I_{C1} = I_{B2}$, respectively, resulting in degraded SCR current regeneration. As a result, it becomes more difficult to drive the SCR device into the ON state for active ESD discharge. Interestingly, if such kind of base-emitter short-circuit is inevitable, increasing R_W and R_{sub} may help to reduce the BE shunting currents and hence makes SCR ESD triggering easier. In math, the intrinsic SCR circuit, i.e., the inner SCR network between *A* and *K*, in Figure 4.36 is similar to that in Figure 4.33. With the inevitable parasitic R_W and R_{sub} included, the avalanche multiplication can be described by the following formulas:

$$I = I_{A'} = I_A + I_{sub} = I_{K'} = I_K + I_{R_w} \tag{4.49}$$

and

$$I = M_p I_p + M_n I_n \tag{4.50}$$

Similar to the previous analysis and assuming $I \gg I_0$, the triggering voltage can be approximated by

$$V_{t1} = BV_{A'K'} \approx BV_{DJ2}\left(1 - \alpha_1 \frac{I_A}{I} - \alpha_2 \frac{I_K}{I}\right)^{\frac{1}{n}} \tag{4.51}$$

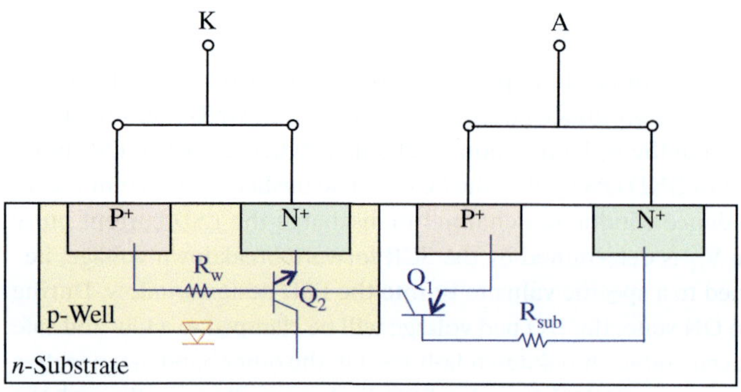

Figure 4.35 A simplified SCR ESD protection structure is a two-terminal SCR device.

This formula suggests that due to the current regeneration degradation effect, the ESD V_{t1} (or, $BV_{A'K'}$) may be higher than that of the generic SCR (BV_{AK}) that has no parasitic R_W and R_{sub}. Moreover, the series resistances in the emitters of Q_1 and Q_2 (i.e., R_{ep} and R_{en}) as shown in Figure 4.36, have influences, which can be approximated by

$$V_{t1} = BV_{A'K'} \approx BV_{AK} + V_{r_{ep}} + V_{r_{en}} \tag{4.52}$$

Due to the transient nature of ESD events, SCR operation under ESD stressing may be different from classic SCR behaviors. First, an ESD pulse introduces a strong $\frac{dV}{dt}$ effect that has strong impact on SCR triggering V_{t1} under ESD zapping. From avalanche analysis, the displacement current of $\frac{d(CV)}{dt}$ flows through the junction capacitances (C) and must be included into the total current estimation as below

$$I = M_p I_p + M_n I_n + I_d \tag{4.53}$$

here $I_d = \frac{d(CV)}{dt}$ is the *displacement current*. Therefore, a reduced triggering voltage is obtained as

$$V_{t1} = BV_{AK} \approx BV_{DJ2}\left(1 - \alpha_1 - \alpha_2 - \frac{I_d}{I_A}\right)^{\frac{1}{n}}. \tag{4.54}$$

Depending upon the magnitude of the displacement current, a SCR ESD protection device, also true to other types of ESD protection structures, may behave very differently from the design specifications if the $\frac{dV}{dt}$ effect is not be considered [14, 15]. Second, ESD pulse induced $\frac{dI}{dt}$ effect may

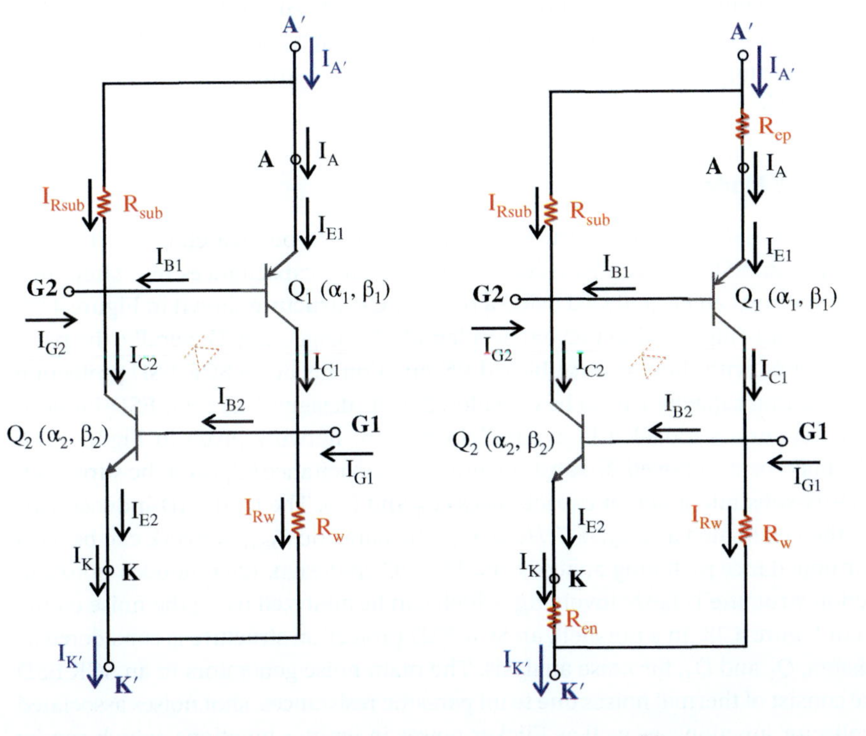

Figure 4.36 Equivalent circuit models for a SCR ESD protection structure include the parasitic resistances: (a) without emitter extension resistances R_{ep} and R_{en}, and (b) with R_{ep} and R_{en}.

set a limit on performance of a SCR ESD protection structure. An actual SCR triggering procedure cannot be a uniform event across the whole SCR junction areas. In reality, aided by the seeding currents, initial SCR turn-on will occur locally in a small area of a SCR junction and then spreads over the whole junction, which is characterized by a spreading velocity. Consequently, the SCR power density will not be the same across a junction during ESD discharges. This results in highly localized overheating in the SCR junctions, i.e., ESD-induced local hot spots, which leads to premature ESD failure, also called early ESD failures. Qualitatively, a relationship of $P \propto \Delta T \propto \frac{dI_A}{dt}$ may exist across the inner SCR junction area, i.e., the actual ESD discharging power dissipation and the ESD heating across the SCR junction area are related to the actual ESD discharge current distribution and its transient behaviors. It is also noteworthy that layout and placement of an SCR ESD protection structure is very critical in practical designs. As discussed previously, current triggering is one method to trigger an SCR device. Therefore, a designer must be cautious about the substrate currents from the surrounding elements and use special techniques, such as double guard-rings, for excellent isolation in order to prevent any accidental SCR triggering in absence of an ESD event, i.e., not to affect normal IC operations.

Obviously, latch-up is a design nightmare for SCR-based ESD protection structures. On the other hand, a snapback-based ESD protection structure is nothing more than a latch-up device. The main difference is that, latch-up in CMOS ICs is unpredictable and uncontrollable, while a snapback-based ESD protection structure is a well-controlled, latch-up device. Therefore, consideration of latch-up in designing SCR-based ESD protection structures has two aspects: First, latch-up should be fully utilized to make an efficient SCR ESD protection device for robust ESD protection. Second, latch-up must be under control in any SCR-based ESD protection structures in order not to interfere with normal IC operations. To achieve this design goal, careful design of the ESD-critical parameters is important for SCR-based ESD protection structures, including β_{NPN}, β_{PNP}, V_{t1}, t_1, R_{ON}, I_h, and V_h.

4.5.3 SCR ESD Parasitic Modeling

While, in general, an SCR ESD protection structure is considered to be area efficient, it is still important to understand its ESD-induced parasitic effects. Parasitic capacitances in a simplified SCR ESD protection structure can be analyzed using a typical SCR structure shown in Figure 4.37, which includes P-well guard-rings used to prevent accidental ESD triggering. Generally, the parasitic capacitances associated with those reverse-biased PN junction inside an SCR ESD protection structure are the dominating capacitances to be considered in IC designs. The total ESD-induced parasitic capacitance network is modeled by a parallel capacitor network given in Figure 4.37, which consists of the main central P-well/N-substrate junction capacitance (C_1) and the surrounding guard-ring P-well/N-substrate junction capacitances (C_2 and C_3). The total ESD-induced parasitic capacitance is then estimated as $C_{ESD} \approx C_1//C_2//C_3$. The parasitic C_{ESD} network can be used in RF IC designs for impedance matching analysis and ESD-IC co-design. ESD-induced noises in an SCR ESD protection structure is fairly involving, which can be analyzed using the noise equivalent circuit shown in Figure 4.38. In a nutshell, an SCR ESD protection structure is considered as a pair of BJT transistors, Q_1 and Q_2, for noise analysis. The main noise generators in an SCR ESD protection structure consist of thermal noises due to all parasitic resistances, shot noises associated with emitter and collector junctions, as well as Flicker noises in emitter junctions, which can be

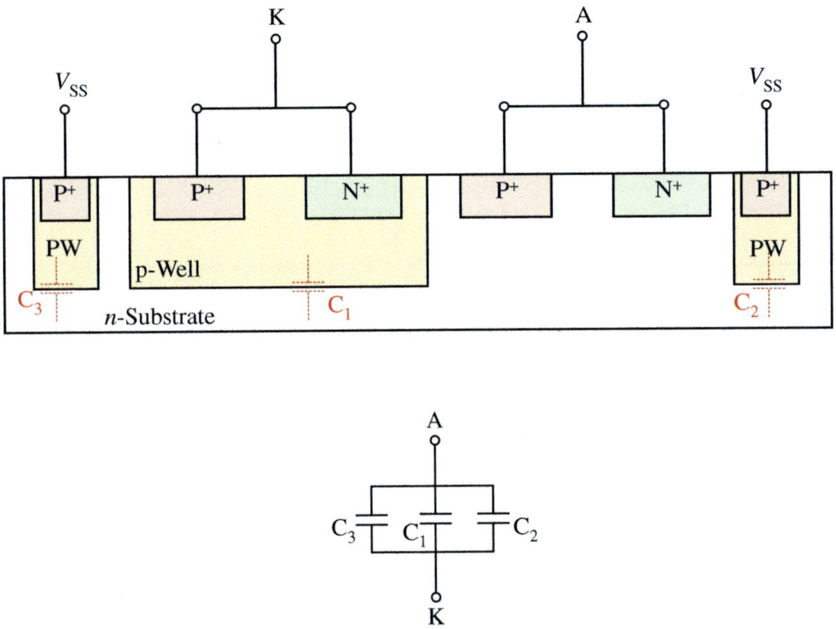

Figure 4.37 A simplified ESD-induced parasitic capacitance model for an SCR ESD protection structure comprises junction capacitances associated with reversely biased junctions.

approximated by the following equations for both Q_1 and Q_2 [10],

$$\overline{v^2_{nR}} = 4kTR\Delta f \tag{4.55}$$

$$\overline{i^2_{nC}} = 2qi_C\Delta f \tag{4.56}$$

and

$$\overline{i^2_{nb}} = 2qi_B\Delta f + K_1\frac{i_B^a}{f}\Delta f \tag{4.57}$$

where R represents all the individual parasitic resistances (i.e., R_W, R_{sub}, r_e, r_b, and r_C for Q_1 and Q_2), i_C and i_B are collector and base currents of Q_1 and Q_2, respectively. In normal IC operations, only ESD-induced leakage currents are considered for noise analysis. Since Q_1 and Q_2 share a collector junction, a strong correlation factor is expected between i_{C1} and i_{C2} in this noise circuit model. In analog and RF IC designs, the equivalent circuit models for ESD-induced parasitic C_{ESD} and noises, as well as extra noise coupling effect due to C_{ESD}, can be included in full-chip circuit analysis by simulation.

4.6 Summary

This chapter discusses the details of classic single-device ESD protection structures including device physics, ESD discharge functions, and ESD-induced parasitic effects. A PN junction diode

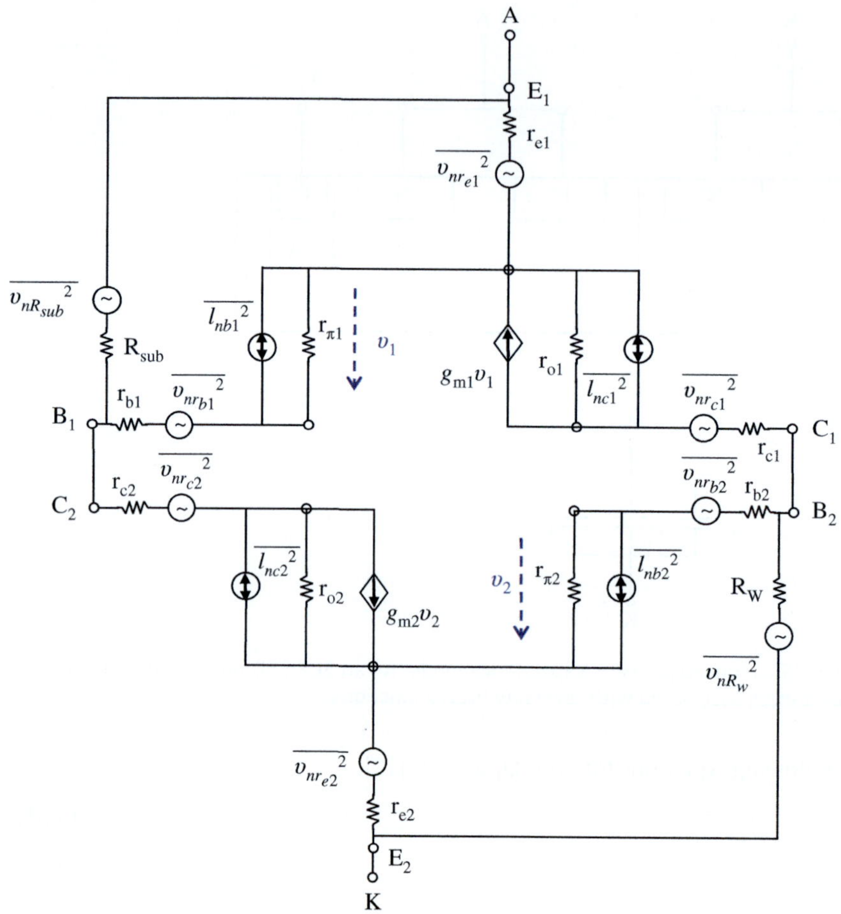

Figure 4.38 A simplified noise equivalent circuit model for an SCR ESD protection structure can be used for analyzing ESD-induced noise effect for ICs.

is the simplest ESD protection structure, which should be used if ever possible. However, generic PN diodes have disadvantages, such as low forward turn-on voltage and high reverse conduction resistance. Zener didoes and diode strings may be used for on-chip ESD protection requiring higher ESD triggering voltages. BJT ESD devices offer high ESD discharge conduction due to amplification effects. One problem with BJT ESD devices is the snapback *I–V* behavior that is not friendly to SPICE circuit simulation. MOSFET ESD protection devices have been widely used for ESD protection in CMOS ICs. Particularly, ggNMOS and ggPMOS devices are used for I/O ESD protection, while gcNMOS is often used as an ESD power clamp. The main disadvantage for MOSFET ESD protection structures is large layout size and low ESD area efficiency. SCR ESD devices are robust ESD protection structures due to the NPN–PNP coupling effect and low-voltage clamping feature due to *I–V* snapback. However, care must be given in designing SCR ESD protection structures to put the latch-up effect under full control. All these classic ESD protection structures reply on PN junction turn-on and conduction. Therefore, a PN junction is the gene of all active ESD protection structures, and BJT operation is the foundational mechanism for most active ESD protection structures, including MOSFET and SCR ESD protection structures.

Since PN junctions are embedded inside various active ESD protection structures, some adverse PN-inherent properties may have serious impacts on ESD protection and core IC performance, such as, junction leakages, junction capacitances, and ESD self-generated noises. To address these problems, advanced ESD protection structures and subcircuits have been developed, which will be discussed in the later chapters.

References

1 Fong, Y. and Hu, C. (1987). The effects of high electric field transients on thin gate oxide MOSFETs. *Proceedings of EOS/ESD Symposium*, pp. 252–257.

2 Wang, A. and Tsay, C. (1999). A low-triggering circuitry for dual-direction *ESD* protection. *Proceedings of IEEE CICC*, pp. 139–142.

3 Wang, A., Tsay, C., Lele, A., and Deane, P. (1998). A study of NMOS behaviors under ESD stress: simulation and characterization. *Microelectron. Reliab.* 38: 1183–1186.

4 Zhan, R., Feng, H., Wu, Q. et al. (2004). ESDInspector: A new layout-level ESD protection circuitry design verification tool using a smart-parametric checking mechanism. *IEEE Trans. Comput. Aided Des. Integr. Circuits Syst.* 23 (10): 1421–1428.

5 Lin, L., Zhang, L., Wang, X. et al. (2011). Novel nanophase-switching ESD protection. *IEEE Electron Device Lett.* 32 (3): 378–380.

6 Ma, R., Chen, Q., Zhang, W. et al. (2016). A dual-polarity graphene NEMS switch ESD protection structure. *IEEE Electron Device Lett.* 37 (5): 674–676.

7 Shockley, T. (1949). The theory of p-n junction in semiconductors and p-n junction transistors. *J. Bell Syst. Tech.* 28: 435; (1950). *Electrons and Holes in Semiconductors*. Princeton, NJ: D. Van Nostrand.

8 Sze, S.M. (1981). *Physics of Semiconductor Devices*, 2e. New York: Wiley.

9 Miller, S.L. (1957). Ionization rates for holes and electrons in silicon. *Phys. Rev.* 105: 1246–1249.

10 Lee, T. (1998). *The Design of CMOS Radio-Frequency Integrated Circuits*. Cambridge University Press.

11 Massobrio, G. and Antognetti, P. (1993). *Semiconductor Device Modelling with SPICE*, 2e. McGraw-Hill.

12 Sze, S.M. and Chang, C.Y. (2000). *ULSI Devices*. New York: Wiley.

13 Estreich, D.B. (1980). The physics and modelling of latch-up and CMOS integrated circuits. PhD dissertation. Stanford University.

14 Wang, A.Z. and Chen, C.H. (2001). On a dual-polarity on-chip electrostatic discharge protection structure. *IEEE Trans. Electron Devices* 48 (5): 978–984.

15 Feng, H.G. (2001). A mixed-mode simulation-design methodology for on-chip ESD protection design. A MS thesis. Illinois Institute of Technology.

5

ESD Protection Circuits

Chapter 4 presents details of single-element electrostatic discharge (ESD) protection devices, including device physics, ESD discharge mechanisms, and ESD-induced parasitic effects. As stated, simplicity is the ultimate beauty. A designer should follow the golden rule of "the simpler, the better," while meeting the design specs. Unfortunately, the real world is never simple. As integrated circuit (IC) technologies continuously advance to 2 nm node, and chip performance and complexity rapidly increase due to the ever increasing demands for system applications, new challenges on ESD protection designs constantly emerge. For example, ultralow-power ICs require low ESD V_{t1}, mixed-signal ICs of multiple power domains prefer localized ESD designs, high-voltage (HV) ICs need scaled V_{t1} and V_h, high-frequency and high-throughput ICs demand for ultralow-parasitic ESD designs, and charged device model (CDM) ESD model requires ultrafast ESD t_1, so on, so forth. Often, these emerging and varying design requirements cannot be met by single-device ESD protection designs. Gradually and as necessary, derivatives and combination of simple ESD devices are needed to achieve complete on-chip ESD protection, leading to ESD protection circuits that, sometimes, can be quite involving. This chapter discusses exemplar ESD protection circuit design techniques.

5.1 I/O ESD Protection

In principle, every pad on an IC chip should be protected against possible ESD failures, including input, output, control, and supply pads. As said, the fundamental principle for on-chip ESD protection is to integrate well-designed ESD switches at pads to swiftly discharge incident ESD transients without overheating and to safely clamp the pad voltage to a sufficiently low level, so as to avoid possible thermal and voltage breakdowns, respectively. On the other hand, there is no one-for-all universal ESD protection solution both for different pads on a chip and for different ICs designed in same or different IC technologies. ESD protection design is truly custom design and IC-specific. For example, from pad perspectives, input signal pads of complementary metal-oxide-semiconductor (CMOS) ICs are very sensitive to gate oxide breakdown, output pads may take advantage of ESD self-protection capability of robust output transistors, and supply bus ESD protection is not concerned about ESD-induced parasitic capacitance. This section presents typical ESD protection subcircuit schemes for different IC pads.

Practical ESD Protection Design, First Edition. Albert Wang.

5.1.1 Two-Stage ESD Protection

Figure 5.1 depicts a classic two-stage primary–secondary ESD protection scheme comprising a primary ESD protection structure (ESD1), a secondary ESD protection unit (ESD2), and an isolation resistor (R) [1–3]. The primary ESD1 structure is typically optimized for high ESD protection level, which however may feature a high ESD V_{t1}, not suitable for low-voltage (LV) ICs. The secondary ESD2 unit serves as a trigger-assisting device that features a lower ESD V_{t1} and fast ESD triggering, which is typically weak in handling large ESD discharge currents. As illustrated in Figure 5.1 for Input-to-GND ESD protection, as a PS mode ESD pulse occurs at the pad, the secondary ESD2 can be turned on immediately and at a low voltage. The large ESD transient current starts to discharge into GND through ESD2, which will produce a large voltage drop across the isolation resistor R. As the transient voltage builds up on R quickly to a given level, the primary ESD1 device will be triggered to discharge the ESD pulse. In general, an ESD1 device is designed for very low ESD discharge resistance R_{ON} and high ESD current-handling capability; hence, the ESD transient will be quickly steered into ESD1 for safe ESD discharge. The isolation R has another role, which is to prevent an ESD pulse from getting into IC core (i.e., stressing the input device) directly, hence avoid possible CMOS gate breakdown. In two-stage ESD protection design, design optimization is critical for both ESD1 and ESD2 devices. The desired specs for ESD2 is low V_{t1} and short t_1, while that for ESD2 include low R_{ON}, low V_h and high I_{t2}. R involves a design trade-off too: large enough for fast voltage build up, but not too large to avoid adverse impact on signal propagation. In practical ESD designs, the primary ESD protection structure can be realized in many ways, e.g., using a thick-gate NMOS, a bipolar junction transistor (BJT), a silicon controlled rectifier (SCR), or diode strings. A ggNMOS has been widely used as the secondary ESD protection unit. The isolation R can be a diffusion resistor or a poly-Si resistor. ESD device layout design plays a key role in ESD design optimization, for example, to round the ESD device corners to avoid ESD current and heat crowding at the corner as discussed in Chapter 2. In general, due to the availability of many robust ESD protection structures in advanced IC technologies, the two-stage ESD protection scheme has lost its popularity some years before. Interestingly, recently, the two-stage ESD protection method is re-gaining attention for CDM ESD protection because it can handle large ESD surges without

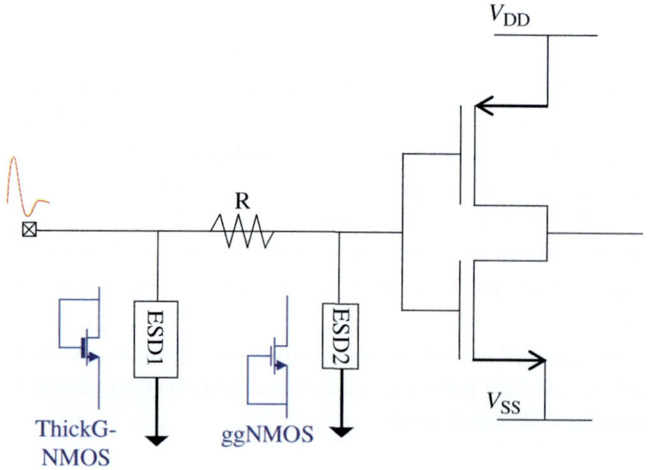

Figure 5.1 A two-stage primary-secondary ESD protection scheme uses a secondary ESD unit (ESD2) for lower V_{t1} and a primary ESD unit (ESD1) for higher I_{t2}.

overheating, while preventing CMOS gate breakdown due to the isolation R (i.e., no direct zapping on the input gate). It is worth to note that one should not expect a single "best" ESD protection solution for all; rather, ESD protection design should always be customized for specific IC needs.

5.1.2 Multiple-Fingers ESD Protection

Layout is always a tricky task in analog IC designs. Typically, multiple-finger layout is used in designing large transistors, such as output buffer metal-oxide-semiconductor field-effect transistors (MOSFETs) in CMOS ICs, which need to handle larger currents. The same layout approach also applies to ESD protection device layout designs because a robust ESD protection structure must handle much larger currents, hence, requires a large size. For example, a ggNMOS ESD protection structure for 8 kV HBM ESD rating may easily require a large device size with a total conduction width (W) of 400 μm, due to its relatively low ESD current-handling capability of ~20 V/μm-W. Apparently, it is not wise to make a very long MOSFET finger of $W = 400$ μm. The required current conduction uniformity cannot be ensured across the long finger of 400 μm because the ESD discharge current is very high, which can readily cause local current crowding, leading to ESD heat crowding, i.e., hot spots, and resulting in low ESD thermal failure threshold. The unavoidable device defects across a long device finger due to either fabrication or operational issues will make the current crowding effect even worse [4]. Therefore, multiple-finger layout is a common practice in practical ESD protection device designs [5, 6]. Typically, a large ESD protection structure contains an array of short fingers with individual finger length around 10–100 μm. The sum of all finger lengths is the total ESD conduction channel width of a large ESD protection structure. Figure 5.2 illustrates a simplified multiple-finger ggNMOS ESD protection device layout, which is equivalent to a parallel ESD device network with each finger being an individual ESD device. There are of course many different multiple-finger layout design variations in practical ESD protection designs, totally up to the designers. However, using multiple-finger layout design approach cannot guarantee that a design of multiple-finger ESD protection structure will be scalable between ESD protection level and ESD device size (i.e., finger numbers). In fact, without careful design

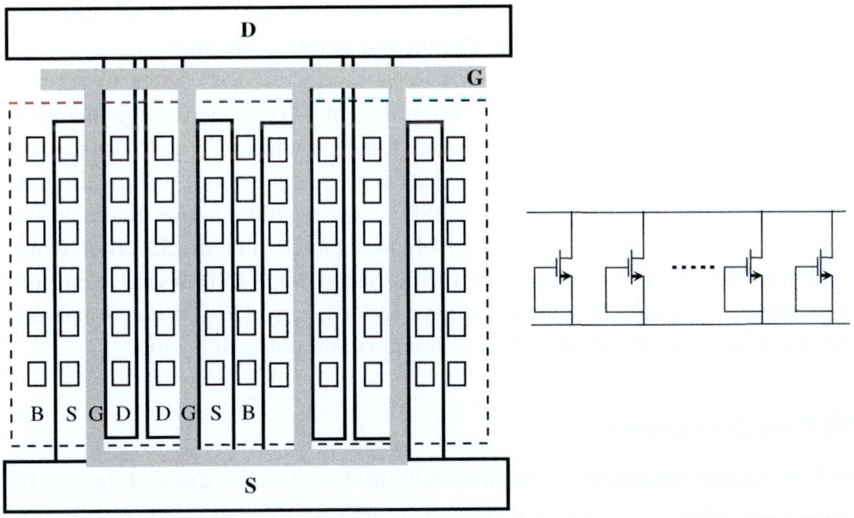

Figure 5.2 A multiple-finger ggNMOS ESD protection structure is equivalent to a parallel ESD protection device network.

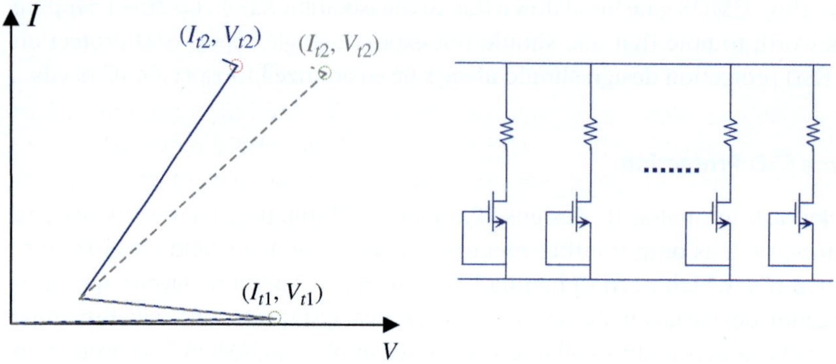

Figure 5.3 ESD discharge I–V characteristics are critical to multiple-finger ESD layout designs: $V_{t1} > V_{t2}$ will break ESD protection scalability. $V_{t1} < V_{t2}$ is preferred to fully utilize all ESD device fingers. Adding a ballasting resistor improves ESD triggering uniformity.

consideration, multiple-finger layout design may not be beneficial at all. Practically, it is quite common that increase in the number of layout fingers may not enhance ESD protection capacity much. One can imagine that, under ESD stressing, one finger may always be turned on first to conduct the ESD transient, which may possibly fail before any more device fingers could be turned on to share the large ESD pulse load. The ESD discharge I–V curves in Figure 5.3 explain the potential problem. If an ESD device follows the ESD discharge I–V curve featuring $V_{t1} > V_{t2}$, then, under ESD stressing, one ESD device finger may be triggered first and starts to discharge the ESD transient, resulting in voltage snapback and then build-up following the I–V curve. In order to trigger another ESD device finger, the transient voltage at anode must reincrease back to V_{t1}. Unfortunately, since $V_{t1} > V_{t2}$, ESD failure may occur to the first ESD device finger before any other ESD device finger(s) may be turned on. Therefore, a multiple-finger ESD protection structure of $V_{t1} > V_{t2}$ will never work as expected, i.e., no scalability between the ESD protection capacity (i.e., I_{t2}) and the number of ESD device fingers (i.e., device size). Instead, the unwanted ESD design overhead, including ESD-induced parasitic effects and ESD device area consumption, will monotonously increase with the number of ESD device fingers. This is particularly true to advanced silicided IC technologies [7]. A straightforward solution to this multiple-finger ESD layout design problem is to redesign an ESD protection structure to make $V_{t1} < V_{t2}$, which means, even though one ESD device finger will be triggered first to discharge the ESD transient, but as the transient ESD anode voltage increases and before any potential ESD failure to the first ESD device finger, it will reach to V_{t1} again and subsequently triggers another ESD device finger(s). This ESD triggering sequence will ensure that all ESD device fingers can be turned on to discharge the ESD surges together, hence, realizing the desired ESD protection device design scalability. Alternatively, as depicted in Figure 5.3, a ballasting resistor can be inserted into each finger branch in series with the ggNMOS ESD device to enforce uniform ESD triggering across all parallel ESD device fingers. Indeed, a good ESD protection structure requires a careful consideration of both electrical parameters and physical layout.

5.1.3 MOSFET ESD Protection Circuits

Though ggNMOS ESD protection structure has been widely used, one major disadvantage is its relatively high ESD triggering voltage V_{t1}, making it often unsuitable for LV CMOS ICs. Also, as discussed previously, if V_{t1} is too high, such that $V_{t1} > V_{t2}$ holds, it will make multiple-finger ggNMOS ESD protection structure practically useless. Therefore, various techniques have been developed to reduce V_{t1} of MOSFET ESD protection structures. *Gate-coupled* MOSFET (gcMOS) was proposed to effectively reduce the V_{t1}. Figure 5.4 depicts the concept of gcNMOS ESD protection subcircuit. In principle, an RC-coupling mechanism is introduced to transiently raise up the gate voltage of an NMOS ESD device under ESD stressing, which leads to a reduction in its V_{t1}. In a PS mode I/O-to-GND ESD protection scheme, an incident positive ESD pulse occurring at the I/O pad will be swiftly coupled to the gate of the NMOS device via the capacitor, quickly and briefly increases V_G of NMOS to reduce the ESD V_{t1} of the gcNMOS ESD subcircuit. Consequently, the desired $V_{t1} < V_{t2}$ design goal will be reached for the ESD protection circuit, good for scalable multiple-finger MOSFET ESD protection structures or LV CMOS ICs. Other than a reduced ESD V_{t1} made possible by the RC coupling effect, ESD discharge mechanism of a gcNMOS ESD protection subcircuit is the same as that for a ggNMOS ESD protection structure. As long as a gcNMOS structure is turned on, the rest of the ESD protection story will be the same as that held for a ggNMOS ESD protection structure, which is that the parasitic lateral NPN transistor inside the NMOS channel will be forced into active amplification mode to efficiently discharge the incoming ESD pulses without overheating. To bias the embedded parasitic NPN BJT into its forward amplification mode, its drain-body junction is reversely biased to breakdown and the holes generated will flows into GND through the Body terminal through the parasitic well-resistor R_w, which quickly builds up a forward voltage across the body-source junction (i.e., NPN V_{BE}), and quickly forward turns on the NPN emitter, forcing the NPN into forward conduction mode to discharge the ESD pulse. This ESD operation mechanism suggests that reduction in ESD V_{t1} can be realized by accelerating the NPN turn-on process. Obviously, increasing the substrate current (I_{sub}) flowing through the parasitic lateral R_W can build up V_{BE} faster to turn on the parasitic NPN. From the classic MOSFET device physics, the I_{sub} can be increased by properly setting the NMOS V_G in various ways as depicted in Figure 5.5. It was found that the DB avalanche breakdown induced I_{sub} is closely related to the NMOS gate bias V_G as shown in Figure 5.6, where the I_{sub} peaks at a certain V_G [8, 9]. For short-channel MOSFETs, the unavoidable hot carrier effect may contribute to ESD triggering. The hot holes will contribute to the I_{sub} and the observed $I_{sub} \sim V_G$ relationship associated with hot carriers also shows a peak I_{sub} at a given V_G, e.g., at $V_G = 0.4–0.5\ V_{DS}$, as shown in Figure 5.6 [10–15]. In addition, the hole current can flow directly from drain to source through the short channel in MOSFET, which also contribute to forward-biasing the BS junction. Collectively, properly increasing the NMOS V_G may help to reduce the ESD V_{t1} as wished, hence came the gcNMOS ESD protection subcircuit. It is interesting to note

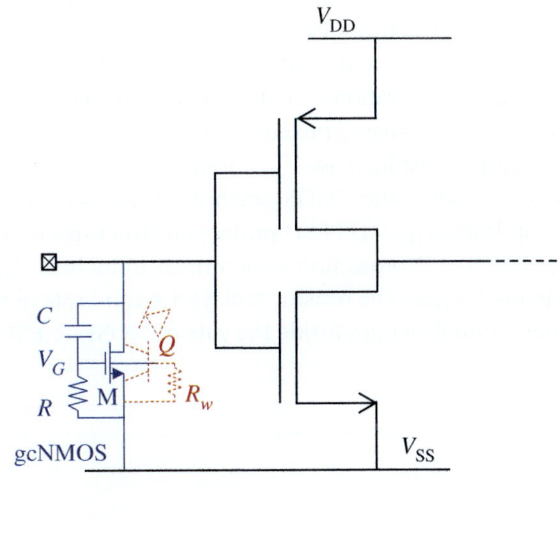

Figure 5.4 A conceptual schematic illustration for gcNMOS ESD protection structure for I/O-to-GND ESD protection in PS stressing mode.

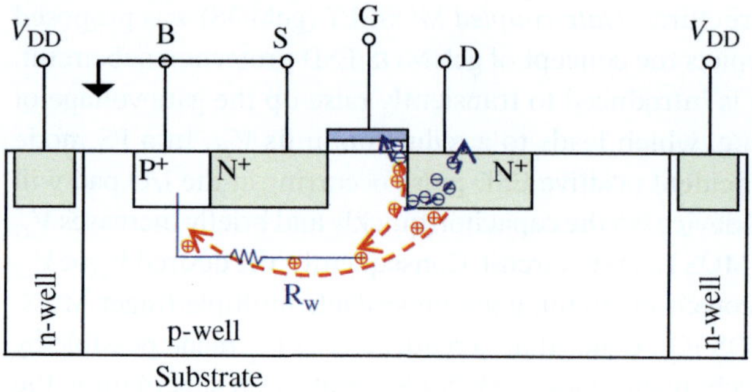

Figure 5.5 A cross-section view of gcNMOS ESD protection structure shows various carrier contributions to the substrate current, which assists in triggering the parasitic NPN for ESD discharge.

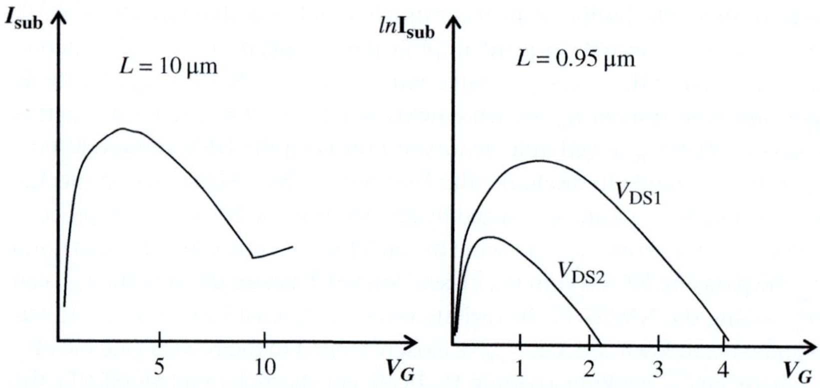

Figure 5.6 Typical bell-shaped $I_{sub} \sim V_G$ characteristics of MOSFETs reported for long-channel and short-channel devices suggest that increasing V_G properly may help to reduce ESD V_{t1}.

that, in normal CMOS IC operations, the substrate current I_{sub} is unwanted because it may cause the latch-up problem. However, the same I_{sub} can be wisely used to reduce the ESD V_{t1} in gcMOS ESD protection structures. Indeed, a good IC circuit designer must have a good understanding of semiconductor device physics, in addition to being able to run SPICE circuit simulation only. Of course, SPICE circuit simulation can help to determine the RC timing in designing gcNMOS ESD protection structures. However, the timing consideration itself by SPCIE simulation is not enough in selecting the R and C values in gcNMOS ESD protection designs because it cannot reveal its ESD discharging operations in details, which must be thoroughly studied by technology computer-aided design (TCAD)-based mixed-mode ESD simulation design, to be discussed in Chapter 8. Obviously, one cannot just increase V_G, aiming to boost the I_{sub}, hence, to reduce the ESD V_{t1} because a high V_G may itself cause CMOS gate breakdown. As always, careful design trade-off must be considered in optimizing gcMOS ESD protection structures by simulation. One important factor in designing gcNMOS ESD protection structures is to include the gate overlap capacitance in ESD protection circuit designs. The resistor R plays a second role in the circuit too, i.e., serves to quickly discharge any residual charges inside the gate of MOSFET ESD devices after the ESD events are over.

5.1.4 BJT ESD Protection Circuits

As discussed previously, almost all existing active ESD protection structures are in-Si PN-junction-based that rely on PN turn-on and conduction to discharge the ESD transients. Further, BJT transistors are the basic building blocks for most ESD protection structures of active amplification, including MOSFET and SCR type ESD protection structures. In BJT-based ESD protection structures, a BJT is certainly the gene. Similarly, MOSFET-based and SCR-based ESD protection structures utilize parasitic lateral or vertical BJT cells to achieve the highly desired active amplification to enhance ESD discharge operations. There are essential countless different ways to use BJT devices to build complex ESD protection structures and subcircuits. This section discusses a few examples of using BJT cells to design ESD protection structures.

Recall the device physics depicted in Figure 4.16 for the basic BJT ESD protection device, the forward conduction for a BJT ESD device starts from forward turn-on of the emitter junction. This is typically realized with the following routines: An incident ESD pulse will reverse-bias the collector junction, causing its avalanche breakdown. The hole current produced will flow into GND. An external trigger-assisting resistor, R, in the conduction path will build up the voltage across the emitter. When the base voltage V_B reaches $V_{BE} > V_{ON}$ for the BE junction, the emitter will be forward ON, which subsequently forces the BJT into forward conduction and amplification mode to discharge the incident ESD surges. The combination of collector junction breakdown and voltage build-up of R makes accurate control of the BJT ESD triggering difficult in practical ESD designs. A conceptual method to enhance the BJT ESD triggering process is depicted in Figure 5.7 for I/O-to-GND PS ESD stressing mode. In concept, a dedicated trigger-assisting current source (I_1) is included in the BJT ESD subnet in a series with the R. The I_1 current source is synchronized with any ESD events. As an ESD pulse comes to the pad, it will swiftly turn on the I_1 source, which provides a stable current flowing through the R to quickly turn on the BJT emitter, forcing the BJT ESD device into forward ESD discharge operation with full amplification. The main benefit of this I_1-assisted BJT ESD protection structure is that one can accurately control the ESD triggering, i.e., V_{t1}, and accelerate the ESD triggering process too, instead of passively relying on PN junction breakdown. In actual ESD designs, there are many ways to realize the I_1-source, of which, the easiest method is to use a Zener diode to serve as the I_1-source. Of course, the Zener diode may be

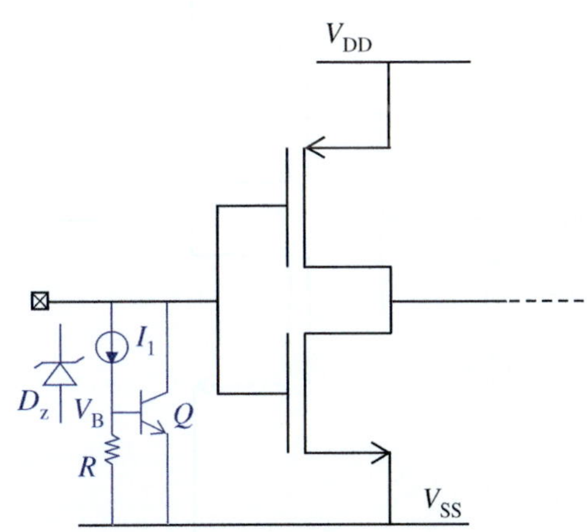

Figure 5.7 A conceptual BJT-based ESD protection sub-circuit for Input-to-GND ESD protection uses a trigger-assisting mechanism to control ESD triggering. The ideal I_1 current source can be realized by a specially designed Zener diode.

optimized and properly selected according to the ESD V_{t1} requirement, i.e., $V_{t1} \approx V_{Dz}$. As such, as soon as an incident ESD pulse turns on the Zener diode, a stable and sufficient I_1 current will flow through the R and trigger the BJT transistor for ESD discharge. The ESD V_{t1} can be accurately controlled by design, which is critical to meeting the ESD Design Window requirements for full-chip ESD protection. In advanced IC technologies, such as BiCMOS processes, there are many PN junctions that can be selected as the Zener diode for ESD triggering. The Zener diode means to assist the ESD triggering not to conduct the large ESD transient current, hence, is a small diode. The large ESD transient current will be discharged through the ESD-optimized BJT device. Alternatively, a gcNMOS can be used to serve as the I_1-source to assist the ESD triggering, as illustrated in Figure 5.8a. Basically, the gcNMOS unit replaces the Zener diode in Figure 5.7. An incident ESD pulse will turn on the gcNMOS ESD cell first, which generate a transient current flowing through

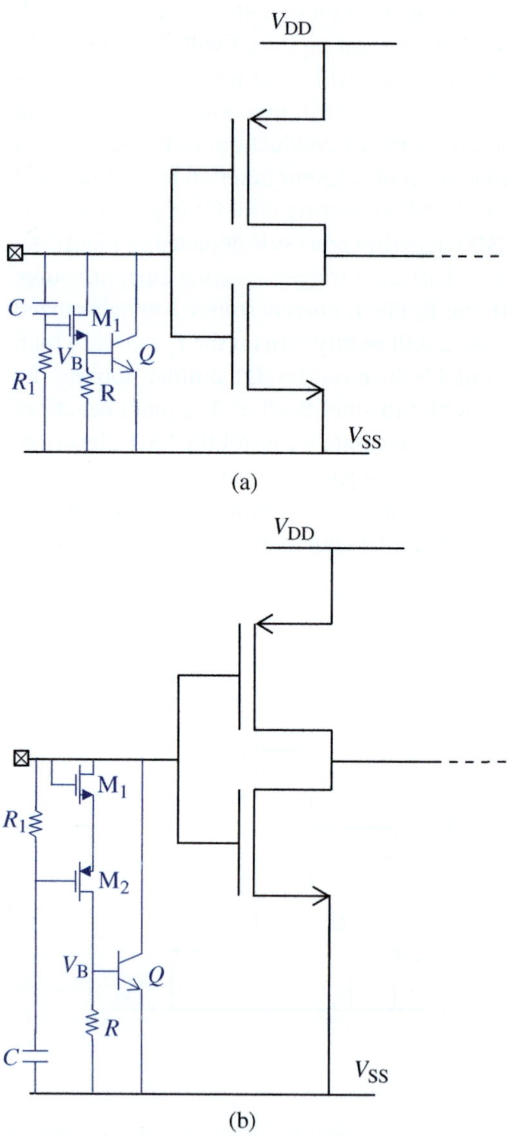

Figure 5.8 Exemplar modified BJT-based ESD protection circuits feature varying ESD trigger-assisting mechanisms: (a) gcNMOS trigger-assisting concept, and (b) a combined RC-NMOS trigger-assisting schematic. Source: Tandan [16].

the R to trigger the BJT ESD device. A gcNMOS unit is selected mainly because it has relatively low ESD V_1 and can be well-controlled in designs. Compared to using a Zener diode for ESD trigger assistance, more design consideration for the RC branch is needed for the gcNMOS unit. The gate overlap capacitance, C_{gd}, of M_1 can be used for RC coupling in the gcNMOS. To address the concern that excessive C_{gd} may cause accidental BJT turn-on in absence of ESD events, i.e., during normal IC operations, a current breaker can be inserted into the R-path, i.e., the PMOS, M_2, connected between M_1 and R, as shown in Figure 5.8b [16]. Under ESD stressing, the gate of M_2 is pulled down and turned on immediately to trigger the ESD protection circuit. During normal IC operations, M_2 stays OFF, and there is no current flowing through the R, even under very noisy conditions. Obviously, careful circuit simulation is required in design for accurate timing control. In addition, the size of the C, ~10 pF as reported, is too large for many advanced ICs, particularly not suitable to protect I/O signal pads. Figure 5.9 shows another BJT-based ESD protection circuit example, which seems to be a real ESD "circuit" due to its schematic complexity [17]. Essentially, the BJT (Q) is still the main ESD discharge device and the diode string (D_1–D_n) serves as the I_1 current source to assist ESD triggering. This BJT ESD protection circuit was designed for typical CMOS technologies that often do not offer suitable Zener diodes; hence, a forward diode-string can be used to properly set the ESD V_1 for ICs. The required ESD V_1 is determined by the combination of the diode string and the PMOS M_1. Under PS ESD stressing, the diode string and the M_1 will be turned on first, and the current will flow through the R; hence, triggering the main ESD structure, BJT Q. PMOS M_2 ensures full turn on of M_1 under PS ESD stressing: M_2 is turned on along with M_1 during the ESD event and its current (I_{D2}) flows through R_1 to turn on NMOS M_3. Further, since PMOS M_4 is always on, turn-on of M_3 will pull V_{G1} to GND and keep M_1 in ON state. Well, this seems to quite a brain-burning game in designing an ESD protection "circuit." The question to

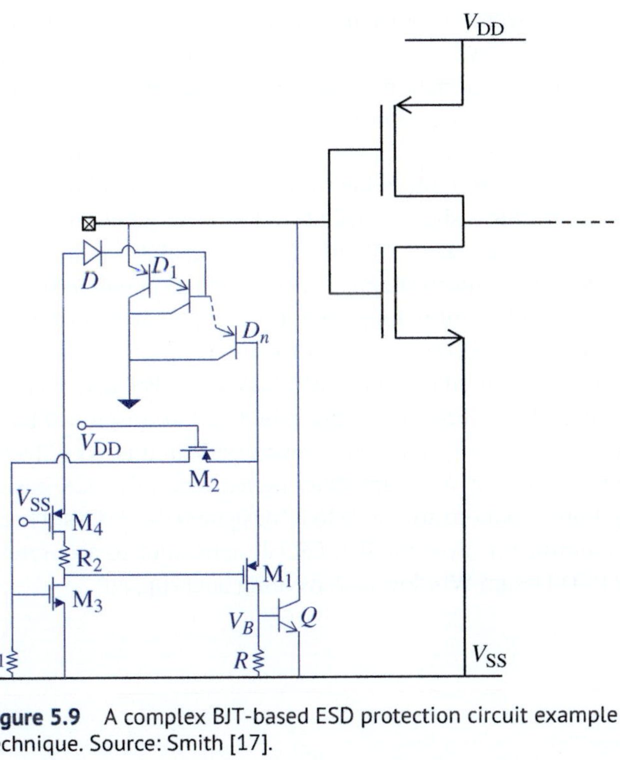

Figure 5.9 A complex BJT-based ESD protection circuit example utilizes a diode-string trigger-assisting technique. Source: Smith [17].

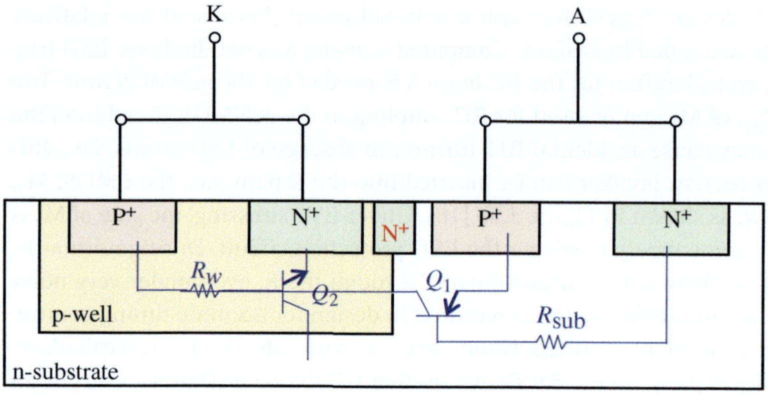

Figure 5.10 A cross-section view for an MVSCR ESD protection structure utilizes an added N+ diffusion plug at the p-well/n-substrate boundary to reduce the SCR ESD triggering voltage V_{t1}.

be asked by an IC designer is WHY making it so complicated?! There are a lot of papers reporting various fancy and complex ESD protection circuit designs. However, a cool-minded IC designer should always follow the golden rule of "the simpler, the better" in designing IC products.

5.1.5 SCR ESD Protection Circuits

SCR has been widely considered to be ESD-robust because of its deep snapback I–V characteristic. One major disadvantage for SCR ESD protection structure is the high ESD triggering voltage V_{t1}, making SCR ESD not suitable for LV CMOS ICs. Various modifications have been proposed to address this SCR ESD design problem. One easy technique to make relatively low-V_{t1} SCR ESD protection structure is depicted in Figure 5.10. From device physics, an SCR ESD protection structure is triggered by avalanche breakdown of the collector junction of the parasitic lateral PNP as shown in Figure 5.10, which is the p-well/n-substrate junction. Unfortunately, both p-well and n-substrate doping concentrations are very low, resulting in very high breakdown voltage, therefore, high ESD V_{t1} for SCR ESD structure. Understand this basic principle, a high-doing N^+ diffusion plug can be added at the p-well/n-substrate boundary, forming a single-sided abrupt N^+/PW junction, which has a lower breakdown voltage under reverse ESD stressing. After the N^+/PW breakdown, the same mechanism applies to force the PNP–NPN pair into deep snapback I–V conduction region to efficiently discharge the ESD pulses. This is a rather beautiful design that uses a small, but clever trick to solve a major design problem. This modified SCR ESD protection structure is commonly referred to as a middle-voltage SCR ESD structure (a.k.a., middle-voltage SCR [MVSCR]) because there is another simple method that can achieve even lower ESD triggering, which will be discussed later along with some other interesting SCR-based ESD protection structures and subcircuits. As IC technologies continue advancing, and chip performance and complexity increase rapidly, SCR-based ESD protection structures are becoming more and more attractive to IC designers these days. There are of course many design challenges remaining unsolved for SCR ESD designs, such as accurately controlling ESD V_{t1} and V_h to meet the ESD Design Window and avoiding latch-up, etc.

5.2 ESD Self-Protection

As discussed previously, ESD protection follows the same general principle for all IC pads, i.e., to discharge the large ESD transient current without overheating and to clamp the pad voltage to a low level to avoid voltage breakdown. In general, the ESD protection schemes for input pads can be used for output pads too. However, as stated before, good ESD protection must be IC-specific, or tailored for each pad per its functional needs. For output pads, ESD protection has its uniqueness, such as, breakdown features and self-protection, which will be discussed in details in this section.

5.2.1 Output ESD Protection

Let us discuss both general and unique attributes for output ESD protection, which are associated with the functionalities and topologies at both device and circuit levels. First, in CMOS ICs, output pads are typically connected to source and drain diffusion regions of the output buffer transistors, different from input pads that are most directly connected to the CMOS gate. Since the vulnerability level of the diffusion regions are different from that for gate oxide at input pads in terms of voltage breakdown; hence, all input ESD protection methods cannot be universally applied to output pads. Second, ESD discharge paths at input pads are typically isolated from the core ICs by the insulating gates. However, since ESD protection structures are connected to sources and drains of output buffer transistors, the ESD structures are often closely tied to the core circuits at output pads, which means many parasitic ESD discharging paths existing at an output pad. This certainly leads to ESD design complexity at chip level, which requires thorough design considerations for output ESD protection at chip level. A few output ESD protection schemes are discussed below. Figure 5.11

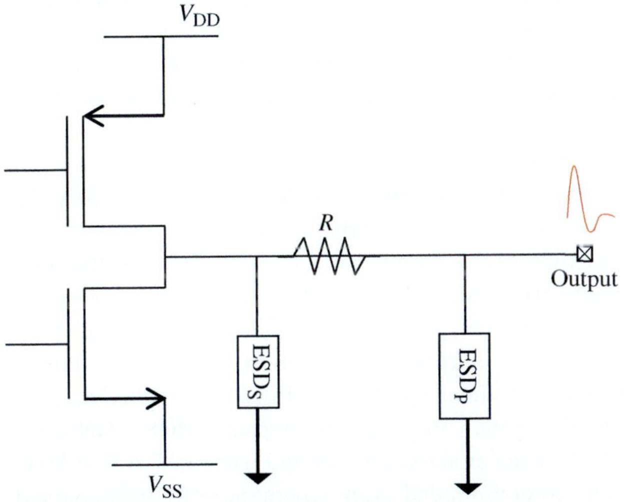

Figure 5.11 A primary-secondary output ESD protection scheme, where the pull-down NMOS may serve as the ESD_S and an isolation resistor, *R*, is optional.

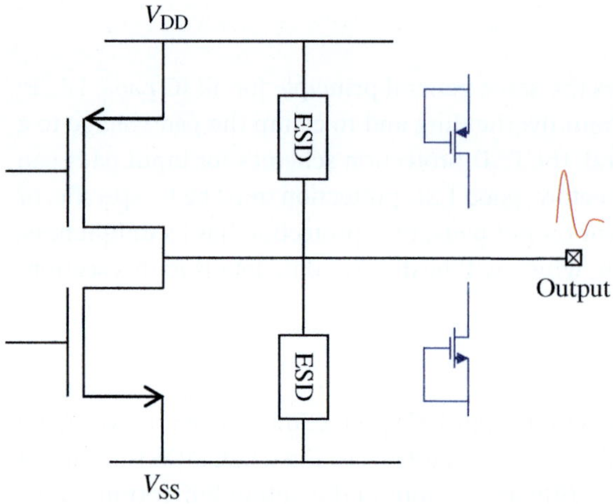

Figure 5.12 A complete output ESD protection scheme uses various pull-up and pull-down ESD protection devices.

shows that the classic two-stage, primary-second ESD protection scheme can be applied to output pads too. Since ESD structures at an output pad is directly tied to IC core, the triggering role of the secondary ESD unit, ESD_S, may be served by the output buffer devices. In addition, the isolation resistor is not preferred at output because it may have a serious effect on the output signals. Figure 5.12 illustrates an exemplar output full ESD protection scheme using a pair of complementary ggNMOS and ggPMOS ESD protection structures to discharge incident ESD pulses in PS, NS, PD, and ND modes. Since an external ESD device and the output buffer transistor are in parallel, one must carefully design the ESD-critical parameters of the ESD device in consideration of the possible parasitic ESD discharging device inside the output buffer transistors in order to ensure that the external ESD protection device, not the internal buffer parasitic device, will be triggered first by an incident ESD pulse. Keep in mind that a buffer parasitic device is not designed to handle large ESD surges, hence, may likely cause early ESD failure if being turned on before the external ESD protection device. This clearly states that ESD protection design is a chip-level design task, not about single ESD device design only, and good on-chip ESD protection design is a quantitative design task in order to accurately design the ESD-critical parameters. For example, how to ensure V_{t1} of the external ESD device will be lower than V_{t1} of the buffer parasitic device? How to reduce the ESD discharging resistance R_{ON} of the external ESD protection structure in comparison with R_{ON} of the internal buffer parasitic device in case that is turned on by an ESD pulse first or simultaneously. A lower R_{ON} will ensure that the large ESD current will mainly flow into the ESD-optimized external ESD protection structure even if the buffer parasitic device may be triggered first. Obviously, diode and SCR ESD devices can also be used for output ESD protection. It is believed that the pull-down MOSFET is usually more ESD-vulnerable, hence, if the pull-down NMOS survives ESD stressing, the whole output buffer would be ESD-safe. Uniquely, a thick-gate NMOS (TGNMOS) can be used for output ESD protection in PD stressing mode as depicted in Figure 5.13. A TGNMOS ESD protection structure can be connected in either grounded-drain schematic or RC well-coupled TGNMOS schematic to reduce the ESD V_{t1} [18]. Similar to a gcNMOS, the ESD transient is coupled into the p-well of the TGNMOS via the RC branch that provides extra body current to turn on the lateral NPN within the TGNMOS. Figure 5.14 shows an exemplar output ESD protection scheme using BJT ESD structure for Output-to-GND PS ESD stressing. Since the NPN ESD device (Q_2) is in parallel with the parasitic NPN device (Q_1) inside the internal NMOS buffer, specially care must be given in ESD design to ensure the external NPN ESD Q_2 have a lower ESD V_{t1} than that of the internal buffer parasitic NPN Q_1, for example, tuning the channel length of the buffer M_1 [19].

Figure 5.13 A TGNMOS ESD protection scheme for output ESD protection using different connections.

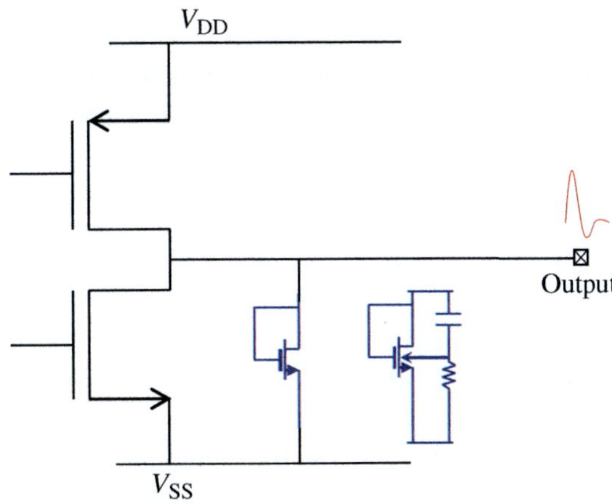

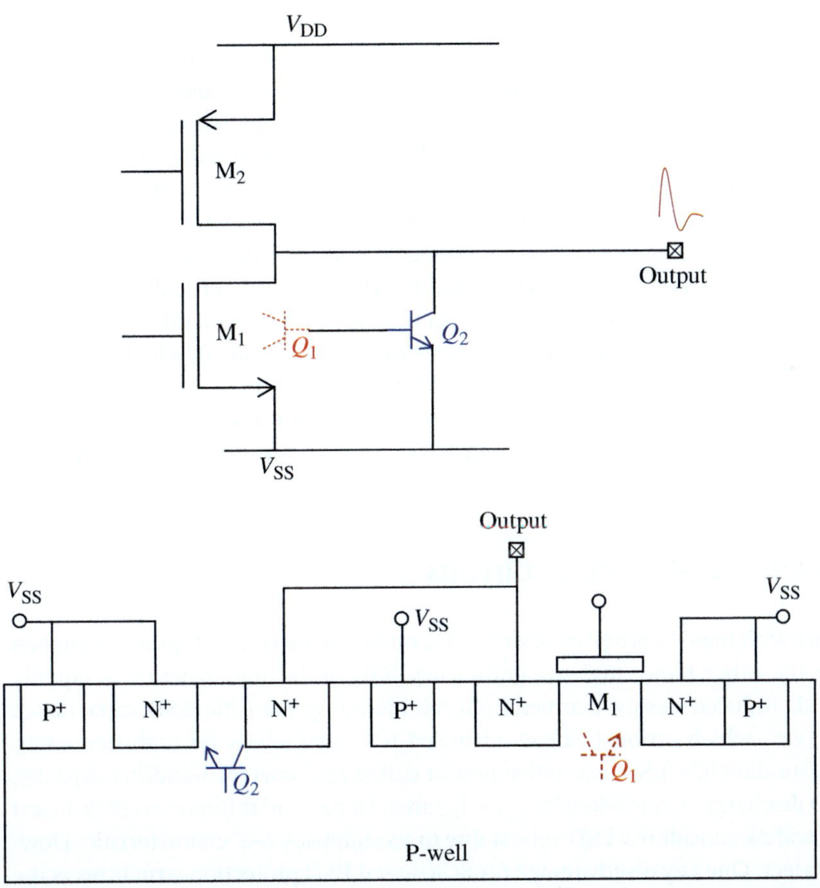

Figure 5.14 A BJT ESD protection scheme for output pad requires careful design of the ESD triggering voltage (V_{t1}) of the external ESD protection structure (Q_2) versus the internal buffer parasitic NPN device (Q_1).

5.2.2 ESD Self-Protection

ESD self-protection is an unique feature for output ESD protection. In principle, an ESD event is a fast transient surge carrying high EM energy, which may cause ESD thermal failures in ICs. Therefore, a good ESD protection structure must be able to handle large ESD surges without overheating. Generally, a more robust ESD protection structure means a larger device size of the same device type. To this end, any larger device inside an IC core can survive relatively stronger ESD stressing. This is particularly true to the output buffer transistors, which typically must be able to process rather strong "useful" signals, which translates into more currents. Accordingly, a large output buffer transistor can provide ESD self-protection. If ESD protection target is moderate, a large output buffer transistor may provide sufficient ESD protection itself. Obviously, careful design trade-off must be considered for ESD self-protection because an output buffer transistor is designed to properly process output signals, though strong, but not optimized for ESD protection that typically involves much higher ESD transient currents. For ESD self-protection, some layout design details have to be considered. For example, a large output buffer transistor is usually large in size and utilizes multiple-finger layout design for uniform conduction. While the conduction uniformity can be ensured for a large output buffer transistor under normal IC operations, localized overheating, i.e., hot spots, is expected under ESD stressing because an ESD pulse involves substantially more transient energy. Similar to the discussion on conduction uniformity across a multiple-finger ESD protection structure, special design attention is needed in dealing with ESD self-protection. For example, the buffer transistor should meet the $V_{t1} < V_{t2}$ criterion in order to trigger all ESD device fingers simultaneously. However, this may not be possible because an output buffer transistor has its own specs to process "useful" signals. Alternatively, a ballasting resistor made in a well extension diffusion can be added to each ESD device finger to enforce uniform ESD triggering; hence, ESD discharge uniformity, as depicted in Figure 5.15, but again may alter the output circuit specs [20]. An n-well diffusion resistor is preferred due to smaller size and good vertical heat dissipation capability. Unfortunately, for advanced silicided IC technologies, a silicided diffusion resistor becomes too low as a ballasting resistor. Polysilicon resistors can be used instead, which, however, has poor thermal conductivity that often results in early ESD thermal failure in the poly-Si resistor itself. Often, dedicated ESD protection structure is still needed for output ESD protection and the buffer ESD self-protection property can be included in ESD protection design to reduce the ESD device size and ESD design overhead effects.

5.3 Low-Triggering ESD Protection Circuits

ESD protection structure with low-V_{t1} is highly desirable for many advanced ICs, typically designed for LV operations. On the other hand, high-area-efficiency ESD protection structure is equally desired for reduced ESD-induced design overhead effects, including parasitic ESD capacitance (C_{ESD}) and layout size, etc., which seriously affect advanced ICs, particularly for high-frequency, broadband, and high-data-date ICs. ESD area efficiency, or called ESD current-handling capacity, represented by the ESD discharge current density, $J_{t2} = I_{t2}$/area. Of particular interest is SCR-based ESD protection that is widely considered ESD-robust due to its snapback I–C characteristic. However, nothing can be perfect. One key disadvantage for SCR-based ESD protection structures is the high ESD triggering voltage V_{t1}. Major design efforts have been devoted to reduce V_{t1} of SCR ESD protection structures. This section discusses a few examples.

MOSFET can be used to assist SCR ESD triggering since MOSFET has lower ESD V_{t1}. Figure 5.16 depicts a simple ggNMOS-triggered SCR ESD protection structure where a small ggNMOS device

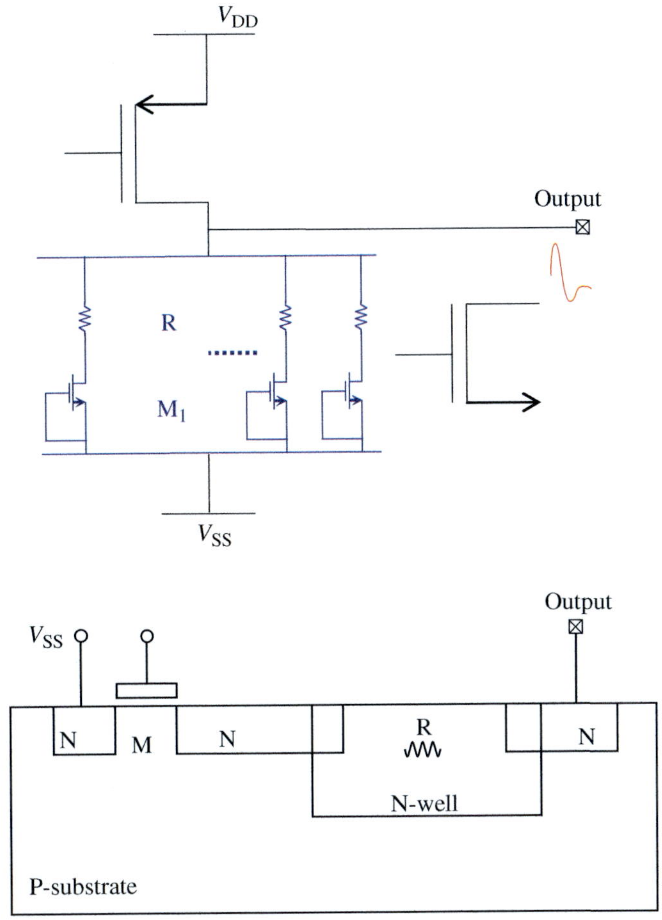

Figure 5.15 ESD self-protection property of output buffer transistors may be included in output ESD protection designs. Ballasting resistor can be added to each finger of the buffer NMOSFET in a multiple-finger layout design, but may affect output circuit specs.

is embedded inside the SCR ESD structure in its cathode region. Under ESD stressing, an incident ESD pulse will bias the ggNMOS device, cause drain-pwell junction breakdown and then turn on the ggNMOS device. The initial ESD discharge current through the ggNMOS device contributes to the substrate current that will turn on the parasitic vertical PNP Q_2, which will further turn on the parasitic lateral NPN Q_1. The SCR ESD protection structure is therefore fully turned on to discharge the ESD pulse with active amplification. The major advantage is the reduced SCR ESD V_{t1} initiated by the lower ggNMOS triggering, instead of the much high reverse breakdown voltage of the pwell/n-substrate junction [21]. Compared to the N^+-triggered MVSCR ESD protection structure depicted in Figure 5.10, this ggNMOS triggered SCR ESD protection structure features even lower ESD V_{t1}, hence, often called an low-voltage SCR (LVSCR) ESD protection structure. It is possible that this ggNMOS-triggered SCR ESD protection structure can be controlled for ESD triggering by varying the short channel length of the MOSEFT. On the other hand, ESD failure may occur to the thin gate of the MOSFET if the SCR ESD structure would not be turned on quickly enough. From ESD V_{t1} reduction view point, a gate-coupled gcMOS ESD device can be integrated into an SCR ESD protection structure to further reduce its ESD triggering voltage. Figure 5.17 illustrates such a

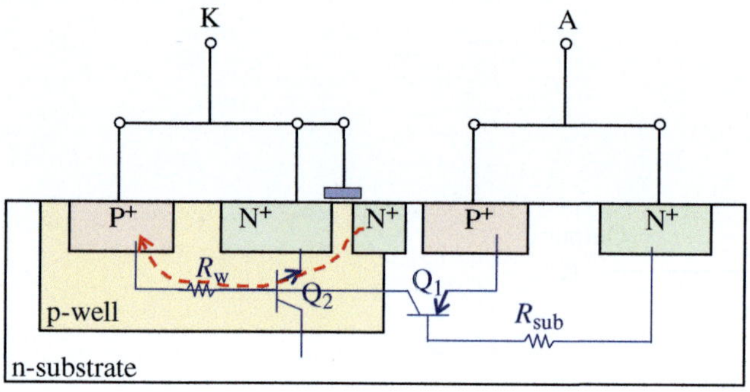

(b)

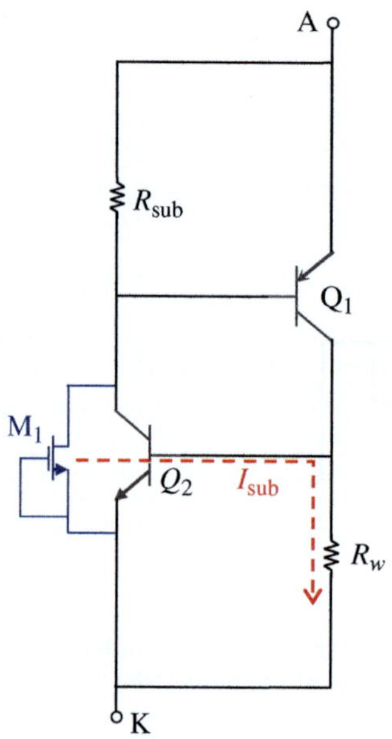

(b)

Figure 5.16 A ggNMOS-triggered SCR (LVSCR) ESD protection structure reduces the triggering voltage V_{t1}: (a) a cross-section, and (b) an equivalent circuit.

gcMOS-trigger SCR ESD protection subcircuit in its complementary mode using a pair of gcNMOS and gcPMOS units for I/O-to-GND and I/O-to-V_{DD} ESD protection on a chip [22]. For I/O-to-GND PS mode ESD protection, the incoming ESD pulse will cause the C_n-R_n-M_1 gcNMOS ESD subnet to turn on first, which then supplies the current needed to trigger the core SCR ESD structure. For I/O-to-V_{DD} ESD zapping, the C_p-R_p-M_2 gcPMOS ESD subnet will be turned on first to trigger the core SCR ESD structure. Obviously, the gcNMOS and gcPMOS are the trigger-assisting units only, while the core SCR device is the main ESD discharge structure. In quantitative designs, the C-R-FET subnets can be simulated by SPICE for timing analysis. However, TCAD simulation is required to simulate transient ESD discharging I–V characteristics including the SCR core device to optimize the ESD discharge performance.

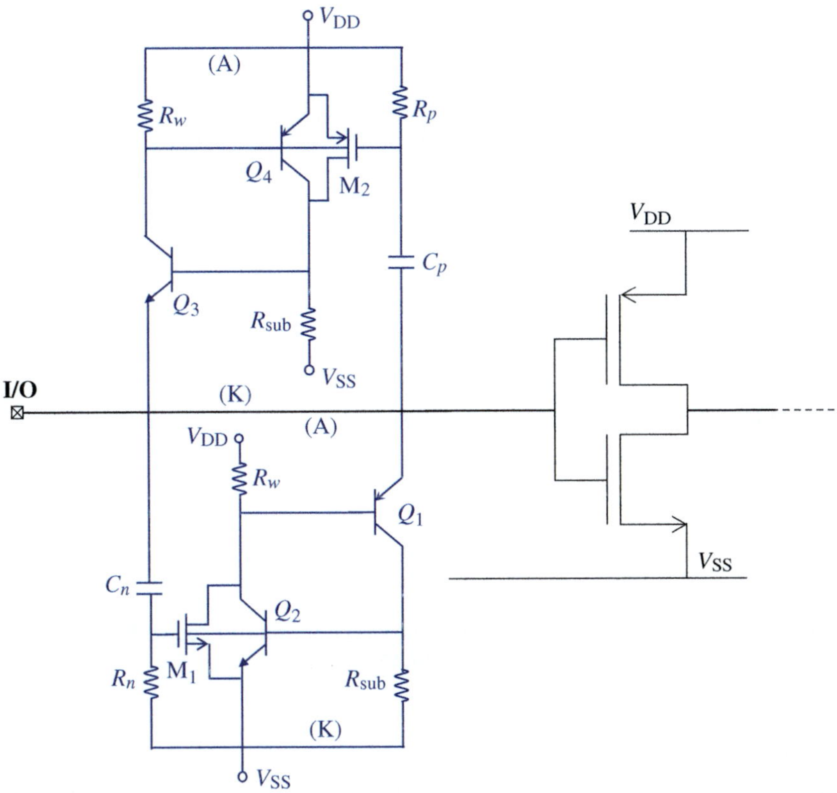

Figure 5.17 A complementary gcMOS-triggered SCR ESD protection circuit, in n-well/p-substrate process, uses gcNMOS/gcPMOS trigger-assisting units to achieve lower trigger voltage.

Figure 5.18a depicts an alternative way to reduce the ESD triggering voltage of SCR-based ESD protection structures where a dedicated current source (I_S) can be swiftly turned on by an incident ESD pulse, which will run through the series resistor (R_S). It will then quickly build up a voltage drop across the emitter of NPN (Q_2) to turn on Q_2 first, which will further turn on PNP Q_1; hence, triggers the NPN–PNP SCR structure to discharge the ESD transient. There are many means to implement this trigger-assisted SCR ESD protection structure. In general, R_S can be an internal well diffusion resistor or an external resistor by design. The conceptual I_S-source may be readily created using diode(s) embedded inside an SCR core structure in either forward or reverse conduction mode to source the current needed, generally referred as diode-triggered SCR (DTSCR) ESD protection structures. Figure 5.18b shows the concept for using a reverse-biased diode to trigger an SCR core structure. The reverse triggering diode is typically an optimized Zener diode (D_z) [23]. The D_z is integrated within the SCR ESD core, connected across the PNP Q_1 and in series with a dedicated resistor (R_S) attached parallelly to the emitter junction of the NPN Q_2. As a positive ESD pulse appears at SCR terminal A, it reverse-biases D_z and will turn on the D_z at the designed voltage ($\sim V_{t1}$). The initial ESD current ($I_{Dz} = I_S$) will flow through the D_z into the R_S and the voltage drop over the R_S will forward turn on the NPN Q_2, which will then turn on the PNP Q_1. Therefore, the SCR ESD core structure will be triggered to discharge the large ESD transient current at a low ESD V_{t1} as designed.

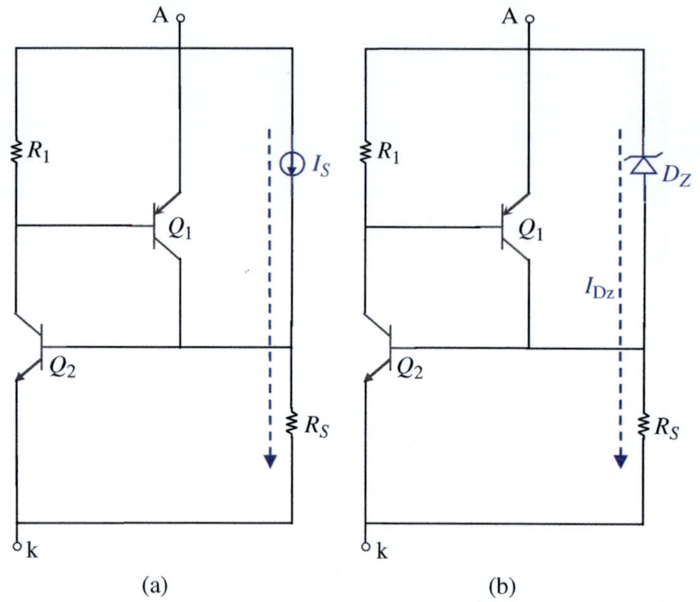

Figure 5.18 A low-triggering DTSCR ESD protection structure: (a) concept of using a *I*-source for trigger-assisting, and (b) using a Zener diode as the *I*-source to reduce V_{t1}.

Alternatively, a forward diode-string can be integrated into an SCR ESD core to form a low-V_{t1} DTSCR ESD protection subcircuit in many mays. Figure 5.19 shows one exemplar DTSCR ESD protection subcircuit using a diode-string of several forward diodes (D_1, D_2, … D_n) [24]. Under ESD stressing, the diode-string serves as the trigger-assisting subnet to the SCR core structure. After the diode-string turns on into forward conduction, it will build up the voltage across R_S to turn on NPN Q_2, which will then turn on PNP Q_1, hence, triggers the SCR structure to form a low-*R* ESD discharge path to protect ICs. Figure 5.20 depicts an exemplar cross-section for the DTSCR implemented in bulk CMOS. Figure 5.21a shows schematic for another DTSCR (Type-I) implementation where a forward diode-string (D_1, D_2, … D_m) is connected across the collector and emitter of NPN Q_2. Upon ESD zapping, the diode-string will be turned on for forward conduction that will turn on PNP Q_1 first, which will then turn on NPN Q_2, hence, triggers the SCR structure to form a low-*R* ESD discharge channel to protect ICs. Figure 5.22 shows cross-section of an exemplar DTSCR implemented in CMOS. In designing the required V_{t1} for the DTSCR ESD protection structure, the emitter voltage drop of Q_1 must be considered. Figure 5.21b depicts a modified diode-triggered SCR (DTSCR) (Type-II) using a forward diode-string to reduce V_{t1}, where D_m in the diode-string is shifted to emitter path of Q_1. The shift of D_m should not affect the V_{t1} of the DTSCR device; however, its hold voltage V_h will be increased due to D_m voltage drop after full ESD discharging starts. This illustrates one way to increase SCR holding voltage to avoid post-ESD SCR latch-up effect. Figure 5.23 presents measured ESD discharge *I*–*V* curves of exemplar DTSCR ESD protection circuits using varying number of gated diodes in the trigger-assisting diode-strings, which were designed and fabricated in a foundry 28 nm CMOS technology. It clearly shows that DTSCR has reduced ESD V_{t1} compared with the SCR ESD core. Further, the measurement also shows that the ESD V_{t1} of DTSCR ESD protection structures increases as the number of didoes in the diode-string unit increases, which is the expected trend [25].

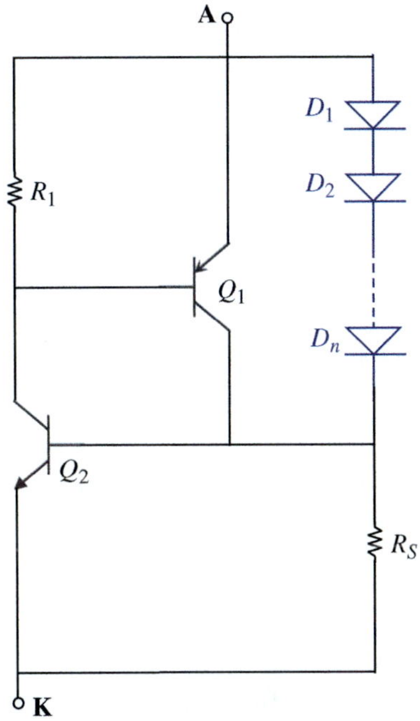

Figure 5.19 An exemplar DTSCR ESD protection structure uses a diode-string in forward mode as *I*-source to reduce V_{t1}.

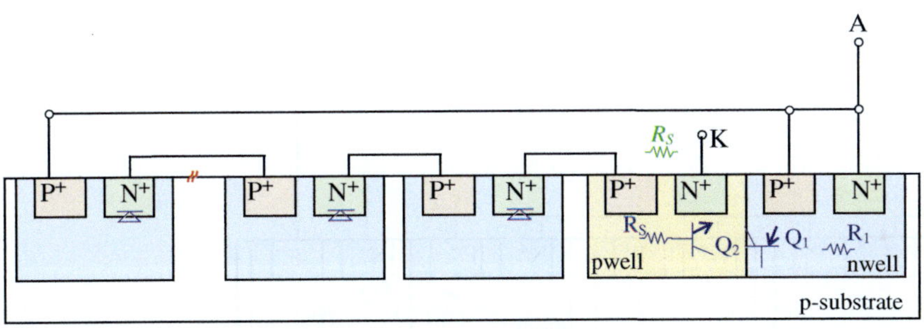

Figure 5.20 Exemplar cross-section view for a forward diode-string triggered DTSCR ESD protection structure, shown in Figure 5.19, implemented in bulk CMOS technology.

Obviously, there can be numerous different ways to build a DTSCR ESD protection structure or subcircuit in practical ESD protection designs. The limit would be a designer's brain power. Regardless any actual schematics to be used for DTSCR ESD protection structures, the SCR core device is always the one to take the main ESD transient currents that should be optimized for R_{ON}, V_h, and I_{t2}, while the trigger-assisting device determines the ESD V_{t1}.

5.4 ESD Power Clamps

During ESD events, an ESD pulse may occur at any pad on a chip; hence, IC supply buses may suffer from ESD failures similar to any I/O pads. In principle, an ESD protection structure is needed at a supply bus pad to protect the power rail, which is called a *power clamp*. In addition to protect a power line on a chip against possible ESD failure, power clamps also serve as key connecting nodes to form a full-chip ESD protection network in such a way that there always exists a low-*R*

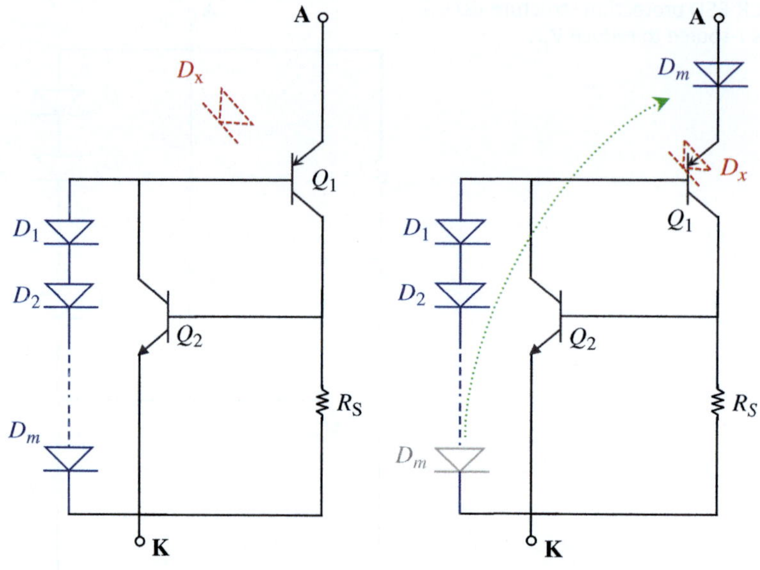

Figure 5.21 An alternative DTSCR ESD protection structure uses a diode-string in forward mode as *I*-source to reduce V_{t1}: (a) Type-I, and (b) Type-II to modify V_h.

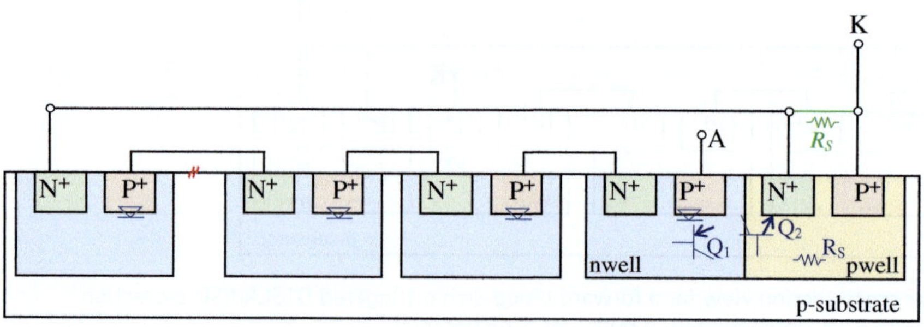

Figure 5.22 Exemplar cross-section view for a forward diode-string triggered DTSCR ESD protection structure, shown in Figure 5.21, implemented in bulk CMOS technology.

conduction path between any pair of pads on a chip to discharge transient ESD current without overheating and to clamp the pad voltage to a safely low level. In case of multiple supply voltages existing on a chip, a power clamp is needed between any two supply buses of different voltages, e.g., V_{DD1}, V_{DD2}, V_{DD3}, V_{SS1}, V_{SS2}, V_{SS3}, etc. One main difference between a power clamp device and an ESD protection device for an I/O pad is that the ESD-induced parasitic capacitance C_{ESD} is not a concern to supply rails. Typically, power clamps are placed at the corners of an IC chip. However, depending on IC schematics and layout floorplans, power clamps may be placed at different sites on a chip. Conceptually, any ESD protection structure can be used as a power clamp. This section discusses common power clamp designs.

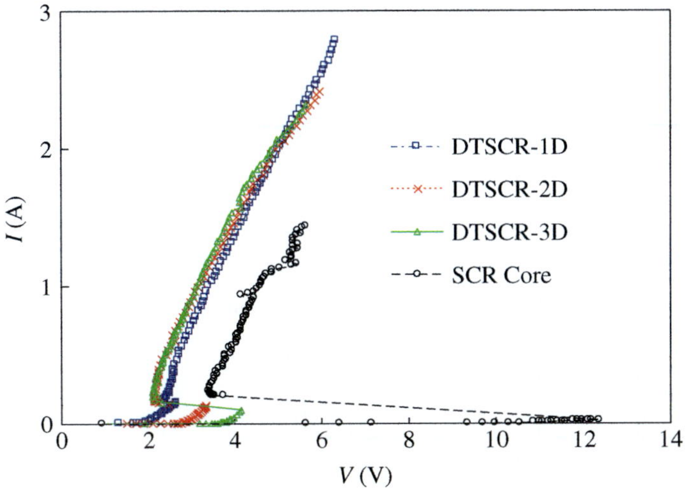

Figure 5.23 Measured ESD discharging $I-V$ curves for exemplar DTSCR ESD protection structures using STI diode-strings of varying number of gated diodes fabricated in a 28 nm CMOS technology [25].

5.4.1 Diode-String Power Clamps

As can be expected, didoes can be used as power clamps. However, single diode cannot serve as a power clamp in forward mode because the low diode turn-on voltage of $V_D \sim 0.65\,\text{V}$ for silicon PN didoes, which is too low for any ICs due to short-circuit risk. A reverse diode, e.g., a Zener diode, may be used as a power clamp if designed properly for a specific supply voltage on a chip. One solution for power clamping is to use a forward diode-string, which consists of a number of diodes in series so that the effective turn-on voltage of the diode-string will be high enough to avoid possible short-circuiting in normal IC operations. Unfortunately, though sounds like simple, designing a good diode-string power clamp may burn many brain cells. Simply stacking up m discrete diodes together and expecting $V_{t1} = mV_D$ for the m-diode-string ESD protection structure will never work. The reason is that a PN junction diode in IC fashion is never a "discrete" diode. Instead, all PN junction diodes in the same Si substrate are connected to each other in certain ways, which leads to some serious parasitic effects that must be carefully considered in designing a diode-string ESD power clamp structure. Take a six-diode-string ESD power clamp shown in Figure 5.24 as an example, the six PN junction didoes (D_1–D_6) are electrically linked as depicted in the cross-sectional view in Figure 5.25 in a p-substrate n-well CMOS technology. Ideally, each diode is a vertical P^+/n-well junction. However, in its IC format, each diode is actually part of a parasitic vertical PNP. Due to the conduction nature of a Si substrate, the six-parasitic PNP transistors (Q_1–Q_6) are connected together similar to a multiple-stage common-collector (CC) amplifier circuit as shown in Figure 5.24. The amplification feature of the PNP improves ESD discharge capability, meanwhile and unfortunately, it also causes design problems, such as reduced total ESD V_{t1} and enlarged total leakage current I_{leak}, etc. The total ESD triggering voltage V_{t1} may be ideally estimated as $V_{t1} = 6 \times V_D$, or generalized for an m-diode-string as follows:

$$V_{t1} = \sum_{i=1}^{6} V_{Di} = 6V_D \cdots = \sum_{i-1}^{m} V_{Di} \tag{5.1}$$

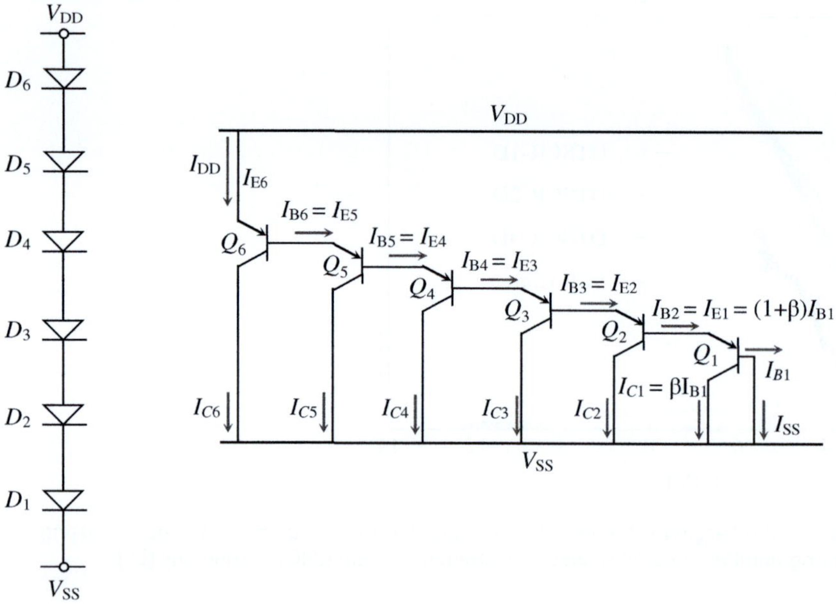

Figure 5.24 Schematic and equivalent circuit for a six-diode diode-string ESD power clamp structure.

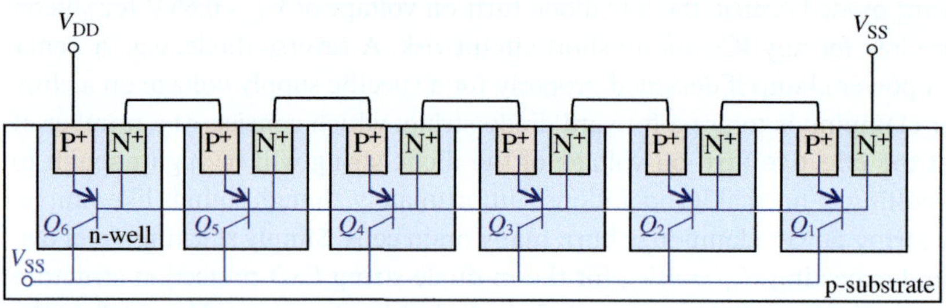

Figure 5.25 Cross-sectional view of a six-diode diode-string ESD power clamp structure implemented in a p-substrate CMOS technology.

where m is the number of diodes in the diode-string. This would result in a $V_{t1} \approx 3.9\,\text{V}$ for the six-diode-string ESD clamp, nicely working for ICs powered by 3.3 V supplies. However, connecting six P^+/n-well junctions in IC fashion brings in the Darlington amplification effect of the Q_1–Q_6 sub-circuit. From the Shockley equation for a PN diode, the forward turn-on voltage follows

$$V_D \cong nV_T \ln \frac{I_D}{I_S} \tag{5.2}$$

where V_T is PN thermal voltage, I_D and I_S are diode current and saturation current, and n is the ideality factor for PN diode, respectively. For the parasitic PNP chain of Q_1–Q_6 shown in Figure 5.24, the following derivation holds,

$$V_{\text{BE1}} = V_{D1} \cong nV_T \ln \frac{I_{D1}}{I_S} \cong nV_T \ln \frac{I_{E1}}{I_S} \tag{5.3}$$

$$V_{\text{BE2}} = V_{D2} \cong nV_T \ln \frac{I_{E2}}{I_S} \cong nV_T \ln \frac{I_{E1}}{(1+\beta)I_S} = V_{D1} - nV_T \ln(1+\beta) \tag{5.4}$$

and

$$V_{BE6} = V_{D6} \cong nV_T \ln \frac{I_{E6}}{I_S} \cong V_{D1} - nV_T \ln (1+\beta)^5 \tag{5.5}$$

where the same V_D for a PN diode and the same current gain of β for the parasitic PNP transistor are assumed. The total diode-string ESD trigger voltage is therefore given by

$$V_{t1} \cong \sum_{i=1}^{6} V_{Di} = 6V_D - 15nV_T \ln(1+\beta) \tag{5.6}$$

which can be generalized for an m-diode-string ESD power clamp as

$$V_{t1} \cong \sum_{i=1}^{m} V_{Di} = mV_D - \frac{m(m-1)}{2} nV_T \ln(1+\beta) \tag{5.7}$$

Equation (5.7) clearly states that the dream of linearly summing up the V_D of a diode-string power clamp will never come to reality in IC designs due to the chain amplification effect of the connected parasitic PNP-chain of Q_1–Q_6. The actual total ESD V_{t1} is seriously affected by the PNP current gain of β. A larger BJT β will dramatically reduce the effective ESD V_{t1} of a diode-string power clamp. It seems that the problem of ESD V_{t1} reduction can be easily resolved by adding more didoes in a diode-string structure. Unfortunately, a longer diode-string inevitably introduces some adverse effects to ICs that must be carefully considered in IC designs. The first such design problem is the increase of the ESD discharge resistance R_{ON} of an enlarged diode-sting. The total R_{ON} is equivalent to the effective input resistance looking into Q_6 in CC mode as shown in Figure 5.24. The total R_{ON} can be derived roughly using the CC input resistance equivalent circuit model depicted in Figure 5.26 starting from the first stage of Q_1. From the BJT T-model for a CC mode amplifier, it follows

$$R_{i1} \cong r_e + \frac{R_w}{1+\beta} \tag{5.8}$$

$$R_{o2} = R_{i1} \tag{5.9}$$

and

$$R_{ON} = R_{i6} \cong r_e + \frac{R_{i5} + R_w}{1+\beta} = r_e \sum_{a=0}^{5} \frac{1}{(1+\beta)^a} + R_w \sum_{b=1}^{6} \frac{1}{(1+\beta)^b} \tag{5.10}$$

where r_e and R_W are series emitter resistance and parasitic lateral n-well (base of PNP) resistance, respectively. Obviously, Eq. (5.10) states that more didoes in a diode-string will increase the total

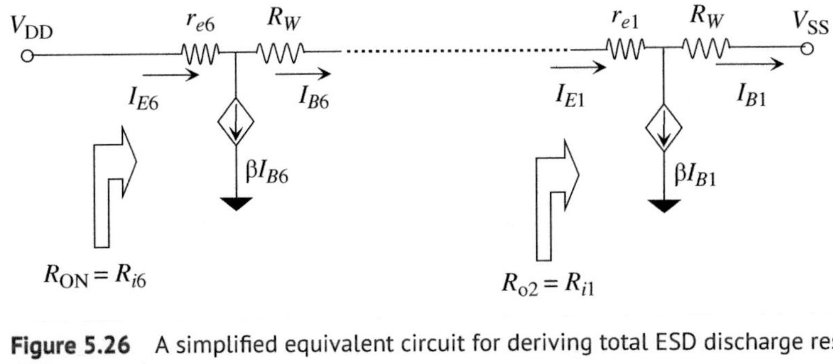

Figure 5.26 A simplified equivalent circuit for deriving total ESD discharge resistance for a diode-string ESD power clamp.

ESD discharge R_{ON}, causing more ESD overheating, while the current amplification gain β of parasitic PNP BJT serves to mitigate the high R_{ON} problem of a large diode-string ESD power clamp. The second major design problem of a diode-string power clamp is the dramatical increase in ESD-induced leakage currents. From Figure 5.24, the total ESD-induced leakage appeared at the V_{DD} pad will be roughly $I_{leak} = I_{DD} = I_{E6}$ of Q_6. Obviously, the Darlington amplification effect of the parasitic Q_1–Q_6 chain will significantly enlarge the total I_{leak}, which is a big problem to ICs. Therefore, the parasitic BJT amplification and Darlington china effect should be minimized in designing good diode-string power clamps. The above discussion clearly states that careful design trade-off must be made in designing good diode-string ESD power clamps by considering ESD-critical parameters, such as, V_{t1}, R_{ON}, I_{leak}, and I_{t2}. A simple ESD protection concept may not always mean design simplicity in real-world IC designs. Interestingly, a diode-string can be used for I/O ESD protection with reduced ESD-induced C_{ESD}.

5.4.2 MOSFET Power Clamps

MOSFET-based ESD protection structures have been widely used for ESD power clamps in CMOS ICs. Figure 5.27a shows one power clamp example using a classic ggNMOS ESD protection device for V_{DD}-to-V_{SS} ESD surge protection in DS mode through the parasitic NPN conduction. For SD mode ESD stressing, the parasitic Drain junction N^+/p-well diode will provide limited ESD protection. Unlike I/O signal pads, the supply pads are not sensitive to ESD-induced parasitic capacitance. However, the ESD triggering voltage for a power clamp must be designed carefully. On the one hand, the ESD V_{t1} should be low enough to ensure ESD protection, i.e., the power clamp can be turned on under ESD stressing before any core IC device may be affected by the ESD transients. On the other hand, a good safety margin is needed for the power clamp V_{t1} to avoid possible ESD mis-triggering because a power bus may have quite a supply voltage fluctuation during normal IC operations. Figure 5.27b depicts another ESD power clamp example formed by a classic two-stage primary–secondary ESD protection subcircuit. The secondary ESD protection device is a ggNMOS featuring a low-triggering voltage. The primary ESD protection device is a thick-gate NMOS with its gate coupled to the supply bus. Typically, multiple-finger layout is used for large MOSFET ESD power clamp structures, normally fairly large in sizes, to ensure ESD discharge scalability, for which $V_{t1} < V_{t2}$ is desirable for uniform ESD triggering across all ESD protection device fingers.

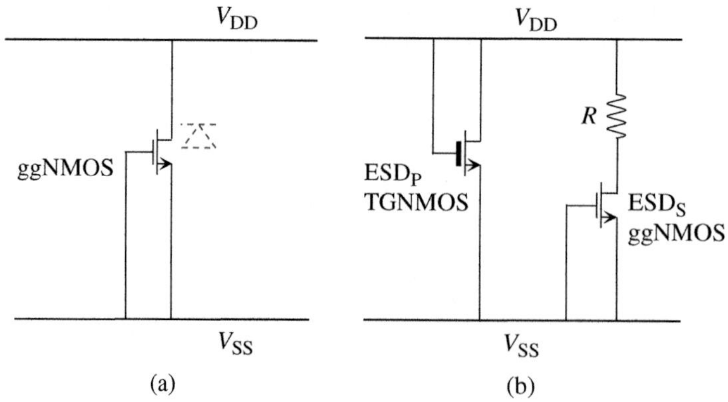

Figure 5.27 Exemplar ESD power clamps using (a) ggNMOS and (b) two-stage ggNMOS-TGNMOS ESD protection structures.

5.4.3 SCR Power Clamps

Due to its high ESD discharge efficiency associated with the snapback ESD discharge I–V characteristics, SCR-based ESD protection structures are commonly used as ESD power clamps. Generally, an original SCR device is not favorable for power clamping, because its high ESD V_{t1} may cause internal device, often parasitic bipolar devices, turn-on ahead of the SCR device under ESD stressing. Typically, lower-V_{t1} SCR structures are used as ESD power clamps, such as MVSCR and LVSCR ESD protection structures. Figure 5.28 illustrates an NMOS-triggered SCR ESD protection subcircuit for power clamping for DS mode ESD discharging, where SD mode ESD discharging relies on a parasitic well-junction diode. For an SCR power clamp, the ESD V_{t1} must be accurately designed with enough safety margin to avoid mis-triggering possibly initiated by normal supply voltage fluctuation, at least 10%. On the other hand, the SCR hold point, V_h and I_h must be carefully designed to prevent post-ESD latch-up lock-in due to the supply.

5.4.4 Any Switch Power Clamps

A non-snapback ESD power clamp structure is highly favored by circuit designers because full-chip SPICE circuit simulation can be readily conducted. Figure 5.29 depicts one such NMOS ESD power clamp circuit [26]. This ESD power clamp contains a main NMOS ESD switch device, M_1, which will be turned on to discharge the ESD pulses in *normal* MOSFET conduction mode. An ESD trigger-assisting subnet is required, which consists of a three-inverter chain and an R_1–C_1 coupling unit. The ESD protection mechanism follows: When an ESD pulse occurs to the V_{DD} pad in DS mode, the input node of the inverter-1 is pulled down by the R_1–C_1 coupling and outputs a logic HIGH, which drives the inverter-2 to logic LOW, which then forces the inverter-3 to produce a logic HIGH. An odd number of inverters are required and designed to produce a high-enough voltage at the inverter-chain output node, which is greater than the threshold voltage V_{th} of the NMOS switch M_1, hence drives the M_1 into active conduction mode with amplification to efficiently discharge the large ESD surge without overheating. Several key design factors must be considered in

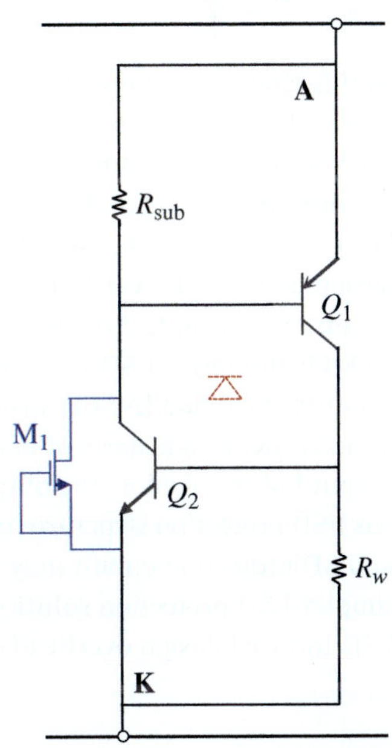

Figure 5.28 A exemplar ESD power clamp uses an NMOS-triggered SCR ESD protection structure.

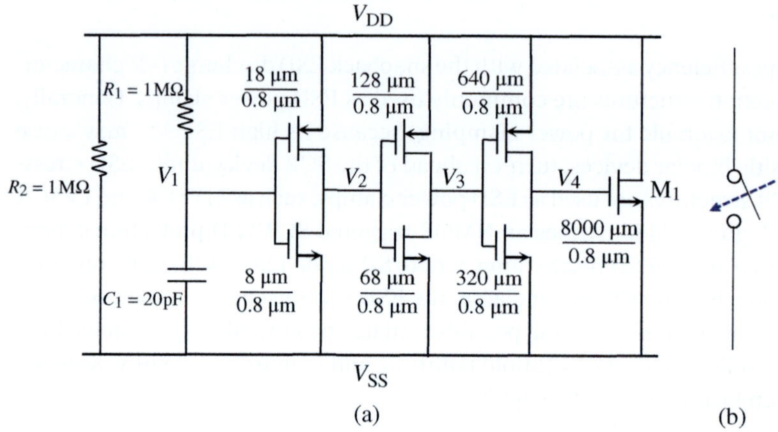

Figure 5.29 (a) A non-snapback NMOS switch serves as an ESD power clamp for power line ESD protection (Merrill and Issaq [26]), and (b) any ideal ESD switch can be used for power clamping.

designing a working NMOS power switch: The R_1–C_1 subnet should be carefully designed for its time constant to ensure fast ESD triggering, e.g., $t_1 < 10$ ns for HBM ESD events. The ESD V_{t1} is determined by tuning the inverter chain and M_1. For a low ESD discharge R_{ON} required to handle large ESD surges, the device size of M_1 will be very large, which is a big disadvantage because it consumes large Si area and make layout floor planning difficult. The reset resistor, R_2, is used to reset the chip after ESD event is over. Analysis of the pros and cons of this NMOS power clamp leads to an interesting point: any ideal ESD switch, as depicted in Figure 5.29b, can be used for ESD power clamping on a chip, for example, a graphene-based NEMS mechanical switch device to be discussed in Chapter 17 [27].

5.5 Summary

In this chapter, various ESD protection circuit schematics are discussed. In general, a more complex ESD protection structure or subcircuit, other than a simple ESD device, is needed to meet specific IC design requirements, e.g., the ESD design window at advanced technology nodes that requires fine-tuning of the ESD-critical parameters. For high ESD robustness, multiple-finger ESD layout designs are commonly used where ESD triggering and discharging uniformity is a main design concern. Lower ESD triggering designs can be achieved by using various trigger-assisting means. For example, MOSFET ESD protection circuits can be realized by a gate-coupling technique (gcMOS) for lower ESD V_{t1}. Low-triggering SCR-based ESD protection sub-circuits can be realized using MVSCR and LVSCR structures. A forward diode-string may be used for I/O ESD protection to meet the design specs of both ESD triggering and low ESD-induced parasitic capacitance. For output ESD protection, the output buffer transistors may be modified for ESD self-protection. Various ESD protection structures can be used as supply line power clamps. It is noteworthy that, while an ESD protection circuit may be needed to meet specific IC specs, one should always try to use a simpler ESD protection solution if ever possible to avoid ESD design complexity and to minimize ESD-induced design overhead effects in real-world IC designs.

References

1 Rountree, R. and Hutchins, C. (1985). NMOS protection circuitry. *IEEE Trans. Electron Devices* ED-32: 910–917.

2 Duvvury, C., Taylor, T., Lindgren, J., and Kumar, S. (1989). Input protection design for overall chip reliability. *Proceedings of EOS/ESD Symposium*, pp. 190–198.

3 Hulett, T. (1981). On chip protection of high density NMOS devices. *Proceedings of EOS/ESD Symposium*, pp. 90.

4 Pan, Z., Li, C., Di, M. et al. (2020). 3D TCAD analysis enabling ESD layout design optimization. *IEEE J. Electron Devices Soc.* 8: 1289–1296.

5 Chen, K.L. (1988). Effect of interconnect process and snapback voltage on the ESD failure threshold of NMOS transistors. *Proceedings of EOS/ESD Symposium*, pp. 212–219.

6 Polgreen, T. and Chatterjee, A. (1989). Improving the ESD failure threshold of silicided nMOS output transistors by ensuring uniform current flow. *Proceedings of EOS/ESD Symposium*, pp. 167–174.

7 Feng, H.G. (2001). A mixed-mode simulation-design methodology for on-chip ESD protection design. A MS thesis. Illinois Institute of Technology.

8 Kamata, T., Tanabashi, K., and Kobayashi, K. (1976). Substrate current due to impact ionization in MOSFET. *Jpn. J. Appl. Phys.* 15: 1127.

9 Sze, S.M. (1981). *Physics of Semiconductor Devices*, 2e. New York: Wiley.

10 Toriumi, A. (1989). Experimental study of hot carriers in small size Si-MOSFETs. *Solid State Electron.* 32 (12): 1519–1525.

11 Tam, S., Ko, P., and Hu, C. (1984). Lucky-electron model of channel electron injection in MOSFET's. *IEEE Trans. Electron Devices* ED-31 (9): 1116.

12 Tam, S. and Hu, C. (1984). Hot-electron-induced photon and photocarrier generation in silicon MOSFET's. *IEEE Trans. Electron Devices* ED-31 (9): 1264.

13 Fischetti, M., Laux, S., and Crabbe, E. (1995). Understanding hot-electron transport in silicon devices: is there a short cut? *J. Appl. Phys.* 78: 1058–1087.

14 Chan, T., Ko, P., and Hu, C. (1984). A simple method to characterize substrate current in MOSFETs. *IEEE Electron Device Lett.* EDL-5: 505.

15 Hu, C. (1989). Hot-carrier effects. In: *Advanced MOS Device Physics*, Chapter 3, VLSI Electronics Microstructure Science, vol. 18 (eds. N.G. Einspruch and G. Gildenblat), 119–160. San Diego, CA: Academic Press.

16 Tandan, N. (1994). ESD trigger circuit. *Proceedings of EOS/ESD Symposium*, pp. 120–124.

17 Smith, J. (1998). A substrate triggered lateral bipolar circuit for high voltage tolerant ESD protection applications. *Proceedings of EOS/ESD Symposium*, pp. 63–71.

18 Wu, C. and Ker, M. (1997). ESD protection for output pad with well-coupled field-oxide device in 0.5 μm CMOS technology. *IEEE Trans. Electron Devices* 44 (3): 503–505.

19 Chan, T. and Culver, D. (1994). ESD protection circuit. US Patent, No. 5, 329, 143.

20 Scott, D., Bosshart, P., and Gallia, J. (1991). Circuit to improve electrostatic discharge protection. US Patent, No. 5, 019, 888.

21 Chatterjee, A. and Polgreen, T. (1911). A low-voltage triggering SCR for on-chip ESD protection at output and input pads. *IEEE Electron Device Lett.* 12 (1): 21–22.

22 Ker, M., Chang, H., and Wu, C. (1997). A gate-coupled PTLSCR/NTLSCR ESD protection circuit for deep-submicron low-voltage CMOS IC's. *IEEE J. Solid-State Circuits* 32 (1): 38–50.

23 Chen, J., Amerasekera, A., and Vrotsos, T. (1995). Bipolar SCR ESD protection circuit for high speed submicron bipolar/BiCMOS circuits. *IEEE IEDM Technical Digest*, pp. 337–340.

24 Mergens, M., Russ, C., Verhaege, K. et al. (2003). Diode-triggered SCR (DTSCR) for RF-ESD protection of BiCMOS SiGe HBTs and CMOS ultra-thin gate oxides. *IEEE Int. Electr. Devices Meeting (IEDM) Tech. Digest.*: 515–518.

25 Research data base, *Wang's Lab*, University of California.

26 Merrill, R. and Issaq, E. (1993). ESD protection methodology. *Proceedings of EOS/ESD Symposium*, pp. 233–237.

27 Ma, R., Chen, Q., Zhang, W. et al. (2016). A dual-polarity graphene NEMS switch ESD protection structure. *IEEE Electron Device Lett.* 37 (5): 674–676.

6

Full-Chip ESD Protection

6.1 Full-Chip ESD Protection Principles

At this moment, we understood that electrostatic discharge (ESD) failure is unavoidable to electronics since ESD event is a natural phenomenon existing everywhere. We also understood the two basic on-chip ESD protection mechanisms, i.e., using an ESD protection structure to form a low-R conduction path to discharge ESD pulses without overheating and to clamp pad voltage to prevent voltage breakdown. We discussed many different ESD protection devices and structures that can facilitate the two ESD protection principles. We further discussed the importance of designing the ESD-critical parameters associated with the triggering, holding, discharging, and failure properties, e.g., V_{t1}, I_{t1}, t_1, V_h, I_h, R_{ON}, V_{t12}, and I_{t12} [1–3]. Nevertheless, many integrated circuit (IC) designers had a bad experience that an ESD protection structure provided by an ESD design "guru" for your IC failed at chip level. It seems that uncertainty always exists for on-chip ESD protection design regardless how wonderful an individual ESD protection device – a gold nugget – seems to be. Here are the explanations for the ESD protection design uncertainty problem in real-world designs. First, an individual/standalone ESD protection device is always characterized for its ESD current handling capability, I_{t2}. However, a working ESD protection device itself does not guarantee on-chip ESD protection, just like a piece of sturdy brick does not always guarantee a strong skyscraper in a real world. Second, ESD protection design is IC-specific that requires quantitative design of ESD-critical parameters, e.g., V_{t1} and V_h, to meet the ESD Design Window for a given IC in a given technology. Third, it is important to know that on-chip ESD protection design is a circuit-level design task that requires full-chip circuit design considerations. Fourth, there exist unavoidable interactions between ESD protection structures and the IC core circuit under protection. The ESD-induced design overhead effects can no longer be overlooked for advanced ICs.

Over years and, unfortunately, still quite often today, a common misconception on ESD protection design exists in the IC design community that considers ESD protection design as a device-level design task, which treats IC designers as customers of an ESD protection "solution" being a "product" delivered by an ESD designer. This ESD protection design approach is fundamentally wrong, which is one main cause to ESD protection design uncertainties and failures in the real world. In one clear statement: on-chip ESD protection design is an IC circuit-level design task, which is IC-specific and requires active involvement of IC designers. To ensure successes of on-chip ESD protection designs, the following full-chip ESD protection principles ought to be considered: Principle-1 is that, per the ESD testing standards (e.g., human body model [HBM] or charged device model [CDM]), every IC pad must be ESD-zapped with respect to other pads on a chip in preset pad combinations. Therefore, the simple and mandatory on-chip ESD protection

Practical ESD Protection Design, First Edition. Albert Wang.

rule is to ensure one low-R ESD discharging path existing between any two pads on a chip for all relevant ESD stressing modes, i.e., PS, NS, PD, ND, DS, and SD, as depicted in Figure 4.4 in Chapter 4 [4]. This principle means a complete *full-chip ESD protection network* is required that guarantees that any pad combination can pass a given ESD zapping routine, hence, realizing full-chip ESD protection. Principle-2 states that each pad-to-pad ESD discharging path mentioned above must be quantitatively designed and optimized for the desired ESD testing models and ESD protection targets. This means that the *equivalent ESD-critical parameters* for a specific ESD discharging path for a given pair of pads, not an individual ESD device, should be evaluated for proper ESD protection functions on a chip. For example, the total equivalent ESD triggering voltage is the sum of that of all ESD devices and voltage drops of any series bus resistors within the given ESD discharging path, i.e., $V_{t1-\text{path}} = \sum_i^n (V_{t1i} + I_{\text{ESD}} R_{\text{bus}i})$, and the total equivalent ESD discharging resistance is the sum of that of all ESD devices and total series bus resistance within the same ESD discharging path, i.e., $R_{\text{ON}-\text{path}} = \sum_i^n (R_{\text{ON}i} + R_{\text{bus}i})$, where i denotes the device number-i of either an ESD device or a series bus resistor. Further, the equivalent ESD current-handling capability of the given ESD discharging path is limited to the minimum (i.e., the weak point) of that of any ESD device within the same ESD conduction path, i.e., $I_{t2-\text{path}} = \text{Min}(I_{t2i})$. Equivalent ESD-critical parameters for an ESD discharging path will be discussed in details in Chapter 15. Principle-3 points out that as long as a discharging path is guaranteed between any pair of pads on a chip, it is unnecessary to use multiple ESD devices per pad, each dedicated to shunt ESD transients of specific ESD stressing mode as depicted in Figure 4.4, i.e., typically up to four single-directional ESD device may be needed for each IC pad. The interconnected circuit schematics containing ESD devices serve to establish a complete ESD protection network on a chip that guarantees an ESD discharging path existing between any two pads on a chip. As such, Principle-3 naturally leads to Principle-4, which offers a smart way to optimize full-chip ESD protection schematics and to minimize the ESD-induced design overhead, including ESD layout size and ESD-induced parasitic effects, such as C_{ESD}, I_{leak}, and noises. Last but not least, Principle-5 enables an IC designer to consider possible interactions between the ESD protection devices and the IC core circuit under ESD protection, which allows IC designers to inspect any potential weak conduction link within the core circuit, adjacent to and in parallel with a specific ESD protection structure, which may be mis-triggered by an incident ESD pulse ahead of the dedicated ESD protection device, which may consequently lead to prematured ESD failure, i.e., early ESD failure. From the above discussions, it becomes apparent that on-chip ESD protection designs must be conducted from the full-chip perspective, never in the simple language of individual standalone ESD protection "device." Full-chip ESD protection design examples will be discussed in Sections 6.3 and 6.4.

6.2 ESD Protection Design Window

Now that we understood that on-chip ESD protection design is a full-chip IC circuit-level design task, not simply about designing an individual standalone ESD protection device, it is equally important to state that ESD protection design must follow a *quantitative* design method, not entirely experience-based. Although rich ESD design experiences help, however, like analog IC design today, success of ESD protection design is all about careful design of the ESD-critical parameters, quantitatively, for a specific IC in a specific technology at a chip level. The traditional experience-based *trial-and-error* ESD protection design approach is no longer suitable for advanced ICs at advanced technology nodes. Rational ESD protection design is truly custom IC design that requires design optimization by CAD simulation aiming for not only design prediction

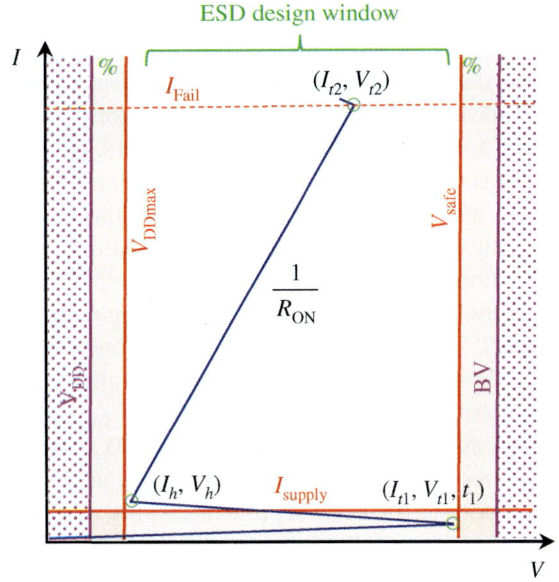

Figure 6.1 Concept of ESD Design Window suggests that ESD-critical parameters should be designed for ESD-safe operations. An ESD design window is dynamic in nature in terms of its boundaries.

but also design optimization. Therefore, practical ESD protection design is all about crunching the numbers for ESD-critical parameters for a specific IC. Recall that the ESD-critical parameters are defined in Figure 4.5 for ESD discharge functions, which include the ESD triggering threshold, ESD hold threshold, ESD discharge resistance, and ESD thermal failure threshold. To optimize the ESD-critical parameters for a specific IC in a given technology means to fine-tune these operational numbers to comply with the *ESD Design Window*, as defined in Figure 6.1 [5]. First, to prevent ESD-induced voltage breakdown, e.g., typically gate or diffusion breakdown in complementary metal-oxide-semiconductor (CMOS), the ESD triggering voltage V_{t1} must be lower than the breakdown voltage of the protected node on a chip. Second, for an ESD protection structure featuring snapback discharge I–V characteristic, e.g., silicon controlled rectifier (SCR) ESD device, its holding voltage V_h and current I_h should be higher than the supply voltage V_{DD} (or, V_{supply}) and the total supply current I_{DD} (or, I_{supply}) on a chip. Therefore, unwanted latch-up will not sustain after an ESD event is over. Third, if $V_{t1} < V_{DD}$ holds for any ESD protection device, then, short-circuiting caused by the ESD protection device will occur, resulting in IC malfunction. Fourth, the top bar, I_{Fail}, is simply limited by the ESD design target in light of, typically, the ESD thermal failure, i.e., $I_{Fail} \approx I_{t2}$, with a safety margin. Consequently, the ESD discharge I–V curve must be designed to fit into the ESD Design Window as shown in Figure 6.1. Further, due to the unavoidable variations in supply voltages and processes (i.e., breakdown voltage), a design *safety margin* of certain percentage is adopted in practical ESD protection designs, which leads to a narrowed ESD Design Window bounded by V_{DDmax} and V_{safe}. Typically, 10–20% is a reasonable number for the design safety margin of IC products. On the other hand, as IC technologies continuously advance, the breakdown voltage decreases substantially, while the supply voltage typically only reduces mildly. Therefore, an *ESD Design Window Shrinking* effect is observed down on the technology roadmap, which makes on-chip ESD protection design more difficult. In practical designs, it is very involving and challenging to understand the exact physical meaning of the ESD design window, and hence, to check design compliance with the ESD design window. The upper limit (V_{safe}) for the ESD design window is often set by the possible voltage breakdown failure threshold at the protected node, where the transient voltage may be built-up during an ESD event, hence being *dynamic*. On the other hand, the lower end (V_{DDmax}) of the ESD design window is limited by the DC supply voltage in normal IC operations (ignore the much smaller *ac* signals) that is static in nature, but fluctuation exists. Clearly, the ESD design window must be considered dynamically and in reference with the protection nodes and process modules (i.e., various BV_{xyz} at pads), ESD stressing modes (i.e., PS, NS, PD, ND, DS, and SD), and the functional and power domains (V_{DDx}, V_{SSy}, V_{CCi}, V_{EEj}, etc.). Hence, the ESD design window is a *dynamic* concept in nature with several unique aspects. On the one hand, the ESD design window boundaries are certainly depending upon the power domains (i.e., different V_{DD}, etc.) and the circuit nodes under protection (i.e., different BV_{xyz}). On the other hand, an ESD design window is also dynamically and closely related to the actual ESD zapping modes at a given protection node (i.e., PS, NS, PD, ND, DS, SD, etc.). It is wrong to expect one single ESD design window for the whole chip. For example, for a power clamp between a positive V_{DD} bus and a negative V_{SS} bus, under the positive rail-to-rail ESD stress mode (DS), $V_{t1} > V_{DD} + |V_{SS}|$ is required to avoid short-circuit in normal operation or latch-up triggered by an ESD pulse; on the other hand, for the negative rail-to-rail ESD stress mode (SD), the reverse V_{t1} (a positive value in terms of ESD triggering voltage) can be reduced to 0 V because, practically, the negative V_{SS} and positive V_{DD} across the ESD clamp prevent any possible short-circuit turn-on of the ESD clamp in the reverse direction. Indeed, comprehensive and quantitative analysis of the dynamic ESD design windows for the whole chip is a complex IC design task, therefore, CAD-based ESD protection design methodology is critical to fine-tuning the ESD-critical parameters to fit into the ESD Design Window, which will be discussed in details in Chapters 8 and 15.

6.3 Advanced ESD Protection: More at Less

In addition to ensure complete pad-to-pad ESD protection, one major benefit for thinking from the full-chip ESD protection perspective is that it helps to achieve the ESD design goal of More at Less. Here, MORE (benefits) means better ESD protection at whole chip level, while, LESS (costs) translates into reduced ESD-induced design overhead including ESD-induced parasitic effects, area consumption and layout floorplan. Recall the classic full-chip ESD protection concept depicted in Figure 6.2, multiple single-polarity ESD protection devices would be required to protect each IC pad against incident ESD pulses of different stressing modes, i.e., PS, NS, PD, ND, DS, and SD. This

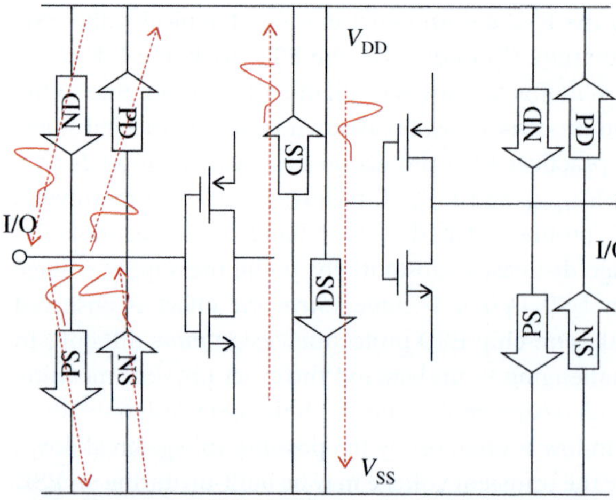

Figure 6.2 Complete on-chip ESD protection requires multiple single-polarity ESD protection devices per pad to construct a full-chip ESD protection network.

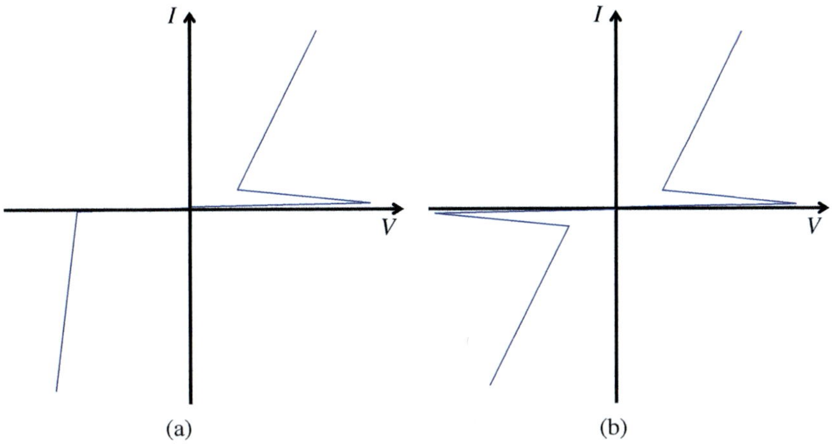

Figure 6.3 Conceptual ESD discharge I–V characteristics for ESD protection structures: (a) a single-polarity asymmetrical ESD discharging, and (b) a dual-polarity symmetrical ESD discharging.

is because a single-polarity ESD protection structure, as shown in Figure 6.3a, for its asymmetrical ESD discharge I–V characteristic, is designed and optimized to be a good active ESD switch in one direction only, while typically a parasitic diode may help to discharge an ESD transient in the opposite direction, often resulting in Early ESD failure in this opposite discharging direction. This is common for MOSFET and SCR-based ESD protection structures, which can readily handle large ESD surges in one direction due to the optimized low-R, however, the parasitic PN junction is an as-is diode that can only conduct limited ESD currents in the reverse direction before being burned out. As an example, Figure 6.4 shows a full-chip ESD protection design using MOS-triggered SCR ESD protection structures devices [6]. Since each ggMOS-SCR ESD protection structure is designed to conduct ESD pulses in one direction only, four such ggMOS-SCR ESD units are required for each pad to actively discharge incident ESD pulses of different stressing modes. While the ESD protection schematic is carefully designed, major adverse effects arise: each ggMOS-SCR ESD unit is very large, and four such ESD units per pad take too much Si area, making layout floor planning a big headache too, and also introducing too much parasitic effects including C_{ESD}, I_{leak}, noises, and global noise coupling. The adverse impact is often unacceptable for advanced ICs. This example clearly states that a good whole-chip ESD protection design cannot be simply adding up ESD protection devices at pads on a chip. One has to burn some brain cells for a well-thought-out smart on-chip ESD protection solution, using either novel multiple-polarity ESD protection devices or full-chip ESD protection schematics. A few such design examples are discussed in Sections 6.3 and 6.4.

6.3.1 Dual-Polarity ESD Protection

Understood that one-directional conduction with traditional single-polarity ESD protection device is inefficient for full-chip ESD protection schematics, it naturally calls for novel ESD protection structure that can simplify ESD protection circuit network at chip level. Figure 6.5 depicts one such novel dual-polarity ESD protection structure based on SCR ESD discharging mechanism. Conceptually, this new dual-polarity ESD protection structure is a two-terminal five-layer ($N_1P_2N_3P_4N_5$) lateral device that can be readily implemented in a BiCMOS technology where an N-type isolation (n-Iso) layer serves as a global decoupling layer [7]. Figure 6.6 illustrates the corresponding intrinsic NPNPN core structure and its equivalent circuit comprising one lateral

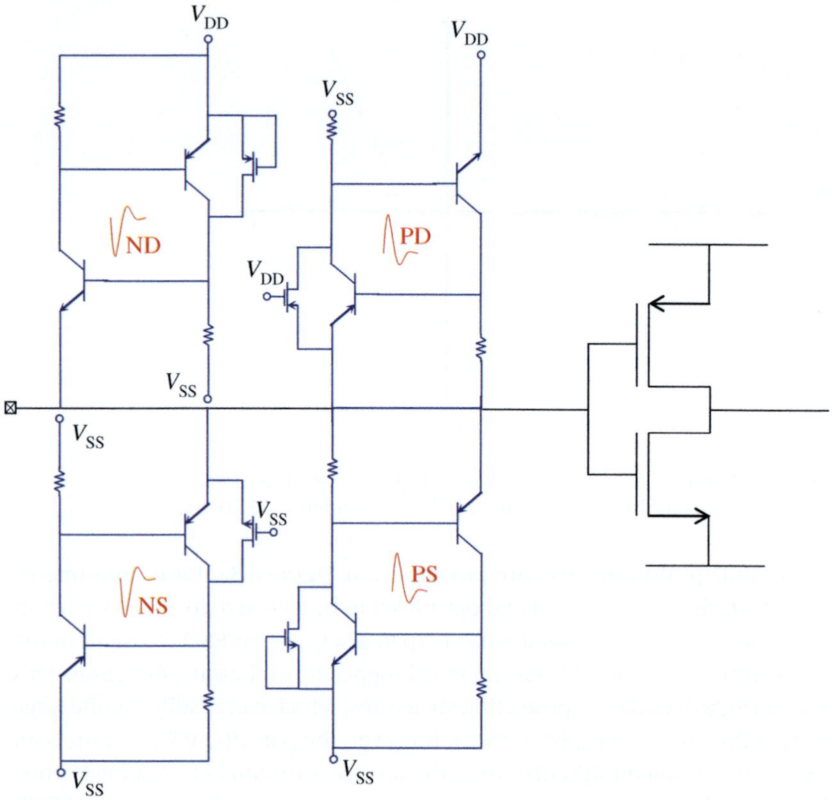

Figure 6.4 A full-chip ESD protection example uses four ggMOS-SCR ESD protection structures per pad to discharge ESD pulses of different stressing modes.

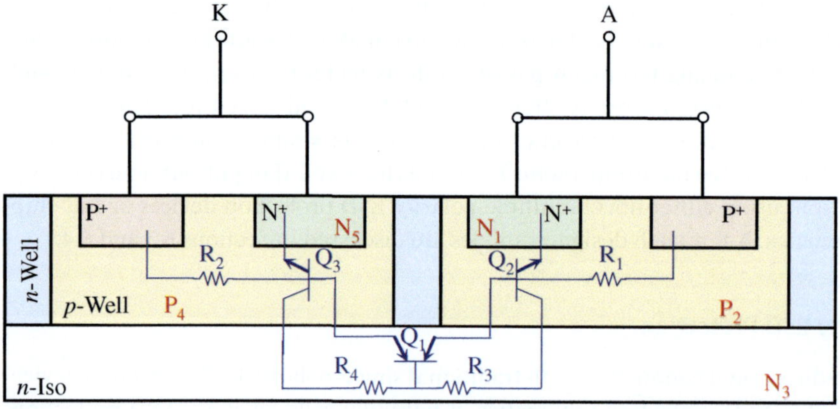

Figure 6.5 A cross-sectional and schematic view for the novel dual-polarity SCR ESD protection structure in BiCMOS technology.

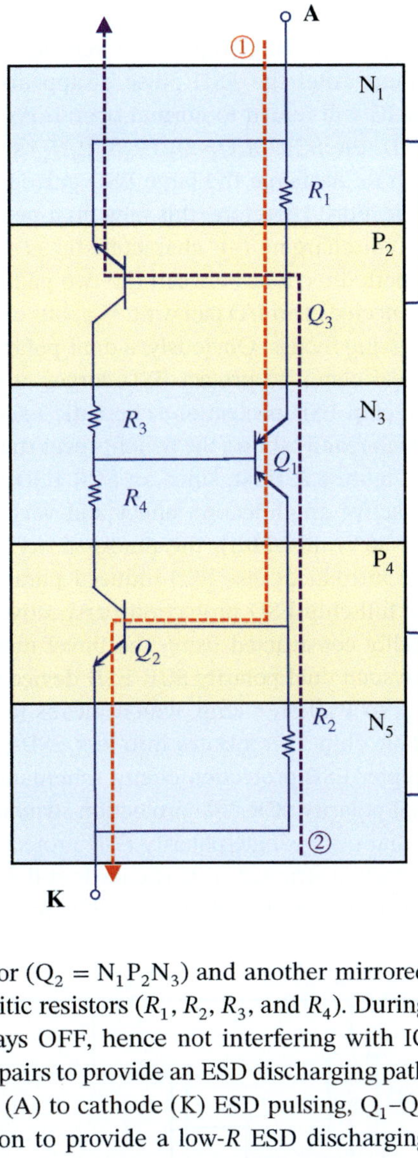

Figure 6.6 Illustration of the intrinsic $N_1P_2N_3P_4N_5$ unit of the novel dual-polarity SCR ESD protection structure in Figure 6.5 and its equivalent circuit shows the two ESD discharging modes, i.e., path ① and path ②.

PNP transistor ($Q_1 = P_2N_3P_4$), one vertical NPN transistor ($Q_2 = N_1P_2N_3$) and another mirrored vertical NPN transistor ($Q_3 = N_3P_4N_5$), and several parasitic resistors (R_1, R_2, R_3, and R_4). During normal IC operations, this ESD protection structure stays OFF, hence not interfering with IC functions. Under ESD stressing, these transistors work in pairs to provide an ESD discharging path between two pads in both directions. For positive anode (A) to cathode (K) ESD pulsing, Q_1–Q_3 forms an SCR ESD protection unit that can be turned on to provide a low-*R* ESD discharging path in one direction. For negative A-to-K ESD zapping, Q_1–Q_2 works as an SCR ESD unit to discharge the ESD surges in the opposite direction. Therefore, this new ESD protection structure can discharge ESD pulses in both directions equally, hence features symmetric ESD discharging *I–V* characteristics as desired. For example, the dual-polarity ESD protection structure can be connected between an I/O (A) pad and the GND pad (K) on a chip. When a PS mode ESD pulse appeals at the I/O w.r.t. GND, the BC junction (N_3P_4) of Q_1 is reverse-biased until avalanche breakdown occurs, which produces a large amount of free electron–hole pairs. Since the K-terminal is grounded through the P_4–P^+ diffusion, excess hole current will flow into GND via K electrode, building up a positive voltage across the parasitic R_2 that eventually forward turns on the BE (P_4N_5) junction of Q_3. Consequently, the SCR ESD device of Q_1–Q_3 ($P_2N_3P_4N_5$) will be triggered off to effectively discharge the large ESD current through a low-*R* without generating much heat. This ensures I/O-to-GND ESD protection in PS stressing mode through one SCR device with active amplification. Due to deep snapback I–V behavior and low SCR holding voltage V_h, the I/O pad will be clamped to a very low level to avoid possible ESD-induced voltage breakdown failure. After the ESD pulse disappears, the SCR device will be automatically turned off and the IC will return to normal operations. In case an NS mode ESD pulse occurs at I/O pad w.r.t. GND, the SCR of Q_1–Q_2 ($P_4N_3P_2N_1$) will work the same way to establish a low-R conduction path to discharge the large ESD current and clamp the pad voltage to a very low level for ESD protection. Therefore, this new dual-polarity SCR ESD protection structure features a symmetric ESD discharging I–V characteristic as shown in Figure 6.3b and provides robust ESD protection in both directions between the two pads. Similarly, a dual-polarity SCR ESD protection structure connected at an I/O pad w.r.t. V_{DD} bus can provide active ESD protection for both PD and ND ESD stressing modes. Obviously, a dual-polarity SCR ESD protection can be used as a dual-directional power clamp to protect ESD surges at supply lines on a chip. Figure 6.7 depicts an exemplar full-chip ESD protection schematic using the new dual-polarity SCR ESD protection structure, which readily shows the benefits over that using single-polarity ESD protection structure as shown in Figure 6.2: First, since an SCR ESD protection device can handle large ESD currents due to its active amplification effect and very low-R (hence, large ESDV-to-Si ratio, readily reaching to ~80 V/μm-width), the dual-polarity SCR ESD protection structure by itself will be smaller and introduces less ESD-induced parasitic effects, e.g., C_{ESD}, $I_{leak,}$ and noises. Second, due to the full-chip ESD protection perspective, a smart whole-chip ESD protection schematic can be readily constructed using the novel dual-polarity SCR ESD protection structures. Hence, only one such dual-polarity SCR ESD device is needed between I/O and GND (also, I/O to V_{DD}, and V_{DD} to V_{SS}) on a chip, which means fewer ESD protection devices needed per pad and for the whole chip. It translates into less ESD-consumed Si area, easier layout and floor planning, and simpler ESD protection circuit schematics, etc. Together, the overall benefits for using the novel dual-polarity SCR ESD protection structures for full-chip ESD protection is substantial compared to that using single-polarity ESD protection devices. Therefore, the novel dual-polarity SCR ESD protection structure is particularly suitable for RF ICs and high-speed mixed-signal ICs. Since the ESD triggering V_{t1} of the SCR ESD protection device is determined by either junction avalanche

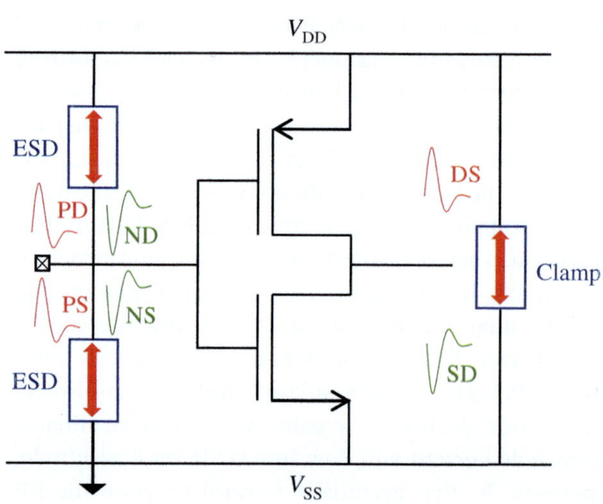

Figure 6.7 A conceptual schematic for full-chip ESD protection scheme using novel dual-polarity SCR ESD protection structures features fewer ESD devices than that when using single-polarity ESD protection structure.

breakdown or punch-through breakdown, the V_{t1} can be handily adjusted for different ICs, for example, by choosing different junction layers or changing the P-well–P-well spacing. Nevertheless, the high-V_{t1} problem of SCR structure remains a design concern, which can be resolved by using various trigger-assisting subnets, to be discussed later. Practically, a wide V_{t1} range of 5–55 V has been reported for such dual-polarity SCR ESD protection structure [8]. On the other hand, one should pay special attention to possible latch-up effect of an SCR ESD protection structure, which is essentially a controlled latch-up device that works in latch-up mode during ESD events, but must be pulled out of latch-up immediately after the ESD stressing in order to not interfering with normal IC operations. Referring to the ESD Design Window shown in Figure 6.1, the holding current I_h of an SCR-based ESD protection structure must be designed higher than the highest on-chip operation current nearby, often being the supply current I_{DD}, in order to remove latch-up after an ESD event is over. In full-chip layout planning, SCR ESD protection structures should not be placed close to any heavy-current device, e.g., a large output buffer transistor, because the large IC current may keep the SCR ESD protection structure in latch-up mode even after the ESD transient disappears. Guard rings are strongly recommended for SCR ESD protection structures for the same reason.

6.3.2 Multiple-Polarity ESD Protection

Comparing the two full-chip ESD protection schemes shown in Figures 6.2 and 6.7, the advantage of using dual-polarity SCR ESD protection structures is obvious due to its active ESD discharge features in both directions: it dramatically reduces the total head count of ESD protection devices required per pad and for the whole chip, while realizing whole-chip ESD protection. From further consideration of all possible ESD stressing modes between pads on a chip, i.e., PS, NS, PD, ND, DS and SD, it appears possible to devise multiple-directional ESD protection structure to further simplify full-chip ESD protection schematic. Figure 6.8 depicts one such a three-terminal multiple-polarity SCR ESD protection structure implemented in BiCMOS technology. Further reduction in protection device count will certainly be beneficial to whole-chip design [9]. The novel multiple-polarity SCR ESD protection can be used for full-chip ESD protection as illustrated in Figure 6.9, where the center electrode (A) is connected at an I/O pad, while the other two terminals (K1 and K2) are connected to supply lines, V_{DD} and GND (or V_{SS}), respectively. By design, this SCR-based ESD protection structure will establish an active low-R ESD discharge path in each direction between any two terminals (A, K_1, and K_2), all featuring deep snapback I–V characteristics like an SCR device. The multiple directional ESD discharge mechanism are

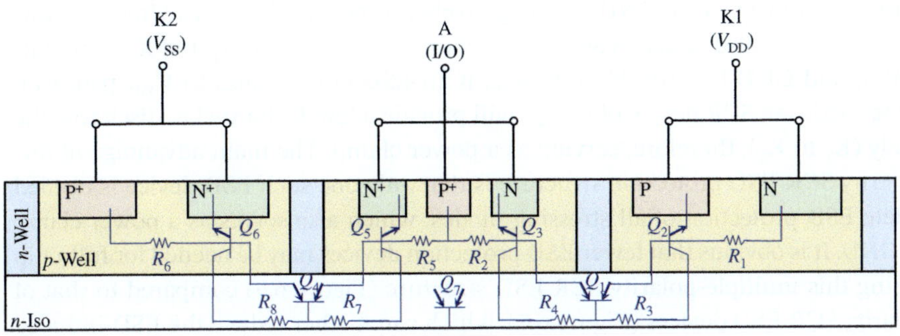

Figure 6.8 A cross-section view of the novel three-terminal multiple-polarity SCR ESD protection structure in BiCMOS.

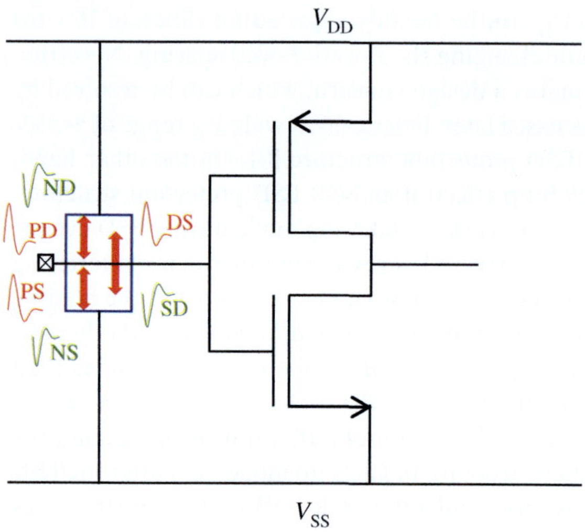

Figure 6.9 A conceptual full-chip ESD protection scheme using the multiple-polarity SCR ESD protection structure results in fewer ESD protection devices for a chip.

discussed below, which can be viewed as a pair of mirrored dual-polarity SCR ESD protection structures described previously with shared device elements. This novel three-terminal SCR structure contains three P-type wells, each has P^+ and N^+ diffusion contacts. The SCR ESD structure comprises six bipolar transistors (Q_1, Q_2, Q_3, Q_4, Q_5, and Q_6) and parasitic resistors (R_1, R_2, R_3, R_4, R_5, R_6, R_7, and R_8), which form two dual-polarity SCR ESD protection units, i.e., Q_1–Q_2–Q_3 and Q_4–Q_5–Q_6, respectively. Figure 6.10 shows the equivalent circuit for the multiple-polarity SCR ESD protection structure. With properly designed ESD triggering V_{t1} for each SCR device inside, this multiple-polarity SCR ESD protection structure will remain OFF during normal IC operations. During ESD events, from Figure 6.9, assume a ND ESD pulse comes to I/O pad (A) w.r.t. to V_{DD} (K_1), the BC junction of Q_1 is reverse-biased to its avalanche breakdown and the substrate current will turn on Q_3, hence, triggers the SCR of Q_1–Q_3 and drives it into deep snapback conduction mode with low holding V_h of ~2 V. It therefore provides an active conduction path to shunt the ND mode ESD surge from A to K_1 negatively (or, positive K_1 to A). In case of PD ESD stressing mode, the positive ESD pulse occurring at I/O w.r.t. V_{DD} will cause BC junction breakdown of Q_1, which then turns on Q_2, therefore, the SCR unit of Q_1–Q_2, and forms an active ESD discharge path from A to K_1 positively to provide ESD protection. Similarly, for PS ESD stressing, the SCR unit of Q_4–Q_6 will be triggered to establish an active ESD discharge path from I/O to GND positively (A–K_2); while for NS ESD stressing mode, the SCR unit of Q_4–Q_5 will be turned on to provide an active ESD shunting channel from I/O (A) to GND (K_2) negatively (or, positive K_2 to A). Uniquely, this SCR ESD protection structure also contains SCR conduction device between K_1 and K_2 terminals from supply bus V_{DD} and GND line (or, V_{SS}). Hence, if an ESD pulse comes to V_{DD} pad w.r.t. GND in DS stressing mode, an SCR device of Q_6–Q_7 will provide a low-R channel to discharge the ESD pulse positively (K_1 to K_2), therefore, serving as a power clamp. The main advantage of this novel multiple-polarity SCR ESD protection structure is that only one such ESD device is needed per pad for complete ESD protection of all stressing modes, which also serves as a power clamp between V_{DD} and GND. It is obvious that fewer ESD protection devices may be needed for full-chip ESD protection using this multiple-polarity SCR ESD structure (Figure 6.9) compared to that of using the dual-polarity SCR ESD devices (Figure 6.7), which can further reduce the ESD-induced design overhead effects discussed before. Another advantage for this multiple-polarity SCR ESD protection structure is that it always provides a deep-snapback ESD unit to discharge ESD pulse of

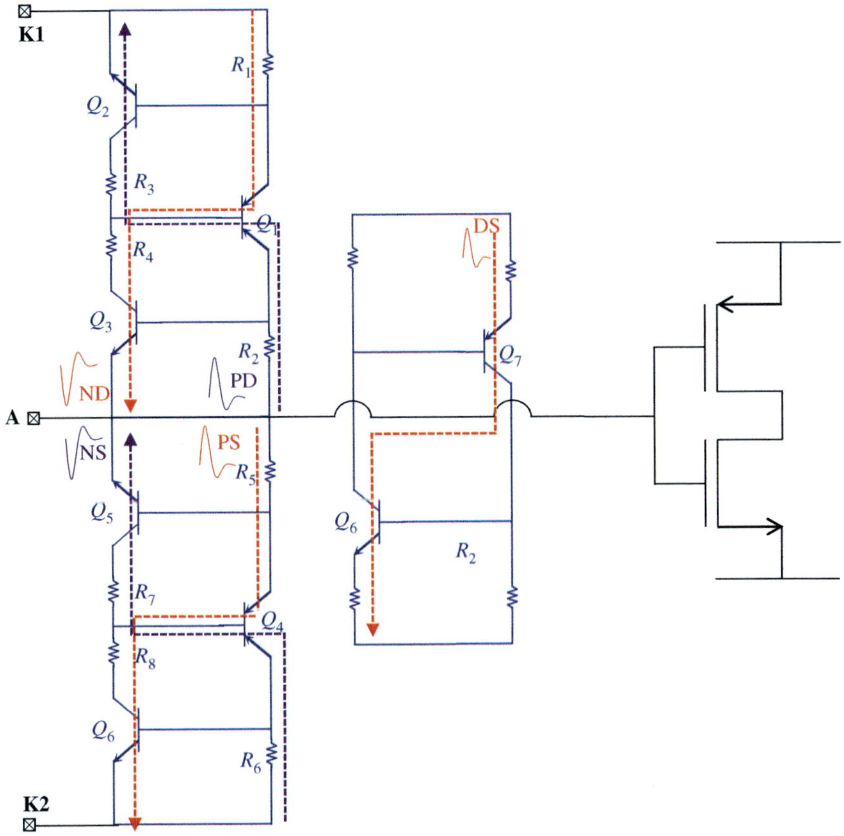

Figure 6.10 Equivalent circuit for the novel multiple-polarity ESD protection structure shows ESD discharge paths for all ESD stressing modes.

any stressing mode, which is very ESD-robust, meaning less ESD-induced parasitic capacitance, very beneficial for high-speed, high-frequency IC cores [10]. Moreover, less ESD-induced parasitic capacitance helps to speed up the ESD response, hence, making it more friendly to ultrafast ESD protection, e.g., IEC and CDM ESD events [11, 12].

The high-V_{t1} problem associated with classic SCR ESD protection structure can be resolved by using certain trigger-assisting techniques. ESD V_{t1} of the multiple-polarity SCR ESD protection structure can be reduced by various means, for example using different diffusion layers as the avalanche breakdown junction, or changing the well-to-well spacing for relatively lower punch-through breakdown. Alternatively, an external trigger-assisting subcircuit cell may be used to reduce ESD V_{t1} of the multiple-polarity SCR ESD protection structure as depicted in Figure 6.11 [13]. The ESD triggering mechanism of the new low-V_{t1} multiple-polarity SCR ESD protection circuit shown in Figure 6.11 follows: The P-well/N-well junction breakdown mechanism is replaced by an external trigger-assisting subcircuit unit consisting of a switch and a current source as shown in Figure 6.11. For example, in the PS ESD stressing mode, a switch ($S3$)-current source ($I3$)-resistor ($R33$) cell is connected to the Q_4–Q_5–Q_6 sub-net. The $S3$ is normally off and can be turned on by the incident ESD pulse. The $I3$ current then flows through $R33$ that will quickly build up the V_{BE} of Q_6 and eventually turns on this NPN transistor; therefore, triggers the SCR of Q_4–Q_6 to discharge the ESD pulse. Many techniques can be used to form the $S3$–$I3$

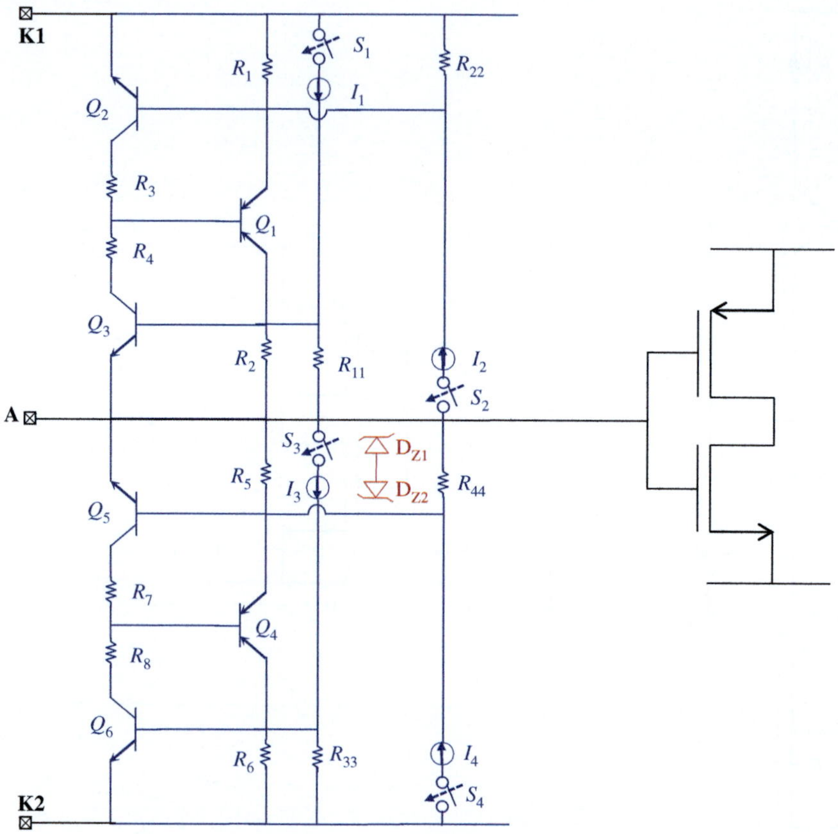

Figure 6.11 An equivalent circuit schematic for an exemplar low-V_{t1} multiple-polarity SCR ESD protection circuit uses a sub-net of switch and current source as the ESD trigger-assisting sub-circuit.

unit. In a simple example, a pair of back-to-back connected Zener diodes (Dz1 and Dz2) shown in Figure 6.11 is used to construct the *S*3–*I*3 network. The Dz1 and Dz2 serve as the switch in the opposite biasing directions, respectively, to ensure the ESD protection circuit stays OFF in normal IC operations. The turn-on voltage of the Dz1 is carefully designed to control the triggering V_{t1} of the SCR ESD protection structure. Similarly, Zener diode pairs form other S-I units for the whole multiple-polarity SCR ESD protection circuit. Since many Zener diodes are usually available in advanced BiCMOS technologies, one can readily adjust the ESD V_{t1} values for the low-V_{t1} multiple-polarity SCR ESD protection circuits for different IC core blocks featuring different local V_{DD} on a chip.

6.4 Full-Chip ESD Protection Schemes

6.4.1 Full-Chip ESD Consideration

At this moment, it is crystal-clear that on-chip ESD protection design is not about only designing individual ESD protection devices or dropping-in an ESD protection device designed by an ESD guru in your company; rather, it is a circuit-level design task requiring full-chip design considerations. The full-chip ESD design perspective not only will ensure passing your ESD protection target in terms of the ESDV, but also allow you to consider the inevitable and complex ESD–IC interactions. The five principles for full-chip ESD protection designs discussed previously lays the foundation for implementing whole-chip ESD protection. The underlying mechanism is that there must always exist a low-*R* ESD discharge path between any two pads on a chip for any ESD stressing modes, while the equivalent ESD-critical parameters (e.g., V_{t1}, V_h, R_{ON}, V_{t2}, and I_{t2}) for each pad-to-pad ESD discharge path must fit into the ESD Design Window shown in Figure 6.1. To this point, utilizing novel ESD protection devices, such as the multiple-polarity SCR ESD protection structure, can significantly improve both ESD protection and IC performance for a chip. A few such full-chip ESD protection examples are discussed below.

6.4.2 Pad-Clamp Scheme

Using traditional single-polarity ESD protection structures and treating each pad as a standalone entity, multiple ESD protection devices are needed per pad, in addition to power clamps, to realize a full-chip all-active ESD discharge network to ensure an ESD discharge path between any two pads on a chip, as depicted in Figure 6.2. However, considering on-chip ESD protection as a circuit design task and following the full-chip ESD protection design principles, it is possible to simplify the ESD protection schematics at whole chip level while ensuring adequate full-chip ESD protection, even if using the traditional single-directional ESD protection structures. Figure 6.12 illustrates an exemplar pad-clamp full-chip ESD protection scheme using conventional single-polarity ESD protection structures for whole-chip ESD protection. For illustration purpose, multiple pads and power supplies are included for a chip. The active ESD discharge path for each ESD protection structure follows the arrow direction for low-*R* conduction. ESD protection is also included between V_{DD} and GND (V_{SS}), as well as any supply buses of different voltages, such as V_{DD1}, V_{DD2}, V_{SS1}, and V_{SS1}. Following the full-chip ESD protection principles discussed, a designer has to ensure that there is always an active ESD discharge path existing between any two pads on a chip. A quick checking follows: In case of PD mode ESD stressing to pulse Pad 1 w.r.t. to V_{DD1} bus, the ESD device, ESD_1, connected between Pad 1 and pad V_{DD1}, will offer an active low-*R* path to discharge

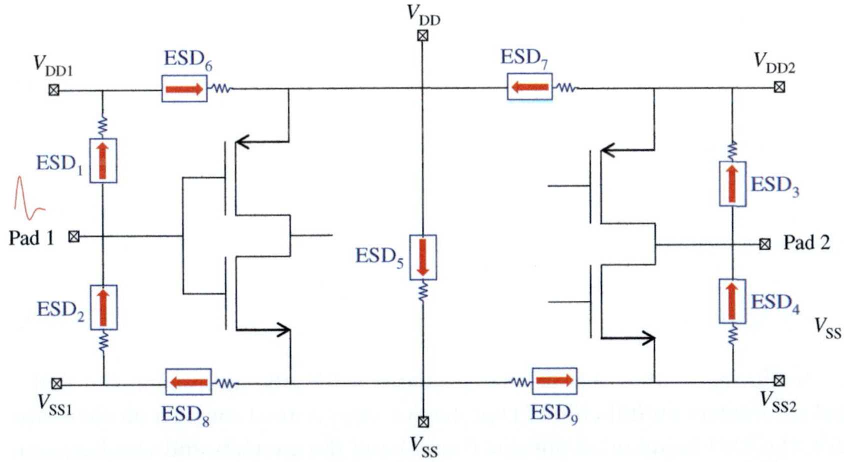

Figure 6.12 Illustration of a pad-clamp whole-chip ESD protection scheme using traditional single-directional active ESD protection structure, denoted by a single-arrowed ESD box. It must ensure an active low-R ESD discharge path tween any two pads on a chip. Any classic ESD protection device can be used, including diodes, BJTs, FETs and SCRs.

the PD ESD pulse. In case of ND mode ESD stressing to Pad 1 w.r.t. V_{DD1}, active ESD discharge will follow a long conduction path of ESD_6-ESD_5-ESD_8-ESD_2, rather than relying on a reverse parasitic conduction device of ESD_1, which is normally ESD-weak. Of course, careful design is needed to ensure the equivalent $R_{\text{ESD-path6582}}$ is much lower than $R_{\text{ESD1-reverse}}$ for the single ESD_1, and the equivalent $V_{t1\text{-path6582}}$ and $V_{h1\text{-path6582}}$ are within the ESD Design Window. Similarly, active low-R ESD discharge path exists for PS and NS ESD stressing cases from Pad 1 to V_{SS1}. The same analysis applies to checking active ESD discharge path between any two pads under any ESD stressing mode, including between different supply buses. Sharply different from device-level ESD protection design approaches, the worst-case analysis is required for full-chip ESD protection schematic design. Three key aspects must be considered in full-chip worst-case ESD analysis. First, one has to identify the worst-case physical ESD discharge path on a chip, which is often causing ESD failure. From Figure 6.12, the worst-case ESD discharge path is the longest one between Pad 1 and Pad 2, which contains the largest number of ESD devices in the path and also features the longest physical conduction distance on the chip, both are undesired for full-chip ESD protection. In case of positive ESD stressing from Pad 1 to Pad 2, the all-active ESD discharge path will be ESD_1-ESD_6-ESD_5-ESD_9-ESD_4. In case of negative ESD stressing from Pad 1 to Pad 2, the all-active ESD charge path will be ESD_3-ESD_7-ESD_5-ESD_8-ESD_2. Second, all contributors must be included in estimating the equivalent ESD-critical parameters for a given ESD discharge path between the two pads. Third, all interconnects resistance (R_{busi}) in the ESD discharge path cannot be ignored due to the large ESD currents, which is often a weak ESD link on a chip. Therefore, the equivalent ESD-critical parameters for the Pad 1 to Pad 2 positive ESD discharging mode can be estimated as following: The total path ESD triggering voltage is

$$V_{t1-\text{path}} = \sum_{i}^{n} (V_{t1i} + I_{\text{ESD}} R_{\text{busi}}) \tag{6.1}$$

the total path ESD discharge resistance is

$$R_{\text{ON}-\text{path}} = \sum_{i}^{n} (R_{\text{ON}i} + R_{\text{busi}}) \tag{6.2}$$

the total path ESD holding voltage is

$$V_{h-\text{path}} \approx \sum_{i}^{n} (V_{hi}) \tag{6.3}$$

the total path ESD failure current is

$$I_{t2-\text{path}} = \text{Min}(I_{t2i}) \tag{6.4}$$

and the total path ESD failure voltage is

$$V_{t2-\text{path}} = \sum_{i}^{n} (V_{t2i} + I_{\text{ESD}} R_{\text{busi}}) \tag{6.5}$$

where i denotes the i-th device in the ESD discharge channel. Clearly, optimizing the path-equivalent ESD-critical parameters for full-chip ESD protection design must consider all elements on a chip. For example, the ESD metal interconnects (buses) and the contacts and vias between different metals layers (together modeled by a total series resistance R_{busi}), which have been typically overlooked, must be included in full-chip ESD protection designs. Third, on-chip ESD failure often occurs at an ESD-weak point, e.g., being a narrow metal line, a lonely via, or a low-I_{t2} ESD protection device on a chip. Further, when considering ESD metal interconnects and contacts/vias in ESD protection designs, it is important that they are accurately characterized for ESD discharging functions by transient TLP and VFTLP stressing tests, not based on their DC and AC aging tests only. One major advantage of using such pad-clamp type full-chip ESD protection schemes is that it can reduce the total head counts of individual ESD protection structures needed per chip, as compared to that shown in Figure 6.2, which translates into reduced ESD design overhead including ESD-induced parasitic effects and die area. Apparently, while full-chip ESD protection schematics can be simplified from a whole chip perspective, comprehensive circuit-level ESD discharging analysis is critical and, unfortunately, rather involving for a chip, which is practically impossible for large and complex ICs. Therefore, new CAD-based methods and new ESD CAD tools are needed for full-chip ESD protection design verification, not just for simply checking a missing device and layout spacings, but also more importantly, to conduct whole-chip ESD function-based smart ESD design verification. The new ESD CAD tools and methods will be discussed in Chapter 15.

6.4.3 Global ESD Bus Scheme

Figure 6.13 depicts another full-chip ESD protection scheme, a global ESD bus scheme, which utilizes chip-wise ESD discharge bus lines in combination with the novel dual-polarity SCR-based ESD protection structures discussed previously. Either one global ESD discharge bus for the whole chip or several domain-wise regional/subglobal ESD discharge buses on a chip can be used. Nicely, only one dual-polarity ESD protection structure is needed per pad connected to the global ESD bus. As depicted in Figure 6.13, a quick check for chip-level ESD discharging paths, following the full-chip ESD protection principles, confirms that there always exists one active ESD discharge channel between any two pads on the chip to ensure efficient ESD discharging for any ESD stressing modes (PS, NS, PD, ND, DS, and SD), hence, forming an active full-chip ESD protection network. Compared with the pad-clamp ESD protection scheme using single-directional ESD protection devices, there are several key advantages for this global ESD bus protection scheme: First, even fewer ESD protection structures are needed per chip for full-chip ESD protection, which ensures even less Si die area consumed by ESD protection devices and much lower ESD-induced

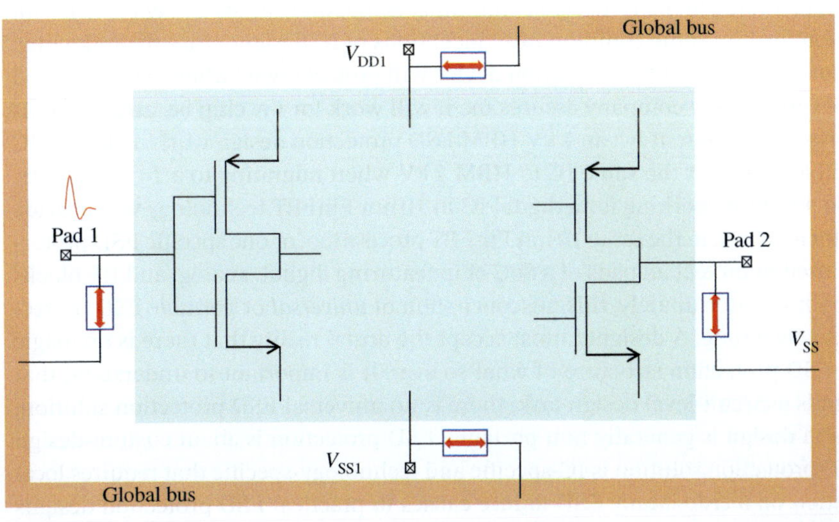

Figure 6.13 Illustration for a whole-chip ESD protection scheme using a global ESD discharging bus and dual-direction ESD protection devices, denoted by the double-arrowed ESD device boxes.

parasitic effects, including C_{ESD}, I_{leak}, self-generated noises, and noise coupling, all of them are real headaches to advanced ICs. Second, the full-chip ESD protection schematics become smarter and simpler, which means less chance of making ESD layout design errors, which is one common ESD design problem for large and complex chips. Third, due to fewer and smaller dual-polarity ESD protection structures, full-chip layout floor planning will become easier. Fourth, since only one dual-polarity ESD protection structure exists in the ESD discharge path between any two pads on a chip, the path-equivalent ESD performance, i.e., the total equivalent path ESD-critical parameters (symmetric in both ESD stressing directions) can be readily optimized for both ESD design procedures (e.g., evaluating all players in one ESD conduction path and estimating the total path-equivalent ESD-critical parameters) and ESD robustness (i.e., $I_{t2\text{-path}}$). It hence compares favorably to the pad-clamp-based ESD protection schemes featuring asymmetric, long, and many-device ESD discharge paths. Certainly, special cares may be needed when using the global bus ESD protection scheme. For example, since the global ESD bus must handle large ESD transient currents, the global ESD metal bus should be wide enough. It is recommended that a highly doped diffusion bus underneath the global metal bus can be parallelly connected together, which will not only enhance the ESD conduction capability but also helps to dissipate the ESD generated heat downward through the Si substrate that may substantially alleviate the ESD metal overheating problem. In principle, single-directional ESD protection devices may be used in this global ESD bus scheme, for example, a diode or MOSFET ESD protection structure, which feature a reverse PN junction diode for ESD discharge in the opposite direction. However, the common problem is that the parasitic PN diodes are not optimized for heavy ESD discharging, hence often results in early ESD failures. Indeed, many novel ESD protection concepts and devices may be used to both simplify chip-level ESD protection schematics and enhance full-chip ESD protection when IC designers apply full-chip ESD protection perspectives to whole-chip ESD protection designs. Every brain cell burned in careful ESD protection designs will be very rewarding.

6.5 No Universal ESD Protection Solution

One common misunderstanding in the IC design community is that there exists a magic *one-for-all* ESD protection solution for ICs. Some common misconceptions of IC designers are the following: a test-verified individual standalone ESD protection device will protect my IC when being dropped onto the chip; the ESD guru in my company assures me it will work for my chip because the ESD protection device works for a different IC; an 2 kV HBM ESD protection design working for my IC in a 22 nm CMOS will also protect the same IC to HBM 2 kV when migrating to a 14 nm CMOS; a 2 kV ESD protection structure working for a digital IC in 10 nm FinFET technology will provide 2 kV ESD protection for an RF IC in the same 10 nm FinFET process too; or one specific ESD protection structure can be used to protect all pads of a SoC chip featuring digital, analog, and RF blocks of varying power supplies. Unfortunately, this misconception of *universal* or *portable* ESD protection solution is completely wrong. A designer must accept the brutal reality that there is no magic one-for-all universal ESD protection structure of what so ever. It is important to understand that ESD protection design is a circuit-level design task, there is no universal ESD protection solution, and any ESD protection design is generally non-portable. ESD protection is about custom-design in nature. A good ESD protection solution is IC-specific and technology-specific that requires local ESD design optimization on a chip. Many ESD failure causes in practical ESD protection designs may be associated with these common ESD design misconceptions. For example because of the complex ESD-IC interactions and the dynamic ESD Design Window requirement, a test-proven individual ESD protection structure may not protect an IC chip when being simply dropped onto the chip, e.g., due to a mis-triggered parasitic bipolar device of guard rings in the core circuit. A working ESD protection structure designed for a 180 nm CMOS will fail in the same ICs migrating into a 28 nm CMOS because the lower gate breakdown voltage will dramatically narrow the ESD Design Window. A functional ESD protection structure working in a digital core may fail in an RF IC implemented in the same technology because the high-frequency RF signals may accidentally trigger the ESD protection device during normal IC operations. A nicely designed robust SCR ESD protection structure of $V_h \sim 1.5\,V$ for a low-voltage analog IC will fail in an HV IC of $V_{DD} = 45\,V$ because of the latch-up problem. There are so many examples attesting that a good ESD protection solution is not universal and cannot be portable. IC designers must always understand the specs of your chips, the functions of an ESD protection structures, and the process technology features in order to design and optimize a suitable ESD protection structure for a given IC with given specs in a given IC technology. There is no shortcut in practical ESD protection designs other than following the full-chip ESD protection perspectives and optimizing ESD protection structures quantitively at chip level.

References

1 Wang, A. (2002). *On-Chip ESD Protection for Integrated Circuits: An IC Design Perspective*. Springer. ISBN: 9780792376477.

2 Wang, A., Tsay, C., Lele, A., and Deane, P. (1998). A study of NMOS behaviors under ESD stress: simulation and characterization. *Microelectron. Reliab.* 38: 1183–1186.

3 Zhan, R., Feng, H., Wu, Q. et al. (2004). ESDInspector: A new layout-level ESD protection circuitry design verification tool using a smart-parametric checking mechanism. *IEEE Trans. Comput. Aided Des. Integr. Circuits Syst.* 23 (10): 1421–1428.

4 Wang, A. and Tsay, C. (1999). A low-triggering circuitry for dual-direction ESD protection. *Proceedings of IEEE CICC*, pp. 139–142.

5 Lin, L., Wang, X., Tang, H. et al. (2009). Whole-chip ESD protection design verification by CAD. *Proceedings of EOS/ESD Symposium,* pp. 28–37.

6 Ker, M. and Wu, M. (1996). CMOS on-chip four-LVTSCR ESD protection scheme. US Patent, No. 5, 572, 394.

7 Wang, A.Z. and Tsay, C. (2001). On a dual-polarity electrostatic discharge protection structure. *IEEE Trans. Electron Devices* 48 (5): 978–984.

8 Wang, A. and Tsay, C. (2001). An on-chip ESD protection circuit with low trigger-voltage in BiCMOS technology. *IEEE J. Solid-State Circuits* 36 (1): 40–45. https://doi.org/10.1109/4.896227.

9 Feng, H.G., Gong, K., and Wang, A.Z. (2001). A novel on-chip electrostatic discharge protection design for RFIC's. *J. Microelectron.* 32 (3): 189–195.

10 Gong, K., Feng, H.G., Zhan, R.Y., and Wang, A.Z. (2001). ESD-induced circuit performance degradation in RFICs. *Microelectron. Reliab.* 41 (9–10): 1379–1383.

11 IEC 61000-4-2 (2008). *Electromagnetic Compatibility, Part 4: Testing and Measurement Techniques, Section 2: Electrostatic Discharge Immunity Test*. the International Electrotechnical Commission (IEC).

12 ANSI/ESDA/JEDEC JS-002-2018 (2018). *For Electrostatic Discharge Sensitivity Testing – Charged Device Model (CDM) – Device Level*. An American National Standard jointly developed by ESD Association and JEDEC.

13 Feng, H.G., Gong, K., and Wang, A. (2001). An ESD protection circuit for mixed-signal ICs. *Proceedings of IEEE CICC*, pp. 493–496.

7

Mixed-Signal and HV ESD Protection

7.1 ESD Protection for Mixed-Signal ICs

Electrostatic discharge (ESD) protection design for homogeneous integrated circuits (ICs) of same or similar type is well understood and the design philosophy is rather straightforward. The key ESD design principles are to create a low-R conduction path to discharge incoming ESD pulses safely without overheating and to clamp the pad voltage to a sufficiently low level to avoid voltage breakdown, which apply to all pad on a chip. On the other hand, ESD protection design for mixed-signal ICs of dissimilar functionalities is often a complicated and challenging design task that requires comprehensive full-chip scale design considerations [1–4]. One can never expect any universal one-for-all-pad ESD protection structure for mixed-signal ICs. Let us revisit the basic concept of ESD design window as depicted in Figure 7.1: a transient ESD discharge I–V curve must fit into the ESD design window for a specific IC designed in a given technology by carefully and quantitatively designing the ESD-critical parameters of an ESD protection structure, including V_{t1}, I_{t1}, V_h, I_h, R_{ON}, V_{t2}, I_{t2} in reference with V_{safe}, V_{DDmax}, I_{supply}, and I_{Fail}. Of course, the ESD response time (t_1) of a specific ESD protection structure must also be carefully designed to meet the different triggering time requirements of different ESD test models, i.e., human body model (HBM), MM, and charged device model (CDM) models. To this end, ESD protection design for heterogeneous mixed-signal ICs is entirely different from that for homogeneous ICs. There are two unique features for mixed-signal ICs. First, a mixed-signal chip often comprises many different circuit domains, such as digital, analog, and RF functional blocks (i.e., functional heterogeneity). Second, a mixed-signal system-on-a-chip (SoC) chip is often powered by different supply voltages in different circuit blocks for whole-chip IC design optimization (i.e., power heterogeneity). These two key differences in mixed-signal ICs make ESD protection designs quite different and challenging. Keep in mind that a good ESD protection design is always IC-specific (or, I/O-specific), *Local ESD Design Optimization* is hence a critical ESD design concept for mixed-signal ICs. This section focuses on mixed-signal ESD protection design considerations associated with different circuit function blocks on a chip. More ESD design considerations related to multiple power supply domains on a chip will be discussed in Section 7.2. The ever-increasing demand for better performance and higher integration comes with larger chip size and complexity of mixed-signal SoC ICs, which directly translates into challenges in ESD protection designs. For example, Figure 7.2 illustrates a typical chip architecture for a 5G smartphone SoC. With more than 15 billions of transistors on a chip fabricated in a 5 nm FinFET IC technology, today's 5G smartphone chipset typically contains many different digital, analog, and RF functional domains, including multiple-core CPUs, GPUs, and NPUs, low-power double data rate (LPDDR) memories and flash memories, various RF circuitry (Wi-Fi, Bluetooth, and mobile links), image signal processor (ISP), artificial intelligence (AI), Hi-Fi

Practical ESD Protection Design, First Edition. Albert Wang.

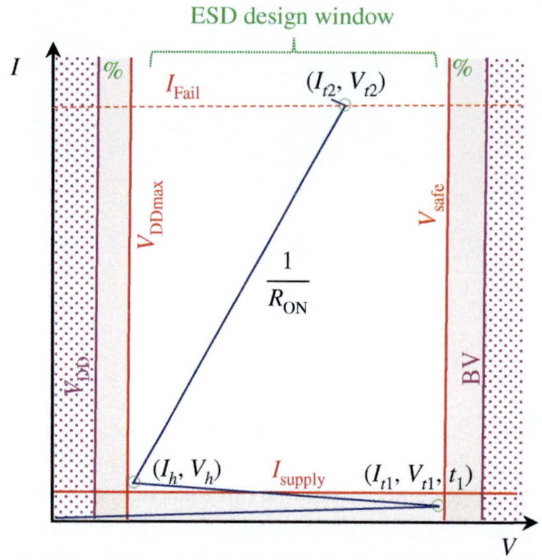

Figure 7.1 Meeting the dynamic ESD Design Window requirement becomes challenging for mixed-signal ICs containing many different functional circuit blocks and power domains, each featuring a different local ESD design window.

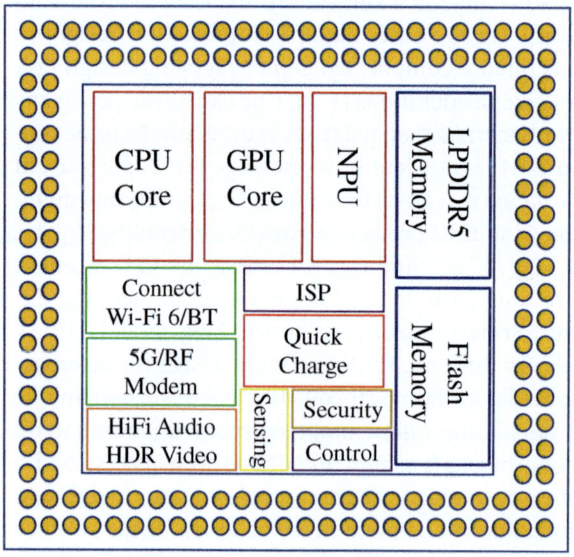

Figure 7.2 Exemplar mixed-signal SoC – a 5G smartphone chipset contains digital, analog, and RF functional domains, each having different ESD protection requirements.

audio and high-dynamic-range (HDR) video, smart charging, and various control, sensing, and security blocks. Typically, SoC optimization requires use of different devices, circuit schematics, and supply voltages for different functional circuit blocks on a chip, which requires careful local ESD design optimization considerations. Because the IC devices and supply voltages are different, the ESD design window for each circuit block can be different across a SoC chip. Consequently, different ESD protection structures may be used for different I/O interfaces and circuit blocks on a mixed-signal IC chip. For instance, in a conceptual mixed-signal SoC shown in Figure 7.3 that contains two digital domains (Domain-I and Domain-II), one analog domain (Domain-III) and one RF domain (Domain-IV). The exemplar ESD protection schemes for the different circuit domains are conceptually depicted in Figure 7.3, which requires special ESD design considerations for each circuit block. In Digital Domain-I, since the input (In-1) goes to a complementary metal-oxide-semiconductor (CMOS) invertor, the key ESD protection consideration is to protect

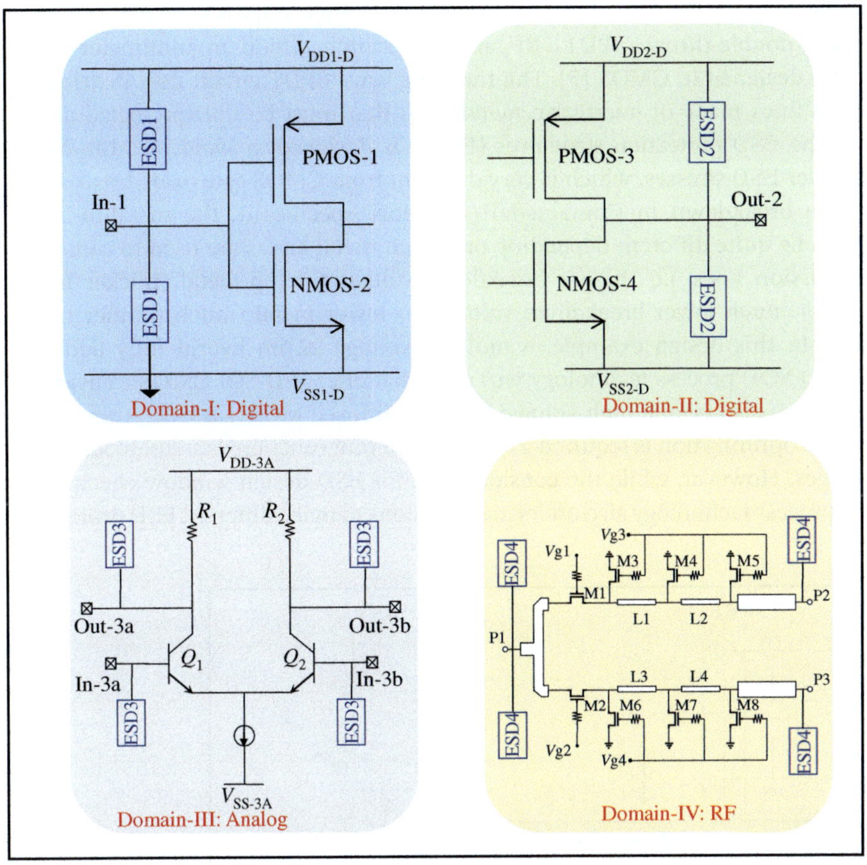

Figure 7.3 A exemplar mixed-signal multiple-domain SoC architecture includes digital, analog, and RF functional circuit domains. The digital, analog, and RF circuit blocks uses different devices for different interfaces. Hence, full-chip ESD protection requires local ESD design optimization for each functional circuit block to protect different device structures under ESD stressing.

the CMOS gate oxide that typically has a low breakdown voltage of as low as BV ~ 0.95 V for a core logic FinFET in 5 nm CMOS technology. Hence, the ESD triggering voltage for the ESD protection device at the input pad (ESD1) should be designed as V_{t1} ~ 0.75 V considering a 20% margin. In addition, CDM ESD protection should be a main design goal for Digital Domain-I, which requires very short ESD triggering time of t_1 for ultrafast CDM ESD response down to 100 ps. In the Digital Domain-II, ESD protection for the output pad (Out-2) is to protect the Drains of PMOS and NMOS transistors in the CMOS output buffer, for which the V_{t1} for the ESD protection structure (ESD2) must be designed to be lower than the source/drain diffusion breakdown voltage that is very different the CMOS gate breakdown voltage as in Digital Domain-I. Further, the key ESD design concern for ESD2 is to prevent ESD-induced thermal failure in the NMOS and PMOS buffer transistors in the diffusion regions. In Analog Domain-III, the interface is with dif ferential inputs (In-3a and In-3b) and outputs (Out-3a and Out-3b) of a bipolar junction transistor (BJT)-based differential amplifier circuit block implemented, where the ESD protection structure (ESD3) is to protect the diffusion regions of bases and collectors of the BJT transistors. Hence, the V_{t1} of ESD3 structures should be set to be lower than the BE and BC junction breakdown voltages of the BJT transistors and the main ESD failure concern is with possible thermal overheating induced by large ESD discharge currents. The RF Domain-IV contains a three-stage distributed traveling wave-based single-pole double-throw (SPDT) RF antenna switch circuit in millimeter wave frequency (28–38 GHz) designed in CMOS [5]. The traveling wave SPDT circuit uses distributed inductive transmission lines made of microstrip metal lines that must be ESD-protected at the I/O pads. The V_{t1} of the ESD protection structures (ESD4) is designed to avoid metal-to-metal voltage breakdown under ESD stresses, which is very different from CMOS gate oxide breakdown and BiCMOS diffusion breakdown in Domains-I/II/III. More specifically, the metal-to-metal breakdown voltage can be quite different depending on which metal layers are used to construct the microstrip transmission lines, i.e., higher breakdown voltage for top metal (thicker metal and dielectric layers) or much lower breakdown voltage for lower metal (much thinner metal and dielectric layers). In this design example, a multiple-voltage 22 nm hybrid fully depleted silicon-on-insulator (FD-SOI) process technology was used that offers FD-SOI CMOS transistors, vertical and lateral BJT transistors, and high-voltage laterally-diffused MOS (LDMOS) transistors. Hence, local ESD design optimization is required to protect different functional circuit blocks with different supply voltages. However, while the consideration for ESD design window checking is involving, the hybrid process technology also offers many options to make different ESD protection

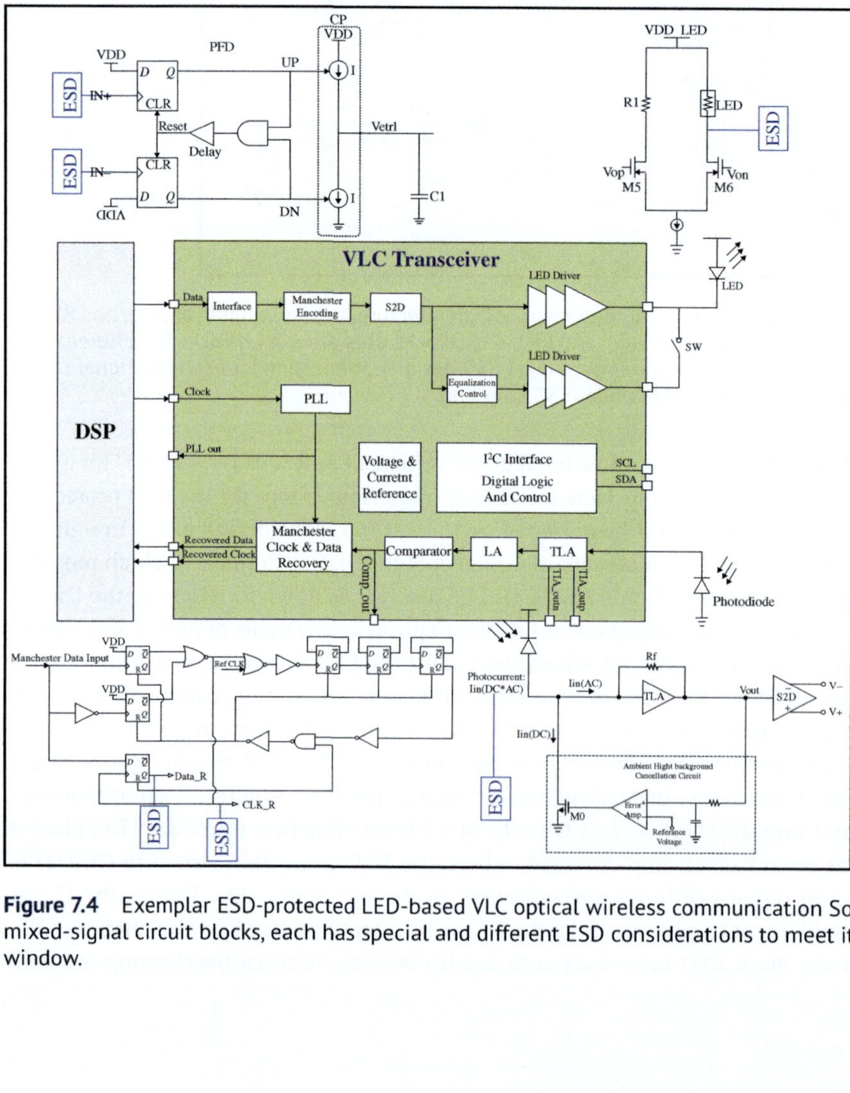

Figure 7.4 Exemplar ESD-protected LED-based VLC optical wireless communication SoC contains many mixed-signal circuit blocks, each has special and different ESD considerations to meet its local ESD design window.

structures, which can be fine-tuned to meet special ESD design specs for local I/O interfaces. Figure 7.4 shows another mixed-signal SoC, which is an LED-based visible light communication (VLC) transceiver chip designed in an 180 nm bipolar-CMOS–DMOS (BCDMOS) technology for high-throughput optical wireless communications [6]. The VLC SoC contains a VLC transmitter to transmit signals via LED light and a VLC receiver to receive incoming signal from an LED photodetector (PD). ESD protection design is carefully optimized for different interfaces (I/O, controls, and supplies) of the VLC SoC IC. Specifically, the multiple-stage LED driver is a CMOS differential amplifier circuit with one output end to drive the LED illuminator device. The ESD protection for the LED driver circuit is mainly to protect the metal-oxide-semiconductor field-effect transistors (MOSFET) Drain diffusion with its V_{t1} designed to avoid the drain junction breakdown. Since the output buffer MOSFET is designed large to provide large current to drive a commercial LED illuminator (off chip), the MOSFET has good ESD self-protection capability. The receiver uses a trans-impedance amplifier (TIA) to amplify the signals from the LED PD (off chip) and an active feedback branch to cancel the background light noises. Therefore, the ESD protection is designed to protect both the input gate of TIA transistor and the drain diffusion of the current sink transistor (M0) in the active feedback loop, and the ESD V_{t1} has to be designed accordingly. The Manchester decoder (digital block) delivers the recovered data and clock signals to the DSP block (separate chip), which requires ESD protection at its output pads that are connected to the output buffer (MOSFET Drain) and input (MOSFET gate) of a D-type flip-flop block. Hence, the ESD V_{t1} must be tuned to meet the ESD design window bounded by either MOSFET drain or gate breakdown. Additionally, ultrafast CDM ESD failure is a main concern in ESD protection design. The phase-locked loop (PLL) circuit provides accurate clocks to synchronize the VLC SoC. The PLL requires ESD protection for its reference clock pad and clock output pad that are connected to the input gates of D-type flip-flop blocks. The ESD V_{t1} is designed to prevent CMOS gate oxide breakdown and ultrafast CDM ESD protection, which is also a main design target.

The second main mixed-signal ESD protection design concern is the varying impacts of ESD-induced parasitic effects on different circuit functional domains. Conventional in-Si PN-junction-based ESD protection structures introduce substantial parasitic effects, including parasitic capacitance (C_{ESD}), leakage current (I_{leak}), noises, and noise coupling, which can seriously affect core mixed-signal IC circuit performance. In general, digital circuitry can tolerate much bigger ESD-induced parasitic effects due to its large logic tolerance margins. For example, an C_{ESD} of a few hundreds of pF is not much a concern for a digital circuit. One the other hand, analog and RF circuits are extremely sensitive to ESD-induced parasitic effects. For example, even a very small C_{ESD} of a few tens of fF will fatally affect a 28–38 GHz broadband RF front-end IC for 5G mobile systems [7, 8]. To complicate the ESD parasitic effect problem, while digital ICs often need low-level ESD protection (e.g., 2 kV for HBM ESD protection), analog and RF ICs for consumer electronics, such as smartphones, always require high ESD protection due to human interfacing (e.g., 8 kV HBM ESD protection for smartphone touch screens). Unfortunately, higher ESD robustness of same type of ESD protection structures always translate into more ESD-induced parasitic effects, which will more seriously affect analog, mixed-signal (AMX)/RF IC performance. ESD-IC codesign may alleviate this ESD parasitic problem to certain extent. However, truly novel, non-traditional ESD protection mechanisms and structures are needed to comprehensively address the ESD parasitic problem for high-data-rate, high-frequency and broadband ICs at advanced technology nodes.

The third major ESD protection design challenge for mixed-signal ICs is the potential ESD mis-triggering effect for mixed-signal SoC with multiple functional circuit domains. Study shows that, in ESD transmission-line-pulsing (TLP) testing, the V_{t1} of an ESD protection structure can be seriously affected by the TLP pulse rise time, t_r. Typically, substantial V_{t1} reduction occurs as t_r of an TLP pulse decreases, i.e., a faster TLP pulse can trigger an ESD protection device at a lower V_{t1}. In general, it is believed that due to the ESD-induced parasitic capacitance and inductance in the discharge channel, a fast-rising ESD pulse (dV/dt or dI/dt) will introduce a significant displacement current, which in turn will accelerate the ESD triggering procedure and result in a reduced ESD triggering V_{t1} for a given ESD protection device. The potential problem is that for mixed-signal/RF ICs designed for very high frequency and very high speed, the ultrafast normal signal waveforms will bring in a large dV/dt or dI/dt. Consequently, an ultrafast and relatively strong normal RF signal running through an ESD protection device featuring a large parasitic capacitance may substantially reduce the ESD V_{t1}, leading to mis-triggering of an ESD protection device under normal IC operations [3]. For example, study finds that a mixed-signal CMOS transceiver IC experiences strong digital noises coupled from the digital block into the analog block, equivalent to a dV/dt of $\sim 1.15 \times 10^7$ V/s. This high level of dV/dt can readily mis-triggers an ESD protection device of moderate parasitic C_{ESD}, causing IC malfunction without ESD stressing. The above discussions clearly state that many new ESD design challenges must be carefully addressed for complex high-performance, mixed-signal SoC ICs designed in advanced technology nodes.

7.2 ESD Protection for Multiple-Voltages ICs

As discussed earlier, functional heterogeneity of mixed-signal ICs certainly makes ESD protection design very complicated and challenging. Further, multiple supply voltages (power heterogeneity) typically used in mixed-signal ICs make ESD protection design even more involving and difficult. Two key design aspects must be carefully considered in designing ESD protection for mixed-signal ICs using multiple supply voltages (e.g., V_{DD}, V_{SS}, and GND). Different circuit blocks using different supply voltages are often referred to as different power domains on a chip. Today's ICs are designed in different process technologies from 5 nm FinFET CMOS to high-voltage (HV) bipolar-CMOS-DMOS (BCD) to wide bandgap (WBG) and ultrawide bandgap (UWBG) semiconductors on silicon (e.g., GaN-on-Si), hence, featuring a widely spread supply voltage ranges, anywhere from 0.75 V (even lower for nanosheet transistors) to above 100 V (much higher in WBG/UWBG on Si). Therefore, success of whole-chip ESD protection must at least consider local ESD V_{t1} tuning for different power domains. For example, an ESD protection device of $V_{t1} = 5$ V is suitable for a circuit domain using $V_{DD} = 3.3$ V considering reasonable power bus fluctuation, and an ESD protection structure of $V_{t1} = 23$ V is good for a circuit domain utilizing $V_{DD} = 15$ V with a safety margin [3]. However, interchanging the two ESD protection structures between the two power domains may cause operation problems. Using $V_{t1} = 5$ V in the $V_{DD} = 15$ V power domain will result in short-circuit. On the other hand, applying $V_{t1} = 23$ V to the circuit block of $V_{DD} = 5$ V may slow down ESD triggering. Depending upon the voltage gap, a much larger than necessary V_{t1} would cause unexpected ESD failure, for example, during ultrafast CDM events due to increased latency in ESD triggering, even though the ESD protection device is very robust in handling large ESD surges. A designer must consider both ESD current/voltage-handling capability and its response time when designing good ESD protection structures, by careful ESD simulation. Unfortunately, the ESD triggering time is often overlooked in practical IC designs. Obviously, a well-thought-out ESD protection design must optimize the V_{t1} quantitatively to accommodate both local circuit functions and supply voltages. A safety margin between the V_{t1} and the local V_{DD} must consider the supply voltage fluctuation in normal and extreme operation conditions in order to prevent possible ESD mis-triggering.

Next, to achieve full-chip ESD protection, power rail-to-rail ESD protection must be carefully considered in practical IC designs in reference with the global noise coupling effects, which becomes increasingly challenging in designing large and complex mixed-signal SoC ICs. As discussed before, full-chip ESD protection requires establishing a low-*R* ESD discharge path between any two pads on a chip, accordingly to the ESD testing standards, which is achieved by connecting an ESD protection device between a pad (signal and control) and a power supply bus (V_{DD}, V_{SS}) or a ground bus (GND), and having a power clamp device between any two power supply buses of different voltages, i.e., rail-to-rail ESD clamp. Under transient ESD stressing, a power rail is equivalent to an *ac* ground. Therefore, per ESD testing standards, every pad under ESD stressing is protected by an ESD protection structure connected at the pad with respect to a transient *ac* GND, which serves to discharge the incoming ESD pulse of any polarities into the chip ground. This full-chip ESD protection scheme is clearly depicted in Figure 7.5, where a low-*R* ESD discharge path always exists between any two pads on a chip. Unfortunately and not infrequently, such an ideal full-chip ESD protection schematic does not always exists on a real-world chip due to several reasons: an ESD protection device may be missing due to a design error; or a pad-to-pad ESD discharge path fails to exist due to poor ESD device design (e.g., incorrect V_{t1}; or a rail-to-rail ESD protection structure cannot be used due to non-ESD design considerations). The latter scenario, i.e., lack of a rail-to-rail ESD clamp between two power supply buses, may occur in practical IC designs for many reasons: First, many functional and power domains on a complex mixed-signal SoC chip are often purposely designed using different and separate power supply buses in order to minimize global noise (or interferences) coupling between digital, analog, and RF blocks. This is true even if circuit blocks are biased by V_{DD} and/or V_{SS} of same voltages where intra-block noise coupling through power rails must be prevented. Second, design of a large complex mixed-signal SoC chip is typically a team design task that often utilizes third-party IPs. Therefore, even if individual circuit blocks were designed with an ideal full-"chip" ESD protection network "locally" within blocks, when constructing the whole mixed-signal SoC chip, some rail-to-rail ESD clamps may likely be missing between circuit domains, resulting in incomplete whole-chip ESD protection network with missing domain-to-domain ESD discharge paths that

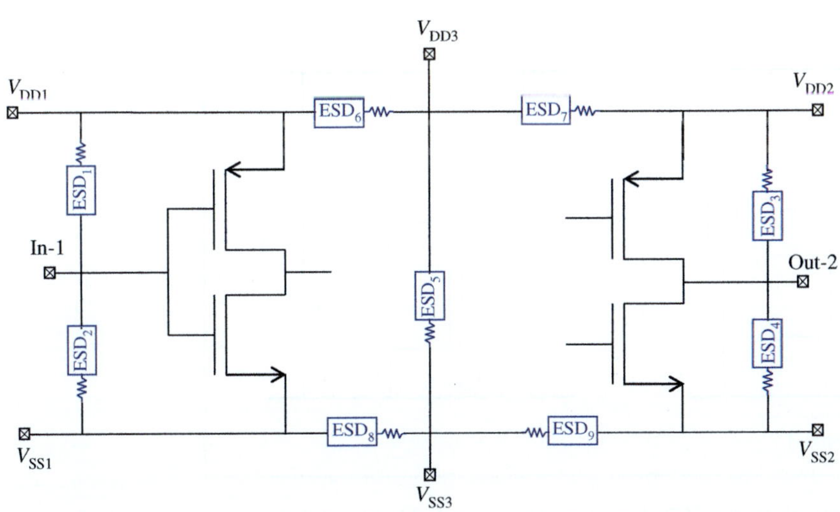

Figure 7.5 Full-chip ESD protection scheme contains power clamp between all supply voltage rails whose series resistances must also be considered.

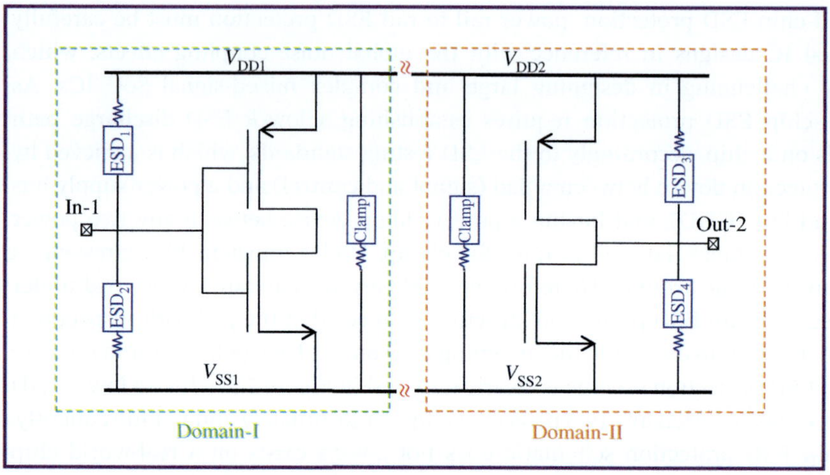

Figure 7.6 Different functional and power domains on a complex mixed-signal chip are often electrically separated in power rails to minimize global noise coupling effect. Hence, the full-chip ESD protection network is incomplete for the whole chip, i.e., missing rail-to-rail ESD clamp between V_{DD1} and V_{DD2}, and between V_{SS1} and V_{SS2}.

will lead to ESD failure at full-chip level. Figure 7.6 illustrates such a case of incomplete full-chip ESD protection network scenario where each power domain (i.e., Domain-I and Domain-II) has full-domain ESD protection locally; however, globally, the full-chip ESD protection network is incomplete due to missing domain-to-domain ESD discharge paths, i.e., no ESD clamp for V_{DD1}-to-V_{DD2} and V_{SS1}-to-V_{SS2} ESD protection. To resolve such a full-chip ESD protection design problem, one has to carefully balance the concerns for ESD protection (e.g., missing rail-to-rail ESD clamps) and global noise coupling (e.g., purposely separating supply buses). Figure 7.7 depicts a conceptual full-chip ESD protection scheme for a reasonable trade-off between noise coupling and full-chip ESD protection. An anti-parallel diode pair can be connected between the two

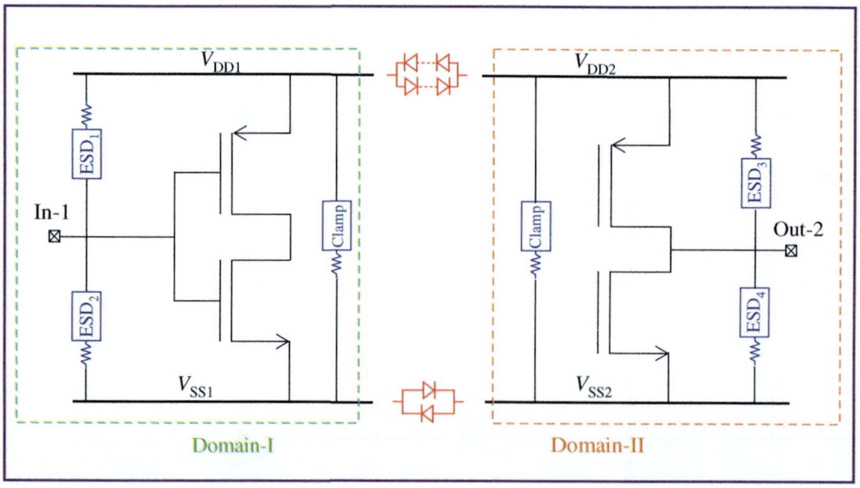

Figure 7.7 Inter-domain rail-to-rail ESD clamps are often inserted between different power buses of different power domains, e.g., using anti-parallel diode or diode-string nets, to re-establish a complete full-chip ESD protection network on an AMX chip.

power rails, which substantially blocks the global noise coupling, while allowing rail-to-rail ESD discharge for power line ESD clamping. In practical design, the equivalent V_{t1} of the inter-domain ESD net must be carefully designed to meet the specific circuit requirements in order to avoid short-circuit errors. For example, an anti-parallel diode-strings may be used for a higher ESD V_{t1} needed, or any other ESD protection devices may be used instead. With the inter-domain ESD clamp devices in place, an ideal full-chip ESD protection scheme can be ensured for a complex mixed-signal SoC chip. Nevertheless, while using inter-domain ESD diode net is popular in IC designs to achieve robust HBM ESD protection, it is noticed that such an ESD protection technique may not improve CDM ESD performance. Let us look at an exemplar chip schematic shown in Figure 7.8, where the CDM-induced electrostatic charges are assumed to accumulate around the V_{DD1} supply bus near the pad. Per CDM zapping procedures, many CDM discharge cases may practically occur in a real world. In the Case-1 depicted in Figure 7.8, assume the V_{DD1} pad is grounded during CDM zapping test, since the charges are close to the V_{DD1} pad, they can be readily discharged locally into the instantaneous (*ac*) GND to realize CDM ESD protection. In Case-2 shown in Figure 7.9, assume the electrostatic charges stay at the same location and the V_{SS1} pad is grounded during CDM zapping test, then the internally stored charges can be discharged into GND via two possible ESD conduction paths, ① and ②, through the V_{DD1}-to-V_{SS1} ESD clamp or the ESD1 and ESD2 devices at the In-1 pad. For a good ESD clamp with low-R, the path-① will function nicely. However, if the path-② dominate the CDM ESD discharge and if the equivalent ESD conduction path resistance from In-1 pad to GND (V_{SS1}) is high (R_{ESD} of ESD2 and bus resistances), a substantial voltage may be built up across the GS of transistor M1, possibly causing gate oxide breakdown, which is typically observed as CDM ESD failure signature. Next, let us look at the even worse Case-3 as illustrated in Figure 7.10, where the Out-2 pad is grounded during CDM zapping test and several possible CDM ESD discharge paths exist on the chip. The path-① will discharge through the V_{DD1} and V_{DD2} buses, the V_{DD1}-to-V_{DD2} inter-rail ESD clamp, and the ESD3 device. Similarly, the path-② will conduct through the V_{DD1}-to-V_{SS1} ESD clamp, the V_{SS1} and V_{SS2} power buses, the V_{SS1}-to-V_{SS2} inter-rail diode clamp and the ESD4 device.

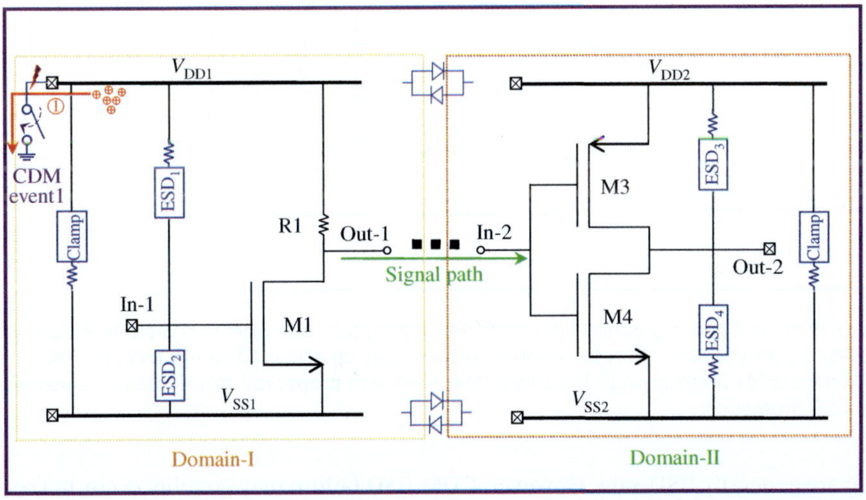

Figure 7.8 It is understood that using inter-domain rail-to-rail ESD clamps may not improve CDM ESD protection. Case-1 assumes that the CDM-induced electrostatic charges are located around the V_{DD1} bus near the V_{DD1} pad, which is grounded during CDM zapping, and CDM ESD discharge occurs via the path-① safely.

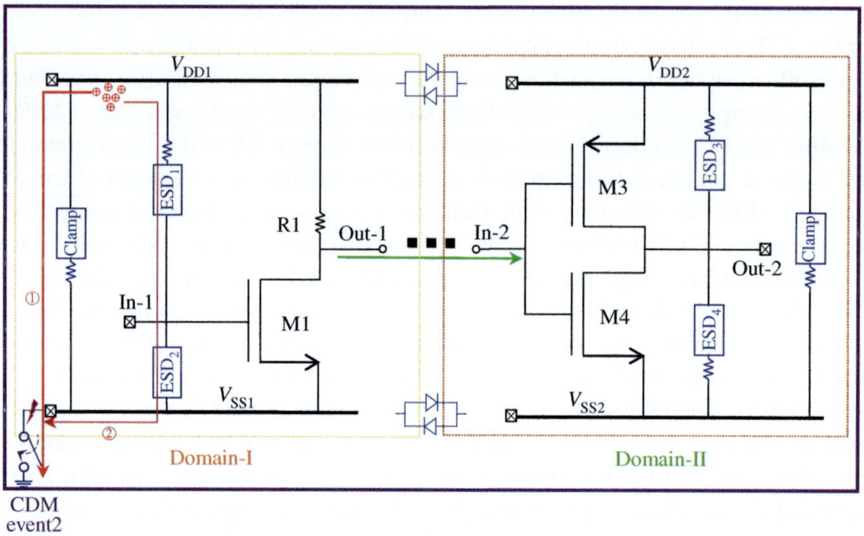

Figure 7.9 In Case-2, the V_{SS1} pad is grounded during CDM zapping test, and CDM ESD discharge occurs possibly in path-① and/or path-②. A substantial voltage built-up across GS of M1 may potentially cause CDM gate oxide breakdown to M1, even with proper rail-to-rail diode clamps in place for good HBM ESD protection.

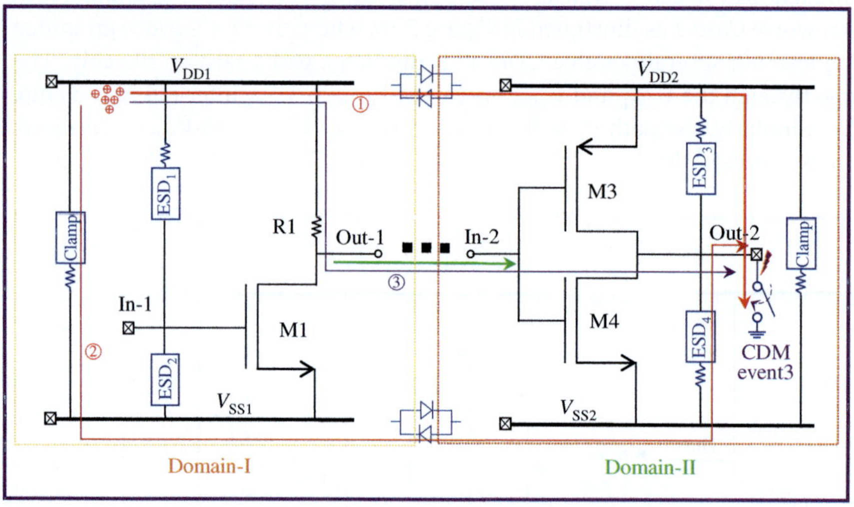

Figure 7.10 In Case-3, the Out-2 pad is grounded during CDM zapping test, and CDM ESD discharge occurs possibly in path-①, path-② and/or path-③. A substantial voltage built-up across GS of M3 and M4 may potentially cause CDM gate oxide breakdown to M3 and/or M4, even with proper rail-to-rail diode clamps in place for robust HBM ESD protection.

Path-① and path-② are generally ESD-safe. However, CDM ESD failure may possibly occur in the long path-③ that discharge via V_{DD1} power bus, the inter-domain signal path (usually via very narrow metal lines), and then through the invertor gate of M3 and M4 by capacitive coupling. If the path is very long and highly resistive, a substantial voltage may be built up across GS of M3 and M4, potentially causing CDM ESD failure to the CMOS gates. The above case analysis

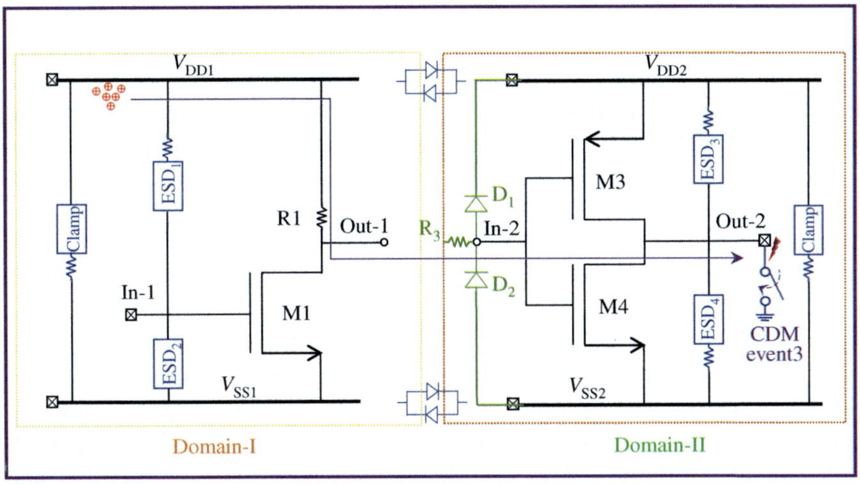

Figure 7.11 Exemplar solution to the internal CDM ESD failure problem utilizes internal and local CDM clamping devices, e.g., D1 and D2 to prevent CDM discharge induced local voltage build-up across M3 and M4.

readily states that even if inter-domain rail-to-rail ESD clamps are in place, whole-chip CDM ESD protection may not be guaranteed at full chip level due to unexpected CDM ESD discharge channels existing on the chip. Figure 7.11 depicts a simple solution to this internal CDM ESD failure problem where *internal* clamping devices (e.g., diodes D1 and D2) can be connected across the vulnerable MOSFETs locally to prevent local voltage build-up due to large discharge resistance. A series resistance (R3) is often inserted into the signal path between Domain-I and Domain-II to avoid direct stressing the MOSFET gate, however, it may seriously affect the circuit performance [9].

7.3 ESD Protection for High-Voltage ICs

As discussed earlier, full-chip ESD protection design for mixed-signal ICs is a very challenging design task, though the design principle is quite straightforward, i.e., domain-specific local ESD protection design optimization by fine-tuning the ESD-critical parameters is critical to ensure full compliance of the ESD design windows across the whole chip. The mixed-signal ESD protection design methods related to heterogeneity of multiple functional domains and inter-domain ESD protection between domains of different power supplies were discussed in Section 7.2. This section focuses on yet another very challenging mixed-signal ESD protection design problem, which is ESD protection design for high (or higher) voltage (HV) ICs that is very different from ESD protection design for low (lower) voltage (LV) ICs.

7.3.1 ESD Design Window Compliance

In addition to functional heterogeneity of typical mixed-signal ICs, biasing heterogeneity due to using multiple supply voltages is another common feature of mixed-signal chips. Specifically, mixed-signal ICs often use lower supply voltages for LV domains and higher biasing voltages for HV domains. For example, typical supply voltages for baseband circuitry in advanced CMOS technologies (e.g., 14 nm FinFET and beyond) can be 0.75 V only; however, high/higher supply voltages are needed for mixed-signal SoC chips using a hybrid bipolar-CMOS-DMOS (BCD) process to power HV circuit domains, e.g., 20 V in a charge pump for power management IC (PMIC) for smartphone chipsets, or 38 V or higher in power converters for automotive SoC IC chips. The supply heterogeneity for mixed-signal ICs implemented in the emerging third-generation semiconductors, e.g., WBG/UWBG (GaN, Ga_2O_3, diamond, etc.) on Si for monolithic ICs, may feature a very wide supply voltage range anywhere from a few volts to a thousand volts. The broad supply voltage range on a chip makes full-chip ESD protection design for complex mixed-signal SoC chips extremely challenging and involving. The key design concern for ESD protection for HV/LV mixed-signal ICs is ESD design window compliance. Refer to the ESD design window shown in Figure 7.1, in HV/LV mixed-signal ICs, the lower-end voltage of V_{DDmax} can vary widely from a sub-1 V to tens of volts; while the upper-end of V_{safe} limited by the breakdown voltage of the protected nodes are also very different due to using different process modules in a given mixed-signal IC technology. Obviously, it becomes very involving and challenging to quantitatively fine-tune the ESD-critical parameters (i.e., V_{t1}, V_h) locally to ensure whole-chip ESD design window compliance. In other word, local ESD design optimization for HV mixed-signal ICs is much harder than that considering functional heterogeneity only. For instance, it may be relatively easier to change the ESD triggering threshold V_{t1} for a HV ESD protection structure according to the higher breakdown voltage (e.g., BV_G, BV_D, or BV_S in CMOS). On the other hand, though a very low ESD holding voltage V_h is often preferred for efficient pad voltage clamping (e.g., an SCR ESD protection device featuring deep snapback I–V characteristics with a low $V_h \sim 1.5$ V in a 0.85 V 28 nm CMOS process), however, such a low-V_h SCR clamp cannot be used in a HV circuit domain featuring a V_{DD} of 10 V supply due to latch-up and short-circuit concerns. Unfortunately, it is very difficult to dramatically increase the V_h of the SCR clamp in a hybrid BCD process technology. In principle, a good full-chip ESD protection solution for HV/LV mixed-signal ICs requires comprehensive local ESD design optimization, including quantitative design of the ESD-critical parameters and careful ESD design window compliance checking for each power domain on a chip. Next, let us use a design example to explain complicated mixed-signal ESD protection designs. Figure 7.12 shows a simplified functional block diagram for a mixed-signal PMIC chip, containing a LV Domain-I and HV Domain-II, designed in a hybrid 30 V 28 nm BCD process technology [10]. The hybrid BCD process is based on a 28 nm CMOS with core supplies of $V_{\text{DD1}} = 0.9$ V and $V_{\text{SS1}} = -0.9$ V, and LV CMOS Gate breakdown of $BV_{G1} = 5.2$ V and source/drain diffusion breakdown of $BV_{S1} = BV_{D1} = 5.6$ V. The added HV process module has supply voltages of $V_{\text{DD2}} = 10$ V and $V_{\text{SS2}} = -10$ V, and offers HV DMOS transistors with Gate breakdown of $BV_{G2} = 52$ V and source/drain diffusion breakdown of $BV_{S2} = BV_{D2} = 55$ V. Roughly, the ESD design windows will be $(V_{\text{DDmax}}, V_{\text{safe}}) \sim (V_{\text{DD1}}, BV_{G1}) \sim (0.9, 5.2\text{ V})$ for I/O blocks and $(V_{\text{DD1}}, BV_{D1}) \sim (0.9, 5.6\text{ V})$ for power buses, respectively, for the LV domains; and $(V_{\text{DDmax}}, V_{\text{safe}}) \sim (V_{\text{DD2}}, BV_{G2D}) \sim (10, 52\text{ V})$ for I/O blocks and $(V_{\text{DD2}}, BV_{DD2}) \sim (10, 55\text{ V})$ for power rails, respectively, in the HV domains. However, one has to understand the dynamic nature of the ESD design window concept: the upper limit of V_{safe} is set by the transient breakdown voltage of the protected nodes during transient ESD stressing events; while the lower boundary of V_{DDmax} is actually determined by the static supply voltage (i.e., DC biasing voltage and ignoring the small *ac* signals). The limit of V_{DDmax} will have two effects: First, possible short-circuit in normal IC operations without ESD events (static in nature) if $V_{t1} < V_{\text{DDmax}}$; Second, possible latch-up after the ESD pulse (transient in nature) is over if $V_h < V_{\text{DDmax}}$ ($I_{\text{supply}} > I_h$ is another factor for latch-up consideration) for a snapback ESD protection device. Hence, the lower limit of the ESD design window related to short-circuit risk under normal IC operations may be varying against the negative supplies (V_{SS}) for a given pad on a chip. This is because it is possible a non-ESD-induced short-circuit risk may be initiated by a voltage across the A–K terminals of the ESD protection device in reference with its V_{t1} in a specific conduction direction. For example, for an ESD protection device with its A-terminal connected to I/O pad and its K-terminal connected to a negative $V_{SS} = -10$ V, and the ESD protection device has a K-to-A $V_{t1} = 3$ V (always a positive value in the respective conduction direction), the lower-end of the related ESD design window will be 0 V, instead of $|V_{SS}| = 10$ V since a negative V_{SS} will never turn on the ESD device from K to A against the I/O pad. Now, let us return to the HV/LV mixed-signal IC depicted in Figure 7.12. The LV Domain-I uses a classic anti-parallel diode (D1, D2, D3, and D4) ESD protection at In-1 pad and a low-V_{t1} diode-triggered silicon controlled rectifier (DTSCR) ESD power clamp. Figures 7.13 and 7.14 depict the ESD discharge I–V curves for the ESD diodes and DTSCR clamp for the LV domain featuring the following ESD-critical

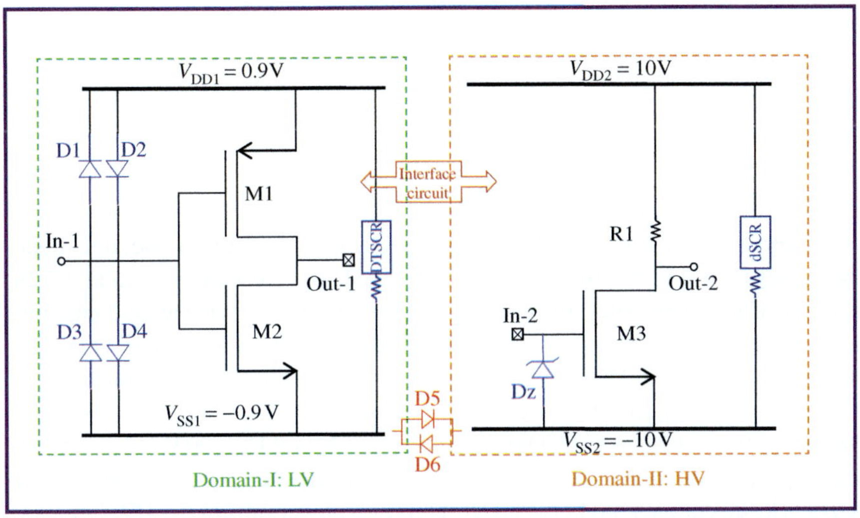

Figure 7.12 A exemplar schematic for a multiple-function/power-domain HV AMX SoC chip implemented in a 30 V hybrid 28 nm BCD process technology. It contains LV circuit domains and HV circuit domains. The LV Domain-I utilizes a classic anti-parallel diode ESD protection sub-net at I/O and a LV DTSCR ESD power clamp. The HV Domain-II uses a Zener ESD diode at I/O and a symmetric dSCR ESD power clamp.

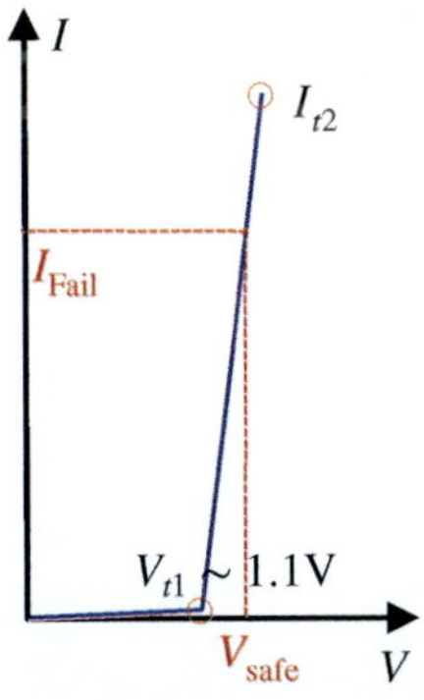

Figure 7.13 A conceptual ESD discharge I–V curve for the anti-parallel diode ESD sub-net used to protect I/O pads in the LV domains. The diodes are in forward conduction to discharge ESD pulses with an estimated turn-on voltage of $V_{t1} \sim 1.1$ V at 10 mA. The V_{safe} value is defined by the ESD Design Window of the mixed-signal IC chip, which sets the limit of ESD current handling capability, I_{Fail}, at chip level.

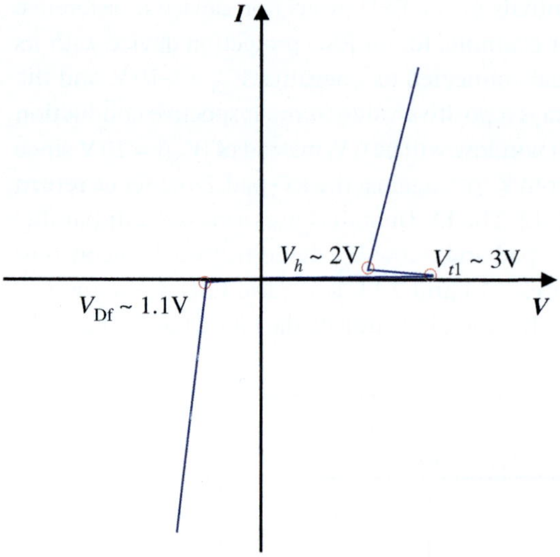

Figure 7.14 A conceptual ESD discharge I–V characteristic for the asymmetric low-V_{t1} diode-triggered SCR ESD power clamp structure (DTSCR) used to protect the power rails in the LV domains. The DTSCR features $V_{t1} \sim 3$ V and $V_h \sim 2$ V, and a parasitic diode of $V_{Df} \sim 1.1$ V.

parameters: forward diode turn-on for ESD diodes (D1–D4), $V_{Df} = V_{t1} \sim 1.1$ V (typically measured at 10 mA), and $V_{t1} \sim 3$ V and $V_h \sim 2$ V for the DTSCR. During ESD stressing at In-1 pad, one diode (D1–D4) will be forward turned on to discharge the incoming ESD pulses. Figures 7.15 and 7.16 show the ESD discharge I–V characteristics for the Zener diode (Dz) ESD device at In-2 pad and a symmetrical dual-directional SCR ESD power clamp (dSCR) used in the HV domain [11]. The ESD-critical parameters for the Dz and dSCR ESD devices are the following: BV_{Dz} (Zener reverse breakdown) = $V_{t1} \sim 15$ V and V_{Dzf} (Zener forward turn-on at 10 mA) = $V_{t1} \sim 1.1$ V (reverse ESD triggering) for the Dz ESD protection device at I/O pad; and $V_{t1} \sim 30$ V and $V_h \sim 20$ V, in both directions, for the dSCR ESD protection structure. The Dz-dSCR ESD protection scheme ensures that there is always an active ESD discharge path between I/O pads and power rails in the HV domains. The ESD design window compliance checking will be analyzed for each pad in the LV and HV domains against different power rails ($V_{DD1} = 0.9$ V, $V_{SS1} = -0.9$ V, $V_{DD2} = 10$ V and $V_{SS2} = -10$ V).

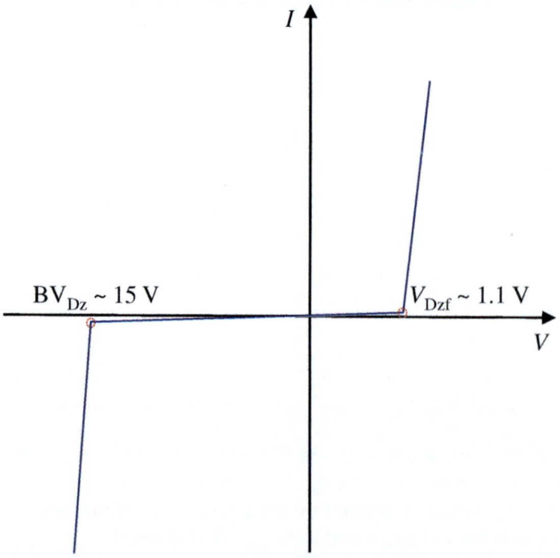

Figure 7.15 A conceptual ESD discharge I–V curve for the Zener diode used to protect I/O pads in the HV domains. The Zener diode features $BV_{Dz} = V_{t1} \sim 15$ V and $V_{Dzf} \sim 1.1$ V.

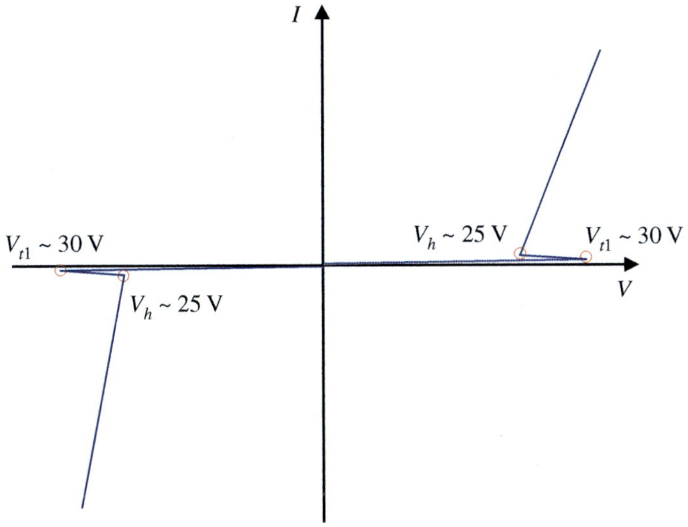

Figure 7.16 A conceptual ESD discharge *I–V* curve for the symmetric dual-direction dSCR ESD power clamp for supply bus ESD protection in the HV domains. The dSCR features $V_{t1} \sim 30$ V and $V_h \sim 25$ V.

In LV domains, the LV I/O ESD design windows are (V_{SS1}/V_{DD1}, BV_{G1}/BV_{D1}) = (0/0.9 V, 5.2/5.6 V) with reference to V_{SS1}/V_{DD1} rails and BV_{G1}/BV_{D1} breakdown, and for PD/ND/PS/NS ESD stress modes, respectively. The LV power bus ESD design windows are ($V_{DD1} + |V_{SS1}|$, BV_{D1}) = (1.8, 5.6 V) for DS ESD stress mode and (V_{SS1}, BV_{D1}) = (0, 5.6 V) for SD ESD stress mode, respectively. Note that, in both windows, when a negative V_{SS1} is referenced, the low-limit will be 0 V, since a negative V_{SS1} cannot forward-trigger the related ESD devices in the direction from V_{SS1} pad to the transient ESD GND. Similarly, in the HV domains, the HV I/O (In-2 pad) ESD design windows are (V_{SS2}/V_{DD2}, BV_{G2}/BV_{D2}) = (0/10 V, 52/55 V) with reference to V_{SS2}/V_{DD2} rails and BV_{G2}/BV_{D2} breakdown, and for PD/ND/PS/NS ESD stress modes, respectively. The HV power bus ESD design windows are (V_{DD2}+| V_{SS2}|, BV_{D2}) = (20V, 55V) for DS ESD stress mode and (V_{SS2}, BV_{D2}) = (0V, 55V) for SD ESD stress mode, respectively. Also note that, in both ESD design windows, when a negative V_{SS2} is referenced, the low-limit will be 0V, since a negative V_{SS2} cannot forward-trigger the related ESD devices in the direction from V_{SS2} pad to the transient ESD GND. Next, let us check ESD design window compliance for individual cases. We start with the LV I/O ESD design window analysis. For ND ESD stress mode (negative ESD zapping at I/O pad w.r.t. V_{DD1}), D2 will be forward turned on to discharge the ESD pulse and the D2 $V_{t1} \sim 1.1$V is within the ESD design window of (V_{DD1}, BV_{G1}) = (0.9V, 5.2V), hence, 0.9V<($V_{t1} \sim 1.1$V)<5.2V passes ESD design window compliance checking. For PD ESD stress mode (positive ESD zapping at I/O pad w.r.t. V_{DD1}), D1 will forward discharge the ESD pulse and the D1 $V_{t1} \sim 1.1$V is within the ESD design window of (V_{DD1}-referenced, BV_{G1}) = (0V, 5.2V), hence, 0V<($V_{t1} \sim 1.1$V)<5.2V passes ESD design window compliance checking. Note that, since V_{DD1} is the reference "GND" during PD ESD stress mode and $V_{DD1} \gg$ any normal input signals that can never forward turn on D1 in the direction from I/O pad w.r.t. V_{DD1} bus, hence, it is safe to reduce D1 V_{t1} to 0V (not V_{DD1} = 0.9V). For PS ESD stress mode (positive ESD zapping at I/O pad w.r.t. V_{SS1}), D4 is forward turned on to discharge the ESD pulse and the D4 $V_{t1} \sim 1.1$V is within the ESD design window of (|V_{SS1}|, BV_{G1}) = (0.9V, 5.2V), hence, 0.9V<($V_{t1} \sim 1.1$V)<5.2V passes ESD design window compliance checking. Note that, since V_{SS1} = –0.9V in the reference "GND" in PS mode, V_{t1}>0.9V is required for D1 to prevent short-circuit without ESD events. For NS ESD stress mode (negative ESD zapping at I/O pad w.r.t. V_{SS1}), D3 will forward discharge the ESD pulse and the D3 V_{t1} ~ 1.1V is within the ESD design window of (V_{SS1}-referenced, BV_{G1}) = (0V, 5.2V), hence, 0V<(V_{t1} ~ 1.1V)<5.2V passes ESD design window compliance checking. For DS ESD stress mode (positive V_{DD1}-to-V_{SS1} ESD zapping from V_{DD1} bus w.r.t. V_{SS1} pad), the DTSCR ESD clamp will be forward turned on to discharge the ESD pulse and the DTSCR V_{t1} ~ 3V and V_h ~ 2V are within the ESD design window of (V_{DD1}+|V_{SS1}|, BV_{D1}) = (1.8V, 5.6V), hence, 1.8V<(V_{t1} ~ 3V and V_h ~ 2V)<5.6V passes rail-to-rail ESD design window compliance checking. For SD ESD stress mode (negative V_{DD1}-to-V_{SS1} ESD zapping from V_{DD1} bus w.r.t. V_{SS1} pad), the DTSCR ESD clamp will be reverse turned on (i.e., a parasitic diode in forward mode) to discharge the ESD pulse and the parasitic diode within DTSCR will conduct in forward mode with $V_{Df} = V_{t1} \sim 1.1$ V that is within the ESD design window of (V_{SS1}-referenced, BV_{D1}) = (0, 5.6 V), hence, 1.8 V< (V_{t1} ~ 1.1 V) <5.6 V passes reverse rail-to-rail ESD design window compliance checking.

We then move to check the HV domain ESD design window compliance checking. As shown in Figure 7.12, a Zener diode ESD device at I/O (In-2 pad) and a dual-direction ESD power clamp (dSCR) together can ensure active ESD discharge path between any two pads within the HV domains. For PS ESD stress mode (positive ESD zapping at I/O pad w.r.t. V_{SS2}), reverse breakdown of Dz will discharge the ESD pulse and the $BV_{Dz} = V_{t1} \sim 15$ V is within the ESD design window of (|V_{SS2}|, BV_{G2}) = (10, 52 V), hence, 10 V< (V_{t1} ~ 15 V) <52 V passes ESD design window compliance checking. For NS ESD stress mode (negative ESD zapping at I/O pad w.r.t. V_{SS2}), forward conduction of Dz will take place to discharge the ESD pulse, and the $V_{Dzf} = V_{t1} \sim 1.1$ V is within the ESD design window of (V_{SS2}-referenced, BV_{G2}) = (0, 52 V), hence, 0 V< (V_{t1} ~ 1.1 V) <52 V passes ESD design window compliance checking.

For PD ESD stress mode (positive ESD zapping at I/O pad In-2 w.r.t. V_{DD2}), since there is no dedicated ESD protection device between I/O pad and V_{DD2} bus, the ESD discharge path will follow reverse Dz conduction (PS) plus reverse dSCR clamp conduction (SD), and the equivalent ESD triggering is $V_{t1\text{-total}} = BV_{Dz} + V_{t1\text{-dSCR}} \sim 15\,V + 30\,V \sim 45\,V$, and the equivalent ESD holding is $V_{h\text{-total}} = BV_{Dz} + V_{h\text{-dSCR}} \sim 15\,V + 25\,V \sim 40\,V$. The PD-related ESD design window is (X, Y) that is actually set by the two segments (parts) of the whole ESD conduction path individually, i.e., part-1 from In-2 to V_{SS2} bus (PS) and part-2 from V_{SS2} bus to V_{DD2} bus (SD), which generate two separate ESD design sub-windows that must be examined coordinately because non-ESD short-circuit may occur in each of the two part of the whole ESD discharge path. The sub-window-1 for the part-1 (PS) is concerned about possible mis-triggering of the Zener ESD diode in breakdown mode; hence, it requires the V_{t1} of Dz (i.e., $BV_{Dz} = V_{t1\text{-Dz}}$) be higher than |$V_{SS2}$| to avoid accident Dz breakdown with ESD stressing; hence, the sub-window-1 is set by (|V_{SS2}|, BV_{G2}) = (10, 52 V). Therefore, 10 V< ($V_{t1} = BV_{Dz} \sim 15$ V) <52 V passes sub-window-1 compliance checking. Similarly, the sub-window-2 for part-2 (SD) is related to the reverse dSCR ESD clamp conduction with $V_{t1\text{-dSCR}} \sim 30$ V and $V_{h\text{-dSCR}} \sim 25$ V, and the sub-window-2 is set as (V_{SS2}-referenced, BV_{D2}) = (0, 55 V). Since a large negative V_{SS2} can never trigger the dSCR clamp w.r.t. a large positive V_{DD2}, the low-limit of the sub-window-2 is 0 V (not |V_{SS2}| = 10 V). Therefore, 0 V< ($V_{t1\text{-dSCR}} \sim 30$ V and $V_{h\text{-dSCR}} \sim 25$ V) <55 V passes the sub-window-2 compliance checking. Together, the overall HV I/O ESD design window compliance checking is validated for the PD ESD stress mode.

For ND ESD stress mode (negative ESD zapping at I/O pad In-2 w.r.t. V_{DD2}), the ESD discharge path will follow the forward dSCR clamp conduction (DS, part-1) plus forward Dz conduction (NS, part-2), and the equivalent ESD triggering is $V_{t1\text{-total}} = V_{t1\text{-dSCR}} + V_{Dzf} \sim 30\,V + 1.1\,V \sim 31.1\,V$, and the equivalent ESD holding is $V_{h\text{-total}} = V_{h\text{-dSCR}} + V_{Dzf} \sim 25\,V + 1.1\,V \sim 26.1\,V$. Similarly, the ND-related ESD design window is (X, Y) that is defined by the two segments (parts) of the whole ESD conduction path individually, i.e., part-1 from V_{DD2} bus to V_{SS2} bus (DS) and part-2 from V_{SS2} bus to In-2 pad (NS), which correspond to two separate ESD design sub-windows. The sub-window-1 for the part-1 (DS) is concerned about the forward dSCR ESD clamp conduction with $V_{t1\text{-dSCR}} \sim 30\,\text{V}$ and $V_{h\text{-dSCR}} \sim 25\,\text{V}$, and the sub-window-1 is set as ($V_{DD2} + |V_{SS2}|$, BV_{D2}) = (20, 55 V). Therefore, 20 V< ($V_{t1\text{-dSCR}} \sim 30\,\text{V}$ and $V_{h\text{-dSCR}} \sim 25\,\text{V}$) <55 V passes the sub-window-1 compliance checking for the dSCR clamp in DS mode. The sub-window-2 is related to possible mis-turn-on of the Zener ESD diode in forward conduction mode without ESD stressing. Since the large negative V_{SS2} can never forward turn on Dz w.r.t. In-2 pad, hence, the sub-windw-2 is set by (V_{SS2}-referenced, BV_{G2}) = (0, 52 V). Therefore, 0 V< ($V_{t1} = V_{Dzf} \sim 1.1\,\text{V}$) <52 V passes sub-window-2 compliance checking. Together, the overall HV I/O ESD design window compliance checking is validated for the ND ESD stress mode.

For DS ESD stress mode (positive V_{DD2}-to-V_{SS2} ESD zapping from V_{DD2} bus w.r.t. V_{SS2} power rail), the dSCR ESD clamp will be forward turned on to discharge the ESD pulse, and the dSCR $V_{t1} \sim 30\,\text{V}$ and $V_h \sim 25\,\text{V}$ are within the ESD design window of ($V_{DD2} + |V_{SS2}|$, BV_{D2}) = (20, 55 V), hence, 20 V< ($V_{t1} \sim 30\,\text{V}$ and $V_h \sim 25\,\text{V}$) <55 V passes rail-to-rail ESD design window compliance checking in DS mode. For SD ESD stress mode (negative V_{DD2}-to-V_{SS2} ESD zapping from V_{DD2} pad w.r.t. V_{SS2} pad), the dSCR ESD clamp will be conduct in "reverse" direction (i.e., same $V_{t1} \sim 30\,\text{V}$ and $V_h \sim 25\,\text{V}$) to discharge the ESD pulses, which again fits into the ESD design window of (V_{SS1}-referenced, BV_{D2}) = (0 V, 55 V), hence, 0 V<($V_{t1} \sim 30\,\text{V}$ and $V_h \sim 25\,\text{V}$)<55 V passes reverse rail-to-rail ESD design window compliance checking in SD mode.

Finally, after going through the very complex ESD design window compliance checking procedures, it confirms that the LV/HV mixed-signal IC shown in Figure 7.12 can provide full-chip ESD protection needed. In fact, the mixed-signal ESD protection design complexity goes beyond the painful procedures discussed previously. Many more design factors must be carefully considered in real-world IC designs. For example, refer to the ESD diode device shown in Figure 7.13, a standalone ESD protection device (typically offered by an ESD design guru in your company) may be very ESD-robust by itself, i.e., a very high ESD current-handling capability represented by a very large I_{t2} that is confirmed in TLP ESD testing. However, often, it may not be able to achieve the expected full-chip ESD protection for mixed-signal ICs due to the ESD design window problem. Specifically, an ESD protection design may has a relatively high ESD discharge resistance R_{ON}, and consequently, the highest ESD failure current (I_{Fail}) is actually limited by the ESD design window set by the supply voltages and breakdown voltages on a mixed-signal IC chip. Hence, one has to first identify the V_{safe} preset by the specific ESD design window in a given functional/power domain, and then extrapolate corresponding I_{Fail} for a specific ESD protection device at the chip level, which is normally substantially lower that the measured I_{t2} of an individual standalone ESD protection device. Therefore, the chip-level ESD protection robustness can then be determined for a mixed-signal IC accordingly. Apparently, your brain would be burning crazily when conducting the ESD design window compliance checking routines even for a moderate-scale, mixed-signal SoC chip. In practical designs, such complex ESD design window-checking routines for mixed-signal ICs can be carried out using CAD-aided full-chip ESD protection design verification techniques and procedures to be discussed in Chapter 15. This will save your brain.

7.3.2 Latch-up Immunity

Latch-up is inherent to CMOS ICs where, illustrated in Figure 7.17, a parasitic NPN and a parasitic PNP nearby, typically existing in a CMOS inverter as shown, form a SCR cell, which can be readily triggered to form a low-R current conduction path between V_{DD} and GND, resulting short-circuit between supply rails on a chip. More dangerously, the deep snapback I–V behavior of a latch-up

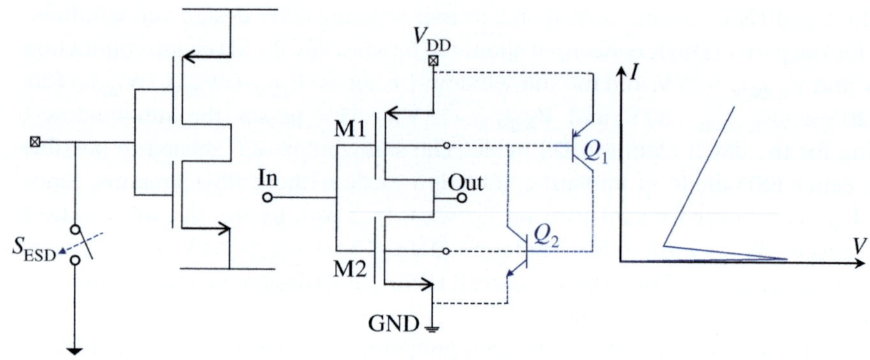

Figure 7.17 In concept, ESD protection utilizes a controlled low-*R* ESD switch that can be considered as a controllable transient latch-up device. In contrast, generic latch-up effect inherent to CMOS is an uncontrollable, disastrous and random current conduction phenomenon. (a) an ESD switch, (b) latch-up in CMOS, and (c) deep snapback *I*–*V* for both ESD protection switch and CMOS latch-up.

device will produce a very large and uncontrollable current that will easily burn the IC. A CMOS latch-up can be initiated by many external factors, such as a current spike injected, single-event radiation, a surge in power supply bus, and perhaps most importantly here, an ESD surge. The possible ESD-induced latch-up is not a surprise because, in principle, ESD protection relies on fast turn-on of a low-*R* electronic switch to efficiently discharge the incident ESD pulses without overheating, as depicted in Figure 7.17. The difference between a generic unwanted latch-up device in CMOS and an intentional ESD switch is that an ESD protection structure is a controllable switch (i.e., controlled latch-up), while CMOS latch-up results in uncontrollable switching that may be triggered randomly for no good. Therefore, in addition to many good features of a well-designed ESD protection structure, such as fast triggering time, high ESD robustness, and low ESD-induced overhead effects, an ideal ESD protection structure must be immune to random latch-up effect. A latch-up-resistant ESD protection device can be swiftly turned OFF after an ESD surge is over so that the ON state of the ESD switch (equivalent to a transient/controlled latch-up effect during an ESD event only) will promptly returns to OFF state; hence, not affecting normal IC operations after the ESD events are over. One key design consideration to ensure ESD switch latch-up immunity is to carefully design the ESD-critical parameters in reference with the ESD design window. In general, it is required to have $V_h > V_{DDmax}$ and $I_h \gg I_{supply}$ on a chip so that the transient latch-up conduction during ESD discharge cannot be sustained after the ESD surge is over because the low V_{DD} and I_{DD} will not be able to keep up the ESD-induced latch-up conduction. This requires careful and quantitative design of the ESD-critical parameters by ESD simulation. Latch-up testing is typically carried out by using the industrial latch-up test standard, JESD78D [12].

7.4 Summary

In summary, full-chip ESD protection design for mixed-signal/RF ICs is a very challenging IC design task. The challenge is directly related to the chip complexity and heterogeneity of a mixed-signal SoC chip that often features multiple functional and power supply domains. It is critical to carefully and quantitatively design the ESD-critical parameters of any ESD protection structures and thoroughly check the dynamic ESD design window compliance across a mixed-signal SoC IC chip. It is vitally important to excise *Local ESD Protection Optimization*, while having a Global ESD protection picture in mind, when designing whole-chip ESD protection for mixed-signal and HV ICs.

References

1 Wang, A. (2002). *On-Chip ESD Protection for Integrated Circuits: An IC Design Perspective*. Springer. ISBN: 9780792376477.

2 Wang, X., Fan, S., Zhao, H. et al. (2010). Whole-chip ESD protection design for RF and AMS ICs. *Tsinghua Sci. Technol.* 15 (3): 265–274.

3 Wang, A., Feng, H., Zhan, R. et al. (2005). A review on RF ESD protection design. *IEEE Trans. Electron Devices* 52 (7): 1304–1311.

4 Feng, H.G., Zhan, R.Y., Chen, G. et al. (2004). Electrostatic discharge protection for RF integrated circuits: new ESD design challenges. *Analog Integr. Circuits Signal Process.* 39 (1): 5–19.

5 Wang, X.S., Wang, X., Lu, F. et al. (2014). Concurrent design analysis of high-linearity SP10T switch with 8.5 kV ESD protection. *IEEE J. Solid-State Circuits* 49 (9): 1927–1941.

6 Dong, Z., Lu, F., Ma, R. et al. (2014). An integrated transmitter for LED-based visible light communication and positioning system in a 180 nm BCD technology. *Proceedings of IEEE BCTM*.

7 Wang, C., Lu, F., Chen, Q. et al. (2017). A study of interferences inside an RF switch array in 45 nm SOI CMOS. *Proceedings of IEEE S3S*.

8 Zhang, F., Li, C., Di, M. et al. (2020). Design and analysis of a 28 GHz 9 KV ESD-protected distributed travelling-wave TRx switch in 22 nm FDSOI. *IEEE J. Electron Devices Soc. (J-EDS)* 8: 655–661.

9 Worley, E. (2004). Distributed gate ESD network architecture for inter-power domain signals. *Proceedings of IEEE EOS/ESD Symposium*.

10 Zhang, F., Wang, C., Lu, F. et al. (2017). Full-chip ESD protection design verification method for HV ICs with multiple power domains. *Proceedings of IEEE ICIEA*.

11 Wang, A. and Tsay, C. (2001). On a dual-polarity on-chip electrostatic discharge protection structure. *IEEE Trans. Electron Devices* 48 (5): 978–984.

12 JESD78D (2010). *IC Latch-Up Test*. JEDEC Solid State Technology Association.

8

TCAD-Based Mixed-Mode ESD Protection Designs

Nowadays, simulation becomes indispensable in integrated circuit (IC) design practices, which is just true for on-chip electrostatic discharge (ESD) protection designs. While design experience is indeed beneficial, today, even analog circuit design, traditionally dubbed as artistic work, cannot be done without comprehensive and quantitative circuit simulation in practical IC designs. Traditionally, ESD protection design has been experience-based where prior success plays a decisive role. However, the experience-based *trial-&-error* design approaches have made ESD protection designs sound more like a black magic, which is certainly unacceptable to the generation of IC designers today. Since 1990s, efforts have been made to develop various CAD methods and simulation techniques for ESD protection designs, ranging from device simulation to circuit simulation to layout verification. This chapter focuses on TCAD-based ESD protection simulation design methods.

8.1 ESD Design Optimization and Prediction

For decades since 1970s, ESD protection designs have been dominated by experience-based trial-&-error design approaches where circuit designers have been relying heavily on the ESD gurus in a company to magically deliver some ESD protection devices to protect IC chips. While many mysteries still exist today, one fact is clear for ESD protection designs, which is that the ESD phenomena to ICs involve complex multiple coupling effects, i.e., materials-process-device-circuit-layout-system-electrical-thermal-transient coupling at all levels. It is this multiple-coupling effect that makes ESD protection designs very challenging, to a certain extent, being mysterious. It is also due to this reason that ESD protection device modeling becomes extremely difficult, which, in turn, makes ESD protection design simulation very challenging, even today. This situation has been compounded by the lack of efficient CAD algorithms and accurate simulation tools for full-chip ESD protection design simulation and verification. ESD protection device modeling will be discussed in Chapter 13. Circuit-level ESD simulation and chip-scale ESD physical design verification will be discussed in Chapters 13 and 15, respectively. This chapter discusses numerical ESD simulation using TCAD techniques.

Figure 8.1 depicts the traditional experience-based trial-&-error ESD protection design approach, which is actually still widely used in industrial ESD protection design practices, believe it or not. In this trial-&-error ESD protection design approach, an IC circuit designer sets the ESD specifications (Specs) for a new IC chip product and then asks an ESD guru in the company to deliver a magic ESD protection solution for the chip. The ESD design master will then search into his/her jewel box full of prior ESD design experiences (i.e., prior design successes) and start the mind

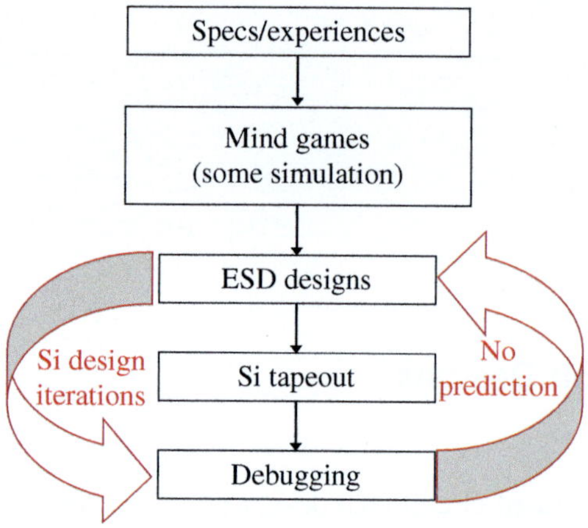

Figure 8.1 Traditional trial-&-error ESD protection design flows are experience-based aided by very limited device or circuit level simulation.

game to find a good ESD protection solution for the IC engineer. The ESD guru's mind game may include some mathematics and/or limited simulation, which delivers an ESD protection solution, but likely not including quantitative ESD protection function details (i.e., the Specs numbers) and without knowing the IC schematic details. The magic ESD protection structures will then be dropped onto the IC layout designed by the IC designer, which will be taped-out for Si fabrication. With the fingers crossed, the IC designer would bet on a good luck with the Si wafer coming in months. Unfortunately, ESD failures occur frequently in chip measurements. The next tedious, painful, time-consuming, and costly task is to go through the long debugging procedures to figure out what was wrong in ESD protection designs. After debugging, an IC design revision and a new tape-out will follow, then waiting anxiously again for the Si and testing results. In industrial design practices, it often takes two to three iterations to make the ESD protection works. This giver-taker style trial-&-error ESD protection design approach is becoming unacceptable to large and complex chips designed in very advanced technologies. Imagine the consequence of an ESD design failure in developing a new 5G smartphone system-on-a-chip (SoC) chipset in 5 nm complementary metal-oxide-semiconductor (CMOS), which typically takes more than one year for a large team of several hundred IC design engineers to complete with typical tape-out costs of anywhere from two-hundreds millions to four-hundreds millions of US dollars today, a company simply may not survive such "fun" design iterations caused by the trial-&-error ESD protection design approach. Indeed, first-Si design success has been a norm in IC design today, thanks to efficient and accurate CAD tools. Therefore, it is just natural to expect the same for on-chip ESD protection designs, which should be quantitative-oriented and CAD-based, with the prior design experiences playing a supportive role only. In summary, it is the "numbers" associated with ESD discharging "functions," not "experiences," that determine successes of on-chip ESD protection designs.

Traditionally, ESD simulation can be conducted at either device level or circuit level [1]. Device level ESD simulation is necessary because on-chip ESD protection relies on low-R discharging through an ESD protection device as the core, for which device physics play a key role. On the other hand, circuit-level ESD simulation is required because, after all, on-chip ESD protection is for IC chips, not about any individual ESD protection device itself.

Device-level ESD simulation is required to investigate the complex ESD phenomena dominated by semiconductor device physics, which must be understood clearly in order to fully model the multiple-coupling effects (transient, electrical, thermal, materials, process, device, circuit, layout, and system-level coupling) of an ESD protection structure on a chip during ESD stressing. The most commonly used ESD protection structures are active devices (e.g., diodes, bipolar junction transistors (BJTs), metal-oxide-semiconductor field-effect transistors (MOSFETs), and silicon-controlled rectifiers (SCRs)) that follow the fundamental semiconductor device physics equations, such as those shown below in their simplified formats [2].

Poisson's equation, coming from Maxwell's equations, defines the electric field (E) as a minus gradient of the electrostatic potential (ϕ), leading to the equations below in both 1D and 3D fashions,

$$\frac{d\phi(x)}{dx} = -E(\mathrm{x}) \tag{8.1}$$

and

$$\vec{\nabla}\phi(x,y,z) = -\vec{E}(x,y,z) \tag{8.2}$$

and the electrostatic potential can be obtained by solving Poisson's equation,

$$\frac{d^2\phi(x)}{dx^2} = -\frac{\rho(x)}{\varepsilon} \tag{8.3}$$

and

$$\nabla^2\phi(x,y,z) = -\frac{\rho(x,y,z)}{\varepsilon} \tag{8.4}$$

where ε is dielectric constant, ρ is charge density of free carriers (with electron density of n, and hole density of p) and fixed charges of ionized impurities (i.e., donor density, N_D^+, and acceptor density, N_A^-), given as

$$\rho = -q(p - n + N_D^+ - N_A^-) \tag{8.5}$$

where q is the elementary electronic charge.

The current density (J) is derived from the carrier transport equations including both electric field induced drift current and carrier gradient originated diffusion current,

$$\vec{J}_n = qn\mu_n\vec{E} + qD_n\vec{\nabla}n \tag{8.6}$$

$$\vec{J}_p = qp\mu_p\vec{E} - qD_p\vec{\nabla}p \tag{8.7}$$

and the total current density is given by

$$J = J_n + J_p = q(n\mu_n + p\mu_p)E + qD_n\frac{dn}{dx} - qD_p\frac{dp}{dx} \tag{8.8}$$

where μ_n and μ_p are mobilities of electron and hole, and D_n and D_p are diffusion constants of electrons and holes, respectively.

Continuity equation states that a change in carrier density over time reflects the balance of incoming and outgoing carrier flux, and carrier generation (G) and recombination (R), following

$$\frac{\partial n(x,t)}{\partial t} = \frac{1}{q}\frac{\partial J_n(x,t)}{\partial x} + G_n(x,t) - R_n(x,t) \tag{8.9}$$

and

$$\frac{\partial p(x,t)}{\partial t} = -\frac{1}{q}\frac{\partial J_p(x,t)}{\partial x} + G_p(x,t) - R_p(x,t) \tag{8.10}$$

Lattice temperature and heat flow relationship follows the heat transfer equation below for heat conduction,

$$mc\frac{\partial T}{\partial t} = H + \nabla \cdot \{\kappa(T)\vec{\nabla}T\} \tag{8.11}$$

where H, m, κ, and c represent total heat source, materials mass density, thermal conductivity, and specific heat, respectively.

Device-level ESD simulation is conducted using numerical simulation, also called TCAD for semiconductor devices. In general, a semiconductor device is partitioned into a mesh of nodes and the semiconductor device physics equations are solved simultaneously at each mesh node across the whole device structure. Device-level ESD simulation is typically used to examine possible ESD-induced thermal failures caused by large ESD transient currents. In such cases, TCAD ESD device simulation will check the maximum lattice temperature (T_{max}) at each mesh node across the ESD protection structure during an ESD discharging event. When TCAD simulation finds the T_{max} at any node exceeding the melting temperature (preset as the ESD thermal failure criterion for a given materials), e.g., 1421 °C for Si, TCAD simulation will report an ESD thermal failure occurrence. It is also possible to use TCAD simulation to study voltage breakdown induced ESD failures, such as those in MOSFET gate oxide layers and PN junctions, by checking the transient electrostatic potential and electric field density at critical device nodes under ESD stressing. Several disadvantages exist for TCAD ESD device simulation though: First, accurate TCAD ESD device simulation requires thorough a understanding of IC process technologies and semiconductor device physics that are often beyond what an ordinary IC circuit design can handle. Second, TCAD ESD simulation at device level can only reveal the ESD discharge behaviors of an individual standalone ESD protection device, which cannot guarantee chip-level ESD protection using the same ESD protection device in an IC. Third, TCAD ESD simulation is very computing intensive because it must handle a large number of nonlinear semiconductor device physics equations for a huge volume of device mesh nodes simultaneously. The main advantage of TCAD ESD device simulation is that it can reveal the insights of ESD discharging characteristics, hence, uncovers the transient ESD discharging functions of an ESD protection device in IC, described by the ESD-critical parameters. TCAD ESD simulation remains the only technique capable of studying full ESD discharging functionalities of ESD protection structures, which cannot be achieved by circuit-level ESD simulation without accurate ESD device models.

On the other hand, ordinary IC designers obviously prefer to conduct circuit-level ESD simulation, if ever possible, just like to simulate a large circuit using SPICE type circuit simulators. A circuit simulator is certainly easier to use compared to its TCAD counterpart. This is because SPICE treats each device in a circuit as a black box with its terminal electrical characteristics being described by a device model, such as BSIM device models. Therefore, a circuit design engineer does not have to understand all complex details of semiconductor device physics as in TCAD simulation. Consequently, circuit simulation can readily handle very large and complex IC chips by simply solving linear circuit equations such as those governed by the Kirchhoff current and voltage laws as below

$$\sum_{x=1}^{m} i_x = 0 \tag{8.12}$$

and

$$\sum_{y=1}^{n} v_y = 0 \tag{8.13}$$

meaning that all branch currents summing at one circuit node is zero and all segment voltages within one circuit loop add up to zero. Obviously, solving a linear equation set is much easier than handling a large number of nonlinear semiconductor device physics equation set as in TCAD cases. Therefore, full-chip ESD simulation using a SPICE-like circuit simulator, theoretically, would be very useful for simulating large IC chips, which is very difficult for whole-chip ESD simulation using TCAD. Of course, this fundamental TCAD ESD roadblock would be nothing, should the quantum supremacy become a realty someday, but not today or tomorrow. Circuit-level ESD simulation using SPICE can be quite informative for circuit-critical properties, such as evaluating the timing of ESD triggering, for instance, in a classic two-stage ESD protection net (Figure 5.1), a gcNMOS ESD protection cell (Figure 5.4), or a nonsnapback NMOS power clamp (Figure 5.26), where both IR drop and RC effects can be readily handled at circuit level. However, to accurately simulate the ESD discharging functions of any ESD protection structures, which is the core of ESD protection designs, accurate ESD protection device models are required, which unfortunately, is not as easy as getting compact device models for SPICE simulation due to several difficulties: First, many ESD protection devices have snapback *I–V* characteristics that may not be easily covered by ESD device models. Second, SPICE-like linear circuit simulation was developed for small-signal circuit analysis, while ESD protection circuit must handle extremely large and ultrafast ESD pulses (i.e., signals), which is unfriendly to normal circuit simulators. Third, thermal effect is uniquely important in ESD discharge operations, which is also ultrafast. While temperature effect may be included in BSIM device models to account for the thermal effects, accurate ESD device thermal modeling is still a huge challenge. Theoretically, as long as an ESD-induced heat source (i.e., hot spot) within an ESD protection structure is known, circuit-level ESD thermal simulation can be performed by simply solving the heat distribution equations. For example, Figure 8.2 illustrates one over-simplified ESD thermal model of an NMOSFET ESD protection device where the ESD heating source is assumed to be located at the edge/corner of the Drain junction that can be modeled by a parallelepiped-shaped heater with physical dimensions of *a*, *b*, and *c* [3]. Enhanced SPICE circuit simulation can then be performed by solving thermal distribution equations to find the device lattice temperatures, which are used as the ESD thermal failure criterion. Unfortunately, it is unrealistic to practically and accurately "locate" the transient ESD discharge-heating source, which may be typically a random "hot spot" or "hot line" due to extremely fast ESD discharging. It is further impossible to accurately "define" the shape and dimensions of an ESD hot spot of whatever shapes, which may also be varying and "flying" over time. In other word, any ESD hot spots are typically unknown for given ESD protection structures. Further, the thermal boundary conditions

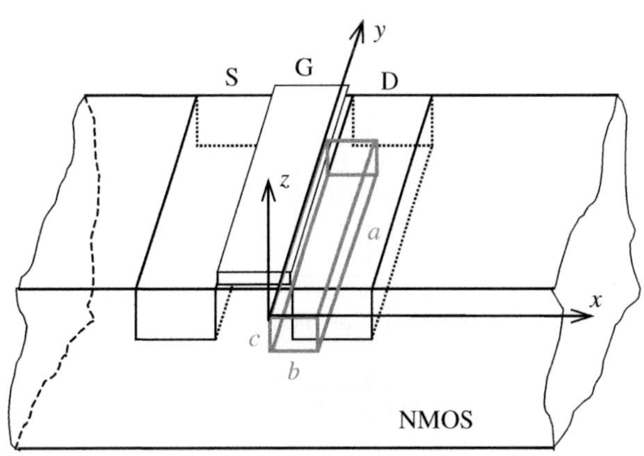

Figure 8.2 Illustration of an over-simplified ESD heating source of an NMOSFET ESD protection device. The ESD-induced heating source is assumed to be a parallelepiped with dimensions of *a*, *b*, and *c* [3].

of any ESD hot spot must be known in order to solve the ESD heat flow equations by a circuit simulator. Unfortunately, these desired thermal boundary conditions, such as thermal resistance and capacitance, cannot be physically measured to calibrate the ESD device thermal models because the ESD hot spot is deeply inside an ESD protection structure that is also varying and floating in a pico-second time scale. Obviously, without accurate ESD device thermal modeling, accurate ESD failure simulation is impractical for any real-world ESD protection designs. Fourth, due to extremely large ESD surge, the layout effect (i.e., ESD device *corner/edge effect*) plays a critical role in ESD discharging and ESD failures, which may not be accurately modeled for ESD protection structures yet. In summary, circuit-level ESD simulation can be informative and theoretically possible, but cannot by itself simulate the critical ESD discharge functionalities without accurate ESD device models, and TCAD ESD simulation is still main technique to reveal the critical ESD discharging functionalities at chip level.

8.2 TCAD-Based Mixed-Mode ESD Simulation-Design Methodology

As discussed previously, the ultimate goal in designing ESD-protected ICs is to achieve design optimization and prediction of on-chip ESD protection designs by CAD simulation. Device-only ESD simulation provides quantitative insights of ESD discharge functions of individual/standalone ESD protection structure, which does not include device–circuit interactions between ESD protection devices and the core circuit under ESD protection. Circuit-only ESD simulation is capable of examining IR drop and RC coupling effects, and allows timing analysis at full-chip level, which however, cannot reveal the critical ESD discharging functionalities of the ESD protection structures due to lack of accurate ESD device models. To accurately address the complex multiple-coupling effects of ESD protection network at full-chip level, a *TCAD-based mixed-mode ESD simulation-design methodology* can be used to achieve design optimization and prediction of ESD protection circuit at chip level [4]. Figure 8.3 depicts the ESD design framework using the mixed-mode ESD simulation-design approach, where the ESD design task starts with user-defined ESD protection Specs and engineer's ESD design experiences. The major design effort is to perform comprehensive TCAD mixed-mode ESD design simulation at chip level, which includes the complex device-circuit interactions during transient ESD discharging. With most of your brain cells burned in the mixed-mode ESD simulation phase, the ESD protection structures can be

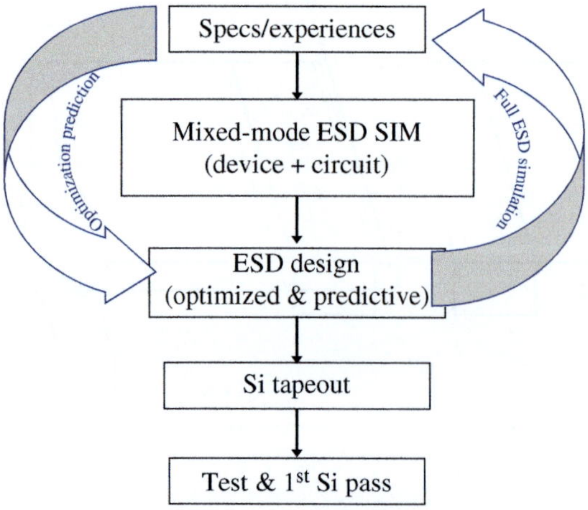

Figure 8.3 A framework for TCAD-based mixed-mode ESD protection simulation-design method relies heavily on integrated and interactive device and circuit simulation at chip level using real ESD stimuli to achieve on-chip ESD protection design optimization and prediction.

optimized with the circuit core (the "user" or "victim") being considered simultaneously; hence, to achieve ESD design prediction at chip level. ESD protection design optimization is as important as ESD design prediction for any real-world IC designs because both under-design (i.e., unexpected ESD failure) and over-design (i.e., too much ESD-induced design overhead such as ESD-induced parasitic capacitance) of ESD protection are harmful to ICs practically. Next, after the full-cycle of TCAD mixed-mode ESD simulation, one can safely "drop off" an optimized ESD protection structure onto the circuit core as a working ESD protection "solution" for a chip and tape-out the IC design for wafer fabrication. It is well known that sturdy bricks do not guarantee a stable skyscraper where the architecture plays a key role. Similarly, a "working" individual/standalone ESD protection device may not be an ESD protection "solution" for a specific IC chip where ESD–IC interactions must be considered at chip level. This is the right approach to realize the desired first-Si pass in practical ESD protection designs.

TCAD mixed-mode ESD simulation-design method can address the complex transient-electro-thermal-materials-process-device-circuit-layout coupling effects within the same TCAD simulation platform in a closed-loop fashion during chip-level ESD simulation. The mixed-mode ESD simulation principle is to virtually "fabricate" the actual ESD protection structures by TCAD process simulation, to examine ESD discharge functionalities of the created ESD protection structures by TCAD ESD device simulation, and to evaluate ESD discharging behaviors at chip level under real-world transient ESD stressing by integrated device-circuit level ESD simulation by taking full account of any ESD-IC interactions. To predict ESD protection at chip level, ESD simulation must reveal the transient ESD discharge I–V characteristics and provide the quantitative ESD-critical parameters at both single ESD device level and circuit level, which include, ESD triggering threshold (V_{t1}, I_{t1}, t_1), ESD holding voltage and current (V_h, I_h), and ESD failure threshold (V_{t2}, I_{t2}), at least. Figure 8.4 describes the design flow of TCAD mixed-mode ESD simulation-design method. The input of the mixed-mode ESD simulation flow include user-defined ESD protection Specs and the process technology Specs (your customers, i.e., chip design and product engineers, choose a specific IC technology to be used), and the ESD designer must thoroughly understand these Specs, as well as the IC circuit schematics and Specs (know your "customer"). Knowing the process technology details is nontrivial for ESD design success. Different processes (e.g., CMOS, BiCMOS, BCD, HV, and FinFET) set different ESD design constraints, while also offering varying options for ESD protection designs. For example, gate oxide thickness and breakdown voltage, doping profiles, PN junction breakdown voltage, diffusion regions, and supply voltages, etc., may critically affect the ESD protection design strategy, choice of ESD protection devices, and ESD performance specs, which practically sets up the stage where an ESD designer can play. For instance, a normal ggNMOS ESD protection device of moderate V_{t1} may not be used for low-voltage circuit domain due to its relatively "high" V_{t1}, and is also not suitable for HV circuit module because its relatively "low" V_{t1} will cause short-circuit under normal IC operations. In another example, a gcNMOS ESD protection should not be used for RF ICs because it brings in too much parasitic capacitance that will seriously affect RF IC performance. Similarly, a "good" SCR ESD protection may not be available in SOI CMOS process due to the lack of vertical diffusion layers. In the next step, a good brain storm will result in a long and rich list of many possible ESD protection structures out of the given process technology. Creating this initial list of possible ESD protection structures is very helpful because one has to dig out the gold from a hidden mine. Yet, most of the available ESD protection structures may not be suitable for the given IC (e.g., circuit schematics and specs), which will be removed during a second round of brain excise, leading to a small set of most-likely ESD protection structures, typically very few of them. Comparative thinking in selecting a best-suited ESD protection structure is always beneficial in practical designs. One then moves to the main

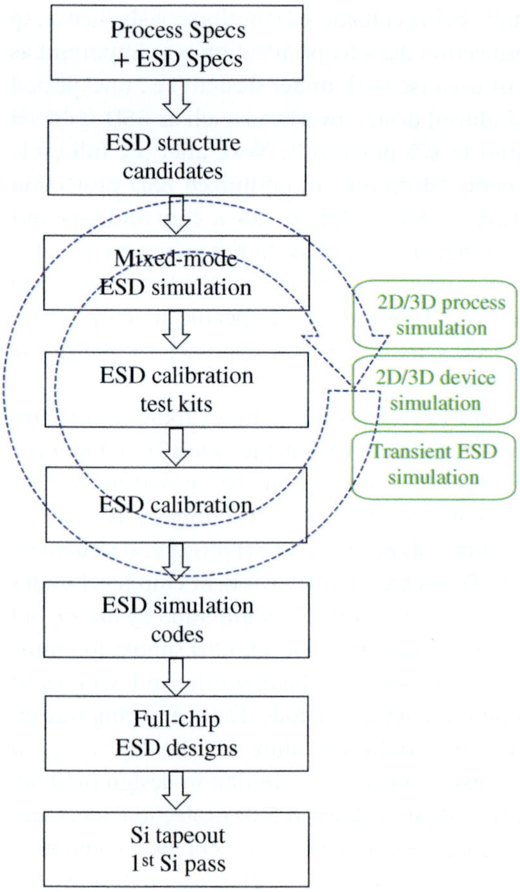

Figure 8.4 A detailed TCAD-based mixed-mode ESD simulation-design flow.

ESD protection design phase, which is to conduct comprehensive TCAD mixed-mode ESD simulation including process simulation, ESD device simulation, and ESD circuit simulation. In process simulation, TCAD is used to create "real" ESD protection structures using "real" technology specs (i.e., process recipes) that include all actual fabrication process steps, such as doping, diffusion, implantation, oxidation, annealing, deposition, etching, as well as the associated time and thermal budgets for temperature ramp-up and cool-down. Using "real" process recipes to create the ESD protection device is important because the device structure details can be very different (e.g., lateral diffusion is controlled by the thermal budget) and the device specs are entirely determined by the fabrication process details (e.g., thin gate oxide quality) and any subtle variations, which eventually affect ESD performance of the ESD protection device fabricated. However, two problems exist here: First, the technology details (process recipes) are highly confidential that is not typically available to a fabless IC design company. Second, full process simulation using all "real" technology recipes is very computing-hungry. Therefore, a balance is needed for TCAD process simulation. As an alternative, an ESD protection device may be created by TCAD by using known doping profiles measured, which dramatically reduces TCAD process simulation time, but may compromise the accuracy of ESD protection device simulation later. Next, device-level ESD simulation will be conducted by TCAD for the ESD protection structure created with the goal being to thoroughly examine its ESD discharging functionalities. The input of TCAD ESD device simulation is an ESD stimulus and the output will be the ESD discharge I–V curves, the ESD-critical parameters (V_{t1}, I_{t1}, t_1, V_h, I_h, V_{t2}, I_{t2}, I_{leak}, C_{ESD}, etc.,) and other critical details, such as lattice temperature (T_{max}) during ESD stressing. To evaluate ESD protection capability of a given ESD protection device, ESD failure criteria need to be preset for TCAD ESD device simulation. The most commonly used ESD failure criterion used is the melting temperature of device materials. For example, $T_{\text{melt}} = 1685\,\text{K}$ for silicon, $T_{\text{melt}} = 823\,\text{K}$ for aluminum, and $T_{\text{melt}} = 1357\,\text{K}$ for copper metal interconnects are typically used in simulating ESD thermal failures. During TCAD ESD simulation, the simulator constantly scans the lattice temperature at all mesh nodes across an ESD protection device to extract the instant maximum lattice temperature (T_{max}) during the ESD discharging procedure. At any time during an ESD discharge simulation, $T_{\text{max}} > T_{\text{melt}}$ occurs for any materials at any location, ESD thermal failure will be reported by TCAD simulation, resulting in the estimated ESD thermal failure level, i.e., I_{t2}, (or, translated into V_{t2} per a given ESD testing model). Further, circuit-level ESD simulation using the given ESD protection structure for a specific circuit core can be conducted by ESD circuit simulation using TCAD too. The circuit-level ESD simulation include both ESD protection devices and core circuit (typically an I/O circuit block) in ESD simulation, where the ESD discharge functionalities of the ESD protection device are simulated by solving the nonlinear device physics equations, while circuit-level ESD protection functions will be evaluated by including ESD device terminal specs (instantly extracted from device simulation) into solving the large set of linear Kirchhoff circuit equations. The device–circuit interactions during transient ESD stressing can be accurately addressed because mixed device and circuit simulation are conducted at the same time, albeit interactively and iteratively, by the same TCAD simulation, therefore, achieving on-chip ESD protection design prediction at chip level (not just simulating a standalone individual ESD protection device without circuit context). It is noteworthy that, though being full-chip ESD simulation in nature, TCAD ESD simulation should be limited to circuit block level that includes the ESD protection device and the I/O circuit block for a practical reason: at least now, TCAD, developed for device simulation, is not realistic for simulating ESD operations of a large IC chip. Will the quantum supremacy eventually solve this TCAD efficiency problem? Time will tell. It is also important to know that any assumption in ESD simulation will compromise the accuracy of ESD simulation results. For example, as discussed earlier, one may assume knowing (impossible though) the ESD heating source being a parallelepiped in a MOSFET ESD protection device as shown in Figure 8.2, then run circuit-level ESD simulation using a SPICE-like simulator. However, the simulation accuracy is questionable due to the assumption of your little brain. Typically, TCAD ESD simulation is performed in both static and transient modes. Since TCAD ESD simulation is very computing-hungry, the common practice is to conduct initial static (DC) ESD simulation to gain a rough picture of ESD discharge behaviors of a given ESD protection structure, which then serves to fine-tune ESD simulation flow in transient ESD simulation that is much more time-consuming. Similarly, TCAD ESD simulation can be conducted in 2D or 3D ways, which certainly have major impacts on ESD simulation accuracy. Particularly, real 3D ESD simulation is critical to accurately addressing the layout effects of ESD protection designs because this may be the only way to take into account the edge/corner effects of ESD protection structures. True 3D simulation requires using real process recipes to create real 3D ESD protection devices in TCAD process simulation and the 3D mesh in 3D ESD device simulation will increase exponentially in 3D TCAD ESD simulation. Again, balance must be considered in TCAD ESD simulation. Assume calibration was completed for mixed-mode ESD protection design simulation, the calibrated ESD simulation codes can then be used to design various ESD protection structures at chip level with design optimization and prediction, heading to first-Si design success in practical IC designs. It is important to know that using real-world ESD stimuli is critical to ESD design simulation in order to achieve ESD design prediction. The famous statement for circuit simulation says "junk in, junk out," which applies to ESD design simulation too. To avoid any assumption, it is suggested that a suitable ESD source equivalent circuit is used to generate an ESD pulse, per any selected ESD test standard, as the input to the mixed-mode ESD simulation circuit deck as illustrated in Figure 8.5. In Figure 8.5, an HBM model circuit is used where the C_{ESD} (electrostatic charge pool) is pre-charged to 2 kV, which then discharge into an ESD-protected IC through R_{ESD} and L_{ESD}. The HBM ESD stressing waveform produced by the HBM ESD source circuit must be exactly the same as that defined in the HBM test standard, which ensures ESD simulation accuracy. Mixed-mode ESD simulation under the HBM ESD pulse is then performed for the ESD-protected IC. Similarly, other ESD source circuits, e.g., per MM, CDM, IEC ESD models, can be used to simulate ESD discharging characteristics under any selected ESD test models. Alternatively, an ESD pulse waveform defined in a given ESD test standard may be used directly as an ESD stimulus for mixed-mode ESD simulation. It is obvious that calibration determines accuracy of ESD simulation, hence playing a vital role in TCAD mixed-mode ESD design simulation. As depicted in Figure 8.6, ESD simulation calibration can be performed in various ways: comparing static ESD simulation with ESD testing by a curve tracer, calibrating transient ESD simulation with transient TLP (or, VFTLP) testing, and matching measured I_{t2} with the ESD protection level obtained by ESD zapping test. In ESD simulation

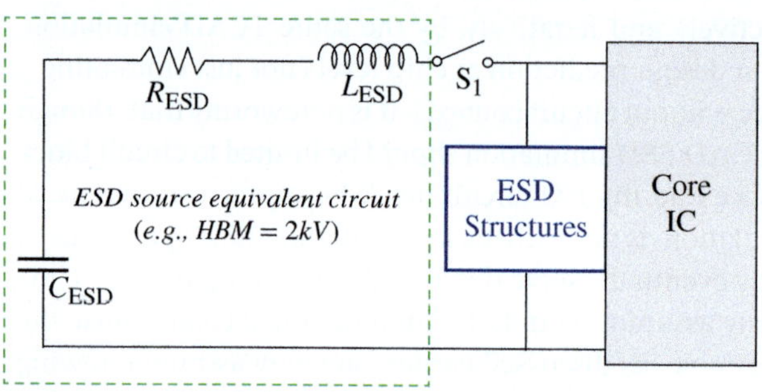

Figure 8.5 An exemplar TCAD-based mixed-mode ESD simulation schematic for HBM ESD zapping simulation.

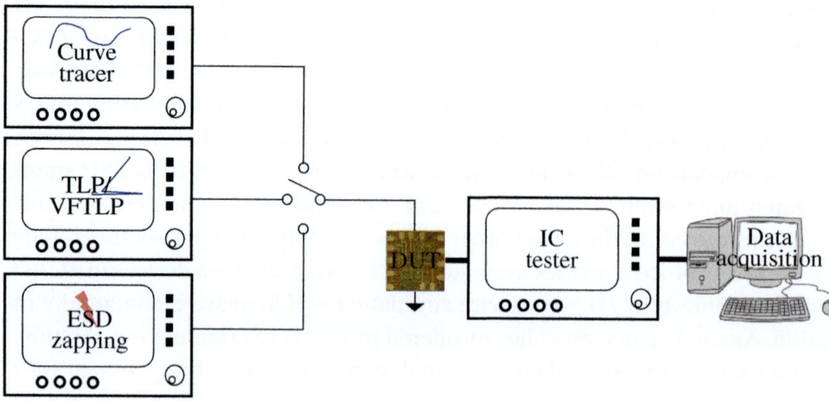

Figure 8.6 A typical ESD simulation calibration framework.

calibration, the first critical thing is to use "good" ESD test kits for both ESD simulation and testing. In general, an ESD test kit should be *simple*, but resemble the ESD protection device to be simulated and tested as much as possible. One should not use the "real" ESD protection structure to be used on a chip for TCAD ESD calibration. The reason is that ESD simulation calibration is very complicated, and a "real" ESD protection structure has too many unknown effects (to be studied by TCAD ESD simulation) and too much complexity affecting actual ESD discharge behaviors. For example, a multi-finger ESD protection structure involves very complex edge/corner layout effects and its transient ESD discharging characteristics can be dominated by the actual metal routing in layout design (not ESD device physics), which will make accurate ESD calibration practically impossible (e.g., calibration is for Si failure, but actual ESD failure is with metal interconnects). As depicted in Figure 8.7, a simple *single-finger* ESD test device should be designed, fabricated, and characterized first, which is then calibrated against TCAD ESD simulation of the *same* simple device. Any device variation during Si testing and TCAD calibration can make the ESD simulation calibration useless or at least misleading. ESD simulation calibration can be very involving, but possible and practical.

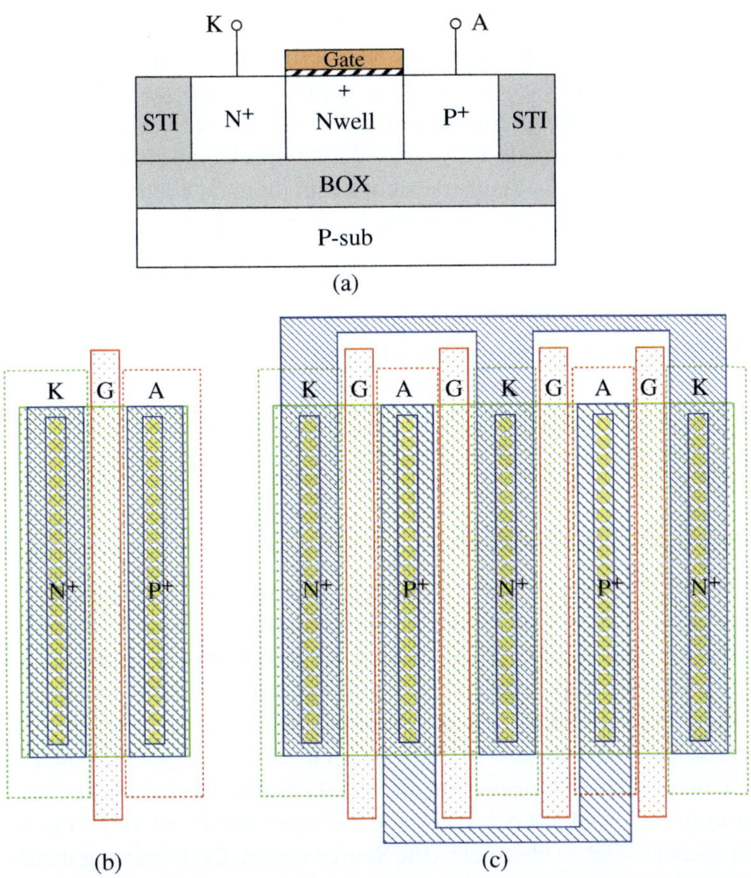

Figure 8.7 Examplar ESD test pattern for TCAD ESD simulation calibration: (a) a cross-section view of a gated P^+/NW ESD protection diode in SOI CMOS, (b) layout view of a single-finger P^+/NW ESD protection diode that is good for TCAD ESD calibration, and (c) a multiple-finger P^+/NW ESD protection diode layout used in real-world ESD protection designs that should not be used for TCAD ESD calibration due to its complexity.

Another key benefit of circuit-level ESD simulation is that it allows to evaluate on-chip ESD performance against other ESD failure mechanisms and criteria. For example, the common ESD voltage breakdown failure criterion induced by the large transient electric field of an ESD event can be evaluated at full-chip level by mixed-mode ESD simulation. During mixed-mode ESD simulation, all node voltage and branch current values induced by ESD stressing can be obtained across the whole chip by TCAD ESD simulation. A map of preset ESD failure voltage at each critical circuit node on a chip can be automatically and instantly compared with the simulated ESD-induced node voltage across the circuit schematics. If any node voltage exceeds the breakdown voltage at the given node during an ESD event per mixed-mode ESD simulation, it reports an ESD voltage breakdown failure for the chip. This is a critical function for ultrafast ESD events, such as CDM ESD events.

Even with accurate ESD calibration, it is cautious that a calibrated ESD simulation deck should only be applied to the same type of circuits in the same process technology using same type of ESD protection structures (i.e., like ESD devices). The reason is that ESD protection solution is not universal and not portable. Rather, ESD protection is highly process- and circuit-specific. ESD calibration may vary significantly for different ESD device types (e.g., diode, MOSFET, SCR) in the same technology, and same ESD protection structures made in different process technologies (e.g., bulk CMOS versus SOI CMOS). It is very important to know that there is no magic one-for-all ESD protection solution for all ICs on Earth. Custom design is required for any good ESD protection designs, which can only be done by comprehensive mixed-mode ESD design simulation. To this end, the main advantage for the mixed-mode TCAD-based ESD simulation-design method is that the seamlessly integrated process-device-circuit-transient simulation, facilitated by TCAD software, makes it feasible to simultaneously address the multiple-coupling effects critical to ESD protection functions at chip level.

8.3 Mixed-Mode ESD Simulation-Design Examples

In this section, several practical ESD protection design examples are discussed using the TCAD-based mixed-mode ESD simulation-design methodology to show its design capability.

8.3.1 Example 1: Understand TCAD ESD Simulation

In this example, a classic ggNMOS ESD protection device designed in a 180 nm CMOS process is used to explain the general meaning of TCAD mixed-mode ESD simulation [5, 6]. Figure 8.8 depicts the ggNMOS ESD protection structure generated by TCAD process simulation in a 180 nm CMOS, which follows a general ggNMOS ESD protection design guideline, i.e., the *source-contact-to-gate spacing* (SCGS) in the layout should be minimized and the *drain-contact-to-gate spacing* (DCGS) should be large enough. Hence, SCGS = 0.36 μm and DCGS = 2.15 μm are chosen per TCAD ESD simulation optimization. In fact, in practical ESD protection designs at advanced technology nodes, one should not choose the minimum-allowed SCGS because the ESD-generated heat at the Drain junction can easily spread over across the short channel to the Source region, likely causing metal melting at the Source contact. Simple DC ESD sweeping simulation was conducted first by directly ramping up a DC bias at the I/O to gain a quick look of the likely ESD discharge function as shown in Figure 8.9, which lead to first-order ESD-critical parameters including ESD triggering (V_{t1}, I_{t1}),

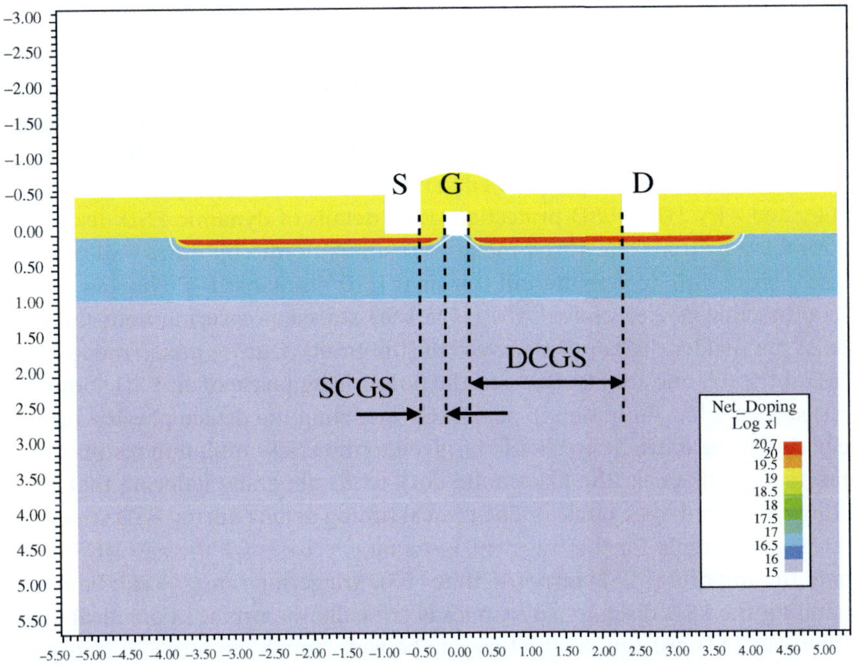

Figure 8.8 A cross-section for a TCAD-created ggNMOS ESD protection device in a 180 nm CMOS process.

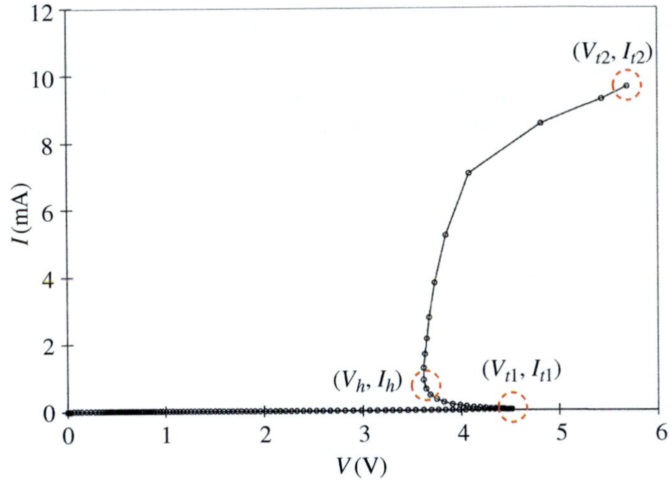

Figure 8.9 DC sweeping ESD discharge I–V curve by TCAD ESD simulation shows ESD-critical parameters: V_{t1}, I_{t1}, V_h, I_h, V_{t2}, I_{t2}, but no timing data.

holding (V_h, I_h), and thermal breakdown (V_{t2}, I_{t2}) values. Due to significant ESD heating, it is observed that the ESD discharge resistance increases substantially as the temperature increases. However, the critical time-domain information (i.e., the transient ESD discharging behaviors) cannot be obtained in DC ESD simulation. The DC ESD simulation is typically used to steer the transient ESD simulation that is usually very time-consuming. One main goal for ESD simulation is to optimize the ESD design size, which was determined to be $W \approx 111\ \mu m$ for the ggNMOS device finger width for the targeted 4 kV HBM ESD protection. Rich details of dynamic ESD discharging behaviors can be revealed by transient mixed-mode ESD simulation to optimize and predict ESD protection designs. Figure 8.10 depicts the full transient ESD discharge I–V characteristics for the ggNMOS ESD protection device under 4 kV HBM ESD stressing covering both the rising and falling phases of the incident ESD pulse waveform (the Inset). During mixed-mode ESD simulation using TCAD software, one should focus on the pulse rising phase of an ESD stimulus because existing TCAD software has limitation in accurately modeling the device physics during the ESD failing pulse period. Hence, attention should be given to the ESD simulation results until the incident ESD pulse reaches its peak (the black solid dot), while generally ignoring that during the ESD pulse falling phase (red open circles). The critical timing details during ESD stressing is given in Figure 8.11 as an example for the transient V–t characteristics for the ggNMOS ESD protection device where the important ESD response time (ESD triggering time, t_1) can be readily obtained. Understanding the ESD discharge dynamics is critically important in practical ESD protection designs. Often, an ESD protection structure can be very "tough" in terms of transient ESD current-handling capability (i.e., bearing the ESD pulse energy), but a chip may still fail at rather lower I_{t2} level that is often because the "tough" ESD protection cannot respond to a fast ESD pulse quickly enough, i.e., during CDM ESD stressing. Another important information from TCAD-based ESD simulation is the insight into the device thermal dynamics during ESD stressing. Figure. 8.12 shows transient maximum lattice temperature (T_{max}) versus ESD discharge current and Figure 8.13 depicts the dynamic T_{max} in the time domain across the ggNMOS ESD protection structure. It clearly shows that the lattice temperature continuously increases during the ESD pulse rising phase and reaches to $T_{max} = 975$ K at the peak of the incident ESD pulse. It is also observed

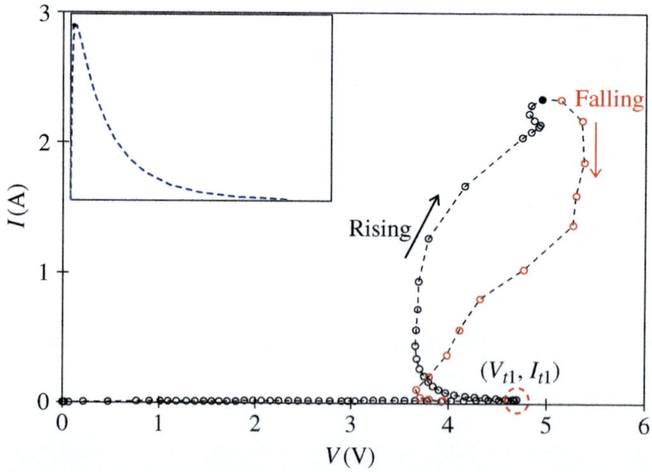

Figure 8.10 Transient ESD discharging I–V curve of a ggNMOS ESD protection device under 4 kV HBM ESD stressing by TCAD ESD simulation. Red circles correspond to the falling phase of the HBM ESD pulse stimulus and the black solid dot represents the peak of the HBM ESD pulse. The Inset shows the incident HBM ESD pulse waveform as the stimulus for TCAD ESD simulation.

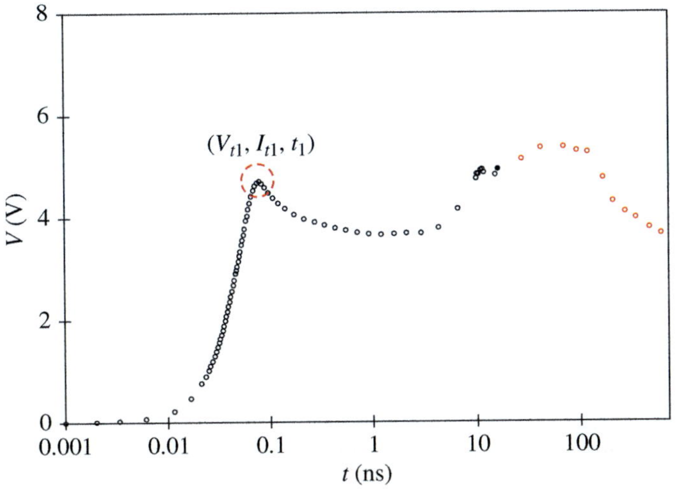

Figure 8.11 Transient ESD discharging $V-t$ curve of a ggNMOS ESD protection device under 4 kV HBM ESD stressing by TCAD ESD simulation. Red circles correspond to the falling phase of the HBM ESD pulse stimulus and the black solid dot represents the peak of the HBM ESD pulse.

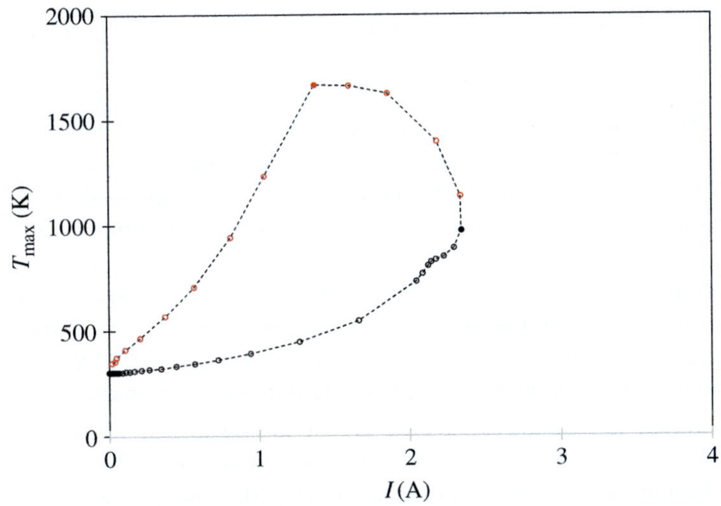

Figure 8.12 Transient ESD discharge $T_{max}-I$ characteristics of a ggNMOS ESD protection device under 4 kV HBM ESD stressing by TCAD ESD simulation. Red circles correspond to the falling phase of the HBM ESD pulse stimulus. The black solid dot represents the maximum lattice temperature ($T_{max} \sim 975$ K) at the peak of the incident HBM ESD pulse, while the red solid dot is the overall-maximum lattice temperature ($T_{max} \sim 1667$ K) in the t-domain that is much higher that the T_{max} at the ESD pulse peak time, which occurs during the ESD pulse falling phase due to slow heat dissipation.

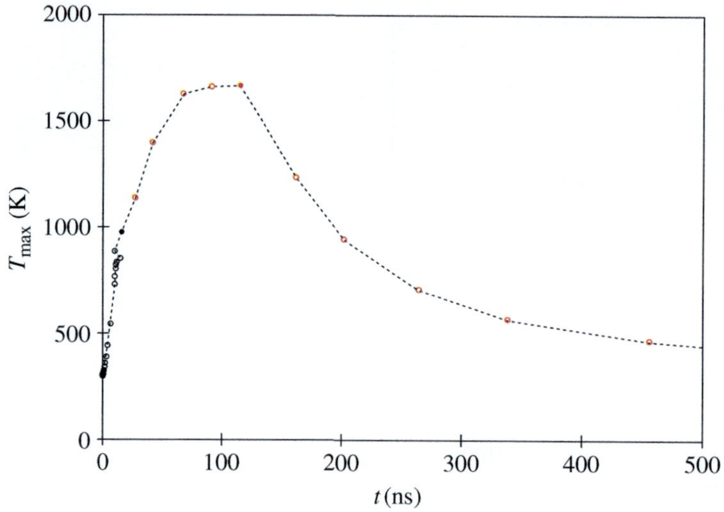

Figure 8.13 Transient ESD discharge T_{max}–t curve of a ggNMOS ESD protection device under 4 kV HBM ESD stressing by TCAD ESD simulation. Red circles correspond to the falling phase of the HBM ESD pulse stimulus. The black solid dot represents $T_{max} \sim 975$ K at the peak of the HBM ESD pulse and the red solid dot shows the overall-maximum lattice temperature $T_{max} \sim 1667$ K.

that as the incident ESD pulse start to fall, the lattice temperature continues to increase and peaks at $T_{max} = 1667$ °K (marked by the red solid dot; still lower than the Si melting temperature of ~1685 K; hence, the ggNMOS ESD device passes 4 kV HBM ESD protection level). This can be easily understood because Si IC is poor in thermal conduction and hence, the ESD-generated local heat will take time to dissipate after the ESD stimulus starts to disappear. This is a very important consideration in practical ESD protection designs, which can only be modeled by TCAD mixed-mode ESD simulation. To further illustrates the ESD thermal failure occurrence by TCAD mixed-mode ESD simulation, a ggNMOS ESD protection device for 2 kV HBM ESD protection was studied with the optimized device width of $W \approx 56$ μm. This ggNMOS device was then down-sized slightly so that it fails at 2 kV HBM ESD stressing by simulation. Figure 8.14 depicts the transient ESD discharge I–V characteristics for the slightly underdesigned ggNMOS ESD protection device under 2 kV ESD stressing and ESD thermal failure was observed during the ESD pulse falling phase as shown in Figure 8.15, where the ESD failure T_{max} (red solid dot) occurs after the ESD pulse peaking (black solid dot) due to heat dissipation latency. The ESD-induced hot spot, corresponding to the overall maximum T_{max} across the underdesigned ggNMOS ESD protection device was readily observed in the lattice temperature maps given in Figures 8.16 and 8.17.

8.3.2 Example-2: ggNMOS versus gcNMOS ESD Protection

In this section, we discuss design comparison of a ggNMOS ESD protection device and a gcNMOS ESD protection structure implemented in a 0.8 μm BiCMOS technology aided by TCAD mixed-mode ESD simulation analysis [7]. Per TCAD ESD simulation, the ggNMOS ESD protection device is optimized for SCGS = 2 μm and DCGS = 4 μm. Figure 8.18 depicts the simulated transient ESD discharge I–V curve for the ggNMOS ESD protection device, which shows an ESD triggering voltage of $V_{t1} \sim 14.68$ V. Figure 8.19 shows simulated V–t characteristics for ggNMOS featuring an ESD response time of $t_1 \sim 0.2$ ns. Obviously, this ggNMOS device is fast enough for not only

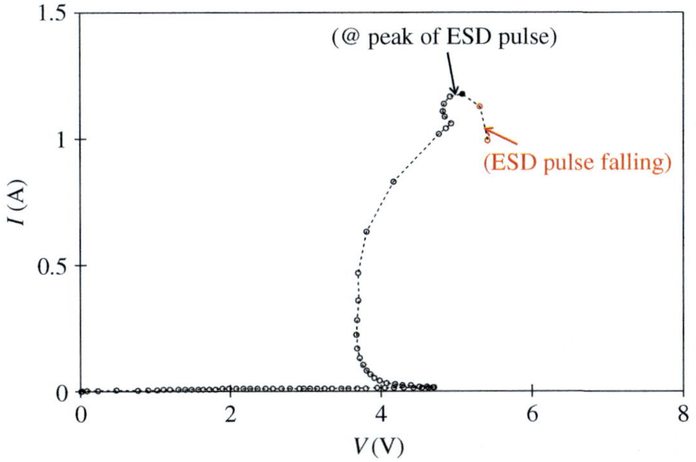

Figure 8.14 Transient ESD simulation for a ggNMOS ESD protection structure shows the ESD discharging I–V curve. The under-designed ggNMOS device fails at 2 kV HBM ESD stressing. The black solid dot corresponds to the peak of the incident HBM ESD pulse and the red circles follow the failing ESD pulse.

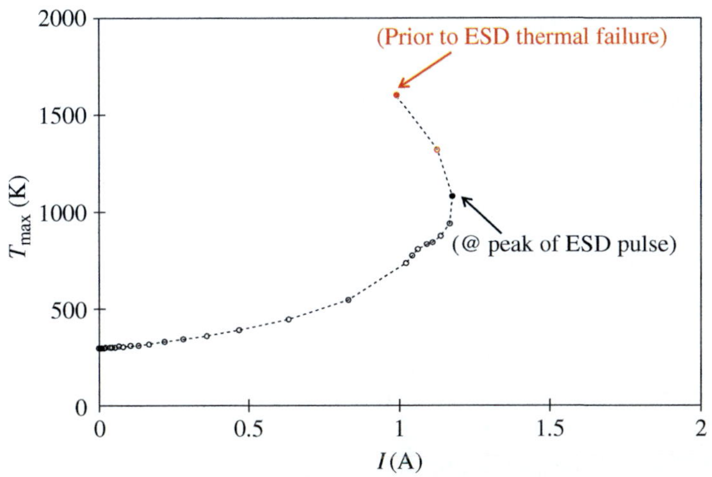

Figure 8.15 Transient $T_{max} \sim I$ characteristics for an under-sized ggNMOS ESD protection device under 2 kV HBM ESD simulation shows $T_{max} \sim 1080$ K at the peak of the HBM ESD pulse (the black solid dot) and thermal failure when $T_{max} > 1685$ K (beyond the red solid dot).

HBM ESD protection but also for ultrafast CDM and IEC ESD protection. Unfortunately, the ggNMOS triggering voltage is too high for many LV ICs. Further, V_{t1} is too close to V_{t2}, which will cause ESD triggering uniformity problem for multiple-finger layout designs. As discussed in Chapter 5, a practical design solution to this ggNMOS problem is to use a gcNMOS ESD protection structure where an RC coupling subnet serves to boost the gate voltage, leading to a reduced ESD triggering voltage. Design optimization of a gcNMOS ESD protection circuit was conducted by TCAD mixed-mode ESD simulation, giving an optimized $R = 10\,\text{k}\Omega$ and $C = 0.1$ pF. Figure 8.20 presents the simulated transient ESD discharge I–V curve for the gcNMOS ESD protection structure, showing a much-reduced ESD triggering voltage of $V_{t1} \sim 7.54$ V, significantly lower than $V_{t1} \sim 14.68$ V of the ggNMOS device. The V_{t1} of gcNMOS device is also lower than its V_{t2} hence,

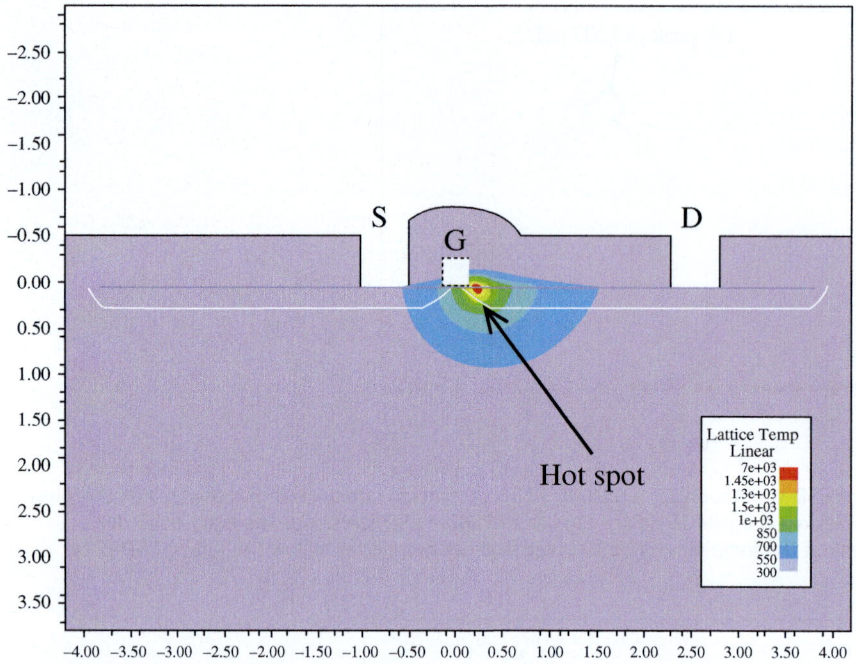

Figure 8.16 Maximum lattice temperature map by transient TCAD ESD simulation shows an ESD-induced hot spot at the drain junction corner under ESD stressing.

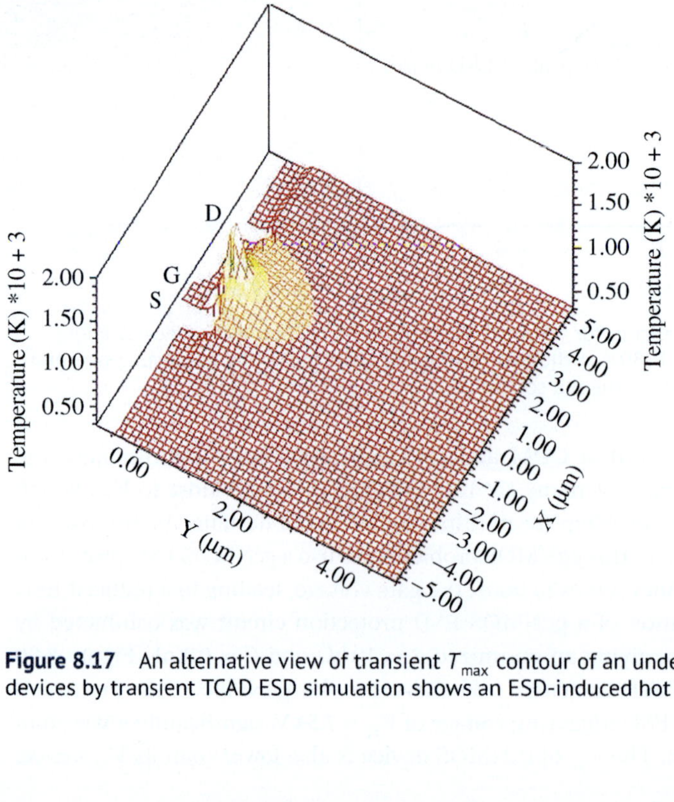

Figure 8.17 An alternative view of transient T_{max} contour of an under-designed ggNMOS ESD protection devices by transient TCAD ESD simulation shows an ESD-induced hot spot causing ESD thermal failure.

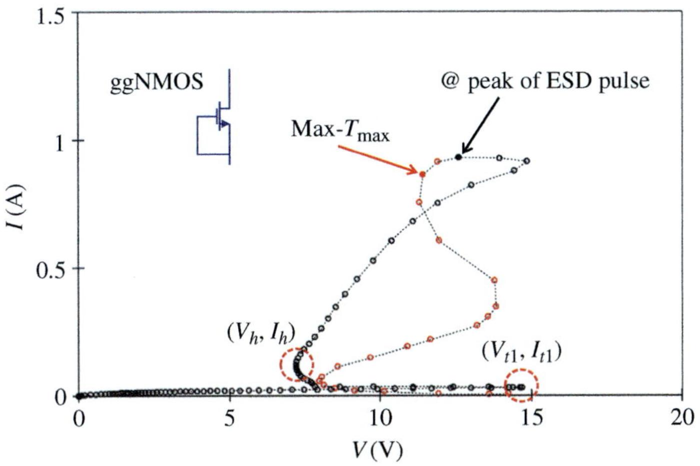

Figure 8.18 Transient ESD discharge I–V curve for a ggNMOS device by TCAD ESD simulation under 1.6 kV HBM ESD stressing shows a high ESD triggering voltage of $V_{t1} \sim 14.68$ V. Black solid dot corresponds to the peak of the incident ESD pulse and red solid dot indicates the peak lattice temperature across the ggNMOS device during ESD stressing.

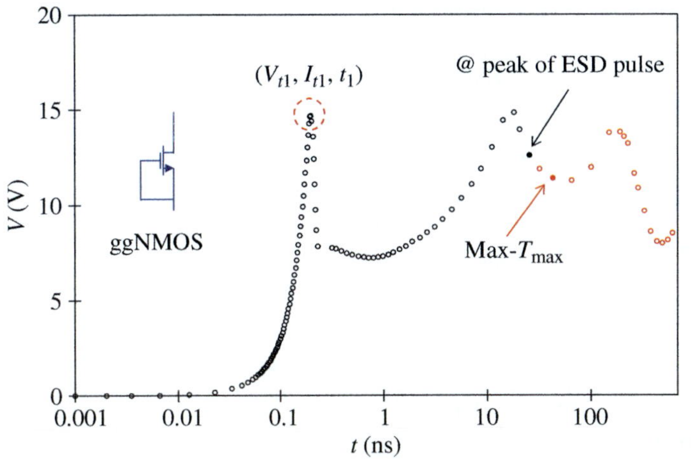

Figure 8.19 Transient ESD discharge V–t curve for a ggNMOS device by TCAD ESD simulation under 1.6 kV HBM ESD stressing shows the ESD triggering time of $t_1 \sim 0.2$ ns, which is fast enough for both HBM and CDM ESD events. Black solid dot corresponds to the peak of the incident ESD pulse and red solid dot indicates the peak lattice temperature across the ggNMOS device during ESD stressing.

making the gcNMOS ESD protection structure not only uniform across multiple fingers but also suitable for LV ICs. Figure 8.21 depicts the simulated transient ESD discharge V–t characteristics that shows an ESD triggering time of $t_1 \sim 0.42$ ns. Apparently, gcNMOS has a slower ESD response than that of ggNMOS. However, $t_1 \sim 0.42$ ns is still very fast for even CDM and IEC ESD protection. Figure 8.22 shows the transient gate voltage behavior during HBM ESD stressing, which is an important ESD design factor, because even though a higher V_G will help to reduce V_{t1} as needed, the design must ensure no voltage breakdown to the MOSFET gate oxide. Table 8.1 summaries the

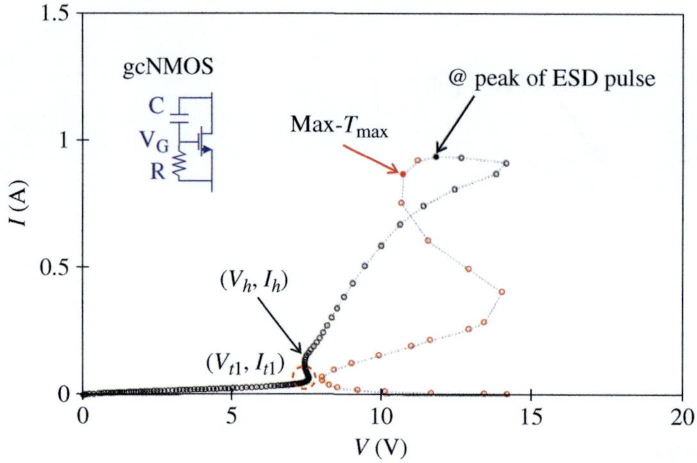

Figure 8.20 Transient ESD discharging $I-V$ curve for a gcNMOS device by TCAD ESD simulation under 1.6 kV HBM ESD stressing shows a reduced ESD triggering voltage of $V_{t1} \sim 7.54$ V. Black solid dot corresponds to the peak of the incident ESD pulse and red solid dot indicates the peak lattice temperature across the gcNMOS device during ESD stressing.

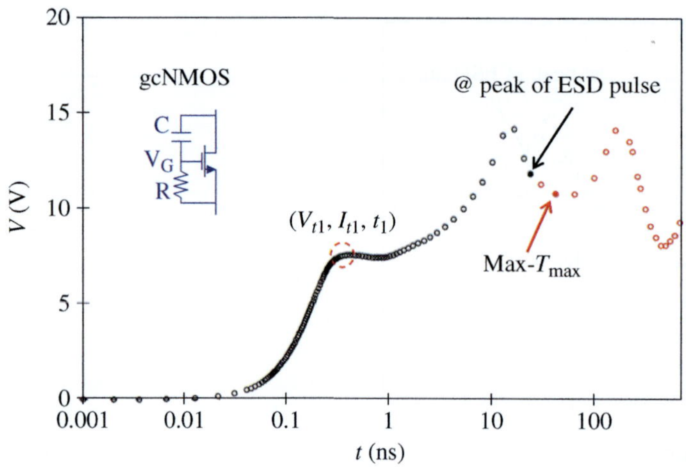

Figure 8.21 Transient ESD discharging $V-t$ curve for a gcNMOS device by TCAD ESD simulation under 1.6 kV HBM ESD stressing shows a longer ESD triggering time of $t_1 \sim 0.42$ ns, which is still fast enough for both HBM and CDM ESD events. Black solid dot corresponds to the peak of the incident ESD pulse and red solid dot indicates the peak lattice temperature across the gcNMOS device during ESD stressing.

simulated and measured ESD specs for both ggNMOS and gcNMOS ESD protection structures. This example demonstrated that TCAD mixed-mode ESD simulation is very useful in quantitative and accurate design of ESD protection structures for both design optimization and prediction.

8.3.3 Example-3: ESD Power Clamp in 0.35 μm CMOS

The mixed-mode nature of TCAD-based ESD simulation can be very useful in optimizing and predicting ESD protection performance at full-chip level. This example discusses design of a gcNMOS ESD power clamp for output circuit block by TCAD ESD simulation. This design was implemented

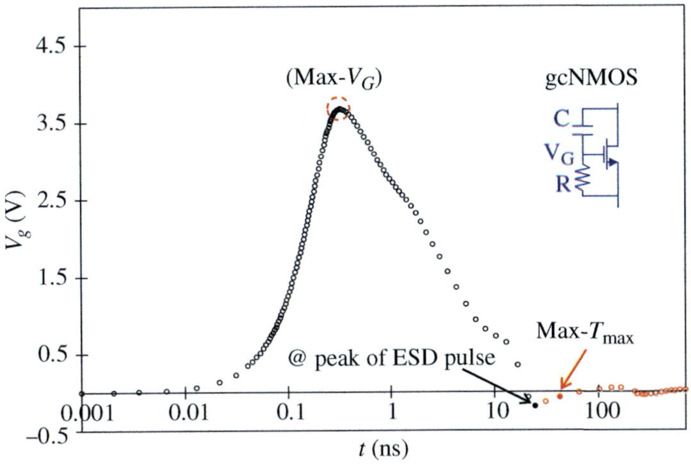

Figure 8.22 Transient ESD discharge V_G–t curve for a gcNMOS device by TCAD ESD simulation under 1.6 kV HBM ESD stressing shows a peak gate voltage of $V_G \sim 3.67$ V due to RC coupling, which is lower than BV_G. Black solid dot corresponds to the peak of the incident ESD pulse and red solid dot indicates the peak lattice temperature across the ggNMOS device during ESD stressing.

Table 8.1 Comparison of key ESD Specs of ggNMOS and gcNMOS ESD Protection structures: simulation versus testing.

ESD Specs	ggNMOS ESD		gcNMOS ESD	
	TCAD	Test	TCAD	Test
V_{t1} (V)	14.68	12.56	7.54	6.66
t_1 (ns)	0.2		0.42	
V_h (V)	6.92	6.48	7.41	6.08
Max-V_G (V)			3.67	

in a 0.35 μm CMOS technology for mixed-signal ICs [6]. Figure 8.23 shows the schematic for an output mixed-signal circuit block, where the power rail is protected by a gcNMOS ESD power clamp structure. While mixed-mode ESD protection circuit simulation is important, TCAD simulation is still very computing-hungry and time-consuming hence, it is wise to have a more efficient TCAD ESD simulation strategy in practical IC designs. In this case, a simplified equivalent circuit shown in Figure 8.23b is used for circuit-level ESD simulation. The general thought is that, even if the gcNMOS power clamp may already be optimized for its own ESD discharge function, when integrating the ESD protection structure into an IC, it will be helpful to check around the full circuit schematic by ESD simulation to identify any potential "danger," possibly overlooked during the design, on a chip. Considering that if an ESD pulse comes to the power bus, M_{P1} will be turned on. Hence, the first stage can be reasonably replaced by a resistor (R_{P2}), which simplifies the schematic as shown in Figure 8.23b. When conducting TCAD ESD simulation on Figure 8.23b, special attention will be given to the transient gate potential to stage-2 (M_{P2} and M_{n2}) during ESD stressing to monitor possible gate breakdown caused by ESD discharge. Figure 8.24 depicts the transient V–t characteristics by TCAD ESD simulation, which reveals the dynamic voltage surges at the Drain of gcNMOS device (V_{D3}) and the Gate of stage-2 (V_{G2}). It is readily observed that, during ESD

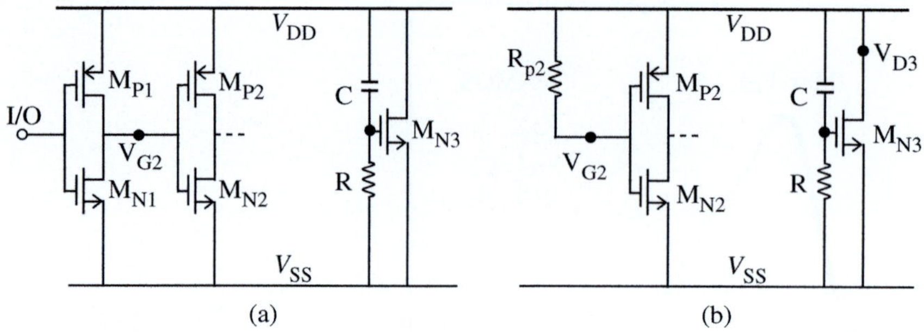

Figure 8.23 Schematic for a gcNMOS power clamp to protect an output block (a) and its simplified circuit (b) for mixed-mode ESD circuit simulation.

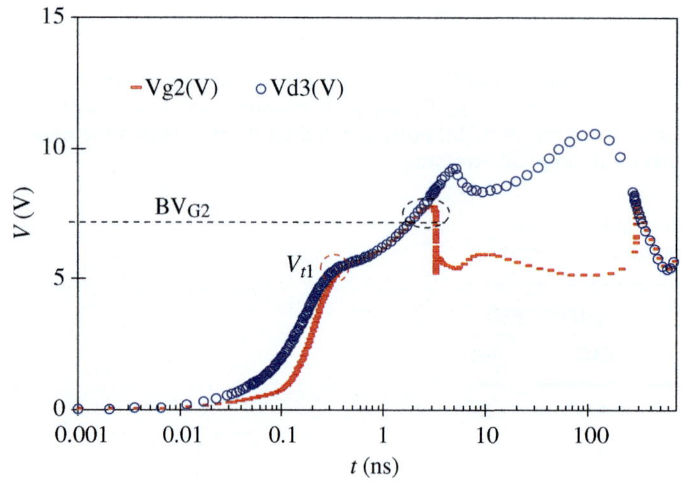

Figure 8.24 Simulated transient ESD discharge V_{D3}–t characteristics for gcNMOS ESD protection structure and dynamic V_{G2}–t curve for internal FETs show that V_{t1} was successfully reduced for gcNMOS ESD power clamp, while the internal V_{G2} slightly exceeds the worst-case BV_{G2} for a very short time during ESD stressing.

stressing, the gcNMOS ESD clamp is turned on at a fairly low triggering voltage of $V_{t1} \sim 6$ V, which is much lower than the gate breakdown voltage of $BV_{G2} \sim 8$ V (typical). On the other hand, the dynamic V_{G2} shows a tiny voltage overshot, slightly above the worst-case breakdown of $BV_{G2} \sim 7$ V (i.e., process corner), for a very brief time (<1 ns). This is certainly a design concern. In general, a designer must fine-tune the design to make sure V_{G2} is always lower than BV_{G2} to avoid any possible gate damage by ESD surges. In this case, simulation shows that the V_{G2} overshot is very brief and small, which meets the reliability specs of the CMOS technology used. Figure 8.25 reveals the transient lattice temperature characteristics of all transistors within the circuit. It clearly shows that, during ESD stressing, all ESD current is discharged through the gcNMOS ESD power clamp, causing $T_{max\text{-}ESD}$ increase within the gcNMOS device; however, T_{max} remains at room temperature for all other internal MOSFETs because almost no ESD current flows through internal circuit. Therefore, circuit-level ESD simulation confirms that this gcNMOS power clamp can protect the output circuit block at chip level.

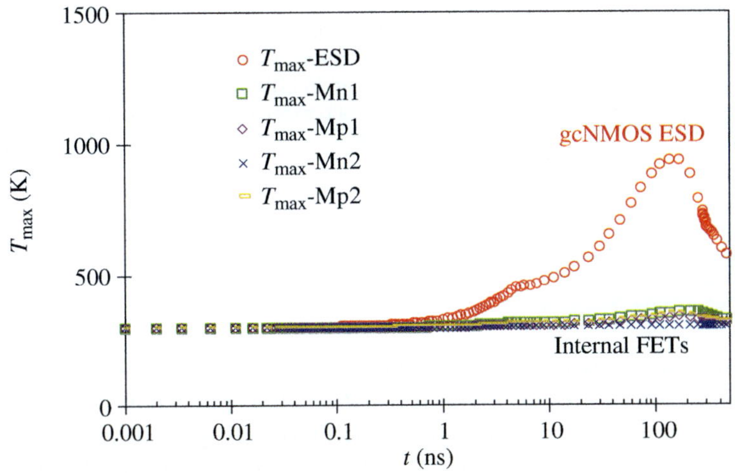

Figure 8.25 Transient ESD simulation reveals lattice temperature for all devices within the IC. It shows that only the gcNMOS ESD protection device experienced over-heating during ESD stressing, while all other internal FETs (Mn1, Mn2, Mp1, and Mp2) remain at close to room temperature, hence realizing full-chip ESD protection.

8.3.4 Example-4: Optimize HV ESD Protection Design

It is recognized that HV ESD protection design is very challenging due to multiple power supplies on a chip that makes it difficult to design ESD-critical parameters to fit into the ESD Design Window. This example shows how to fine-tune HV ESD protection structures against the HV ESD design window. A grounded-gate HV LDMOS FET (HVggNMOS) ESD protection structure is designed and implemented in a foundry 30 V bipolar-CMOS-DMOS (BCD) technology [8], which establishes an ESD Design Window of (35, 45 V) as illustrated in Figure 8.26. Figure 8.27 depicts

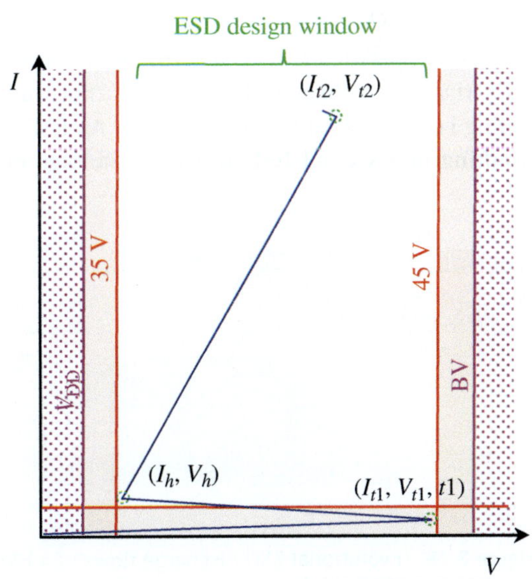

Figure 8.26 A 30 V BCD process establishes the HV ESD Design Window of (35, 45 V).

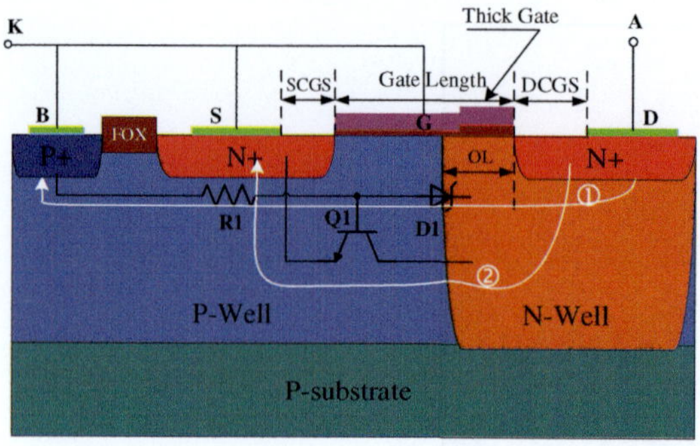

Figure 8.27 A cross-section view of HVggNMOS ESD protection structure in a BCD process with embedded equivalent circuit shown. 30 V high voltage is realized in LDMOS featuring a lightly-doped drain extension region covered by thick gate oxide.

the HVggNMOS ESD protection structure, which relies on a lightly doped Drain extension and a thick oxide gate to achieve high breakdown voltage. The HVggNMOS ESD protection structure consists of BJT ($Q1$), P-well/N-well diode ($D1$), and P-well resistor ($R1$). The HVggNMOS functions in that when an ESD pulse occurs at the Drain anode (A), D1 is driven into avalanche breakdown across the N-well/P-well junction that creates an initial ESD discharge channel of Path-①. The current flowing through $R1$ quickly builds up the potential at the Source junction and then turns on $Q1$. Consequently, a low-R ESD discharge channel of Path-② is formed through $Q1$ to effectively discharge the large ESD current. ESD discharge through Path-①and Path-② can be readily understood by TCAD ESD simulation as shown in Figure 8.28. The initial HVggNMOS ESD protection devices were designed, fabricated, and characterized. Figure 8.29a presents the measured transient ESD discharge I–V curve of a HVggNMOS ESD device of $W = 30\,\mu m$, showing $V_{t1} \sim 67\,V$ and $V_h \sim 12.7\,V$. Figure 8.29b depicts the DC ESD discharge I–V curve from a curve tracer, giving a $V_{t1} \sim 58\,V$. Obviously, this initial HVggNMOS ESD device does not meet the ESD Design Window of (35, 45 V). Design optimization for HVggNMOS ESD protection structures was conducted by aid of careful TCAD ESD simulation. The HVggNMOS ESD discharge characteristics were first studied by TCAD ESD simulation with its ESD discharge I–V curve shown in Figure 8.30, which readily reveals the complicated multiple-step ESD triggering procedures associated with the complex device structure of an HVggNMOS device. At point P1, avalanche breakdown of D1 is initiated across the N-well/P-well junction leading to the first ESD triggering voltage of $V_{t1}{}^1 \sim 58.5\,V$ that was confirmed in DC testing, but not caught in TLP testing because the ultralow $I_{t1}{}^1$ at P1. As ESD discharge continues, $Q1$ is turned on at point P3, resulting in a second ESD triggering voltage of $V_{t1}{}^2 \sim 66.3\,V$ that is confirmed by TLP testing.

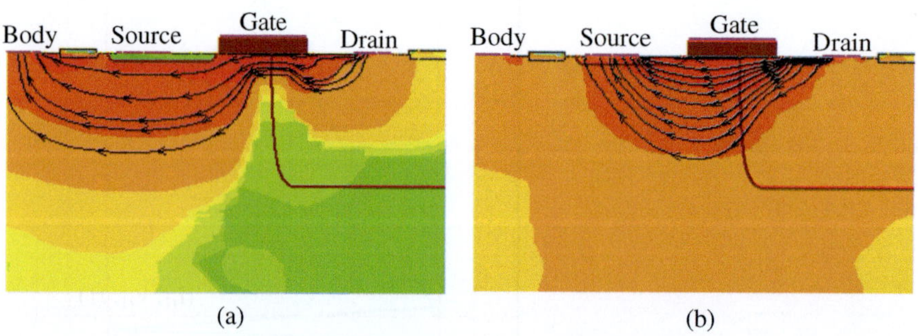

Figure 8.28 Evolutional ESD discharge flows of a HVggNMOS ESD protection structure by TCAD ESD simulation: (a) ESD discharging via path-① with current collected by the Body under small ESD current condition, and (b) ESD discharging via path-② with current collected by the Source under large ESD current condition.

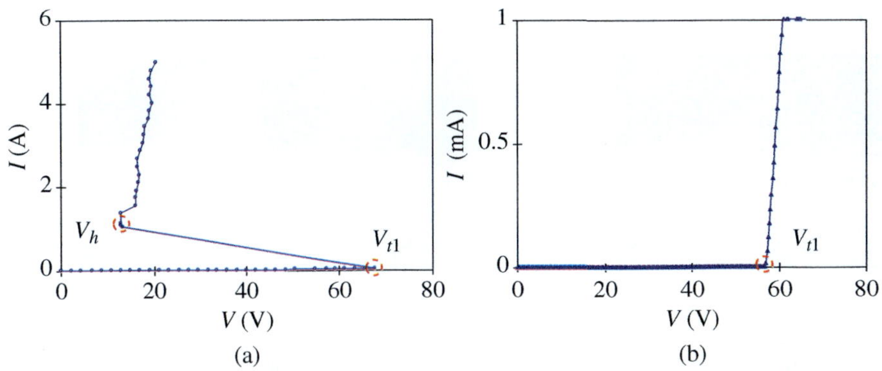

Figure 8.29 Measured ESD discharging I–V characteristics for the initial HVggNMOS ESD protection devices do not fit into the ESD Design Window: (a) TLP testing, and (b) DC testing.

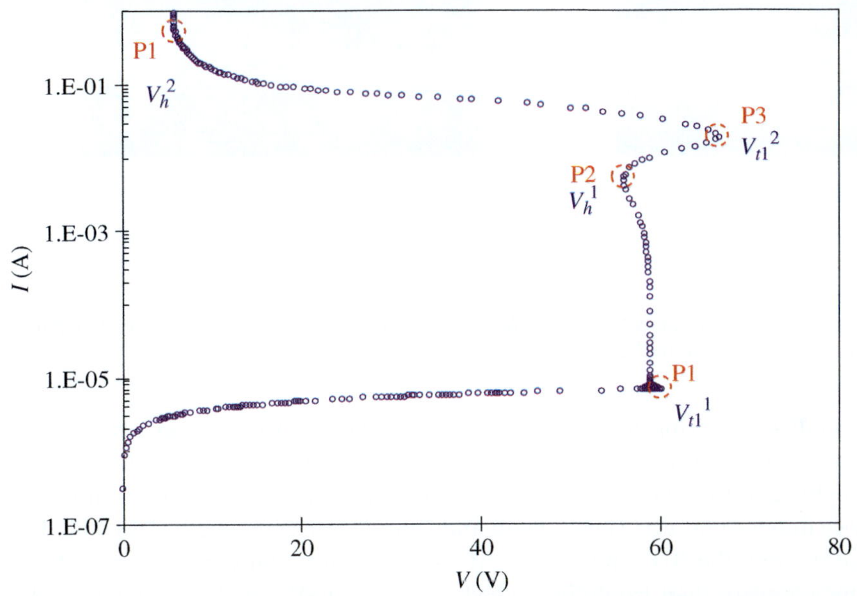

Figure 8.30 Simulated ESD discharging I–V curve for the initial HVggNMOS ESD protection device by TCAD reveals four key points corresponding to internal ESD discharge evolution.

Active ESD discharge through $Q1$ leads to dramatic snapback I–V behavior, resulting in a very low ESD holding voltage a $V_h{}^2 \sim 4.8\,\text{V}$ as shown in TLP testing. Therefore, this HVggNMOS ESD protection structure features transient $V_{t1} = V_{t1}{}^2 \sim 66.3\,\text{V}$ and $V_h = V_{t1}{}^2 \sim 4.8\,\text{V}$ in TCAD simulation. TLP testing shows $V_{t1} \sim 67\,\text{V}$ and $V_h \sim 12.7\,\text{V}$, which falls outside of the ESD design window of (35, 45 V). Discrepancy between ESD simulation and testing was related to calibration at the time of the design. Figure 8.31 depicts the evolutional flow of the transient electric field density within the HVggNMOS structure during ESD stressing from TACD ESD simulation, which corresponds to the four critical points of P1, P2, P3, and P4. As ESD discharge evolves from initially low current to very high current, Q1 internal conduction structure changes from a BJT of N^+/P-well/N-well-N^+ to a BJT of N^+/P-well/N^+ because the lightly doping drain extension region is gradually depleted under very large current and electric field related to the Kirk effect.

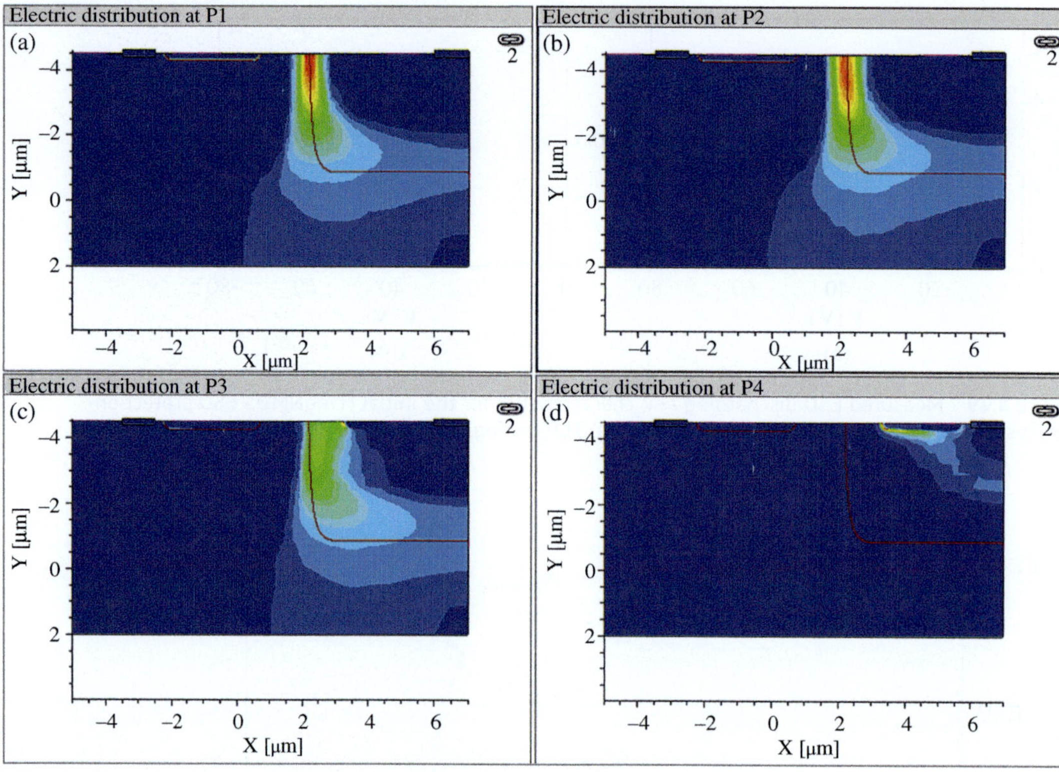

Figure 8.31 Evolution in electric field density of the HVggNMOS by TCAD ESD simulation shows dynamic ESD discharge behaviors corresponding to the four critical points (P1, P2, P3, P4).

Understanding this double-triggering mechanism of HVggNMOS ESD protection structure is important to optimizing the overall V_{t1} and V_h of the HVggNMOS in design. It is understood that P-well and N-well doping densities will affect V_{t1} and V_h of HVggNMOS in different conduction phases, and the device dimensions, e.g., Gate Length and Drain Extension Length (OL) will also affect ESD discharge through the BJT amplification (β). Design splits and comparison analysis of key device parameters were then carefully studied by TCAD ESD simulation. Figure 8.32 presents simulated transient ESD discharge I–V characteristics for HVggNMOS with varying P-well/N-well doping densities, i.e., from a baseline doping density to increased dosage of 1.3×, 1.6×, and 3.1×. It is readily observed that the doping can substantially change the ESD triggering and holding internally, i.e., $V_{t1}{}^1$ (P1), $V_h{}^1$ (P2), and $V_{t1}{}^2$ (P3). Figure 8.33 depicts the simulated transient ESD discharge I–V curves for HVggNMOS devices with varying OL splits (0.5, 1, 2, and 3 μm), which clearly the variation of internal ESD triggering ($V_{t1}{}^2$) because the β of Q1 changes as the equivalent base width of BJT changes. Figure 8.34 describes ESD discharge behaviors of HVggNMOS versus the Gate Length, which also shows clear shift in $V_{t1}{}^1$, $V_h{}^1$, and $V_{t1}{}^2$ of the ESD protection structure. With the thorough understanding of the ESD discharge mechanism, the HVggNMOS ESD protection structure was then carefully optimized to realize the required $V_{t1} = V_{t1}{}^2 \sim 42.6\,\text{V}$ and $V_h = V_h{}^1 \sim 35.6\,\text{V}$ for the HVggNMOS, which fits into its ESD design window of (35, 45 V), as shown in Figure 8.35. This HV ESD design example demonstrates how to use TCAD-based mixed-mode ESD simulation to accurately design ESD protection structures, particularly in HV ESD protection designs.

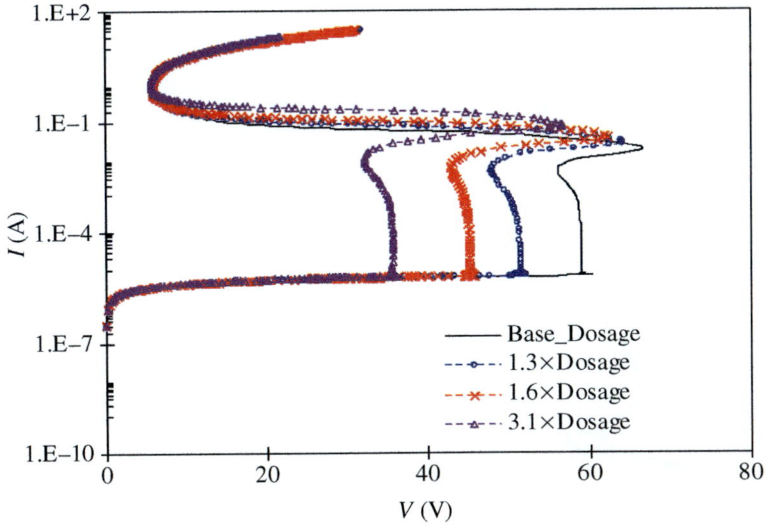

Figure 8.32 Simulated ESD discharging I–V curve for HVggNMOS ESD protection devices by TCAD reveals significant impact of Nwell/Pwell doping density on ESD triggering and holding behaviors (V_{t1}^{1}, V_{h}^{1}, V_{t1}^{2}).

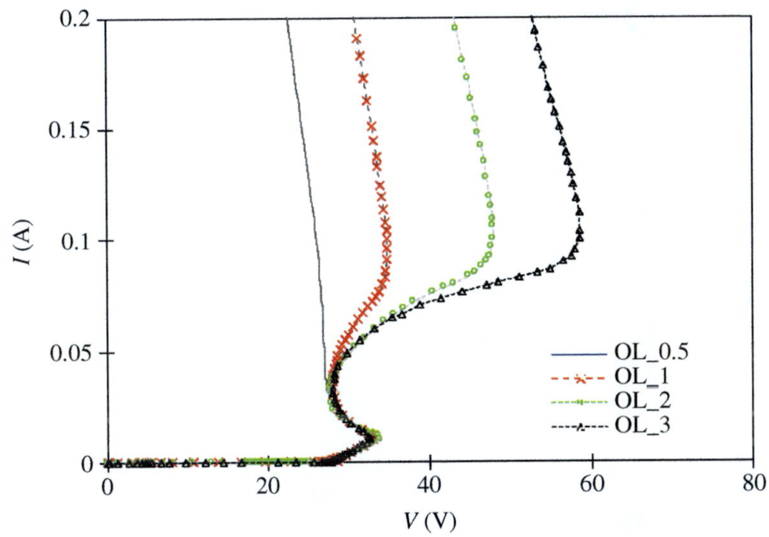

Figure 8.33 Simulated ESD discharging I–V curve for HVggNMOS ESD protection devices by TCAD reveals significant impact of Drain extension length (OL splits) on the secondary ESD triggering behaviors (V_{t1}^{2}).

8.3.5 Example-5: ESD Layout Analysis by 3D TCAD

Design optimization and prediction of ESD protection structures are highly desirable, but extremely difficult in practical IC designs. The main challenge is associated with the nature of ESD phenomena on a chip: ESD pulses are ultrafast (down to picosecond scale), large ESD transient currents produce substantial heat, and Si wafer has poor thermal conductivity. As such, on-chip ESD events always result in internal overheating (i.e., tiny hot spots) and local heat crowding at the edges and

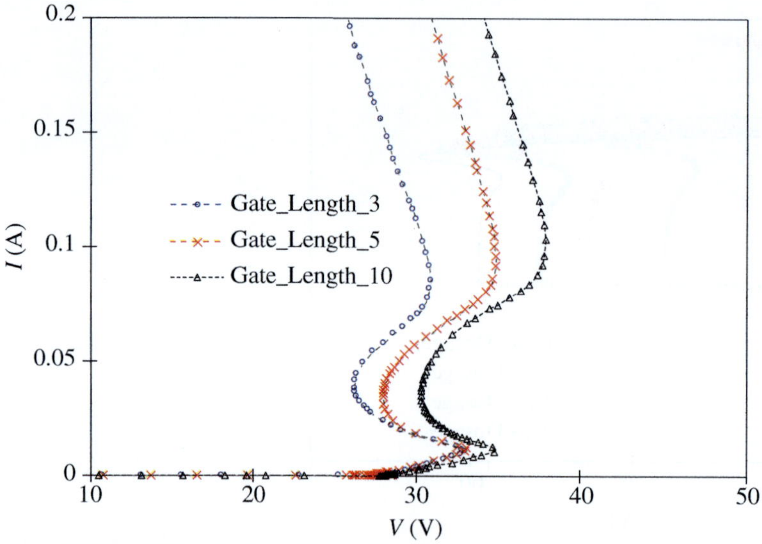

Figure 8.34 Simulated ESD discharging I–V curve for HVggNMOS ESD protection devices by TCAD reveals impact of Gate length on ESD triggering and holding behaviors (V_{t1}^{1}, V_{h}^{1}, V_{t1}^{2}).

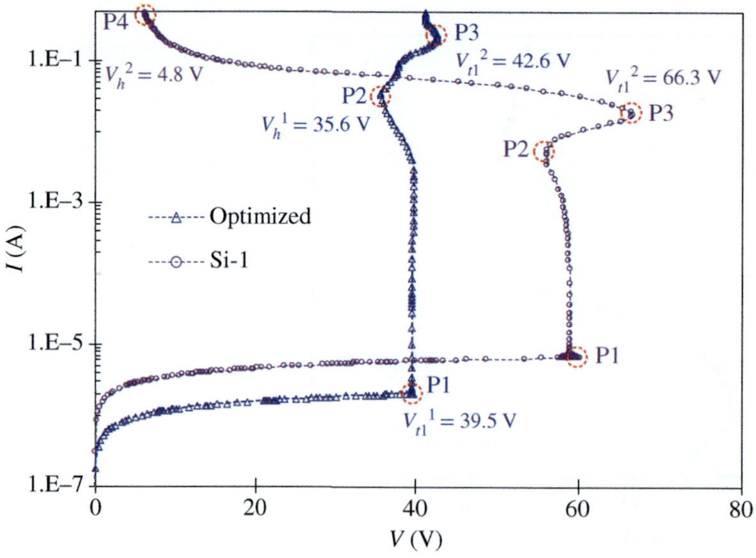

Figure 8.35 TCAD ESD simulation assists to tune the ESD discharging I–V curve of the HVggNMOS ESD protection device into the ESD Design Window: Purple line for the initial device (Si-1) and Blue line for the optimized device.

corners of an ESD protection structure (i.e., corner/edge effect). Consequently, physical layout design of an ESD protection structure plays a crucial role in real-world ESD protection designs, which cannot be accurately addressed by your "rich" experiences. Unfortunately, such complex practical design issues cannot be accurately handled by 2D TCAD ESD simulation, not even if one uniformly extends the 2D ESD structure in the third direction (i.e., in Z-direction) to construct a

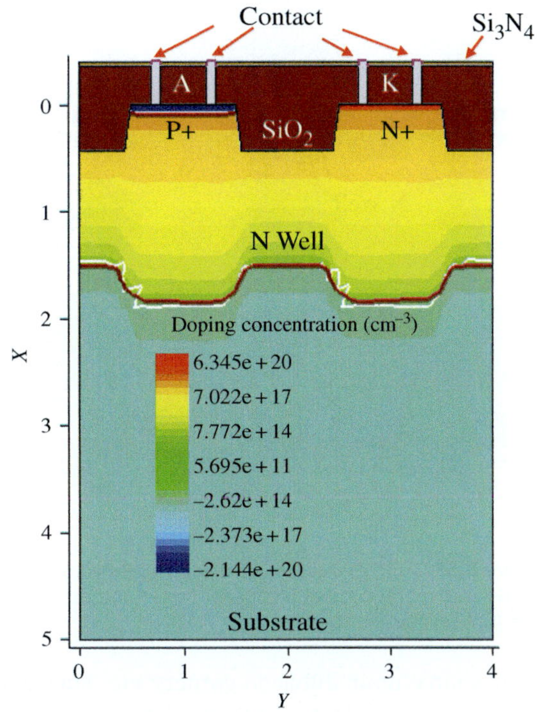

Figure 8.36 A cross-section ($X-Y$) view of P^+/Nwell ESD protection diode created by 2D TCAD process simulation.

"real-sized" ESD device, which is often referred to as 2.5D TCAD simulation (i.e., pseudo-3D). *True 3D* TCAD ESD simulation is required to accurately and reliably study the local overheating and edge/corner heat-crowding effects of any ESD protection structures, which is discussed in this section. In this example, 3D TCAD ESD simulation is used to analyze layout effects and optimize layout design of P^+/N-well (PPNW) ESD protection diodes that were implemented in a foundry 55 nm CMOS technology [9]. Figure 8.36 shows a cross-section for a PPNW ESD protection diode created by 2D TCAD simulation. As a comparison, Figure 8.37 depicts the equivalent PPNW ESD diode structure generated by true 3D TCAD process simulation. Obviously, the 3D structural details for the PPNW ESD diode structure is clearly revealed by 3D TCAD, which consists of two lines of evenly distributed metal contact arrays for the P^+ (A) and N^+ (K) regions, respectively, i.e., total six metal contacts (tungsten plugs) for both A and K terminals. To ensure "real-world" ESD diode structures, true 3D TCAD process simulation was conducted using actual fabrication process recipes as in a foundry, including photoresistor masks, shallow trench isolation (STI) etching, deposition, chemical-mechanical polishing (CMP), diffusions, ion implantation and annealing, SiO_2 and Si_3N_4 isolation, and metal contacts, etc. It is noteworthy that any assumption from your little "smart" brain may cause errors in 3D TCAD simulation, resulting in inaccuracy in transient 3D ESD simulation. It is well known that mesh generation plays an important role in TCAD simulation accuracy. To study the impact of device meshes on TCAD accuracy, three-device mesh grid splits were used for the same PPNW ESD diode structure in 3D TCAD ESD simulation, i.e., *Fine* mesh, *Finer* mesh, and *Finest* mesh. Since the mesh density has significant weight on TCAD run time, a mesh should be created smartly for a 3D ESD protection structure. Typically, a sparse global mesh is applied to the whole ESD protection structure, while fine meshes are created locally at the spots that may critically affect ESD discharge characteristics, such as the conduction channel, gate

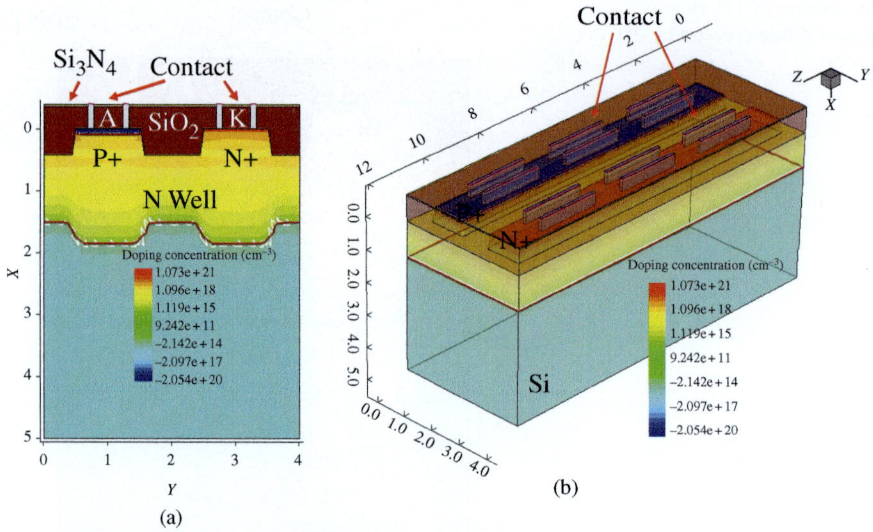

Figure 8.37 A P^+/Nwell ESD protection diode created by true 3D TCAD process simulation features evenly distributed metal contact layout for the P^+ anode (A) and the N^+ cathode (K): (a) $X-Y$ cross-section view, and (b) transparent 3D view.

layer, source/drain diffusion corners, etc. Transient HBM ESD pulse waveforms are then used as the stimuli in 3D TCAD ESD simulation. Figure 8.38 depicts the transient ESD discharge details, including voltage, current, lattice temperature (T_{max}), and time, for the PPNW ESD diodes (Fine, Finer, and Finest mesh splits) by 3D TCAD ESD simulation under 500 V HBM ESD zapping. The transient ESD discharge I–V curves by 2D and 3D TCAD ESD simulation, given in Figure 8.38a–c, clearly shows the impact of device meshes on simulation accuracy where a denser mesh grid is necessary to improve simulation accuracy of 3D TCAD over 2D TCAD. Similarly, a comparison of the transient $T_{max} \sim t$ characteristics for PPNW ESD diodes of varying meshes also confirms that a denser mesh is needed for accuracy in 3D TCAD ESD simulation. Importantly, a sizable difference in T_{max} is observed between Fine mesh and Finest mesh in 3D TCAD ESD simulation for the same ESD diode, i.e., $\Delta T_{max} \sim 30$ K. In addition, a significant variation in lattice temperature is discovered between 2D and 3D TCAD ESD simulation for the same ESD diode of Finest mesh, i.e., $\Delta T_{max} > 110$ K. This is an important observation, which states that 3D TCAD ESD simulation using good mesh grid is vital to ensure ESD simulation accuracy, hence affecting ESD design optimization and prediction. Further, the transient ESD discharge thermal maps for the PPNW ESD diode by 2D (Figure 8.39) and 3D (Figure 8.40) TCAD ESD simulation readily reveals the device geometrical impact on possible ESD thermal failures. The 3D TCAD ESD simulation clearly shows that the ESD-induced thermal distribution is not uniform across the ESD protection structure, i.e., dramatical overheating at the center and edge of the ESD diode reflecting the concerned layout impact on ESD thermal failures. The critical local overheating and edge/corner heat-crowding details can only be obtained by true 3D TCAD ESD simulation. As an example of using 3D TCAD ESD simulation for layout design optimization in practical ESD protection designs, four layout design splits were used for the same PPNW ESD diode as shown in Figure 8.41. Split-1 (Base-4 × 3) serves as a reference (i.e., the baseline device) that features two lines of evenly distributed metal contacts for the P^+ (A) and N^+ (K) terminals, respectively, where the metal contacts were made of tungsten (W) plugs and the "pink" cap indicates that a specific contact is actually electrically connected for ESD discharge conduction. Split-1 represents the common ESD device layout practices. Split-2 (Edge-4 × 1)

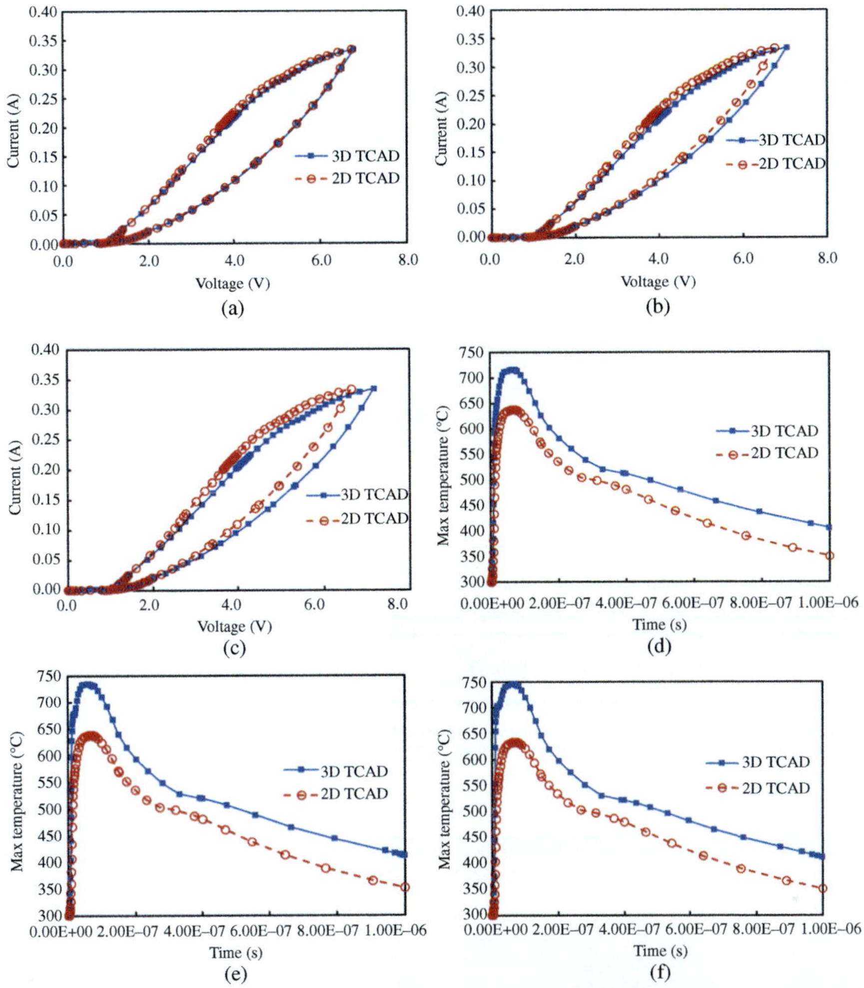

Figure 8.38 Comparison of transient ESD discharging $I-V$ and $T_{max}-t$ characteristics of the PPNW ESD diode by 2D and 3D TCAD ESD simulation under 500 V HBM ESD stressing shows impact of device mesh grids on simulated ESD discharge behaviors: (a) $I-V$ curves for *Fine* mesh, (b) $I-V$ curves for *Finer* mesh, (c) $I-V$ curves for *Finest* mesh, (d) $T_{max}-t$ curves for Fine mesh, (e) $T_{max}-t$ curves for Finer mesh, and (f) $T_{max}-t$ curves for Finest mesh.

has only four-edge metal contacts (with pink caps) being electrically connected for ESD discharge conduction, which is used to study the edge crowding effect in ESD discharge since the large ESD current can only be conducted through these metal plugs (pink-capped) at the edge of the PPNW ESD diode. Split-3 (Even – 4 × 6) contains 2 × 6 evenly distributed "smaller" metal contacts in A and K terminals, respectively, which is similar to Split-1 that has 4 × 3 evenly distributed "larger" metal plugs. Split-4 (Uneven – 4 × 3) is similar to Split-1 except that the edge metal contacts are larger and the central metal plugs are smaller; hence, to purposely redistribute the ESD discharge current for a more balanced thermal map across the PPNW ESD diode structure. The four-layout design splits are studied by 3D TCAD ESD simulation under 500 HBM ESD stressing. Figure 8.42 compares Split-1 and Split-2 for simulated ESD discharge $I-V$ curves, $T_{max} \sim t$ curves and temperature maps, which clearly shows that a poor layout design (i.e., only edge metal contacts can discharge ESD

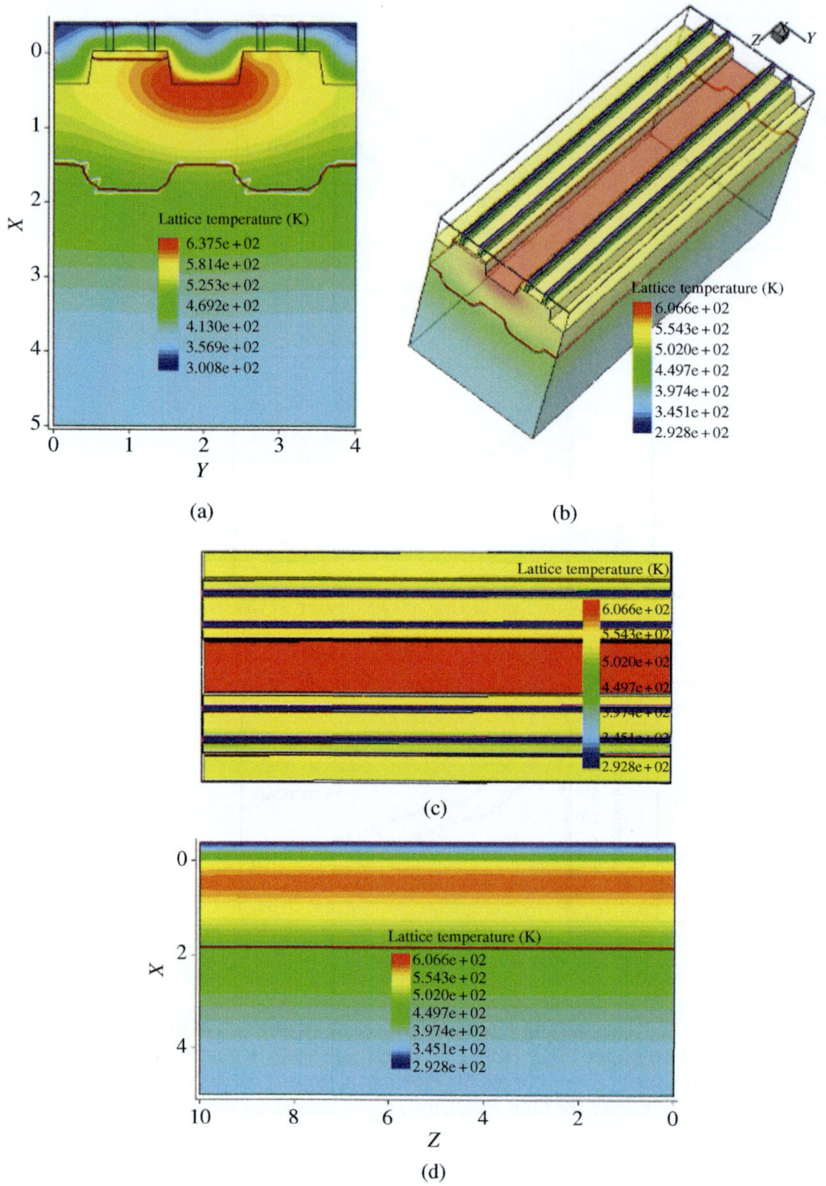

Figure 8.39 Temperature map for the PPNW ESD diode by 2D TCAD simulation: (a) $X-Y$ cross-section, (b) 2.5D view with uniform extension in Z-axis, (c) top view ($Y-Z$), and (d) $X-Z$ cross-section view.

current in Split-2) will cause severe imbalance in ESD discharge current and thermal distribution, often resulting in early ESD thermal failures. Figure 8.43 compares Split-1 and Split-3 for simulated ESD discharge I–V curves, $T_{max} \sim t$ curves and temperature maps, which indicates that a carefully balanced layout design can dramatically improve ESD discharge performance of an ESD protection structure, i.e., more uniform ESD thermal map, lower T_{max} (less a concern on hot spot), lower

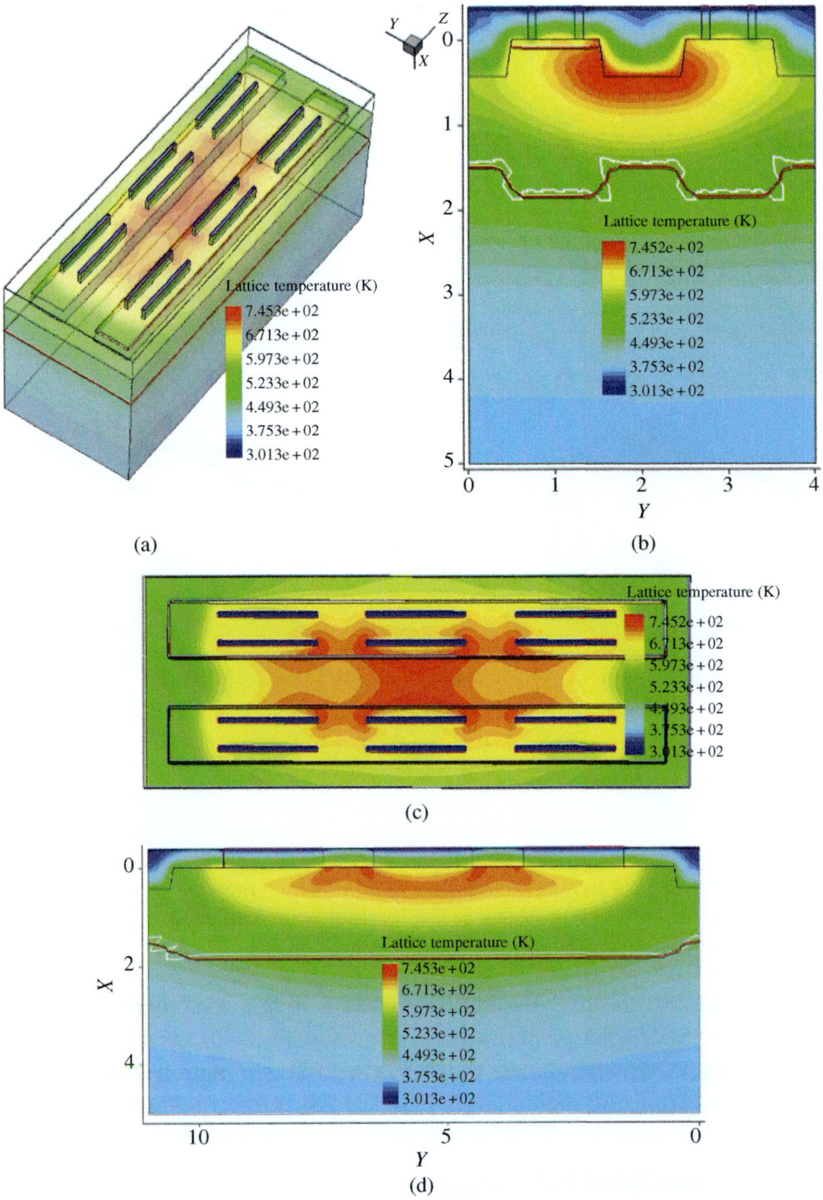

Figure 8.40 Temperature map of the PPNW ESD diode by 3D TCAD simulation: (a) 3D transparent view, (b) $X-Y$ cross-section view, (c) top view, and (d) $X-Z$ cross-section view.

ESD discharge $R_{ON,}$ hence, higher ESD protection level. Figure 8.44 compares Split-1 and Split-4 for simulated ESD discharge $I–V$ curves, $T_{max} \sim t$ curves and temperature maps, which shows that a uniform metal contact layout is not preferred in routing the large ESD transient current because it always cause ESD overheating in the center of the ESD protection structure, and a good layout optimization is to design an unique metal contact array to smartly re-route the large ESD current

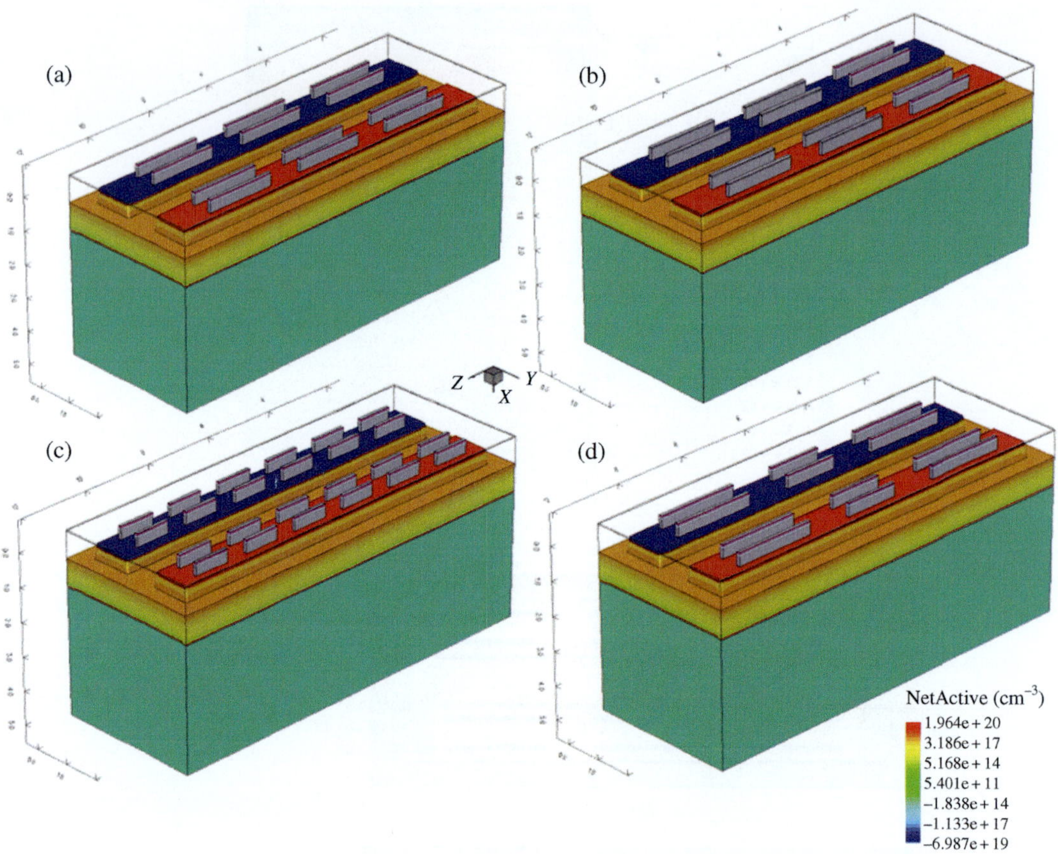

Figure 8.41 Four layout design splits of 3D PPNW ESD diodes by 3D TCAD simulation: (a) Split-1: Base-4 × 3, (b) Edge-4 × 1, (c) Even-4 × 6, and (d) Uneven-4 × 3. The tungsten plug marked by a pink bar on top is actually connected electrically for ESD discharging.

to realize a much-balanced global ESD thermal distribution, hence, a higher ESD thermal failure threshold. The layout impact on ESD discharge performance can be readily observed in the ESD thermal maps depicted in Figure 8.45, which indicates that 3D TCAD ESD simulation is a powerful tool to achieve ESD design optimization and prediction in practical ESD protection designs.

8.3.6 Example-6: Multiple-Stimuli TCAD ESD Simulation

While TCAD-based mixed-mode ESD simulation method is a powerful technique for practical ESD protection designs, many factors may affect the accuracy of TCAD ESD simulation. This Section discusses correlation between TCAD ESD simulation and ESD testing, and a multiple-stimuli TCAD ESD simulation approach for real-world ESD protection designs [10]. Obviously, ESD simulation means to provide guidelines for ESD design optimization and prediction of ESD protection structures. Therefore, correlation between TCAD mixed-mode ESD simulation with various ESD test methods is important. As discussed before, TLP ESD testing is very useful because it is generally non-destructive and offers transient ESD discharge details, including instantaneous

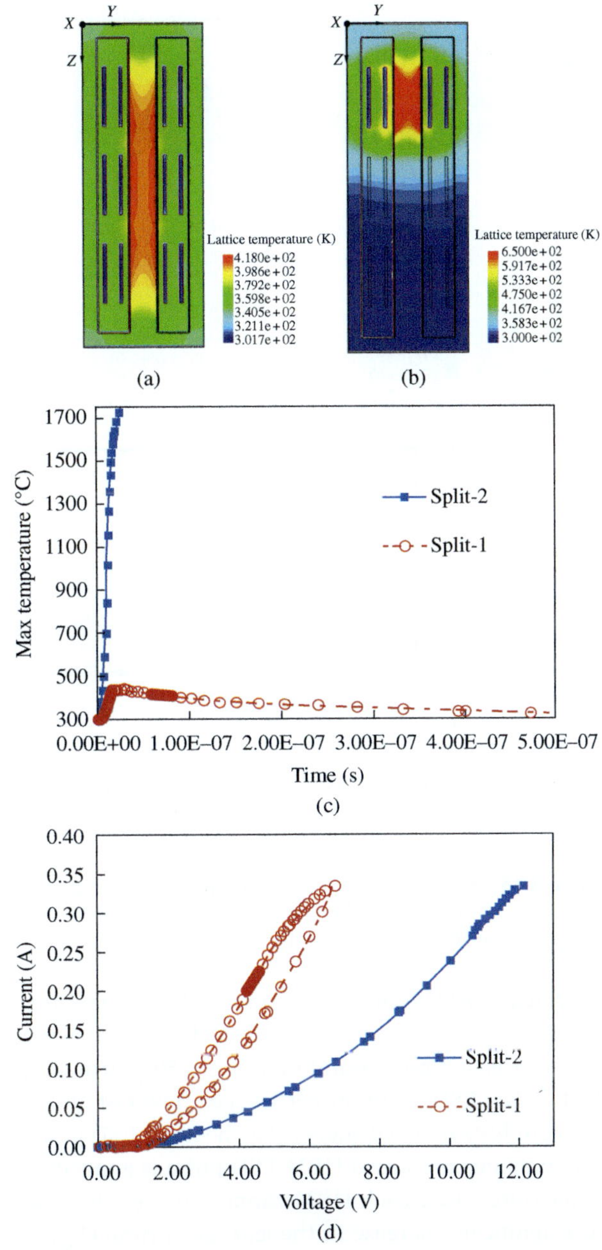

Figure 8.42 Split-1 and Split-2 comparison of PPNW ESD didoes by 3D TCAD under 500 V HBM ESD stressing reveals strong edge overheating due to layout effect: (a) Top view for thermal map of Split-1, (b) Top view for thermal map of Split-2, (c) T_{max}–t variation, and (d) ESD discharge I–V difference.

ESD discharging I–V and leakage information during ESD stressing. On the other hand, all IC products must be evaluated by ESD zapping test, which is developed to accurately mimic real-world ESD events, such as the HBM ESD test standard. Indeed, HBM ESD test is destructive and does not offer any useful design insights to improve ESD protection structures other than generating an ESDV number for an IC datasheet. It is recognized that TLP testing is an emulation of, but does not 100%-ly model the real-world HBM ESD events. On the other hand, the input ESD stimuli used in ESD simulation may also affect the accuracy of TCAD ESD simulation results.

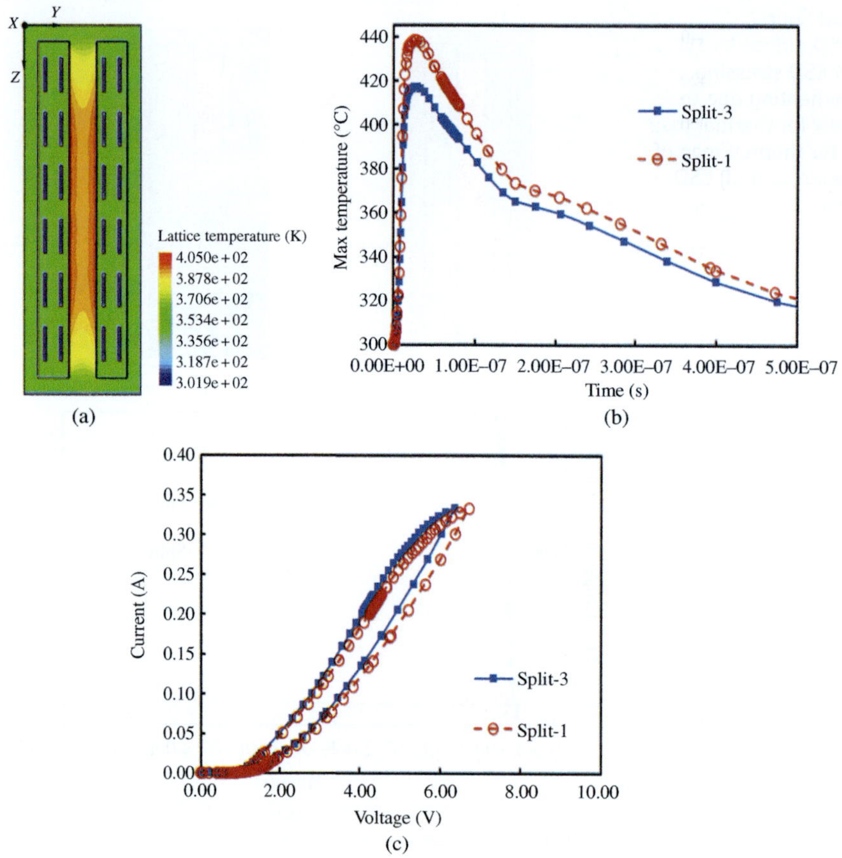

Figure 8.43 Split-1 versus Split-3 ESD discharge comparison by 3D TCAD under 500 V HBM ESD stressing: (a) top view (Y–Z) of temperature map for Split-3, (b) T_{max}–t compasiron for Split-1 and Split-3, and (c) ESD discharge I–V comparison for Split-1 and Split-3.

In a real world, human body-induced ESD events are modeled by HBM ESD test standard, which defines a typical HBM ESD pulse waveform as shown in Figure 8.46 that features a pulse rise time of $t_r \sim 10$ ns and pulse duration of $t_d \sim 150 \pm 20$ ns [11]. The pulse rise time and duration specify the two key ESD event properties, i.e., how fast an ESD response time is needed and how much transient ESD-induced energy (i.e., heat) an ESD protection structure has to handle. During HBM ESD zapping, a required HBM ESD pulse is applied to the DUT device and the key IC Specs will be measured after each HBM zapping run to check if any ESD failure occurs, typically indicted by a significant increase in the leakage current (I_{leak}). In TLP ESD testing, a square waveform is produced by a transmission line pulse generator, which will be used to stress a DUT [10]. The TLP pulse waveform is defined per its HBM counterpart as depicted in Figure 8.47. Typically, to "equate" TLP and HBM ESD tests, a TLP square waveform is set as $t_r \sim 10$ ns and $t_d \sim 100$ ns for HBM pulse equivalence in terms of pulse response time and ESD energy involved. Figure 8.48 depicts the TLP ESD testing mechanism and procedures. After one TLP pulse is applied to a DUT device, the incident and reflected pulse signals will be monitored, and the instant voltage and current of the DUT device will be recorded. In addition, the leakage current (I_{leak}) of the DUT after one TLP stress is measured at normal IC operation bias. In a real world, the incident and reflected pulse waveforms

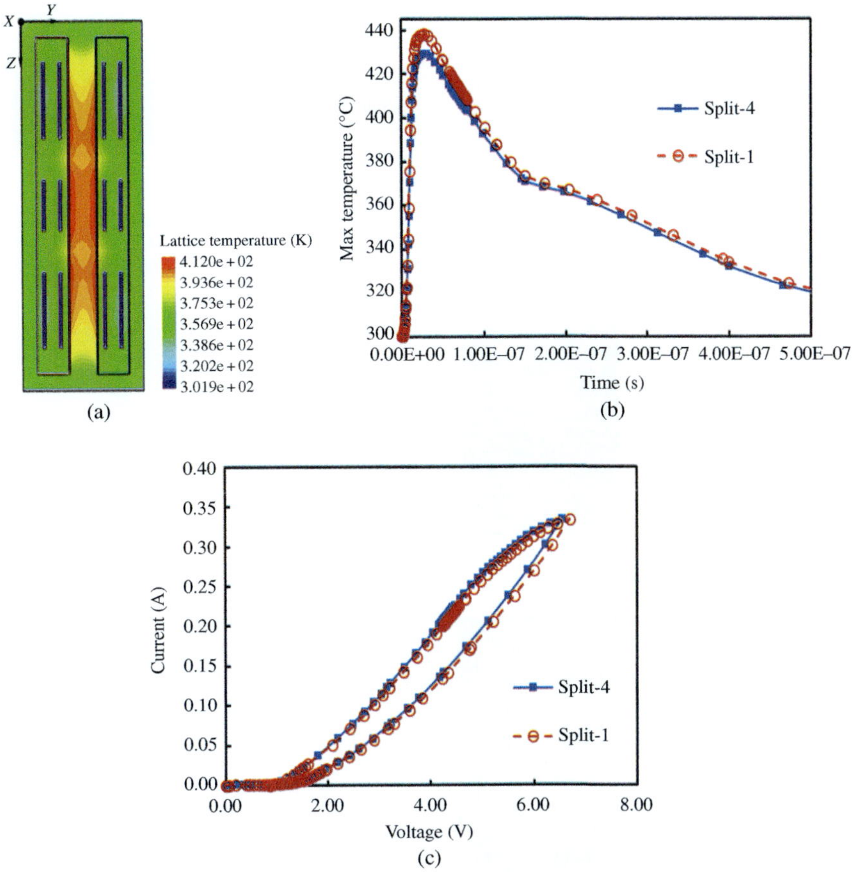

Figure 8.44 Split-1 versus Split-4 ESD discharge comparison by 3D TCAD under 500 V HBM ESD stressing: (a) top view (*Y–Z*) of temperature map for Split-4, (b) $T_{max}-t$ compasiron for Split-4 and Split-1, and (c) ESD discharge *I–V* comparison for Split-4 and Split-1.

in TLP testing mostly are not flat, hence, a typical TLP tester is designed to take an integration of the monitored TLP waveforms within a narrow pulse window of 70–90% to estimate a pair of *I* and *V* values for the DUT during each TLP pulse. A complete TLP ESD test routine uses *a pulse train* to stress the DUT device, with the pulse height stepping up gradually one pulse after another, therefore, produces a transient ESD discharging *I–V* curve and the dynamic leakage current curve for the DUT tested.

In TCAD ESD simulation, various incident ESD pulses can be used as the input ESD stressing signals to zap an ESD protection structure (DUT), typically, an HBM ESD pulse waveform, one single TLP square pulse waveform, and a TLP pulse train. In this example, an ESD diode with shallow trench isolation (STI) designed and fabricated in a foundry 28 nm CMOS technology, shown in Figure 8.49, is studied by TCAD ESD simulation and TLP testing. Figure 8.50 illustrates a multiple-stimuli TCAD ESD simulation set-up that uses different incident ESD pulses as the input signals, i.e., TLP *single pulse*, HBM *single pulse*, and TLP *pulse train*. Figure 8.51 compares the transient ESD discharging *I–V* curves for the STI ESD diode obtained from both TCAD ESD simulation and TLP measurements. Several interesting observations follow: First, there is a good

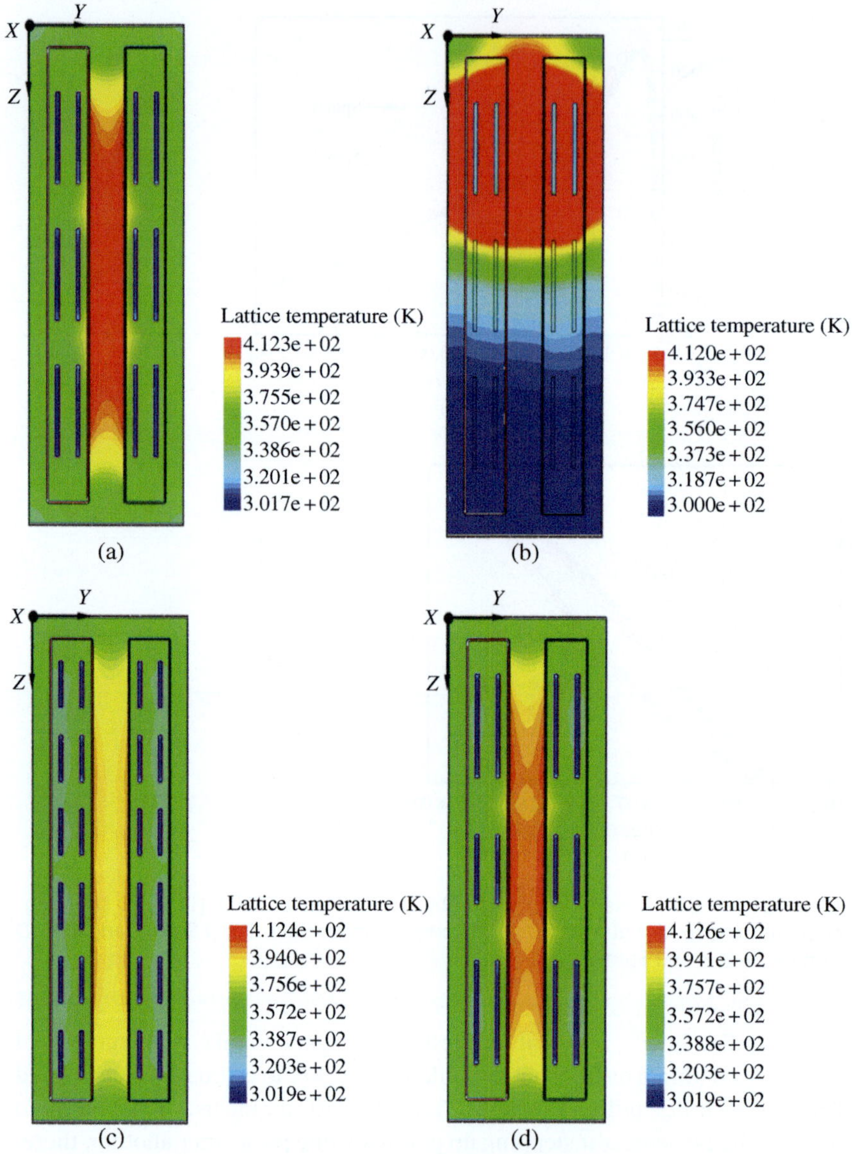

Figure 8.45 Top view temperature contour comparison for PPNW ESD diodes by 3D TCAD clearly reveals severe ESD-induced local overheating and edge/corner heat-crowding effects due to layout variations (in same 300 to 412 °C scale): (a) Split-1, (b) Split-2, (c) Split-3, and (d) Split-4.

match between TLP testing and TCAD ESD simulation using a TLP pulse train as the stimuli, given a good calibration for ESD simulation in place. Second, there is substantial discrepancy in the ESD discharging I–V behaviors between TCAD ESD simulation using single TLP pulse and a TLP pulse train, even though the ESD thermal failure levels (I_{t2}) seems to be equivalent. This observation suggests that TCAD ESD simulation using single TLP pulse is not preferred for ESD protection design. Third, there exists significant difference between TLP testing and ESD

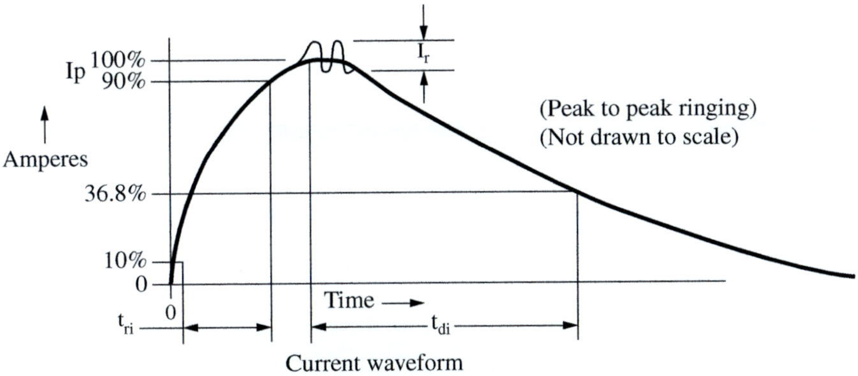

Figure 8.46 Original HBM ESD pulse waveform per *Method 3015.7, MIL-STD-883E* [11].

Figure 8.47 Correlating TLP and HBM waveforms in ESD testing for equivalent ESD pulse responses and ESD energy dissipation.

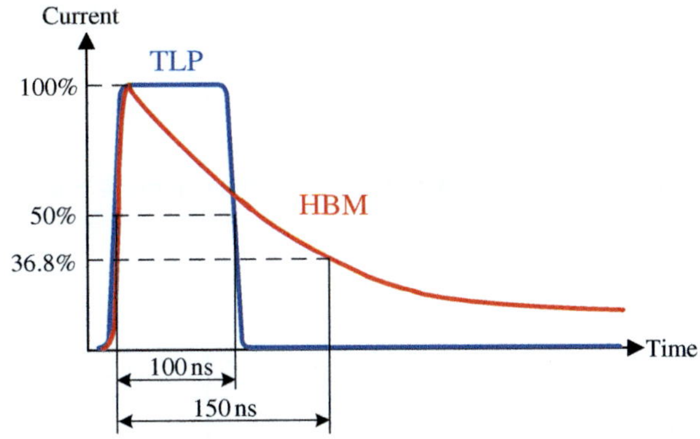

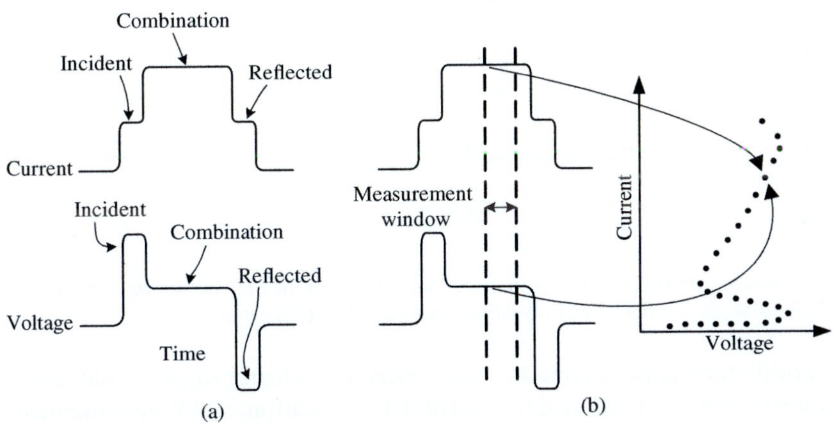

Figure 8.48 TLP ESD testing applies a TLP pulse train across a DUT: (a) the incident and reflected TLP pulses, and (b) the obtained transient ESD discharge $I-V$ curve.

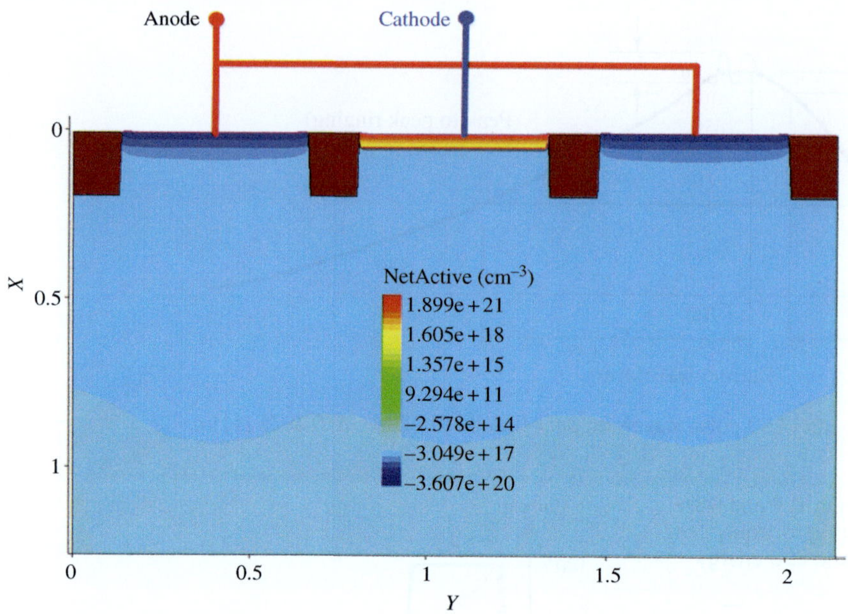

Figure 8.49 Cross-section for a STI ESD diode made in a 28 nm CMOS by TCAD simulation.

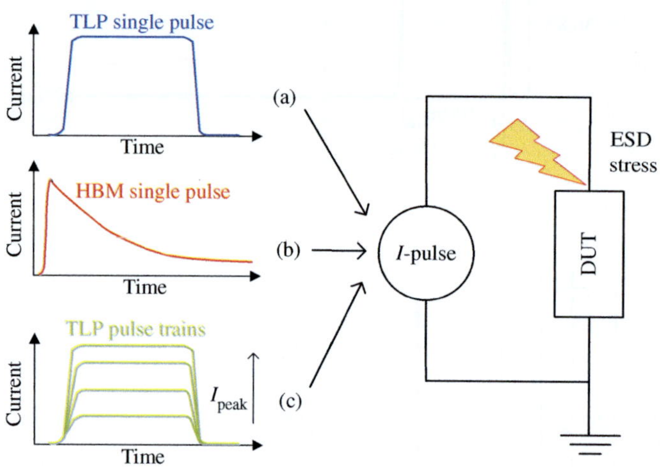

Figure 8.50 A TCAD mixed-mode ESD simulation set-up uses multiple different ESD stimuli to stress a DUT device: (a) a single TLP pulse, (b) a single HBM waveform, and (c) a TLP pulse train.

simulation using real-world HBM pulse waveform. This observation clearly suggests that, even though TLP testing is developed to mimic real-world HBM ESD event and TLP measurement provides rich transient ESD discharging information for ESD design optimization, it is not realistic to use TLP ESD stimulus "only" to accurately model real-world HBM ESD events, e.g., trying to equate the I_{t2} value by TLP stressing test with the ESD failure result from HBM zapping

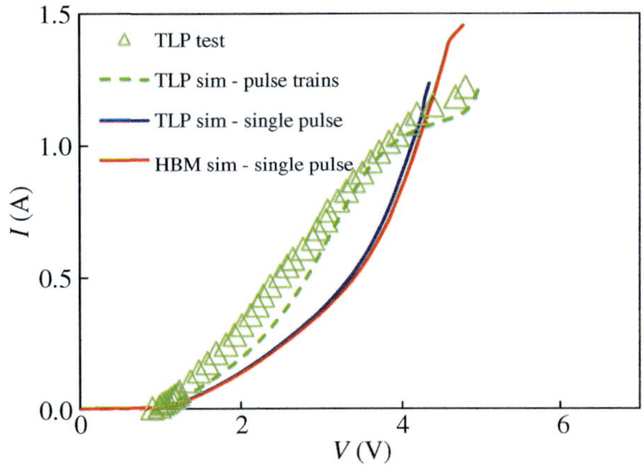

Figure 8.51 Transient ESD discharging $I-V$ curves for a STI ESD diode fabricated in a 28 nm CMOS obtained by TLP testing, and TCAD simulation using different stimuli: a single TLP pulse, a single HBM pulse and a TLP pulse trains. Transient ESD simulation ends at the ESD thermal failure threshold, I_{t2}.

measurement. Fourth, it is observed that the I_{t2} value from ESD simulation using an HBM ESD pulse stimulus is substantially higher that that predicted by TCAD simulation using TLP ESD pulse (both single pulse and a pulse train) as the ESD stimuli. This observation again states that TLP ESD stressing and HBM ESD zapping are not exactly the same ESD test equivalents. The observed differences in ESD discharging characteristics for the STI ESD diode by ESD simulation using different ESD stimuli can be understood by examining the incident ESD pulse I–t and ESD-induced transient $T_{max} - t$ behaviors given in Figure 8.52. For the same level of ESD inputs of TLP and HBM pulses, the ESD-induced transient heating characteristics are quite different in the time domain. It is readily observed that, in HBM zapping case, the ESD-induced lattice temperature (ESD heating) trend synchronizes well with the ESD input stimulus, i.e., $T_{max} - t$ increases, peaks, and falls, pretty much following the incident HBM ESD I–t waveform. However, in TLP stressing case, though the $T_{max} - t$ curve generally follows the incident ESD I–t waveform in terms of pulse rising and falling, the T_{max} peaks well after the peak of the input TLP ESD stimulus and T_{max} stays at a very high level for a substantial long period. In addition, it is found that the ESD-induced peak lattice temperature in HBM zapping case is $T_{max} \sim 1163$ K, which is significantly lower than that in TLP stressing case, i.e., $T_{max} \sim 1683$ K. This is attributed to the fact that the TLP square waveform has a long/flat pulse duration of $t_d \sim 100$ ns (i.e., at the peak current), during which the ESD stimulus keeps heating up the STI ESD diode internally, with the maximum flat and long stress. Hence, even though the incident ESD stimuli of the HBM and TLP pulses are intentionally designed to be "equivalent" in terms of its ESD rise time and ESD-induced pulse energy, the two different ESD stimuli lead to different ESD discharge thermal behaviors, which is very critical to accurately estimating ESD thermal failures in practical designs. Unfortunately, the HBM zapping test result was not available in this study. In summary, it is important to understand the mechanisms of TCAD mixed-mode ESD simulation. A *multiple-stimuli TCAD ESD simulation approach* is recommended in ESD simulation in order to properly guide design optimization and accurately predict ESD performance in practical ESD protection designs. It is cautious to simply equate the ESD testing results from TLP stressing and HBM zapping measurements.

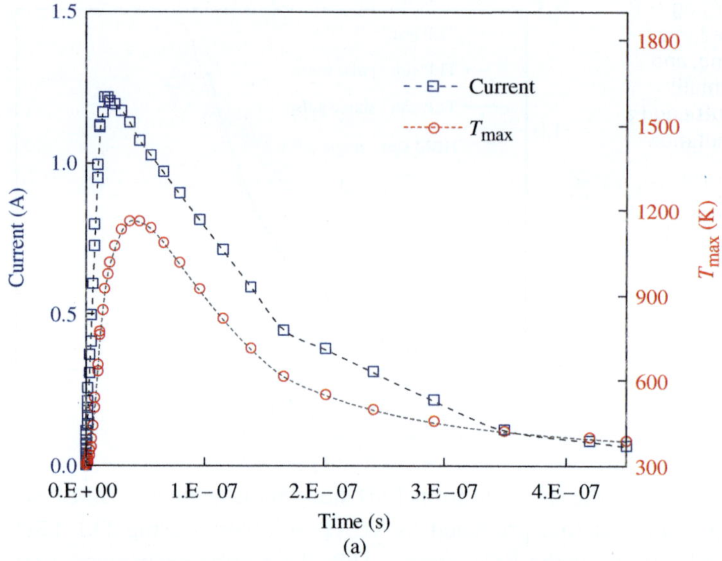

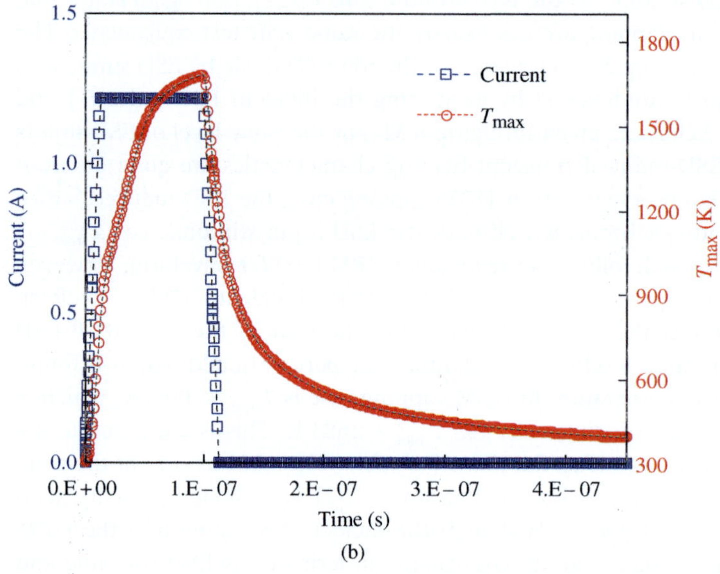

Figure 8.52 Transient incident ESD I–t curve (blue) and ESD-induced T_{max}–t curve (red) by (a) HBM ESD zapping simulation and (b) TLP pulse stressing simulation for a STI ESD diode fabricated in a 28 nm CMOS shows different details in ESD heating and thermal failure behaviors.

8.4 Summary

This chapter discusses details of TCAD-based mixed-mode ESD simulation-design methodology from principles and mechanisms to procedures and know-hows, which is a powerful ESD design technique for design optimization and prediction of practical on-chip ESD protection designs. TCAD is powerful that can facilitate not only process- and device-level simulation, but also circuit-level simulation including ESD protection simulation through integrated and interactive device–circuit-level interaction at chip level. Practical ESD design simulation has to address several important factors. First, ESD discharge phenomena involve complex multiple-coupling effects including materials, process, device, circuit, layout, electrical, thermal, and transient characteristics. Second, common circuit-level simulation is developed to handle small-signal circuit functions by solving linear circuit equation set, while ESD discharge phenomena involve ultralarge signals. Third, good circuit simulator requires accurate device models, which unfortunately is still a technical challenge for ESD protection device modeling that is entirely different from the common SPICE-like device modeling. Therefore, TCAD-based mixed-mode ESD simulation, which integrates numerical simulation to address ESD discharge details at device physics level and circuit simulation to cover the complex ESD-IC interactions becomes essential for accurate ESD design optimization and ESD protection prediction at chip level. The TCAD-based mixed-mode ESD simulation-design method has been comprehensively validated in practical on-chip ESD protection designs. Yet, it is noteworthy that TCAD ESD protection simulation requires comprehensive ESD knowledge and design know-hows in order to realize ESD protection design optimization and prediction in a real world. On the other hand, alternative ESD protection simulation methods that are more designer-friendly, such as, pure circuit-level ESD simulation using ESD device models, and new ESD CAD algorithms and software for whole-chip ESD protection physical design verification, etc., have been actively investigated and will be discussed in later chapters.

References

1 Wang, A. (2002). *On-Chip ESD Protection for Integrated Circuits – An IC Design Perspective.* Kluwer. ISBN: 0-7923-7647-1.

2 Sze, S.M. (1981). *Physics of Semiconductor Devices*, 2e. Wiley.

3 Diaz, C., Kang, S.M., and Duvvury, C. (1994). Circuit-level electricalthermal simulation of electrical overstress failures in advanced MOS I/O protection devices. *IEEE Trans. CAD* 13 (4): 482–493.

4 Feng, H., Chen, G., Zhan, R. et al. (2003). A mixed-mode ESD protection circuit simulation-design methodology. *IEEE J. Solid-State Circuits* 38 (6): 995–1006. https://doi.org/10.1109/JSSC.2003.811978.

5 Feng, H.G., Gong, K., and Wang, A. (2000). *ESD Protection Design Using Copper Interconnects: More Robustness and Less Parasitics.* SRC Publication: 2000 Publications in Copper Design Challenge, Pub. P000375.

6 Feng, H. (2001). A mixed-mode simulation-design methodology for on-chip ESD protection design. A MS thesis. Illinois Institute of Technology.

7 Wang, A., Tsay, C., and Deane, P. (1998). A study of NMOS behaviours under ESD stress: simulation and characterization. *Microelectron. Reliab.* 38: 1183–1186.

8 Wang, S.J., Yao, F., Qin, B. et al. (2010). Analysis and optimization of HV ESD protection. *Proceedings of IEEE International Conference on Electron Devices and Solid-State Circuits (EDSSC).* https://doi.org/10.1109/EDSSC.2010.5714031.

9 Pan, Z., Li, C., Di, M. et al. (2020). 3D TCAD analysis enabling ESD layout design optimization. *IEEE J. Electron Devices Soc. (J-EDS)* 8: 1289–1296. https://doi.org/10.1109/JEDS.2020.3027034.

10 Di, M., Pan, Z., Li, C., and Wang, A. (2021). A new multi-stimuli-based simulation method for ESD design verification. *Proceedings of IEEE Electron Devices Technology and Manufacturing Conference (EDTM).*

11 MIL-STD-883E, Method 3015.7 (1989). *Electrostatic Discharge Sensitivity Classification.* U.S. Dept. of Defense.

9

RF ESD Protection

9.1 What Is Special for RF ESD Protection?

The past two decades witnessed unprecedented proliferation and prosperity of wireless communications, which has been made possible by amazing advances in radio-frequency (RF) integrated circuits (ICs). No surprise, RF ICs require on-chip electrostatic discharge (ESD) protection too. In fact, the nature of wireless products, such as smartphones and tablets, imposes higher ESD protection requirements on RF ICs because handheld-oriented, high-frequency, broadband RF ICs are more vulnerable to ESD risks, while also more sensitive to any ESD-induced parasitic effects. Therefore, on-chip ESD protection for RF ICs has become a major challenge for advanced RF ICs. Before working on RF ESD protection designs, the first question to ask is what is unique in ESD protection for RF ICs compared to ESD protection for normal ICs? The fundamental answer is that there exist interactions between any ESD protection structures and the core circuits under ESD protection, and such ESD–IC interaction effects are much severe for RF ICs and the interactive problem keeps getting worse as the RF IC frequency, bandwidth, and data rate continue to increase [1–3]. This is basically why RF ESD protection design has been so challenging and mounting.

The *ESD–IC interaction* is a two-way problem. In one direction, the core circuitry can affect ESD protection, which is referred as the *Circuit-to-ESD Influence*. In the opposite direction, any ESD protection structure will affect core circuit performance, which is defined as the *ESD-to-Circuit Influence*. There are two major circuit-to-ESD influences. First, on-chip ESD protection involves an IC chip, which has various parasitic devices within the core circuit, some of them may be unexpectedly and randomly turned on by an incident ESD pulse that will likely by-pass the designed on-chip ESD protection structure (i.e., *intentional* ESD protection device) at I/O to discharge the ESD transients unexpectedly. Since any internal parasitic structures may be unknown and, certainly, are not designed to handle large ESD transients, therefore, premature ESD failure (i.e., *Early ESD failure*) may occur regardless how well an intentional I/O ESD protection structure is designed. As discussed before, this is the reason that on-chip ESD protection design is considered as a chip-level design task, not just about an individual standalone ESD protection device at a pad. Second and more specifically for RF ESD protection, an I/O ESD protection structure may be accidentally triggered by a non-ESD normal RF input signal, which is often very fast and strong. RF ESD *mis-triggering* in normal IC operations will cause short-circuit malfunction of a chip in absence of any ESD events. Studies show that there exists a clear correlation between ESD triggering voltage (V_{t1}) and the rise time (t_r) of an incident ESD pulse for common ESD protection structures [1–11]. Figure 9.1 depicts the measured V_{t1} and t_r data by transmission-line-pulsing (TLP) ESD testing for some popular ESD protection structures, including NMOS and dual-polarity silicon controlled

Practical ESD Protection Design, First Edition. Albert Wang.

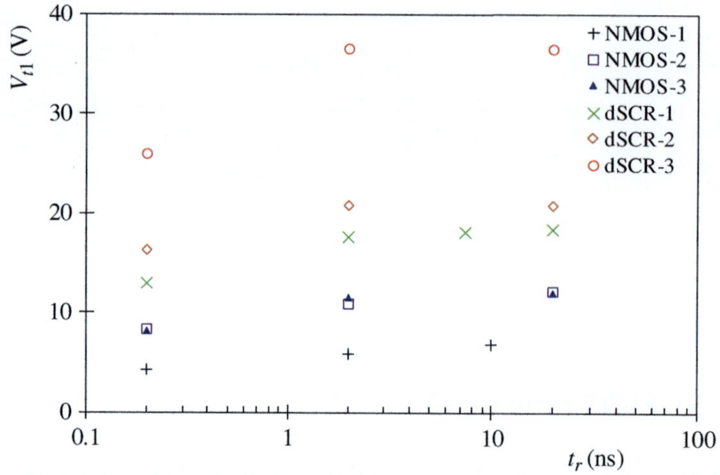

Figure 9.1 Sample TLP-measured triggering voltage V_{t1} data for commonly used ESD protection devices, including NMOS and dual-polarity SCR ESD protection structures, show strong dependence of V_{t1} on TLP pulse rise time t_r. As t_r decreases for a faster ESD pulse, V_{t1} can drop substantially.

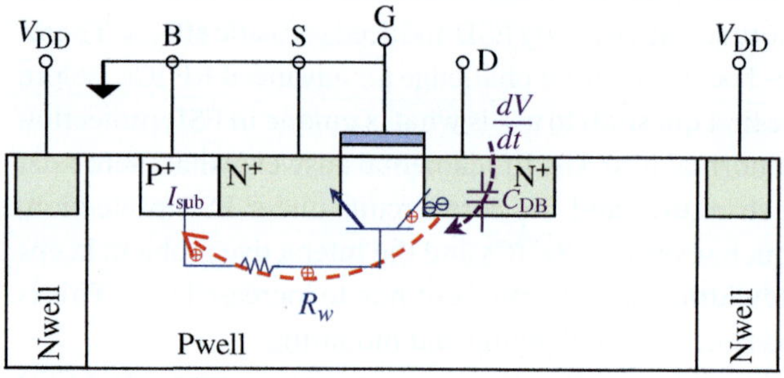

Figure 9.2 A ggNMOS ESD protection device relies on the ESD-induced Drain avalanche current running through the Pwell resistance (R_w) to build up a potential to trigger the parasitic BJT for ESD discharge. With a large $\frac{dV}{dt}$ of the incident ESD pulse across a sizable parasitic drain capacitance (C_{DB}), a substantial ESD-induced displacement current $C_{DB}\frac{dV}{dt}$ will dramatically increase the total substrate current (I_{sub}), which will reduce the effective ESD triggering voltage, V_{t1}.

rectifier (dSCR) ESD protection structures fabricated in various IC processes. A strong V_{t1}–t_r relationship is readily observed, i.e., typically, V_{t1} decreases as t_r becomes shorter (from 200 to 20 ns) for faster TLP pulses. It reveals a major design headache, which is that, no matter how accurately an ESD protection device is designed for its preferred V_{t1}, in a real world, the actual ESD triggering V_{t1} may vary substantially depending upon the incident ESD pulses. This V_{t1}–t_r relationship may be explained by the *displacement current effect* in the following ESD design examples. Figure 9.2 shows a cross-section of a classic ggNMOS ESD protection structure, whose ESD triggering is initiated by the avalanche breakdown of the Drain junction, upon which the avalanche current (I_{sub}) is collected by the grounded body terminal. As I_{sub} flows through the P-well resistance (R_w), it builds up the voltage potential across the source junction and will turn on the parasitic lateral bipolar junction transistor (BJT) to discharge the ESD transient. In real-world ESD events, a very fast ESD pulse has a large $\frac{dV}{dt}$ ratio, which runs through a sizable, revised-biased drain junction capacitance (C_{DB}) and produces a *displacement current* ($C_{DB} \times \frac{dV}{dt}$) that adds up to the total substrate current

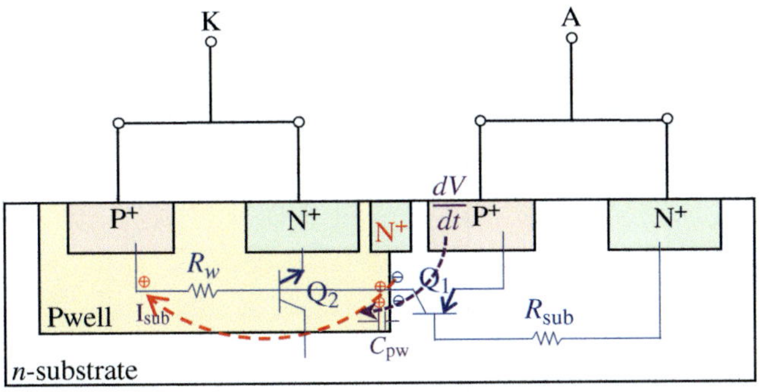

Figure 9.3 An SCR ESD protection device relies on the ESD-induced Pwell/N-sub avalanche current running through the Pwell resistance (R_w) to build up a potential to trigger the parasitic BJT pair for ESD discharge. With a large $\frac{dV}{dt}$ of the incident ESD pulse across a sizable parasitic Pwell/N-sub capacitance (C_{pw}), a substantial ESD-induced displacement current $C_{pw}\frac{dV}{dt}$ will significantly increase the total substrate current (I_{sub}), which will reduce the effective ESD triggering voltage, V_{t1}.

($I_{sub} + C_{DB}\frac{dV}{dt}$). Consequently, V_{t1} of the ggNMOS may be reduced, depending upon the $\frac{dV}{dt}$ ratio of incident ESD pulses. In a second example, Figure 9.3 shows a cross-section of a common SCR ESD protection structure, whose ESD triggering is initiated by the avalanche breakdown of the P-well/N-sub junction, upon which the avalanche-induced I_{sub} is collected by the cathode (K). As I_{sub} flows through R_w in the P-well, it builds up the potential across the N^+/P-well junction, which turns on the parasitic NPN–PNP BJT pair, hence triggering the SCR ESD protection structure. Similarly, a large $\frac{dV}{dt}$ ratio of an incident ESD pulse combining a sizable P-well/N-sub junction capacitance (C_{pw}) will generate a displacement current ($C_{DB}\frac{dV}{dt}$) that will increase the total substrate current ($I_{sub} + C_{pw}\frac{dV}{dt}$). As a result, V_{t1} of the SCR ESD protection device will decrease according to the $\frac{dV}{dt}$ ratio of incident ESD pulses. It is therefore obvious that the ESD triggering can be an unexpected varying factor in real-world ESD events. Next, we will consider if a normal RF signal may mis-trigger an ESD protection structure at I/O. From the ggNMOS and SCR ESD protection structure examples, a forward bias of 0.65 V is needed to forward turn on the PN junction in Si in order to eventually trigger the ggNMOS and SCR ESD protection devices. How much a $\frac{dV}{dt}$ ratio would be needed to establish a 0.65 V bias across a PN junction in order to trigger an ESD protection device solely by an ESD-induced displacement current? Studies suggest that the needed $\frac{dV}{dt}$ ratio to trigger the common ESD protection structures are estimated to be 3×10^{10} to 1×10^{11} V/s [8, 9], which are shown (Blue points) in Figure 9.4. Studies also estimate the $\frac{dV}{dt}$ ratios for typical ESD stimulus waveforms in real-world ESD testing, including standard human body model (HBM) ESD waveforms, industrial HBM ESD zapping testers, and TLP stressing testers, which range from 7×10^8 to 1×10^{11} V/s as marked (open symbols) in Figure 9.4 [8–10]. In comparing these two groups of $\frac{dV}{dt}$ ratio data, it is observed that these data points are at the similar level, suggesting the needed $\frac{dV}{dt}$ ratio to trigger the common PN-based ESD protection structures, such as ggNMOS and SCR ESD protection devices. Concerning the possible ESD mis-triggering induced by normal RF signals, a few normal RF and high-speed signals in IC design papers are used to calculate the $\frac{dV}{dt}$ values, resulting in ~2.5×10^8 V/s for a 2.5 GHz complementary metal-oxide-semiconductor (CMOS) clock recovery circuit [12], ~4.3×10^7 V/s for a 1 GHz CMOS clock synthesizer [13] and ~1.23×10^7 V/s in a 7.1 MHz noise coupling circuit on a mixed-signal CMOS receiver chip [14]. These $\frac{dV}{dt}$ data points from the early day papers are added to Figure 9.4 (Red points), which seem to be much smaller than the $\frac{dV}{dt}$ values estimated to trigger a PN-based ESD protection structure. Figure 9.4 suggests, rather

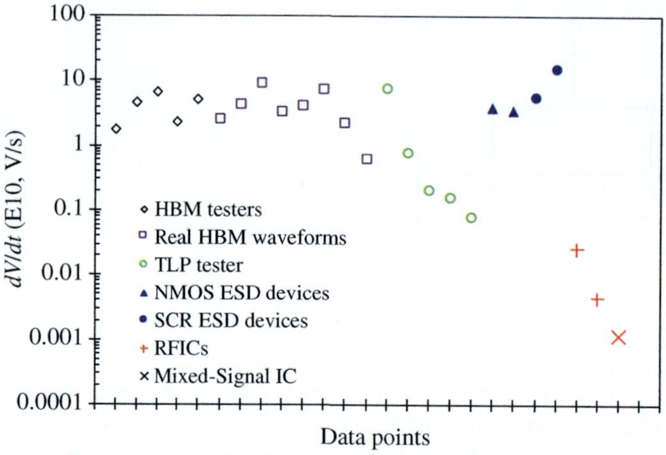

Figure 9.4 Measured data show that the $\frac{dV}{dt}$-induced displacement current plays a role in ESD triggering. The extracted $\frac{dV}{dt}$ data for real-world HBM ESD waveforms, HBM zapping pulses and TLP stressing waveforms are at the same level as that corresponding to forward-PN-induced ESD triggering of common ESD protection structures (NMOS, SCR, etc.). The $\frac{dV}{dt}$ values extracted from some reported RF ICs were much lower than that for ESD triggering, which may change for ultrahigh frequency RF ICs, possibly causing mix-triggering of ESD protection structures by normal RF signals.

intuitively, that normal RF input signals may have been generally "safe" to ESD protection structures, i.e., no ESD mis-triggering. However, the recent relentless pursuit for ultrahigh frequencies, ultrawide frequency bandwidth, and extremely high data rate for RF ICs, now already in millimeter wave spectrum, suggests that, at some moment, an ultrafast very strong normal RF signal may induce a sizable displacement current that may mis-trigger an ESD protection device at I/O in absence of any ESD events, resulting in short-circuit malfunction in normal RF IC operations. It is noteworthy that the actual ESD mis-triggering phenomenon is rather complicated for real-world ICs. Nevertheless, the risk of ESD mis-triggering does exist.

On the other hand, the ESD-to-Circuit Influence is always a concern to ICs because any ESD protection structure, mostly in-Si PN-based active devices, will inevitably produce ESD-induced parasitic effects, including parasitic capacitance (C_{ESD}) and resistance (R_{ESD}), leakage (I_{leak}) currents, and ESD self-generated noises and ESD-induced noise coupling through C_{ESD}. Such ESD-induced parasitic effects are much more detrimental to RF ICs that are extremely sensitive to and become more and more intolerable to any ESD parasitic effects. For example, RF ICs often rely on resistor-inductor-capacitor (RLC) matching networks to ensure input and output impedance matching, which may be easily destroyed by the inherent parasitic C_{ESD}, hence, resulting in severe RF IC performance degradation. Similarly, low-noise design is a main goal for most RF ICs. A sizable PN-induced leakage current in an ESD protection structure will not only cause large standby current but also generate extra thermal and shot noises in RF ICs. The ESD self-generated noises were discussed in Chapter 4 for common ESD protection structures. In addition, a large ESD-induced C_{ESD} will substantially increase the noise coupling effect on a chip, both locally (within a power/function domain) and globally (across the full chip), which is also deadly for RF ICs. Figure 9.5 depicts a chip scenario showing ESD self-generated noises and C_{ESD}-induced noise coupling between digital and RF circuit domains [15]. All these factors will become even worse in RF ICs because RF chips are widely used in handheld devices, such as smartphones and tablets, which require much higher ESD protection that typically translates into

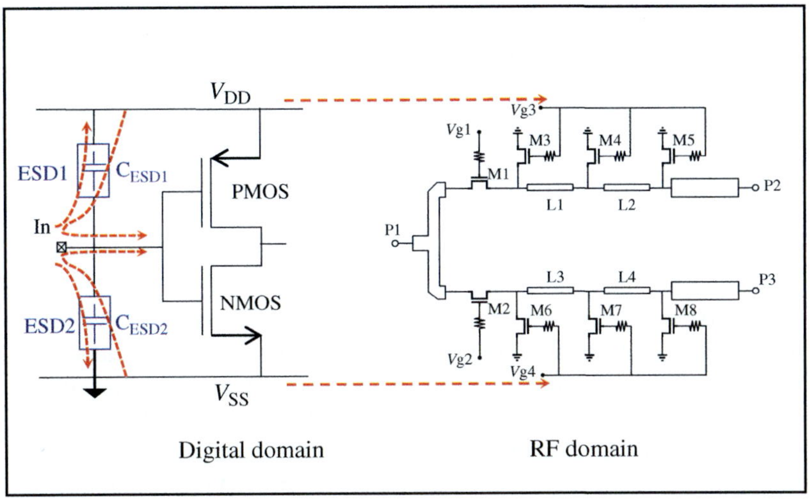

(a)

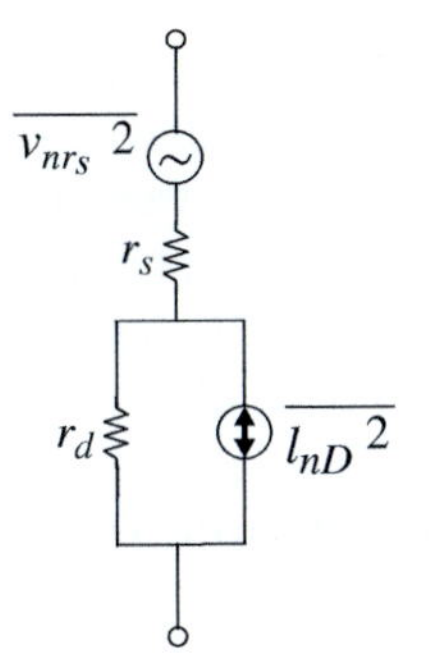

(b)

Figure 9.5 Illustration of ESD-induced IC noise degradation in two possible ways: (a) a parasitic C_{ESD} can cause local and global noise coupling (red dashed arrows) on a chip between different circuit blocks, and (b) an ESD protection structure can self-generate extra noises due to its leakage current and physical resistance as shown for a PN diode, further complicating the global noise coupling problem.

large ESD protection device size of the same type of ESD protection structure, hence having more ESD-induced parasitic effects. For these reasons, one can easily find that high-performance RF ICs often do not have enough on-chip ESD protection and, sometimes, do not have any ESD protection for the high-speed signal pads, in order to achieve the desired high speed or keep up with the high RF frequencies. This problem becomes much worse for higher frequency and broader band RF ICs because of the severe fluctuation in performance degradation due to varying capacitive and noise effects across a wide frequency band.

In summary, the core challenge for RF ESD protection design is the inherent ESD–IC interactions that become severely worsen in RF ICs (i.e., *ESD–RFIC Interaction Effect*), which must be comprehensively handled in RF ESD protection designs. Specifically, there are four unique challenges in RF ESD protection: first, the RF-induced ESD mis-triggering (i.e., V_{t1} reduction due to the displacement current induced by normal RF signals); second, the ESD-induced parasitic ESD capacitance (i.e., C_{ESD}); third, the ESD-induced noise effect (i.e., ESD self-generated noises and C_{ESD}-induced global noise coupling); and fourth, the substantial fluctuation in these negative ESD effects (e.g., ΔC_{ESD}, ΔR_{ESD}, and ΔNF associated with variations in process, voltage, and temperature, as well as frequency bandwidth). More details on the complex ESD–RFIC interactions will be discussed with examples in Chapter 10.

9.2 RF ESD Protection Characterization

Now that we understood that any ESD protection structure inevitably introduces parasitic effects, which can seriously affect RF IC performance, an engineer should handle the RF ESD design problem in two general ways: The first is to minimize ESD-induced parasitic effects when designing an ESD protection structure, which can be achieved by aid of the TCAD-based mixed-mode ESD simulation technique. The second is to accurately characterize the ESD-induced parasitic effects, including C_{ESD} and noises, which can then be used for ESD–RFIC codesign that will be discussed in Chapter 10. This section discusses how to characterize RF ESD protection structures.

Arguably, the most concerned ESD-induced parasitic parameters to RF ESD protection designs are the parasitic C_{ESD}, noise and leakage, which can seriously affect RF IC performance. There are different ways to measure the ESD-induced C_{ESD} for an ESD protection structure. The traditional method of direct measuring capacitance of an ESD protection structure using an ordinary capacitance meter has its limitation in achieving the required testing accuracy, broadband frequency responses, and differentiating contributions to C_{ESD} from an ESD protection device itself (e.g., in-Si structure) and its peripherals (e.g., contact and via plugs, metal interconnects and bonding pads). These C_{ESD} testing concerns are important to ESD design optimization and prediction, as well as the ESD–RFIC co-design approaches to be detailed in Chapter 10. In general, an RF approach (i.e., treat an RF ESD protection device as an RF element) is required to accurately characterize C_{ESD} of RF ESD protection structures, which typically treats an ESD protection structure (i.e., *device under test,* DUT) as a two-port subnet consisting the C_{ESD} and R_{ESD} in series or parallel formats. Correspondingly, typical RF-like ESD test patterns are used, e.g., ground-signal-ground (GSG) or ground-signal (GS) layout patterns, to eliminate the unwanted peripheral "noises." Further and more importantly, a dummy test structure is used in a pair of RF test pattern of the ESD protection structure, as depicted in Figure 9.6, which allows separating the parasitic capacitance contributions from the in-Si ESD protection device itself (often including the contact plugs) and its unavoidable ancillary parts (i.e., via plugs, metal lines, and pads). A well-proven method for characterizing the C_{ESD} is the *S*-parameter measurement technique where the *S*-parameters are measured for an ESD protection structure across the concerned frequency spectrum. With the aid of GSG/GS and dummy test patterns, the C_{ESD} value can be accurately extracted for any ESD protection structures [1, 3, 6, 7, 16]. Figure 9.7 depicts one of the C_{ESD} extraction methods, which utilizes a *Y*-parameter model to extract the ESD-induced parasitic capacitance of a real-world ESD protection layout cell (e.g., an ESD P-cell in an IC technology cell library), where the mathematical comparison of the ESD and dummy test pattern pair allows to separate the Si ESD C_{ESD} from its metal interconnects (C_{metal}) and pad (C_{pad}) capacitance [16]. This *S*-parameter C_{ESD} characterization technique has been widely used in the field, enabling very accurate C_{ESD} measurement results. In one practical design example for comparing the ESD-induced C_{ESD} for various common ESD protection structures, a set of ggNOMS, diode, diode-string, SRC, and dSCR ESD protection structures were designed and fabricated in a 0.35 μm BiCMOS technology [3, 8]. These ESD protection structures were first optimized by TCAD ESD simulation to minimize their parasitic C_{ESD}. *S*-parameter measurement was conducted for all the fabricated ESD protection structures across a wide frequency bandwidth from DC to 9 GHz, and subsequently, the C_{ESD} was extracted for these ESD protection

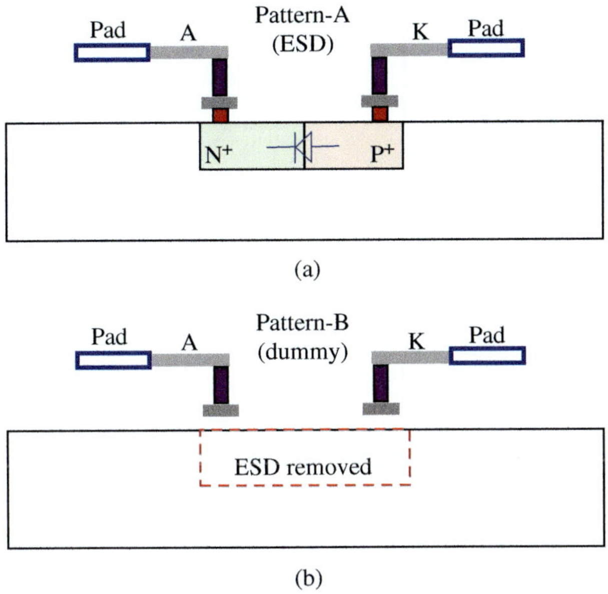

Figure 9.6 Cross-section view for de-embedded ESD test pattern pair for accurately measuring ESD-induced C_{ESD} for the in-Si device structure only: (a) a full ESD protection diode, and (b) a dummy (metal + pad only) for (a) with the in-Si PN diode removed.

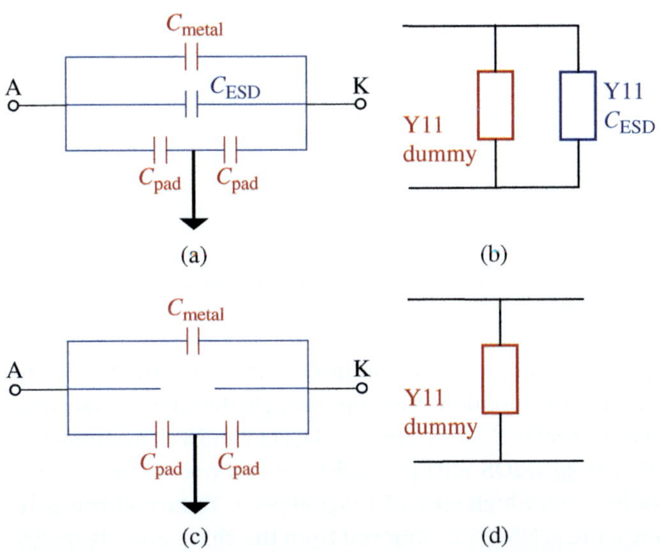

Figure 9.7 Equivalent two-port network models for de-embedded ESD C_{ESD} test patterns: (a) full ESD protection structure, (b) Y-model for full ESD protection structure, (c) dummy ESD test pattern, and (d) Y-model for the dummy pattern. C_{ESD} (or, $C_{ESD\text{-}Si}$) is the ESD-induced capacitance for the in-Si ESD structure only, C_{pad} (or, $C_{ESD\text{-}pad}$) is the pad capacitance and C_{metal} (or, $C_{ESD\text{-}metal}$) is the total ESD metal interconnects capacitance.

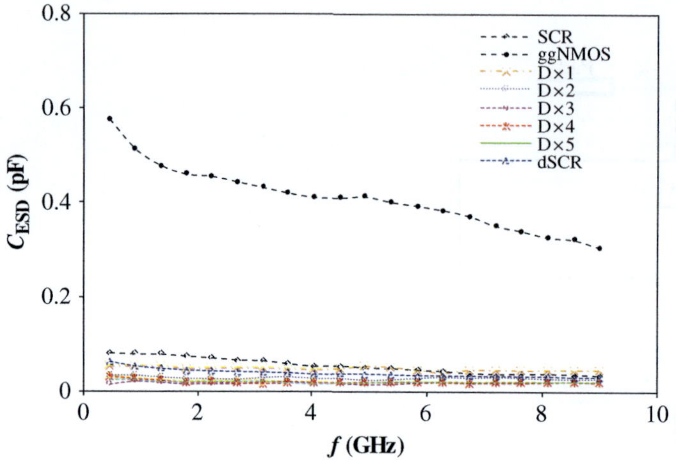

Figure 9.8 Measured C_{ESD} across a 9 GHz bandwidth for various 2 kV ESD protection structures fabricated in a 0.35 μm BiCMOS technology.

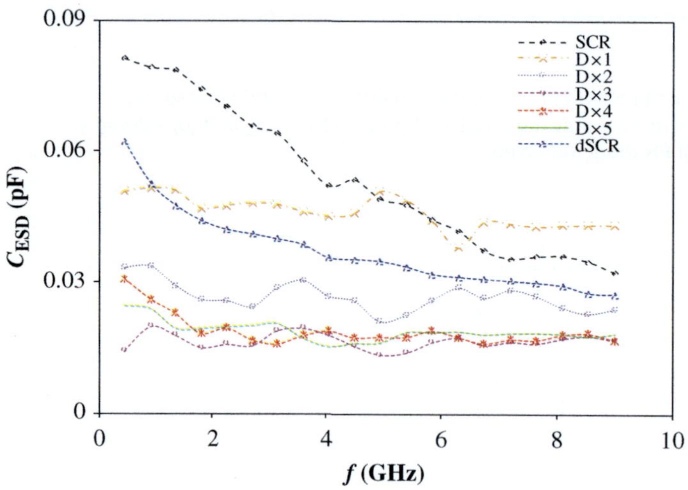

Figure 9.9 Measured C_{ESD} across a 9 GHz bandwidth for various 2 kV ESD protection structures fabricated in a 0.35 μm BiCMOS technology (ggNMOS removed).

devices, designed for 2 kV HBM ESD protection target, as shown in Figure 9.8. Figure 9.8 clearly shows that, though optimized by TCAD ESD simulation, the popular ggNMOS ESD protection device still introduces significantly more parasitic C_{ESD}, across the 9 GHz bandwidth, compared to its other counterparts. This suggests that ggNMOS ESD protection should not be used for RF ESD protection, at least not for the high-frequency high-speed RF signal ports. To further compare other common ESD protection structures, the ggNMOS is removed from the chart, and Figure 9.9 depicts the measured $C_{ESD} \sim f$ curves for other ESD protection structures. It is clearly observed that SCR has fairly low parasitic C_{ESD}, especially at higher frequency, and the dual-polarity SRC ESD protection structure (dSCR) has substantially lower C_{ESD} than that for SCR ESD device. It is interesting to further study the C_{ESD} behaviors of the diode and diode-string ESD protection series.

In general, the C_{ESD} of an ESD diode is considered coming from the PN junction (C_j). So, to the first order, using a diode-string of n diodes in series connection would substantially and monotonically reduce the total C_{ESD} of a diode-string, i.e., $C_{ESD-total} \approx \frac{C_j}{n}$. Interestingly, Figure 9.9 tells a rather different story about an ESD diode-string in real-world measurements. From one diode (D × 1) to a two-diode string (D × 2), C_{ESD} drops almost in half. Though the three-diode string (D × 3) does show a sizable reduction in its C_{ESD}, however, the reduction is not as expected per the formula above. When adding more diodes to the diode-string, i.e., four-diode (D × 4) and five-diode (D × 5) in a diode-string, it does not further reduce the total C_{ESD} of a diode-string ESD protection structure. This phenomenon can be explained through the fact that the parasitic capacitor network for a large diode-string ESD protection structure in IC formats actually contains many ancillary capacitances (e.g., mental interconnects), not simply being n number of $C_{j\text{-Si}}$ in series. Figure 9.10 gives the C_{ESD} values measured at $f = 2.4$ GHz for all ESD protection structures fabricated, revealing interesting insights: First, ggNMOS has the highest C_{ESD} in the design group and should not be used for RF ESD protection. Second, a diode-string ESD protection offers reduced total C_{ESD} to certain extent. An SCR ESD protection device has rather low C_{ESD}, which can be further reduced in its dSCR revision. The observations are certainly very informative for RF ESD design optimization to minimize the parasitic C_{ESD}. Unfortunately, practical IC design is much more involving that requires more comprehensive design considerations, beyond looking at only one specs parameter (e.g., C_{ESD}). As discussed before, the ESD design overhead effects include not only the ESD-induced parasitic C_{ESD} and noises, but also other design factors such layout sizes and layout floor planning. Figure 9.11 presents the layout sizes of the ESD protection structures in this study as another design factor, which clearly shows that ggNMOS is fairly large, SCR is very compact, dSCR is much smaller, and a diode-string increases its size linearly, all assuming for the same 2 kV HBM ESD protection. Obviously, different factors have different impacts for the ESD protection structures, and the corresponding "trends" for different ESD parameters may be in opposite directions. Therefore, as in general IC design practices, some kind of *figure-of-merit* (FOM) parameter may be used to evaluate

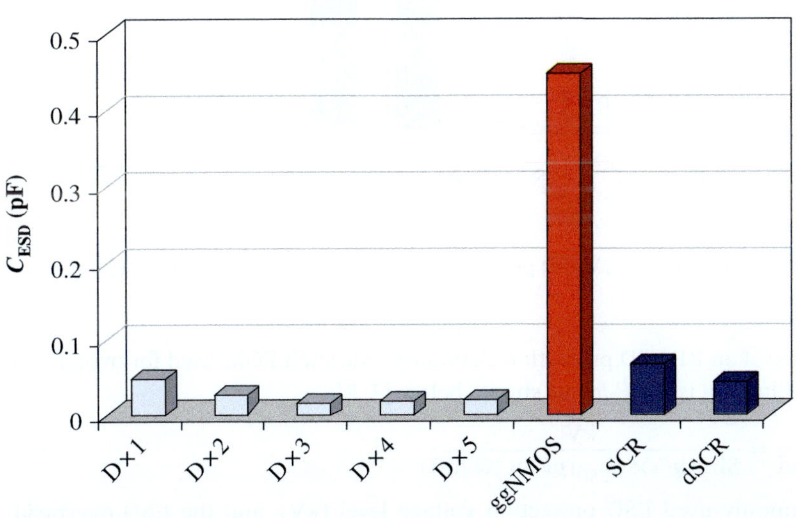

Figure 9.10 Measured C_{ESD} at 2.4 GHz for various 2 kV ESD protection structures fabricated in a 0.35 μm BiCMOS technology.

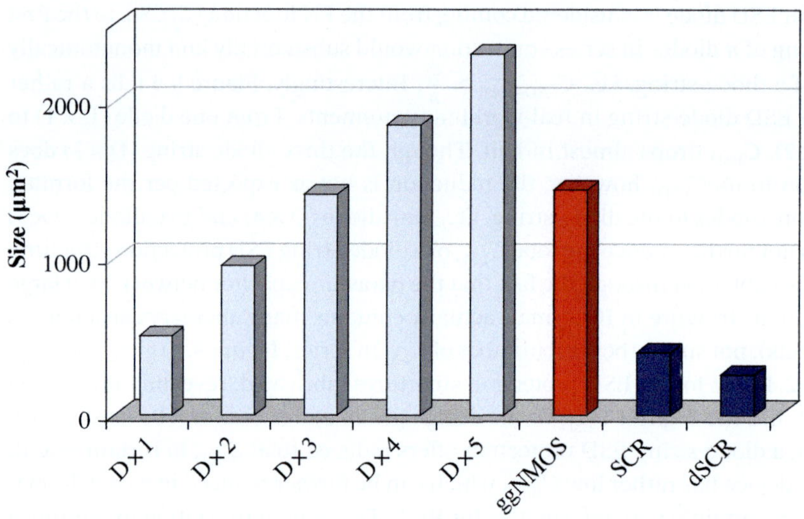

Figure 9.11 Layout sizes are different for various 2 kV ESD protection structures implemented in a 0.35 μm BiCMOS technology.

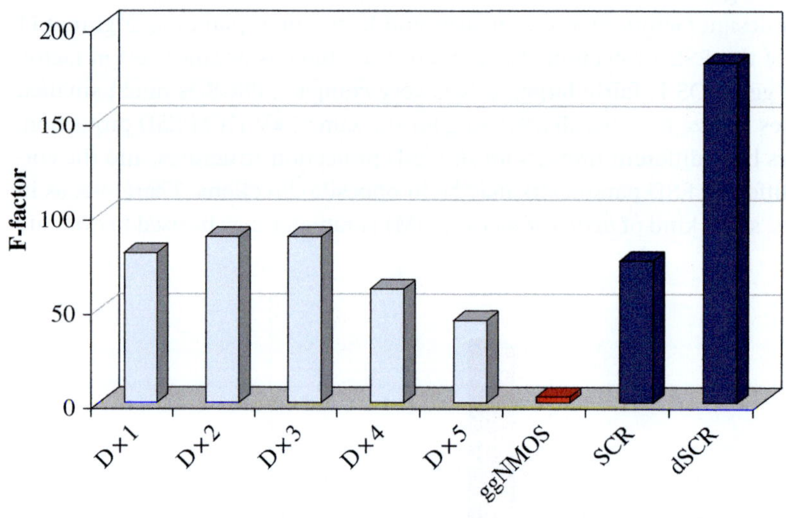

Figure 9.12 F-factors extracted for various 2 kV ESD protection structures implemented in a 0.35 μm BiCMOS technology.

overall design performance of an RF ESD protection structure. One such FOM used for overall RF ESD protection design evaluation is the *F-factor* shown below [3, 8],

$$F = \frac{\text{ESDV}}{\text{ESD Overhead}} = \frac{\text{kV}}{\text{Size}(\mu^2) \times C_{\text{ESD}}(pF) \times \text{NF(dB)}} \tag{9.1}$$

where *ESDV* is the commonly used ESD protection voltage level (kV), and the ESD overhead includes all unwanted ESD-induced design factors, such as parasitic C_{ESD}, noises (in noise figure, NF), and layout size. Indeed, this F-factor can be an open-boundary evaluation parameter that may be modified per special needs for ESD design specs evaluation in practical designs, for example adding the ESD discharging resistance (R_{ON}) and thermal conduction property. Nevertheless,

F-factor is defined so that a larger value is always preferred for an ESD protection structure to ensure the best overall ESD protection design performance. Figure 9.12 depicts the extracted F-factor for various ESD protection structures in this study. It readily shows that ggNMOS has very low F-factor, hence not good for RF ESD protection; SCR has high F-factor and dSCR has very high F-factor, hence good for RF ESD protection. Interestingly, a diode-string increases its F-factor when adding more diodes in the diode string structure but only to certain number. One critical message of this study is that a good ESD protection design must be optimized and evaluated quantitatively by simulation and measurement, and an optimum RF ESD protection must be characterized for its overall performance specs in real-world RF IC designs. A second design example is used to show how accurately the C_{ESD} of the Si ESD protection structures can be measured using the combined Y-model and ESD dummy test pattern approach. In this study, STI-isolated ESD protection diodes and low-V_{t1} diode-triggered silicon controlled rectifier (DTSCR) ESD protection structures designed and fabricated in a foundry 28 nm CMOS technology were characterized using the S-parameter method [16]. The goal was to accurately extract the C_{ESD} values for the in-Si ESD protection devices by de-embedding the parasitic C_{metal} and C_{pad} of the complex layout cells. Figure 9.13 depicts the measured Si-only C_{ESD} for the STI ESD diodes made in the I/O process module in the 28 nm CMOS technology, and Figure 9.14 presents the extracted C_{ESD} results for the Si-only DTSCR ESD protection structures made in the core process module in the same 28 nm CMOS. Very accurate C_{ESD} results are obtained for the in-Si ESD protection structures, which are used for ESD-RFIC codesign later.

In addition to extract the ESD-induced parasitic C_{ESD}, S-parameters for ESD protection structure can also be directly used in ESD-RFIC co-design in real-world IC designs, which will be discussed in Chapter 10. Characterization of ESD-induced noise effects, including the ESD self-generated noises and global noise coupling due to the C_{ESD}, is also important in practical RF ESD protection design [5, 9].

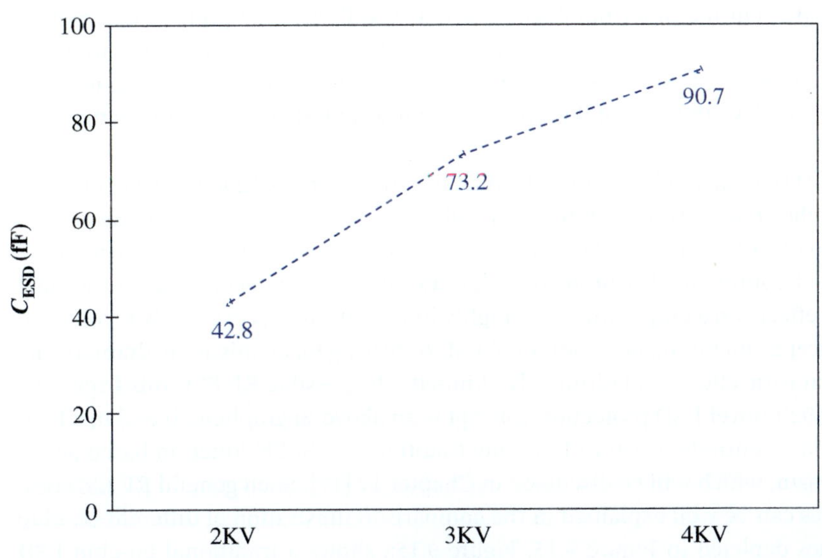

Figure 9.13 Measured C_{ESD} for STI P^+ ESD protection diodes in the I/O process module fabricated in a foundry 28 nm CMOS technology.

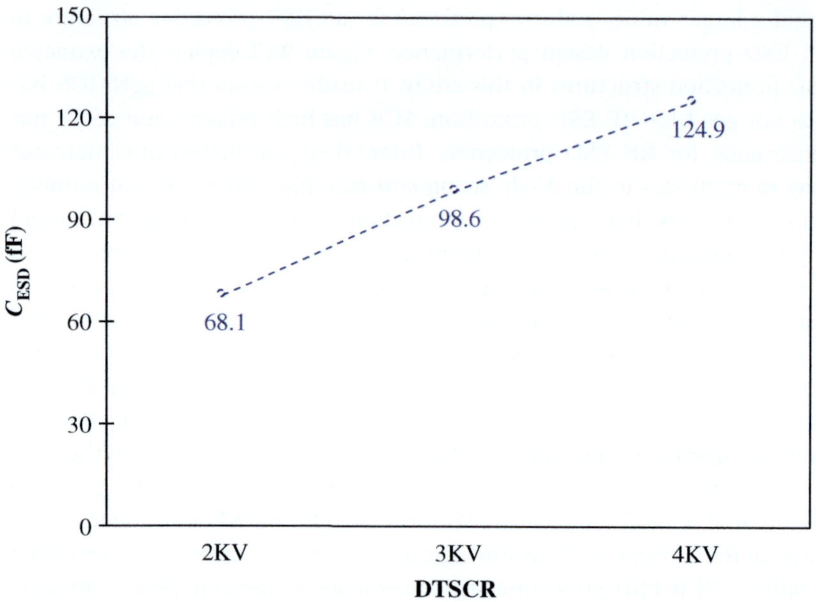

Figure 9.14 Measured C_{ESD} for single-diode-triggered DTSCR ESD protection structures in the core process module fabricated in a foundry 28 nm CMOS technology.

9.3 Low-Parasitic ESD Protection Solutions

The next question is therefore "what is an ideal RF ESD protection solution(s)?" Unfortunately, the answer is that there is NO "ideal, universal, magic and one-for-all" RF ESD protection solution in practical RF IC designs. On the other hand, generally, a "good" RF ESD protection structure must be a low-parasitic design, which minimizes the complex ESD–RFIC interactions and hence has minimized negative impacts on RF IC performance. It is obvious up to you, the IC designer, to burn your brain cells to achieve the goal of designing low-parasitic ESD protection structures for RF ICs.

In principle, there exist several highly desirable technical features for any "good" RF ESD protection solutions: First, higher ESD current-handling capability and smaller layout size are needed to achieve higher *ESDV/Si ratio*. Second, novel multiple-mode ESD protection structures are required to reduce the total head counts of ESD protection devices on a chip, hence, reduced the total ESD-induced parasitic effects on a chip. Third, it is highly imperative to discover truly revolutionary ESD protection concepts, including nontraditional ESD triggering mechanisms, to dramatically reduce ESD-induced parasitic effects and ultimately eliminate the possible RF ESD mis-triggering risk. One example for such novel ESD protection concept is an above-Si graphene-based mechanical switch structure that is entirely different from any traditional in-Si PN-junction-based active ESD discharge mechanism, which will be discussed in Chapter 17 [17]. Such general RF ESD protection design principles can be well explained in the comparison illustration of different on-chip ESD protection schemes depicted in Figure 9.15. Figure 9.15a shows a traditional on-chip ESD protection scheme utilizing *single-directional* (i.e., single-mode or single-polarity) ESD protection structure, which requires multiple ESD protection devices at each pad on a chip, hence resulting in significant overall ESD-induced parasitic effects. As discussed in Chapter 6, *dual-polarity*

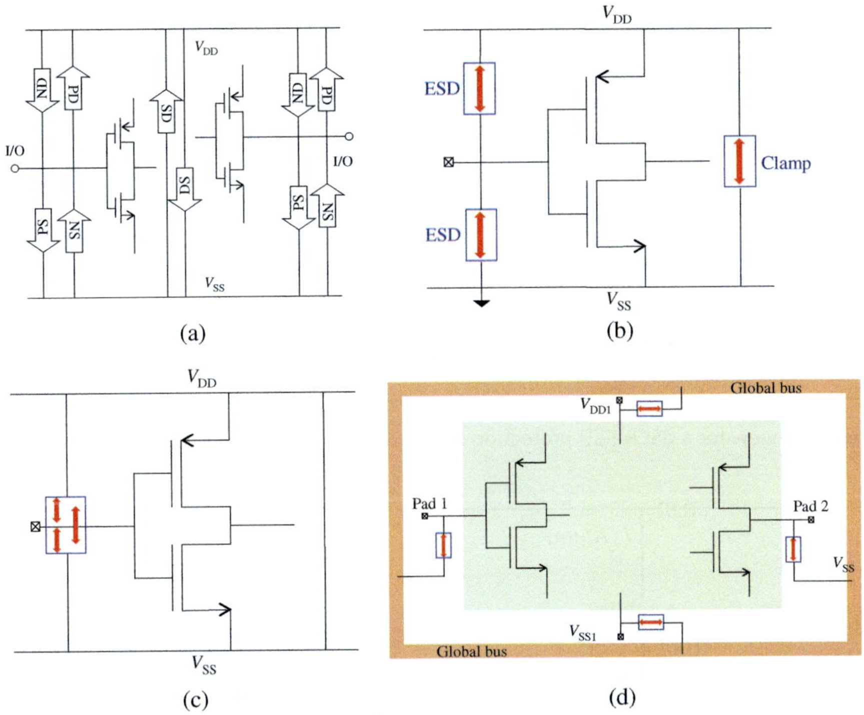

Figure 9.15 A comparison illustration for various on-chip ESD protection schemes using different ESD protection devices and having different ESD-induced parasitic effects: (a) multiple ESD protection devices needed per I/O pad if using single-mode ESD protection structure, (b) fewer ESD protection devices needed per pad if using dual-directional ESD protection structures, (c) one ESD protection device needed per pad if using an all-mode ESD protection structure, and (d) much simplified whole-chip ESD protection schematics when using dual-polarity ESD protection devices and global ESD bus in planning whole-chip ESD protection.

(i.e., dual-mode, Figure 9.15b) and *multiple-mode* (i.e., multiple-directional, Figure 9.15c) ESD protection structures, such as dSCR and all-mode SCR ESD protection structures, can be used to substantially simplify full-chip ESD protection schematics, leading to much reduced overall ESD-induced parasitic effects on a chip. Further, with a global perspective, multiple-mode ESD protection structures can be used to dramatically simplify the whole-chip ESD protection schemes, as depicted in Figure 9.15d, which translates into not only much reduced overall ESD-induced ESD design overhead, but also greatly simplified on-chip ESD protection schematics. Similarly, as discussed in Chapter 6, the pad-clamp full-chip ESD protection scheme can be used to simplify the whole-chip ESD protection circuit schematics and to reduce the total head count of ESD protection devices on a chip, even if using single-direction ESD protection devices, therefore, to minimize the overall ESD-induced parasitic effects for a chip.

9.4 RF ESD Protection Design Example

Now that we understood the harmfulness of ESD-induced parasitic effects on RF ICs and the methods of characterizing RF ESD protection structures, as well as the principles for RF ESD protection designs, this section discusses a design example of low-parasitic RF ESD protection structure.

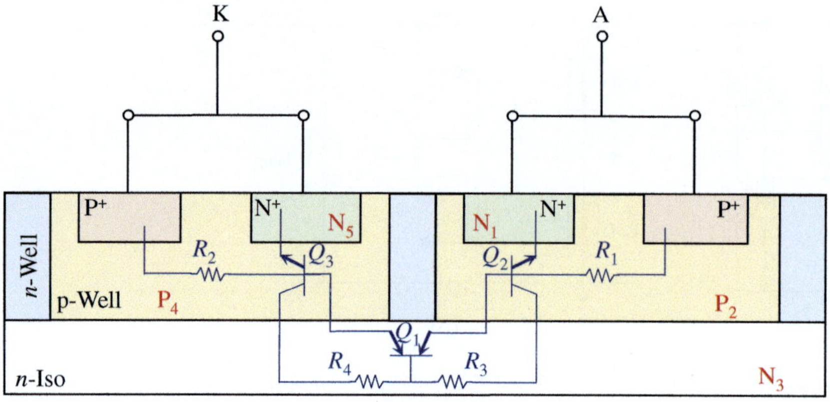

Figure 9.16 A cross-section view for a dSCR ESD protection structure implemented in a BiCMOS process.

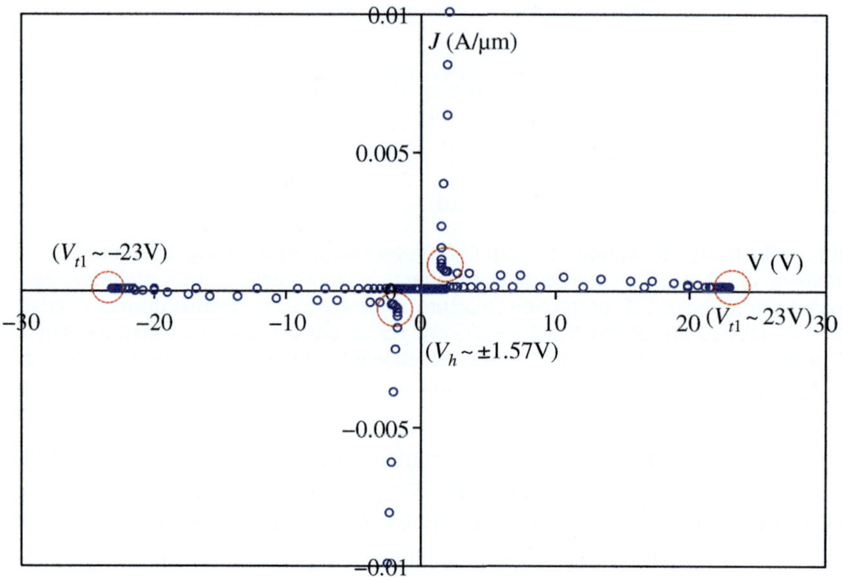

Figure 9.17 DC ESD discharging *I–V* curve for the dSCR ESD protection structure by TCAD ESD simulation shows ideal symmetric deep-snapback *I–V* characteristics.

This example illustrates the TCAD ESD design flow for a novel compact low-parasitic, dSCR ESD protection structure that was presented in Chapter 6. Figure 9.16 shows the cross-section for the dSCR ESD protection device, which is a two-terminal device consisting of two vertical NPN (Q_2, Q_3) and one lateral PNP (Q_1) transistors. This novel dSCR ESD protection device was created completely by TCAD-based mixed-mode ESD simulation and was demonstrated in a 0.8 μm BiCMOS technology [18–20]. In the design, TCAD DC ESD simulation was conducted first and the desired symmetric deep-snapback *I–V* characteristic was confirmed by ESD simulation as given in Figure 9.17, showing $V_{t1} \sim 23$ V and $V_h \sim 1.57$ V. In the next step, the dynamic ESD discharge functionality was proven by transient TCAD ESD simulation using HBM ESD model circuit. Figure 9.18 depicts the simulated transient *I–V* characteristic in one direction from the anode to cathode terminal, showing a $V_{t1} \sim 24.7$ V. With complete confirmation by TCAD mixed-mode ESD simulation,

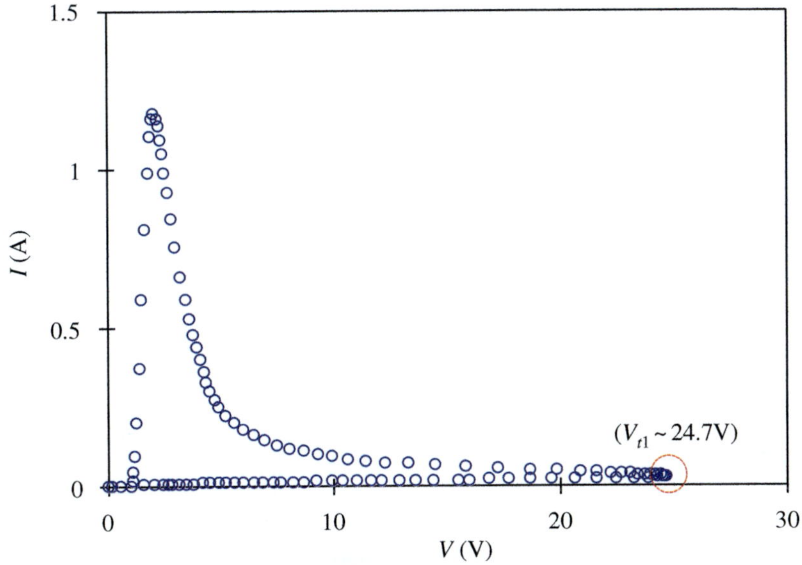

Figure 9.18 Transient ESD discharging I–V curve in one direction for the dual-polarity dSCR ESD protection structure is obtained by TCAD ESD simulation under 2 kV HBM ESD stressing. Deep I–V snapback is incomplete in TLP testing within the TLP pulse period due to SCR capacitive effect.

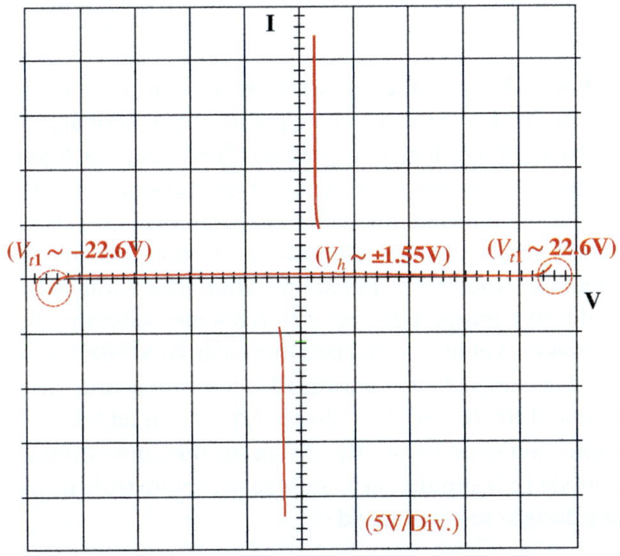

Figure 9.19 Measured symmetric ESD discharge I–V curve by a Curve Tracer matches ESD DC simulation well for the dSCR ESD protection device.

the dSCR design was implemented in silicon in a foundry 0.8 μm BiCMOS. Finally, Si measurements were conducted using both a curve tracer and a transient TLP ESD tester, with its I–V characteristics presented in Figures 9.19 and 9.20, respectively. Figure 9.19 shows the DC ESD triggering of V_{t1} ~ 22.6 V and ESD holding of V_h ~ 1.55 V. Figure 9.20 shows the transient HBM ESD triggering of V_{t1} ~ 21.8 V, ESD holding of V_h ~ 2.53 V, and ESD thermal failure at V_{t2} ~ 12 V

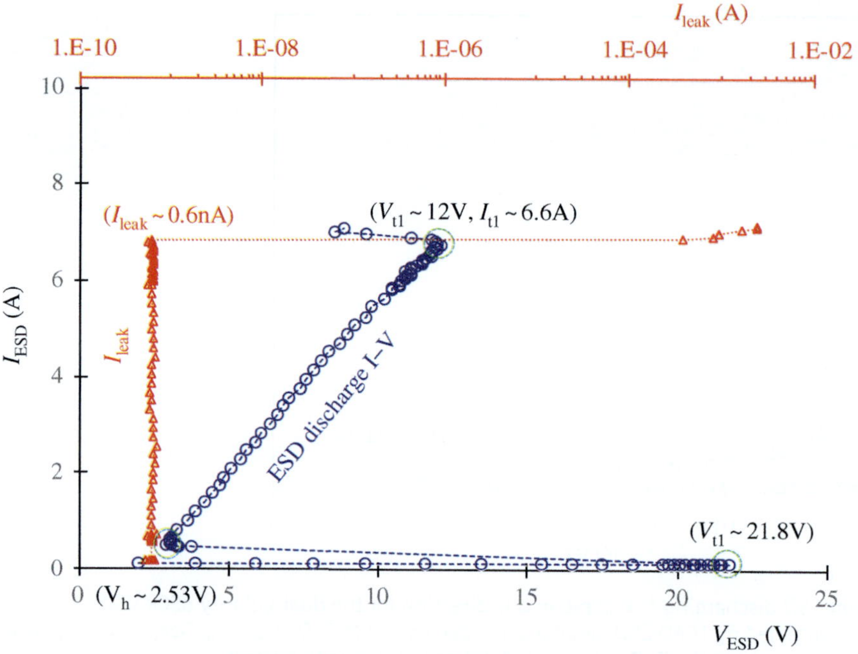

Figure 9.20 Measured transient ESD discharge *I–V* curve (blue), in one direction, by TLP testing matches transient TCAD HBM ESD simulation well. Measured leakage current (red) increases suddenly at the ESD thermal failure threshold. The dSCR ESD structure conducts ESD currents symmetrically in both directions.

and $I_{t2} \sim 6.6$ A. Figure 9.20 also shows very low leakage for the dSCR ESD protection structure of $I_{leak} \sim 0.6$ nA. In addition, Figure 9.20 readily shows that at the threshold point of ESD thermal failure, the leakage current suddenly jumps up by several orders of magnitude. Therefore, this novel dSCR features very high ESDV/Si ratio, resulting in very low parasitic C_{ESD} that is desirable for RF ESD protection. Further, as discussed in Chapter 6, applying this symmetric dual-directional ESD protection structure for a chip will dramatically simplify the full-chip ESD protection schematics, i.e., reducing the total number of ESD protection structures required on a chip, which translates into not only much-reduced overall ESD-induced parasitic effect on a chip but also substantially less Si area needed for ESD protection structures on a chip. Created by a thorough TCAD ESD simulation, this dSCR has resulted in a new breed of compact low-parasitic ESD protection structures that have been widely used for mixed-signal and RF ICs by the industry [21–27]. In addition to demonstrating a low-parasitic dSCR ESD protection structure for RF ESD protection, this example also shows how to use TCAD-based mixed-mode ESD simulation technique to explore truly novel ESD protection concepts and to optimize the designs in a real world.

9.5 Summary

This chapter comprehensively discusses RF ESD protection designs as an emerging RF IC design challenge. In general, the common ESD–IC interactions become much worse, which makes RF ESD protection designs very challenging. For RF ESD protection, ESD–RFIC interactions become severe in many ways, resulting in several unique RF ESD protection design challenges. First, the RF-induced ESD mis-triggering (i.e., V_{t1} reduction) may cause non-ESD short-circuit malfunction in ultrahigh frequency RF ICs. Second, the ESD-induced parasitic C_{ESD} can seriously affect RF IC performance, particularly ruining the I/O impedance matching. Third, the ESD-induced noise effect, including the ESD self-generated noises and C_{ESD}-induced global noise coupling, will substantially affect noise specs of RF ICs. Fourth, these ESD-induced parasitic effects can be fluctuating, making it even more difficult to find a good remedy for good RF ESD protection for high-frequency broadband RF ICs. In principle, any good RF ESD protection designs must carefully address these unique RF ESD design challenges. It is recognized that there is no "ideal" and "universal" RF ESD protection solution for all RF ICs. Instead, the principle for good RF ESD protection designs is to maximize the ESDV/Si ratio to minimize ESD-induced parasitic effects of individual ESD protection devices and to utilize multiple-mode ESD protection structures to simplify full-chip ESD protection schematics for reduced whole-chip ESD parasitic effects. Ultimately, revolutionary ESD protection concepts and mechanisms are desired for future RF ESD protection. In practical RF IC designs, it is important to accurately characterize the ESD-induced parasitic effects, such as S-parameter, C_{ESD}, ESD noises, and noise coupling, and ESD I_{leak}, which can be applied to ESD–RFIC codesign practices.

References

1 Wang, A., Feng, H., Zhan, R. et al. (2005). A review on RF ESD protection design. *IEEE Trans. Electron Devices* 52 (7): 1304–1311. https://doi.org/10.1109/TED.2005.850652.

2 Wang, A., Lin, L., Wang, X., and Liu, H. (2008). Emerging challenges in ESD protection for RF ICs in CMOS. *J. Semicond.* 29 (4): 628–636.

3 Chen, G., Feng, H., Xie, H. et al. (2004). RF characterization of ESD protection structures. *Proceedings of IEEE Radio Frequency Integrated Circuits Symposium (RFIC)*, pp. 379–382.

4 Wang, A. (2002). *On-Chip ESD Protection for Integrated Circuits – An IC Design Perspective*. Boston, MA: Kluwer Academic Publishers. ISBN: 0-7923-7647-1.

5 Feng, H. (2001). A mixed-mode simulation-design methodology for on-chip ESD protection design. An MS thesis. Illinois Institute of Technology.

6 Gong, K. (2001). ESD protection in copper interconnect and ESD-to-circuit performance influences. An MS thesis. Illinois Institute of Technology.

7 Chen, G. (2003). Design and characterization of ESD protection for RFICs. An MS thesis. Illinois Institute of Technology.

8 Wang, A. (1998–2007). *ESD Design Database*. The Integrated Electronics Laboratory, Illinois Institute of Technology.

9 Wang, A.Z., Feng, H., Zhan, R. et al. (2002). ESD protection design for RF integrated circuits: new challenges. *Proceedings of IEEE Custom IC Conference (CICC)*, pp. 411–418. https://doi.org/10.1109/CICC.2002.1012860.

10 Barth, J. and Richner, J. (2001). Correlation consideration: real HBM to TLP and HBM testers. *Proceedings of EOS/ESD Symposium*, pp. 453–460.

11 Barth, J., Verhaege, K., Henry, L., and Richner, J. (2000). TLP calibration, correlation, standards, and new techniques. *Proceedings of EOS/ESD Symposium*, pp. 85–96.

12 Anand, S. and Razavi, B. (2001). A CMOS clock recovery circuit for 2.5 Gb/s NRZ data. *IEEE J. Solid-State Circuits* 36 (3): 432–439.

13 Foley, D. and Flynn, M. (2001). CMOS DLL-based 2V 3.2ps jitter 1GHz clock synthesizer and temperature-compensated tunable oscillator. *IEEE J. Solid-State Circuits* 36 (3): 417–423.

14 Xu, M., Su, D., Shaeffer, D. et al. (2001). Measuring and modelling the effects of substrate noise on the LNA for a CMOS GPS receiver. *IEEE J. Solid-State Circuits* 36 (3): 473–485.

15 Zhang, F., Li, C., Di, M. et al. (2020). Design and analysis of a 28 GHz 9 KV ESD-protected distributed travelling-wave TRx switch in 22 nm FDSOI. *IEEE J. Electron Devices Soc. (J-EDS)* 8: 655–661. https://doi.org/10.1109/JEDS.2020.2975598.

16 Lu, F., Ma, R., Dong, Z. et al. (2016). A systematic study of ESD protection co-design with high-speed and high-frequency ICs in 28 nm CMOS. *IEEE Trans. Circuits Syst. I* 63 (10): 1746–1757. https://doi.org/10.1109/TCSI.2016.2581839.

17 Ma, R., Chen, Q., Zhang, W. et al. (2016). A dual-polarity graphene NEMS switch ESD protection structure. *IEEE Electron Device Lett.* 37 (5): 674–676. https://doi.org/10.1109/LED.2016.2544343.

18 Wang, A. and Tsay, C.H. (2001). On a dual-polarity on-chip electrostatic discharge protection structure. *IEEE Trans. Electron Devices* 48 (5): 978–984. https://doi.org/10.1109/16.918246.

19 Wang, A.Z., Tsay, C.H., and Deane, P. (2002). Dual-direction over-voltage and over-current IC protection device and its cell structure. US Patent No. 6, 365, 924.

20 Wang, A.Z., Tsay, C.H., and Deane, P. (2001). Method for manufacturing a dual-direction over-voltage and over-current IC protection device and its cell structure. US Patent 6, 258, 634.

21 Wang, A.Z. (2003). Bonding pad-oriented all-mode ESD protection structure. US Patent No. 6, 635, 931.

22 Wang, A.Z. (2003). Single structure all-direction *ESD* protection for integrated circuits. US Patent No. 6, 512, 662.

23 Wang, A.Z., Tsay, C.H., and Deane, P. (2012). Dual-directional electrostatic discharge protection method. US Patent No. 8, 305, 722 B2.

24 Wang, A.Z., Tsay, C.H., and Deane, P. (2011). Dual-directional electrostatic discharge protection device. US Patent No. 7, 936, 020 B1.

25 Wang, A. and Tsay, C.H. (2001). An on-chip ESD protection circuit with low trigger-voltage in BiCMOS technology. *IEEE J. Solid-State Circuits* 36 (1): 40–45. https://doi.org/10.1109/4.896227.

26 Wang, X., Liu, J., Fan, S. et al. (2010). Cross-coupling low-triggering dual-polarity CLTdSCR ESD protection in CMOS. *IEEE Electron Device Letters* 31 (10): 1143–1145. https://doi.org/10.1109/LED.2010.2058842.

27 Liu, J., Wang, X., Zhao, H. et al. (2011). Design and analysis of low-voltage low-parasitic ESD protection for RF ICs in CMOS. *IEEE J. Solid-State Circuits* 46 (5): 1100–1110. https://doi.org/10.1109/JSSC.2011.2118290.

10

ESD-RFIC Co-Design

10.1 ESD-IC Interactions

As discussed earlier, the electrostatic discharge–integrated circuit (ESD–IC) interactions are inherent to any ESD-protected ICs, which simply become much worse for RF ICs due to two key reasons: First, high-performance RF ICs are very sensitive to any ESD-induced parasitic effects, especially as RF ICs continue to increase operating frequencies (e.g., millimeter wave), frequency bandwidth (e.g., 7.5 GHz for single-band impulse-radio ultrawide band, i.e., IR-UWB radios), and data rates (e.g., beyond 10 giga bits per second). Second, higher ESD protection is required for RF ICs used in handheld wireless electronics, e.g., smartphones and tablets, which generate more ESD-induced parasitic effects. Although RF ESD protection structures can be optimized by TCAD-based ESD simulation and ESD protection structures can be accurately characterized for their parasitic effects, e.g., C_{ESD}, leakage, and noises, it is practically impossible to completely eliminate the ESD-induced parasitic effects in any real-world RF ESD protection designs. Therefore, *ESD–RFIC co-design* becomes a critical design concept for advanced RF ICs with on-chip ESD protection. Before discussing the ESD–RFIC co-design methodology, the following two sections discuss the IC-to-ESD influences and the ESD-to-IC influences, respectively, which comprise the ESD–IC interactions (a.k.a., *ESD–RFIC interactions*).

10.1.1 IC Affects ESD Protection

In practical IC designs, it is not unusual that an IC protected by a well-designed ESD protection device fails in field at an ESD zapping test level much lower than expected, which is the widely known *Early ESD Failure* Phenomenon (i.e., pre-mature ESD failure). Early ESD failure is often attributed to the IC-to-ESD influences. One main cause to such early ESD failure is the unexpected conduction through any unwanted inner *ESD-like* devices, mostly parasitic structures, within the core circuit, which are mis-triggered by an incident ESD pulse. The inner devices mis-triggered under ESD stressing can be a legitimate device, e.g., a NMOSFET in an inverter right next to a ggNMOS ESD protection device, or a parasitic structure, e.g., a parasitic lateral BJT across a CMOS guard-ring, inside the IC core. The unexpected early ESD triggering will by-pass the intentionally designed ESD protection structure to discharge the large ESD transient, resulting in early ESD failure, regardless how well an intentional ESD protection structure is designed and tested as a standalone ESD protection device. This is because a parasitic non-ESD protection structure is never designed to conduct a huge ESD transient. On the other hand, even if the unwanted ESD mis-triggering of an internal device occurs concurrently with the adjacent intentional ESD protection device, if a sizable ESD current discharges through the internal parasitic device, depending

Practical ESD Protection Design, First Edition. Albert Wang.

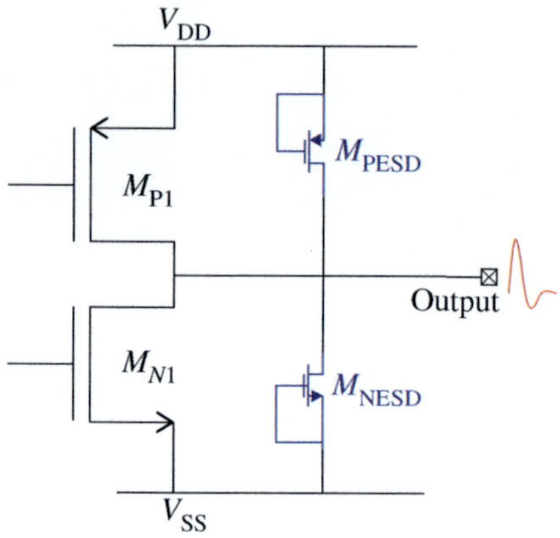

Figure 10.1 The complementary ggNMOS and ggPMOS ESD protection transistors can be similar to the core CMOS transistors under ESD protection, possibly causing ESD discharge competition between the ESD protection devices and the internal transistors, resulting in unexpected early ESD failure.

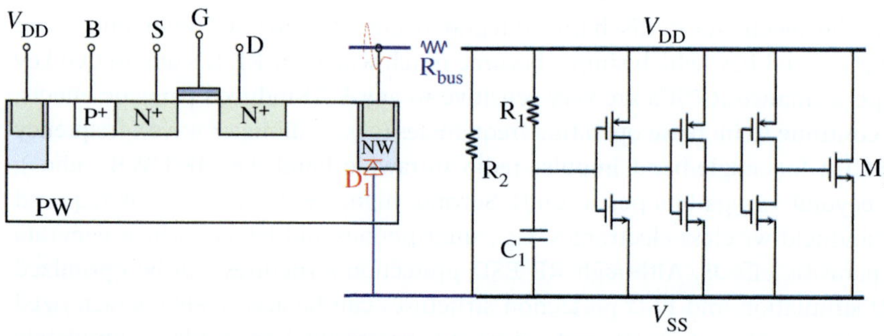

Figure 10.2 This design uses a CMOS power clamp for supply bus ESD protection, which is placed far away from the power pad, but adjacent to an internal NMOSFET. Under ESD stressing, it is possible to unexpectedly turn on the parasitic guard-ring diode of the internal NMOSFET, creating an ESD discharge competition and resulting in early ESD failure.

upon its conduction resistance, it likely will destroy the parasitic ESD-like device because it never intends to handle a large ESD current. The following examples explain these scenarios.

In the first example shown in Figure 10.1, the output circuit block is protected by complementary ggNMOS and ggPMOS ESD protection structures, which are physically located next to each other. Likely, the core MOSFETs are similar to the ggMOS ESD protection devices structurally and dimensionally, which means that the core M_{N1} and the ESD device M_{NESD} have a good chance of being turned on simultaneously by an incident ESD pulse, creating an ESD discharge competition that likely destroys the core CMOS transistors being protected since the core MOSFET can handle very little currents. Early ESD failure therefore will occur even if the intentional ggNMOS ESD protection device is well designed and experimentally validated as a standalone ESD protection device.

In the second example depicted in Figure 10.2, a non-snapback MOSFET ESD power clamp, described in Chapter 5, is used. Assume the MOSFET power clamp is placed far away from the V_{DD} pad, but adjacent to a core NMOSFET, when an incident ESD pulse appears as the V_{DD} pad, there is a good chance the PN junction of the guard ring plug will be turn on before the MOSFET power clamp can be triggered. Should this happen, likely the power clamp will fail under ESD stressing on the power bus, resulting in early ESD failure even if the ESD power clamp is well designed. Hence, the circuit-to-ESD influence must be addressed for full-chip ESD protection.

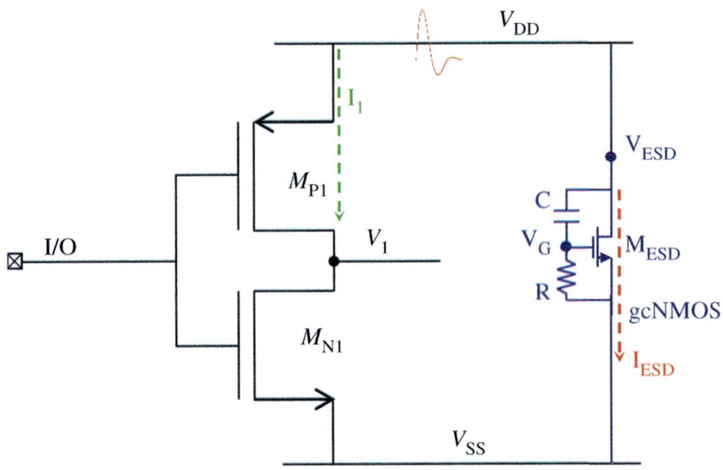

Figure 10.3 This example uses a gcNMOS power clamp for supply bus ESD protection, which is placed near the output CMOS buffer. Under ESD stressing, it is likely the CMOS buffer will be turned on to compete with the gcNMOS ESD protection structure, leading to possibly early ESD failure in the CMOS buffer.

In the third example as illustrated in Figure 10.3, a gcNMOS ESD power clamp is used, which is physically placed adjacent to an output buffer block [1]. Full-chip ESD protection design verification is needed to validate that the output buffer transistors, M_{P1} and M_{N1} will be actually protected, which is conducted using TCAD-based mixed-mode ESD simulation. Designed in a foundry 0.35 μm CMOS technology, the gcNMOS ESD is designed as $W/L = 80/0.35$ μm, and the core transistors M_{P1} and M_{P1} have W/L ratios of 2/0.35 μm and 1.5/0.35 μm, respectively. During normal IC operations, when the invertor is in logic "H" state, M_{N1} is ON and M_{P1} is OFF. However, if an ESD pulse appears at V_{DD} pad, it may force M_{P1} ON as well, and the whole invertor will be in conduction mode, which creates ESD discharge competition with the gcNMOS power clamp, and possibly causing early ESD failure due to ESD damage to M_{P1} and/or M_{N1}. This is studied by TCAD ESD simulation. Figure 10.4 depicts the simulated I–V–t curves for the concerned transistors, which clearly shows that the gcNMOS dominates in ESD discharge and transistor M_{ESD} conducts the majority of the ESD transient, while M_{P1} and M_{N1} are slightly ON, but conduct negligible current. Figure 10.5 presents simulated lattice temperature in time domain, $T_{MAX} \sim t$, for the transistors involved, which readily shows that the M_{ESD} is heated up substantially during ESD discharge, while M_{P1} and M_{N1} remain close to room temperature, hence being properly protected against ESD stressing. This example also serves to demonstrate how to quantitatively analyze the inevitable internal ESD discharge competition by TCAD ESD simulation to address the troublesome circuit-to-ESD influences in practical RF ESD protection designs.

In addition, the circuit-to-ESD influences may also cause unexpected RF signal induced ESD mis-triggering in absence of any ESD events, resulting in short-circuit malfunction of RF ICs. This RF-induced ESD mis-triggering may become a real threat to advanced RF ICs running ultrahigh frequencies, which is a major emerging RF ESD protection design challenge. Revolutionary ESD protection mechanisms are therefore required for advanced RF ESD protection in near future.

10.1.2 ESD Affects IC Performance

There is never free lunch, so is on-chip ESD protection, which inevitably introduces design overhead in practical IC designs. As discussed previously, the ESD-induced design overhead effects

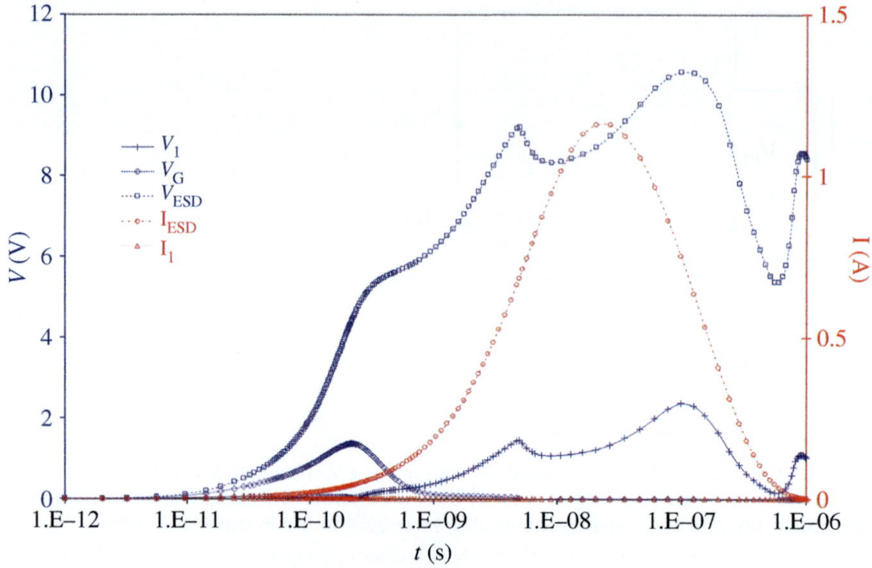

Figure 10.4 TCAD ESD simulation for the circuit in Figure 10.3 shows that, under ESD stressing, the gcNMOS power clamp is triggered properly to discharge the majority of the ESD current, hence, protect the CMOS buffer.

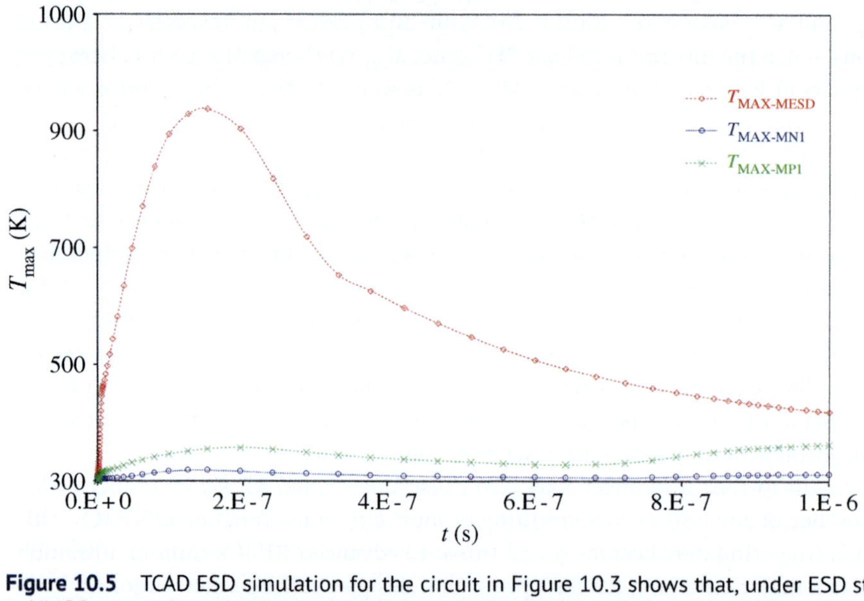

Figure 10.5 TCAD ESD simulation for the circuit in Figure 10.3 shows that, under ESD stressing, the gcNMOS power clamp is triggered properly to discharge the majority of the ESD current, leading to ESD-induced heating in the gcNMOS device. The CMOS buffer transistors are properly protected as shown by the low variation in lattice temperature during ESD discharge.

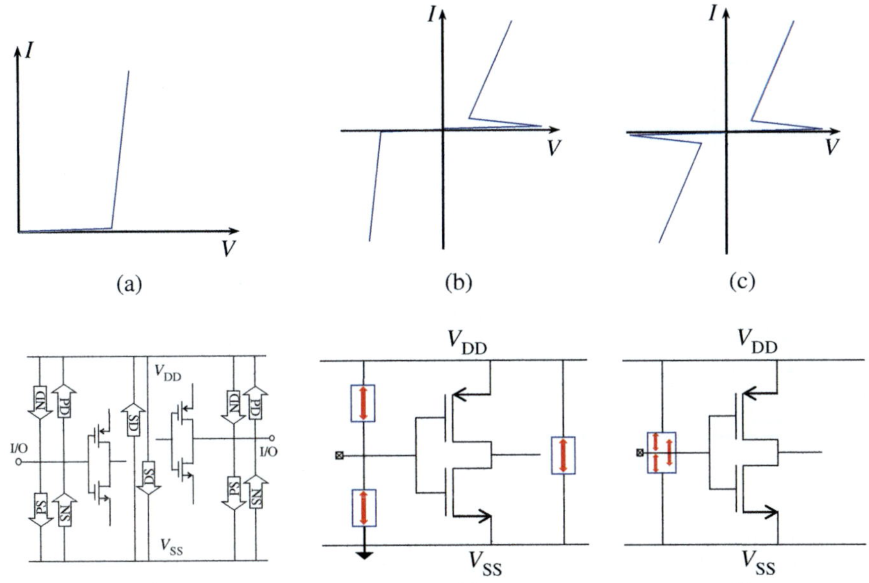

Figure 10.6 General principle in reducing ESD impacts on RF ICs: (a) & (d) use of traditional single-direction ESD protection structures has severe ESD-induced design overhead including ESD parasitic effects and layout complexity; (b), (e), (c) & (f) use of novel dual/multiple-directional ESD protection structures can dramatically simplify the whole-chip ESD protection circuit schemes and significantly reduce ESD influences on RF ICs.

typically include both physical design headache, e.g., layout floor-planning and Si consumption, and circuit performance problem such as ESD-induced parasitic C_{ESD}, leakage, and noises. Such ESD-to-circuit influences become much severe and unacceptable to high-frequency broadband and high-data-rate RF ICs, which must be carefully addressed in designing advanced RF ICs [2–4]. In principle, the general solution to this design challenge is to minimize the ESD-induced parasitic effects, e.g., C_{ESD}, or more wisely, to explore novel ESD protection structures to reduce the overall ESD-induced parasitic effects on a full chip level. As depicted in Figure 10.6a,d, using traditional single-directional ESD protection devices requires too many ESD protection units on a chip, which not only brings in significant parasitic effects but also consumes substantial Si asset. On the other hand, if novel ESD protection structures are utilized, e.g., dual-polarity or all-mode ESD protection structures, the whole-chip ESD protection solutions will lead to much simplified full-chip ESD protection circuit schematics, as illustrated in Figure 10.6b,c,e,f. It is extremely important to keep this RF ESD design principle in mind in practical RF IC designs. This section discusses several design examples to show how seriously ESD-induced parasitic effects can affect RF ICs.

In the first example implemented in a foundry 180 nm BiCMOS technology featuring six copper metal interconnects layers, an Op Amp circuit is protected by different ESD protection structures to study the C_{ESD} impacts on general IC performance of the Op Amp chip [3]. The two ESD protection structures for targeted 2 kV HBM ESD protection are traditional single-direction ggNMOS and dual-polarity dSCR ESD protection structures. The extracted ESD-induced parasitic C_{ESD} for the whole ESD cell is 0.84 pF for the ggNMOS, including both Si and metals, which is much higher than that of 0.12 pF for the dSCR. Figure 10.7 shows the Op Amp schematic, which has ESD protection at the output port in design splits with the key performance specs listed in Table 10.1, including low power of ~0.43 mW, high slew rate of ~116 mV/ns, short settling time of ~3.7 ns (at 1%), wide

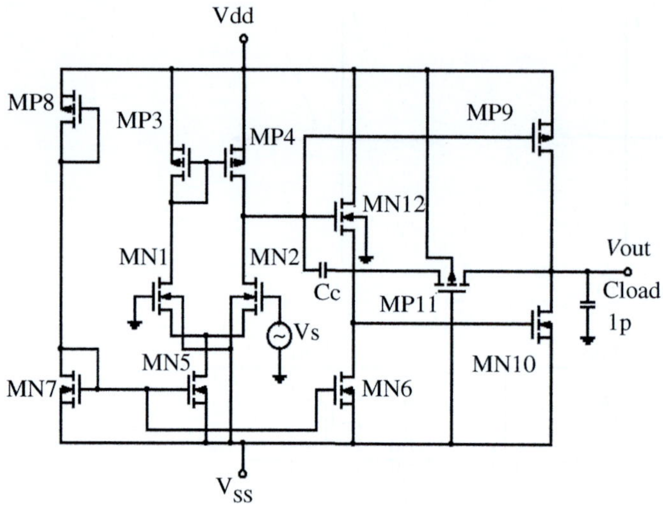

Figure 10.7 A schematic for a low-power Op Amp designed in a 180 nm BiCMOS technology has ESD protection at the output, which introduces parasitic ESD capacitive load.

Table 10.1 Op Amp IC specs.

Technology	180 nm, 3 V 6-Cu BiCMOS
Gain	73.07 dB
Phase margin	70.1°
f_T	120.7 mHz
f_{-3dB} (kHz)	29.7 kHz
V-swing	0.96 V (@80%)
Settling time	3.77 ns (@1%)
Slew rRate	115.9 mV/ns
Power dissipation	0.43 mW

output voltage swing of 0–96 V (at 80% gain), and large unity-gain bandwidth of ~121 MHz. In circuit simulation, the ESD influences on the Op Amp is readily observed. Figure 10.8 depicts the phase Bode plot, which clearly shows that the parasitic-heavy ggNMOS ESD protection structure seriously affects the Op Amp, compared to the split without ESD protection, which recovers substantially when using the parasitic-light dSCR ESD protection device. Figure 10.9 presents the Op Amp settling time that is also severely affected by the ggNMOS ESD protection device, which again, almost completely recovered when using the dSCR ESD protection device. Figure 10.10 presents the Op Amp slew rate that shows large degradation due to the ggNMOS-induced parasitic effect, which can be effectively eliminated when using the low-parasitic dSCR ESD device. As summarized in Table 10.2, this design example readily shows that ESD-induced parasitic effects can severely affect general circuit performance, while using low-parasitic ESD protection solution is very beneficial to high-performance ICs.

Noise impacts of ESD protection on RF ICs is another major design concern. As discussed previously, ESD-induced noise effects can appear in two formats: extra noises due to ESD

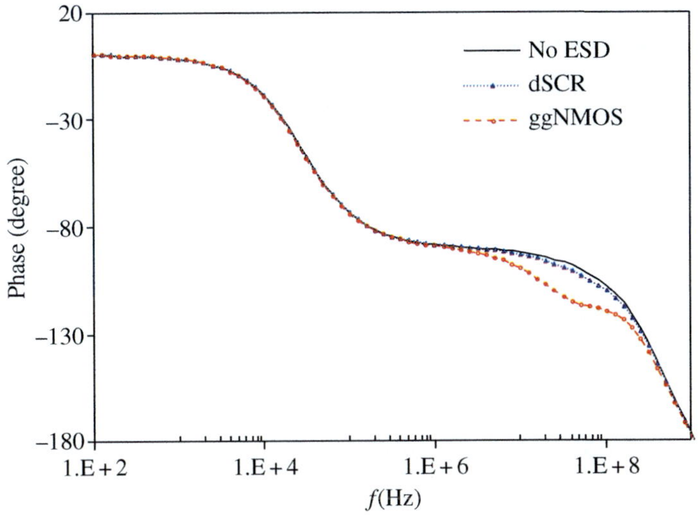

Figure 10.8 The phase Bode plot for the Op Amp ICs with no ESD protection, and ggNMOS and dSCR ESD protection structures.

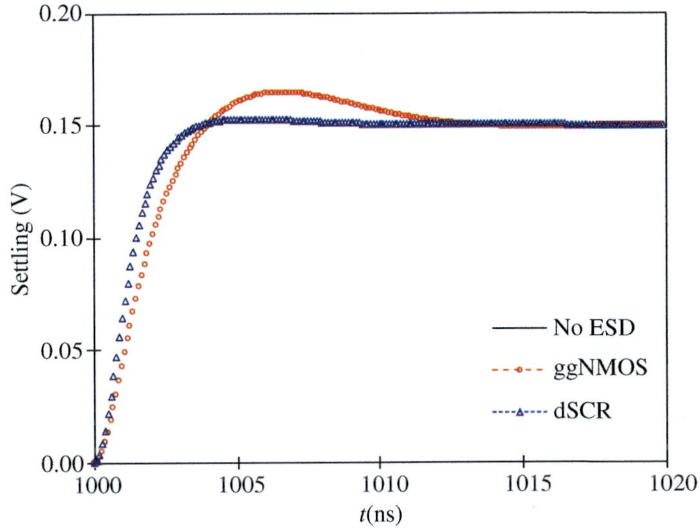

Figure 10.9 The settling time for the Op Amp ICs with no ESD protection, and ggNMOS and dSCR ESD protection structures.

noise self-generation and noise coupling due to the C_{ESD}. In the second example, the ESD self-noise generation is evaluated using a 2.4 GHz LNA designed in a 3 V 35 μm CMOS technology [3]. Figure 10.11 depicts the LNA circuit featuring a two-stage topology. This LNA uses a current-sharing biasing technique to achieve low power dissipation, and on-chip inductors and capacitive source degeneration to realize 50 Ω on-chip impedance matching for the I/O ports. Simulation shows key LNA specs: a gain of ~23.4 dB, a low dissipating current of ~8.5 mA, and a noise figure of NF = 1.76 dB at the center frequency of 2.4 GHz and a supply of 3 V. ESD protection is provided at the input port. For noise analysis, the ESD equivalent noise circuit models discussed

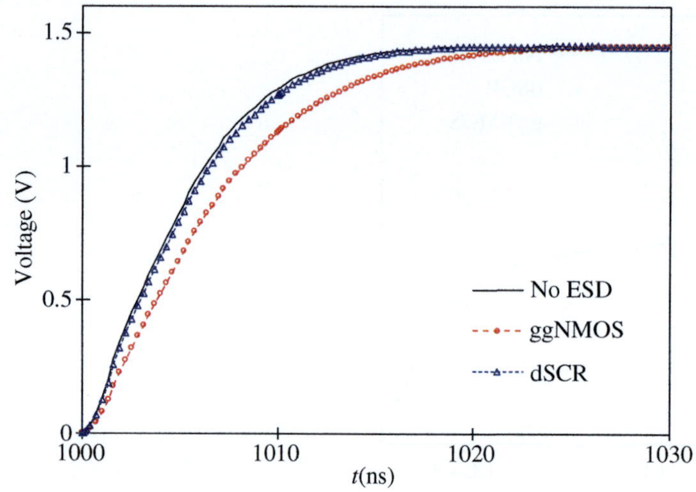

Figure 10.10 The slew rate for the Op Amp ICs with no ESD protection, and ggNMOS and dSCR ESD protection structures.

Table 10.2 Impacts of ESD protection on Op Amp Specs.

Specs	No C_{ESD}	ggNMOS	dSCR
f_T	120.7MHz	−35.8%	−8.3%
		Recovery +77%	
Phase margin	70.1°	−12.7%	−1.9%
		Recovery +85%	
Slew rate	115.9 mV/ns	−27.2%	−4.7%
		Recovery +83%	
$t_{settling}$		−216%	−98%
		Recovery +55%	

in Chapter 4 are used in noise simulation for different ESD protection structures. First, to evaluate the sensitivity of noise self-generation in ESD protection, multiple-finger ggNMOS ESD protection structures of varying sizes were used in LNA noise analysis. Figure 10.12 presents the simulated LNA NF against the sizes of ggNMOS ESD protection devices, i.e., the finger width (W), which readily shows the strong dependence of LNA NF degradation on ggNMOS device sizes, hence the ESD protection level. Next, ggNMOS and dual-polarity SCR (dSCR) ESD protection structures, both at 4 kV HBM ESD protection, were used to protect the LNA input port in two design splits. Table 10.3 summaries the extracted NF values for the LNA of different ESD protection splits. It is observed that ggNMOS introduces significant self-generated noise that degraded the LNS NF by

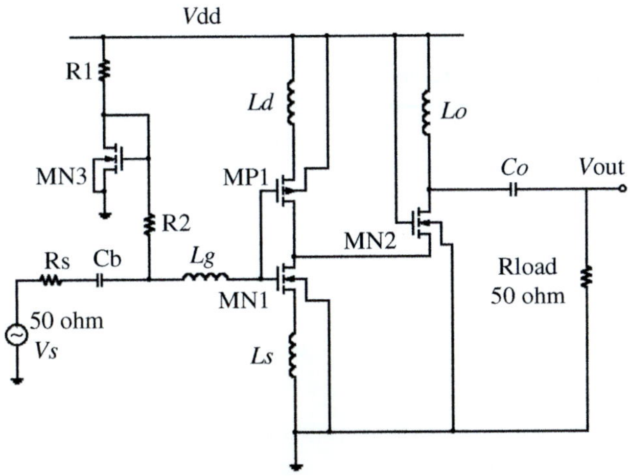

Figure 10.11 Schematic for a 2.4 GHz LNA circuit with ESD protection at the input port designed in a foundry 0.35 μm CMOS technology.

Figure 10.12 Simulation for the ESD-protected 2.4 GHz LNA shows strong noise impact on the LNA noise figure, depending the finger width (W) of the ggNMOS ESD protection structures.

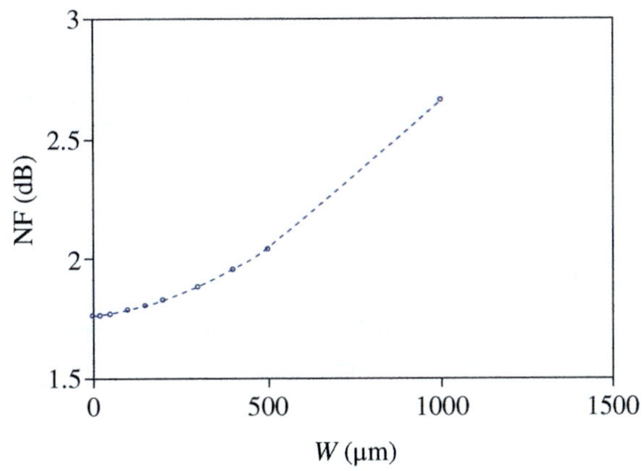

~3.8%; however, dSCR ESD protection has negligible self-generated noise. Thus, using dSCR ESD protection structure can almost recover all NF degradation in the LNA caused by using the noisy ggNMOS ESD protection device, i.e., ~97% recovery. The ESD-induced noise degradation is further validated in Si measurement of an ESD-protected 5.5 GHz LNA circuit designed and fabricated in a foundry 180 nm SiGe BiCMOS technology [5]. Figure 10.13a depicts the schematic for the 5.5 GHz LNA featuring a CE–CB cascade topology for high gain, fine isolation, and suppressed Miller effect. The high/low-gain switching is controlled by the transistor T3 and a double-shutdown function for dual-band 2.4 GHz/5.5 GHz switching is realized by using the M1/M2 pair. Input impedance and noise matching is realized by on-chip inductor, L_b and L_e. 2 kV HBM ESD protection is provided at the LNA input port using a foundry-provided N^+/Pwell ESD diode with a finger width of $W = 48$ μm. Figure 10.13b gives the die photo of the fabricated LNA chip showing the G-S-G test pattern designed for S-parameter and NF measurements. The measured LNA specs are summarized in Table 10.4 where substantial ESD-induced LNA performance degradation, i.e., gain (S_{21}), reflection (S_{11}) and NF, are clearly observed. Figure 10.14 depicts the measured NF curves in frequency domain for the design splits of LNA without and with 2 kV ESD protection diode. Obviously, the 2 kV ESD diode affects LNA NF across a wide frequency spectrum, which

Table 10.3 Comparison of ESD-induced NF degradation of LNA.

ESD protection	NF (dB)	Degradation	Recovery
No ESD	1.758	—	—
ggNMOS	1.825	+3.8%	97%
dSCR	1.760	+0.1%	

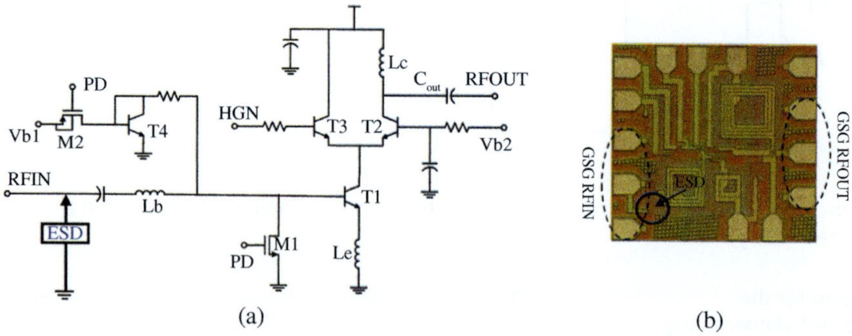

Figure 10.13 A 2KV ESD protected dual-band 2.4 5.5GHz LNA circuit implemented in a foundry 180 nm BiCMOS technology: (a) schematic, and (b) die photo.

Table 10.4 Measured NF for LNA design splits.

LNA splits	LNA performance		
	S_{21}(dB)	S_{11}(dB)	NF(dB)
w/o ESD	18.11	-8.3	2.99
w/o ESD	15.08	-7.2	3.19
Degradation	16.73%	17.25%	6.8%

is attributed to ESD-induced noise effects including both self-generated noise and C_{ESD} noise coupling. The NF of the standalone 2 kV ESD protection diode was also extracted as shown in Figure 10.14. These two design examples, in both simulation and measurement, clearly show that the ESD-induced noise degradation cannot be ignored in practical RF IC design, which is a major design challenge for practical RF ESD protection designs.

10.2 ESD-RFIC Co-Design

From previous discussions, it is obvious that any on-chip ESD protection solution will bring in ESD-induced parasitic effects that can significantly affect RF IC performance. It is hence important to minimize the ESD-induced parasitic effects by ESD design optimization, typically through TCAD ESD simulation, and to accurately characterize the ESD-induced parasitic parameters such as C_{ESD}, I_{leak}, and NF. Nevertheless, an ESD protection structure exists physically. No matter

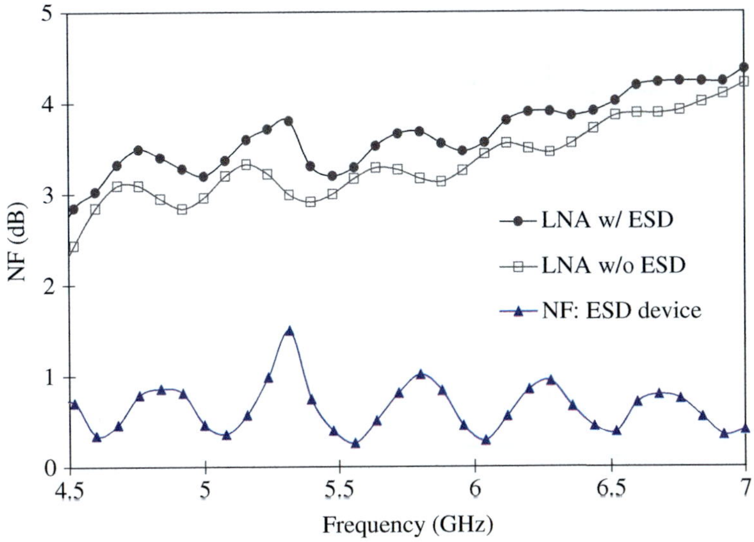

Figure 10.14 Measured NF behaviors for the LNA circuits without and with ESD protection, and the NF of the standalone 2 kV ESD protection diode structure.

how well an ESD protection device may be optimized, its parasitic effect cannot be reduced to zero. Depending upon a specific RF IC, even very "tiny" ESD-induced parasitic effect, e.g., C_{ESD}, may still have non-negligible impact on circuit performance, which is particularly true as RF ICs moving into extremely high frequency (e.g., millimeter wave band), ultrawide band (e.g., 7.5 GHz UWB band), and very high data rate (e.g., 10–130 Gbps and beyond) domain. This RF ESD design challenge can be addressed by a novel ESD-RFIC co-design technique to be discussed in this section [6].

10.2.1 ESD-RFIC Co-Design Principle

The principle for the ESD-RFIC co-design technique is that the known ESD-induced parasitic effects must be included in RF IC design in order to *co-optimize* the ESD-protected RF IC, i.e., simultaneously achieving the best possible performance of both ESD protection and RF core circuit. This is a tough design task, yet very possible and already validated in practical RF IC designs.

The *ESD–RFIC co-design* method can be simplified as following for the RF IC design procedures, as depicted in Figure 10.15 [6–11]:

Step-1: Set the ESD protection target for an RF IC.

Step-2: Complete ESD design optimization for suitable ESD protection structure(s) by TCAD ESD simulation. The goal is to minimize any ESD-induced parasitic effects while achieving the ESD protection target.

Step-3: Fabricate the optimized ESD protection structure(s).

Step-4: Characterize the ESD protection structure(s) to obtain accurate ESD-induced parasitic effects, including C_{ESD}, I_{leak}, and NF. The characterization is typically done by measuring the S-parameters and noises of the ESD protection structure(s) using GSG de-embedding ESD test patterns as discussed in Chapter 9. Note that, if a suitable ESD protection structure and its accurate ESD-induced parasitic parameters are available, i.e., from a foundry PDK package, Steps-2/3/4 can then be by-passed.

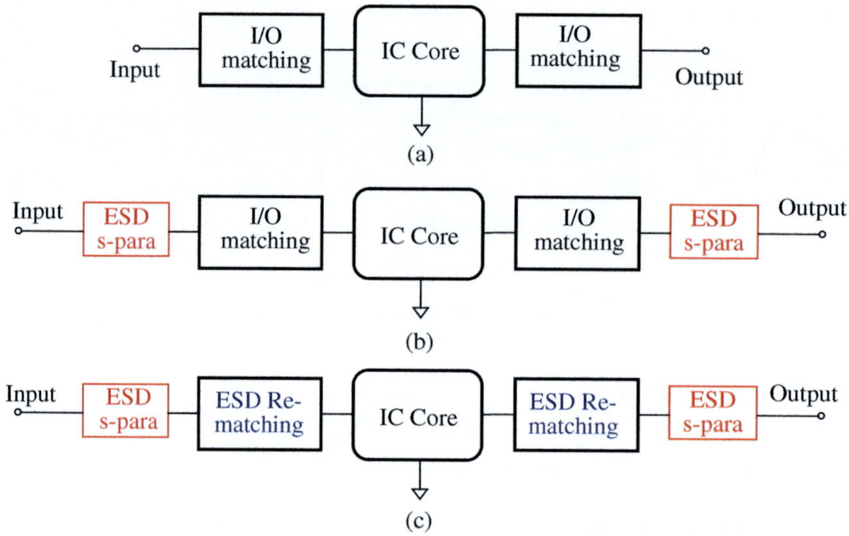

Figure 10.15 Illustration of the ESD-RFIC co-design flow: (a) normal RF IC design with I/O impedance matching, (b) injecting ESD parasitic parameters in RF IC simulation, observing corruption in I/O matching, and (c) I/O impedance re-matching for RF IC with ESD parasitic parameters included.

Step-5: Complete design and optimization of the RF IC planned. In this step, a critical design task is to design the I/O impedance matching networks for both input and output ports, as shown in Figure 10.15a. Good I/O impedance matching ensures almost all RF IC specs, including gain, noise figure, data rate, power dissipation, center frequency, frequency bandwidth, performance flatness across the whole frequency bandwidth, etc. Normally, after post-simulation validation, an RF IC design can be taped-out for Si fabrication, if no ESD protection design is involved.

Step-6: Apply the measured ESD-induced parasitic parameters into the RF IC designed in Step-5 to evaluate any ESD impacts on RF IC performance specs. It is guaranteed that substantial performance degradation in the RF IC specs will be observed, often very severe, depending upon the type of RF ICs. In general, this is reflected in corruption in I/O impedance matching, which, in turn, will affect almost every single RF IC specs parameter. When integrating ESD-induced parameters into RF IC specs evaluation, one can use the measured ESD-induced C_{ESD}, R_{ESD}, I_{leak}, and noises, if they are available and accurate. Unfortunately, accurate measurement of ESD-induced parameters is fairly challenging and the ESD-induced parasitic parameters may not be accurate. One alternative and practical technique to overcome this problem is to directly apply the measured S-parameter and noise data files into the CAD simulation deck to simulate the RF IC. There are two main benefits for using this unique *S-parameter-insertion* technique in RF IC simulation: first, avoid the complexity of extracting C_{ESD} from the measured S-parameters; second, be able to include ALL ESD-induced parasitic effects that are lump-summed into the measured S-parameter data file; third, avoid possible inaccuracy induced by converting the S-parameters into C_{ESD}. This S-parameter-insertion technique has been proven to be very useful in practical RF IC designs including on-chip ESD protection. Step-6 is illustrated in Figure 10.15b.

Step-7: Now that the unwanted ESD-induced parameters are injected into RF IC simulation, careful design revision and balance can be excised for both the ESD protection structure and the I/O impedance matching networks. Most commonly, ESD-aware I/O *re-matching* tuning will be conducted to recover the corrupted I/O matching caused by ESD-induced parasitic effects. The

I/O impedance matching networks and/or ESD protection structures may therefore be revised to re-gain any performance degradation of the RF ICs. Step-7 is depicted in Figure 10.15c.

10.2.2 ESD–RFIC Co-Design Examples

The following RF IC examples illustrate the ESD-RFIC co-design flow and its benefits. The first example is a design of a fully-integrated 5 GHz LNA with 5 kV HBM ESD protection in a foundry 180 nm RFCMOS technology [6]. Figure 10.16 depicts the schematic for the 5 GHz LNA with ESD protection at the input port. This LNA circuit is designed with power-constrained noise optimization and features a fine-tuned 50 Ω input matching network comprising C1, L1, L3, L4, and L5. The LNA circuit is optimized for a gain of 15.82 and NF of 3.12 dB at center frequency and draws a 6.4 mA current from a 1.8 V supply. In design phase one, a foundry-provided 5.7 kV ESD protection diode (N^+/P-sub) available in the PDK package is selected to protect the LNA. The SPICE device model for this ESD protection diode in the PDK package is used to simulate the ESD-protected LNA. Figure 10.17 depicts simulated gain of the 5.7 kV ESD-protected LNA IC showing negligible impact of ESD protection, i.e., ~0.08 dB (−0.5%) drop in the worst case. Figure 10.18 presents simulated NF of the 5.7 kV ESD protected LNA circuit, which again shows negligible NF degradation induced by ESD protection, i.e., ~0.13 dB (+4%) increase in the worst case. The simulation is actually mis-leading in suggesting that a 5.7 kV ESD protection structure would have almost no substantial negative impacts on LNA circuit performance, which was proven wrong in LNA measurements. In design phase two, a custom-designed ESD protection diode, optimized by TCAD ESD simulation, is used to protect the same LNA circuit. For comparison, two LNA design splits, i.e., LNA without ESD and LAN with 5 kV ESD diode, were designed, fabricated, and characterized. Figure 10.19 depicts the measured gain for the two LNA splits that readily show the substantial degradation induced by the ESD protection, i.e., ~0.25 dB (−1.7%) drop in gain at 5 GHz (~3 dB drop at 7 GHz). Figure 10.20 presents the measured NF for the two LNA splits, which also clearly shows a sizable NF degradation associated with the ESD protection, i.e., ~0.6 dB (+12.56%) increase in NF at 5 GHz. Obviously, the foundry-provided SPICE models for ESD protection devices are inaccurate, mainly

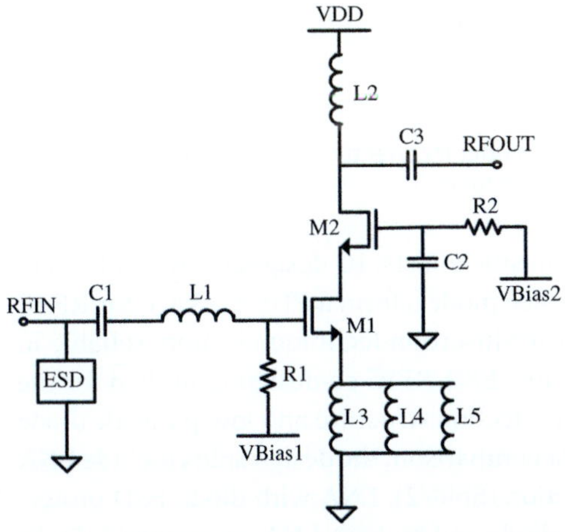

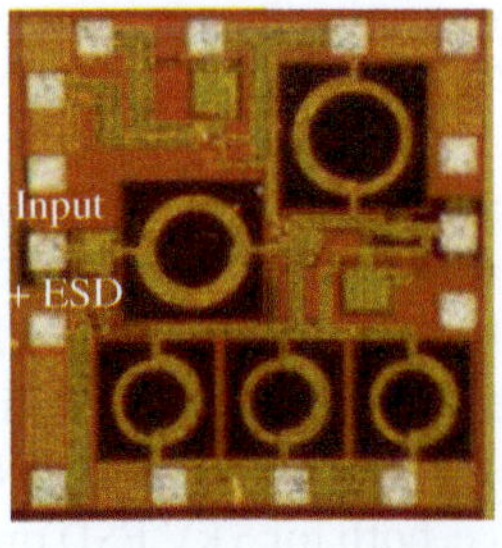

Figure 10.16 Schematic (a) and die photo (b) of a 5 GHz LNA implemented in a 180 nm RFCMOS technology. 5 kV ESD protection is provided at the input.

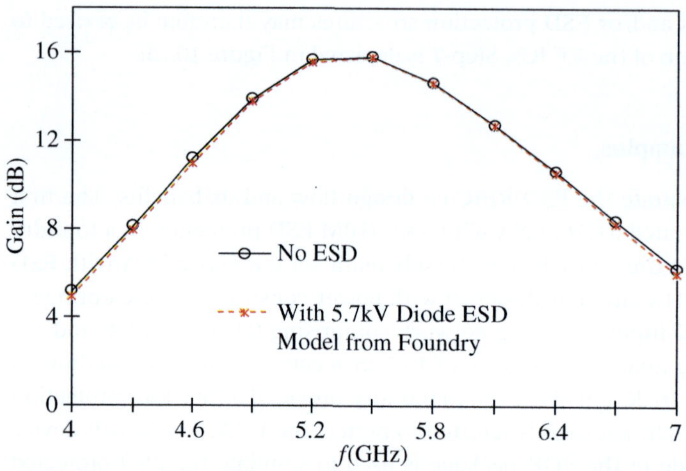

Figure 10.17 Simulation using foundry-provided ESD device model incorrectly shows negligible gain degradation for LNA with a 5.7 kV ESD protection diode from the foundry.

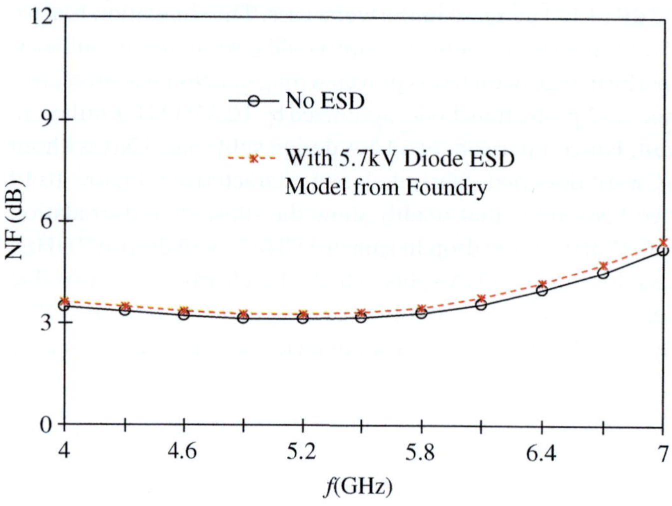

Figure 10.18 Simulation using foundry-provided ESD device model incorrectly shows negligible NF degradation for LNA with a 5.7 KV ESD protection diode from the foundry.

due to lack of inclusion of all possible ESD-induced parasitic effects. IC designers should be very cautious in using any foundry-provided SPICE ESD device models from a PDK package for RF IC + ESD simulation. For this reason, using the S-parameter-insertion technique is more reliable in evaluating ESD impacts on RF ICs. In design phase three, ESD-RFIC co-design is studied for the LNA IC splits using both high-parasitic ggNMOS ESD protection structure and low-parasitic diode ESD protection device, both for 5 kV ESD protection. For comparison, the design splits include LNA without ESD (Split-1), LNA with ggNMOS ESD protection (Split-2), LNA with diode ESD protection (Split-3), and LNA with the same ESD protection diode and by ESD-LNA co-design (Split-4). The LNA in Split-1 (LNA + No ESD) was optimized for I/O impedance matching. Split-2 (LNA + 5 kV ggNMOS ESD) means to show that the ggNMOS has heavy ESD parasitic impacts. Split-3 (LNA + 5 kV Diode ESD) intends to show a much less (still nontrivial) ESD parasitic effect of the

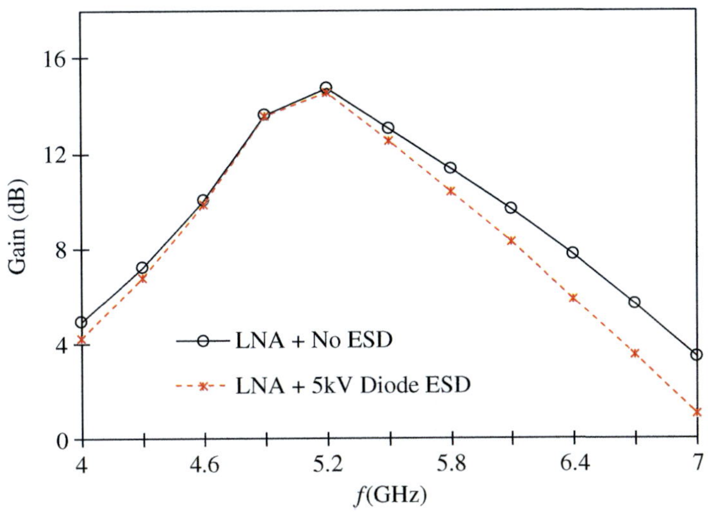

Figure 10.19 Measured gain curves show significant degradation for LNA with a 5 kV ESD protection diode due to ESD-induced noise effect.

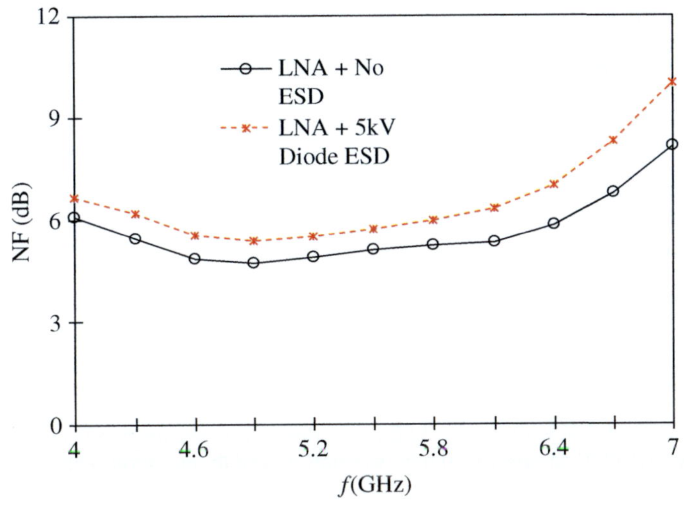

Figure 10.20 Measured NF curves show significant degradation for LNA with a 5 kV ESD protection diode due to ESD-induced noise effect.

ESD protection diode. Split-4 (LNA + 5 kV Diode ESD Rem) aims to demonstrate the benefits of ESD-LNA co-design. The ggNMOS and diode ESD protection structures are optimized first for minimum ESD parasitic C_{ESD} by TCAD ESD simulation first [12]. Following the ESD-RFIC co-design flow, the LNA circuit is first optimized for its I/O impedance matching. The ESD-induced parasitic parameters, i.e., S-parameters, are then applied to LNA circuit simulation to study the ESD-induced performance degradation, e.g., corruption in I/O matching. Finally, ESD-LNA co-design is conducted to recover the ESD-corrupted LNA I/O matching through ESD-I/O re-matching. To fully account for all layout effects, post-simulation is conducted for ESD-LNA co-design, which is then compared with the measurements. Figure 10.21a depicts the simulated gain for the LNA splits from post-simulation, which clearly shows that the high-parasitic ggNMOS severely affects the LNA

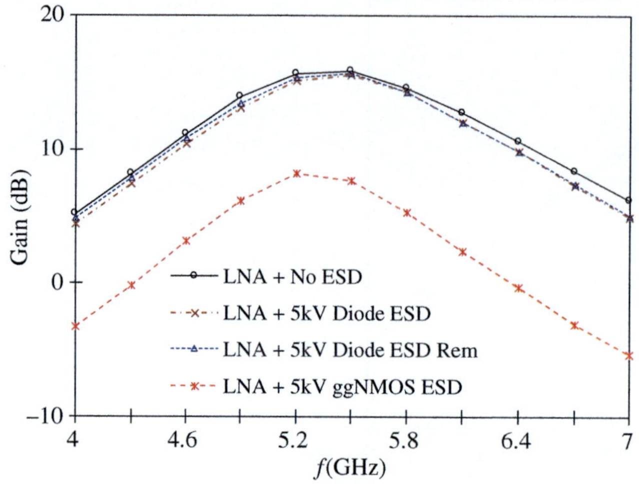

(a)

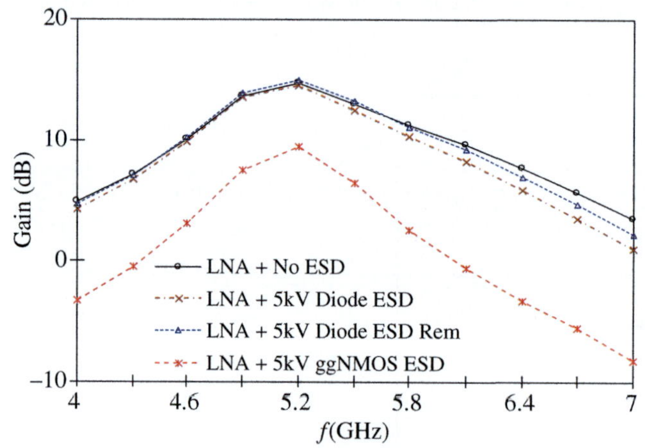

(b)

Figure 10.21 ESD-induced impacts on gain of LNA with various ESD protection structures are significant, which can be substantially alleviated by ESD-RFIC co-design: (a) post-simulation, and (b) measurement (Rem = re-matching).

gain. On the other hand, the low-parasitic ESD diode shows much less degradation (still not trivial across the full bandwidth) in gain drop (~4.2%), which is still sizable though minimized C_{ESD} is realized for the ESD protection diode via TCAD ESD simulation. It is observed that ESD-LNA co-design alleviates the ESD diode induced LNA gain drop across the full frequency spectrum. Figure 10.22a presents the simulated NF for LNA splits by post-simulation, which shows similar observation: ggNMOS ESD protection dramatically increases the LNA NF and low-parasitic ESD diode has much less NF degradation, yet still observable. On the other hand, ESD-LNA co-design further reduces the ESD diode induced NF degradation across the full spectrum. The co-design impact is more obvious in the simulated input reflection parameter (S_{11}) of the LNA splits by post-simulation as given in Figure 10.23a. It is clearly observed that the high-parasitic ggNMOS ESD protection deadly degrades the LNA input reflection, and even the optimized low-parasitic ESD diode can cause significant input reflection, particularly at the center frequency (>7 dB input signal loss).

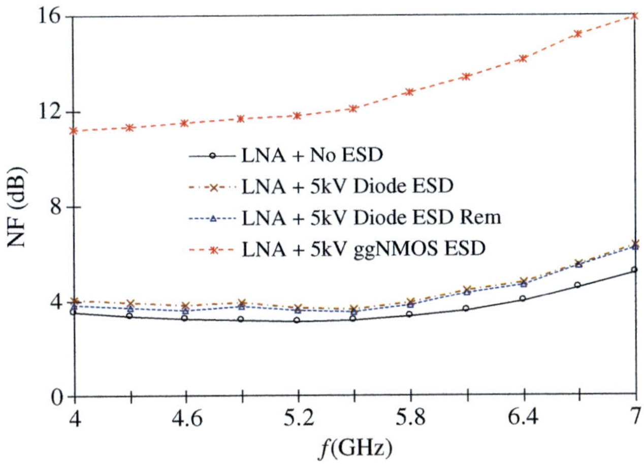

(a)

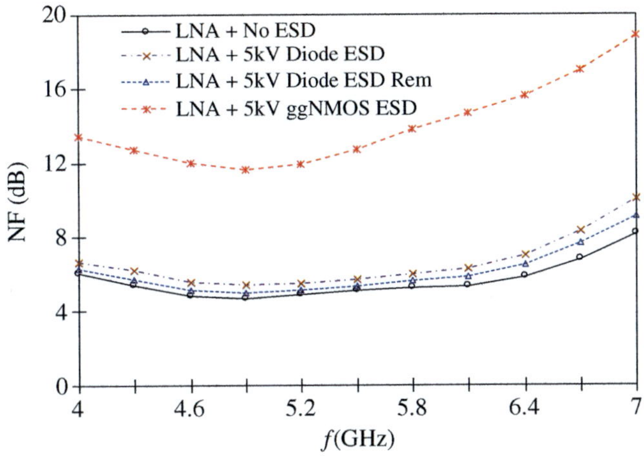

(b)

Figure 10.22 ESD-induced impacts on NF of LNA with various ESD protection structures are significant, which can be substantially alleviated by ESD-RFIC co-design: (a) post-simulation, and (b) measurement (Rem = re-matching).

Post-simulation reveals that ESD-LNA co-design can substantially reduce the ESD diode induced degradation in input reflection. Figure 10.16b depicts the die photo of a sample LNA fabricated. The measurements, shown in Figures 10.21b, 10.22b, and 10.23b, fully validated the observation in post-simulation in the curve trends. Further, Si measurements show a more clearer co-design benefits in terms of the scale of LNA specs degradation and its recovery by co-design. It is noteworthy that this example not only validates the benefits of ESD-RFIC co-design, but also clearly states the importance of TCAD ESD simulation and post-simulation in ESD-RFIC co-design practices. It is also worth to note that the ESD parasitic impacts on RF IC performance may be different for different specs parameters relevant to the circuit schematics, nevertheless, the ESD-induced RF IC performance degradation always exists.

The ESD–RFIC co-design technique was successfully used to design the world's first 8.5 kV ESD protected high-linear single-pole ten-throw (SP10T) transmitting and receiving (T/R) antenna

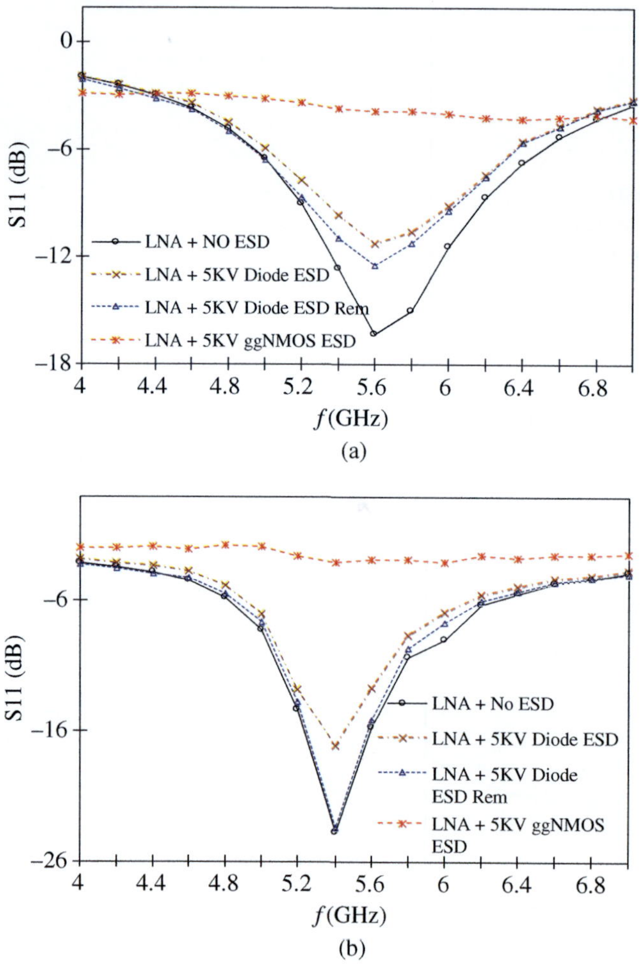

Figure 10.23 ESD-induced impacts on input reflection loss of LNA with various ESD protection structures are significant, which can be substantially alleviated by ESD-RFIC co-design: (a) post-simulation, and (b) measurement (Rem = re-matching).

switch IC for quad-band (850/900 MHz and 1.8/1.9 GHz) GSM and multiple-band WCDMA smartphones [13]. Figure 10.24a depicts the architecture for the SP10T T/R switch chip. As shown in Figure 10.24b, the SP10T IC features a series-shunt circuit topology for time-division duplex (TDD) transmitting (T_x) and receiving (R_x) channels and frequency-division duplex (FDD) Tx/Rx channels to handle high GSM transmitter power. This SP10T switch must handle 10 GSM T_x/R_x and WCDMA TR_x bands. One main design challenge is to achieve high linearity while handling large output power in GSM T_x mode, e.g., ~35 dBm in the low bands (800–900 MHz) and ~33 dBm in the hand bands (1.8–1.9 GHz). In off-state mode, the MOSFET transistors in the SP10T experience very high voltage drop, for example, in GSM T_x mode, an output power of 35 dBm results in a peak AC voltage up to 30.5 V, which will readily cause gate breakdown of MOSFETs in the foundry 180 nm SOI CMOS technology. The solution is to use stacked FETs in each circuit branch to withstand the high RF signal power, ranging from 6 to 12 MOSFETs in circuit branches as illustrated in Figure 10.24c. However, the downside of using MOSFET stacks is the uneven voltage distribution across each MOSFET in a circuit branch due to imperfect gate isolation,

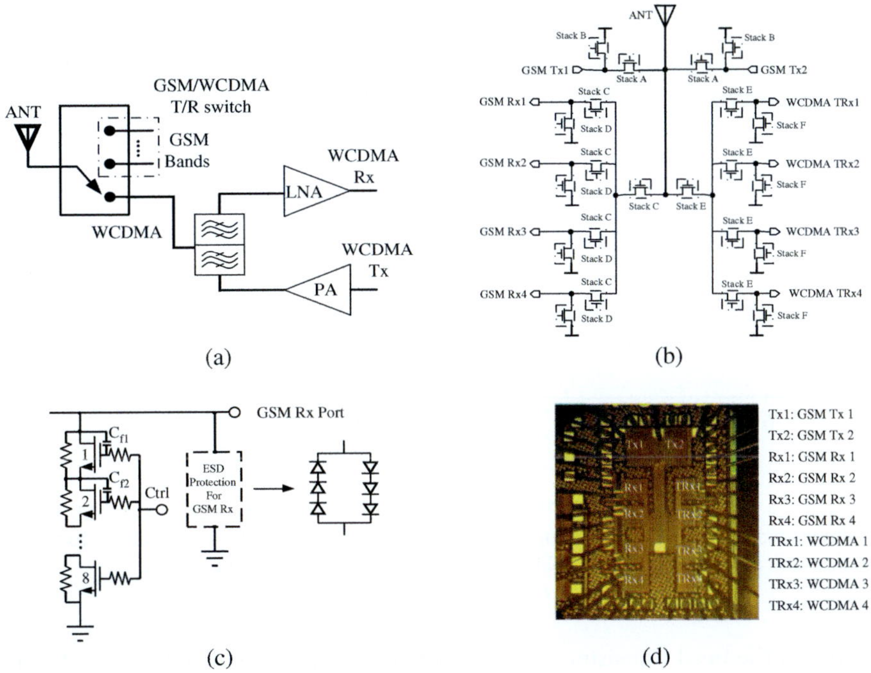

Figure 10.24 Design of a SP10T T/R switch in 180 nm SOI CMOS: (a) quad-band GSM/WCDMA Tx/Rx architecture, (b) series-shunt switch topology, (c) exemplar stacked-FET switch branch with anti-parallel diode-string ESD protection, and (d) a die photo.

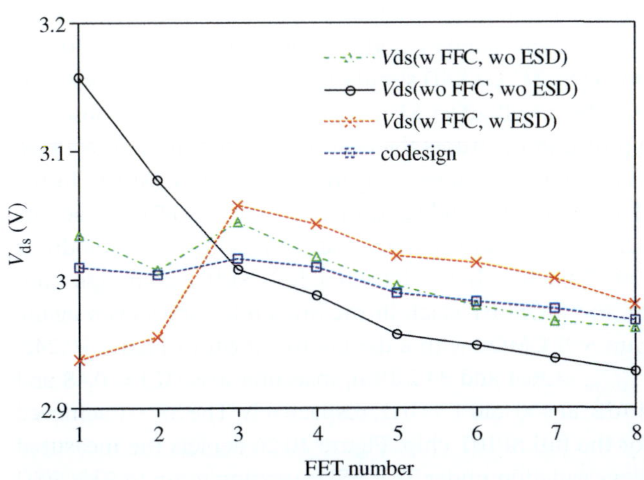

Figure 10.25 Evaluation of voltage distribution across MOSFETs in one 8-FET stack in SP10T shows strong dependence of voltage drop on FFC and ESD effects. ESD-SP10T co-design ensures even voltage drop across the stacked FETs.

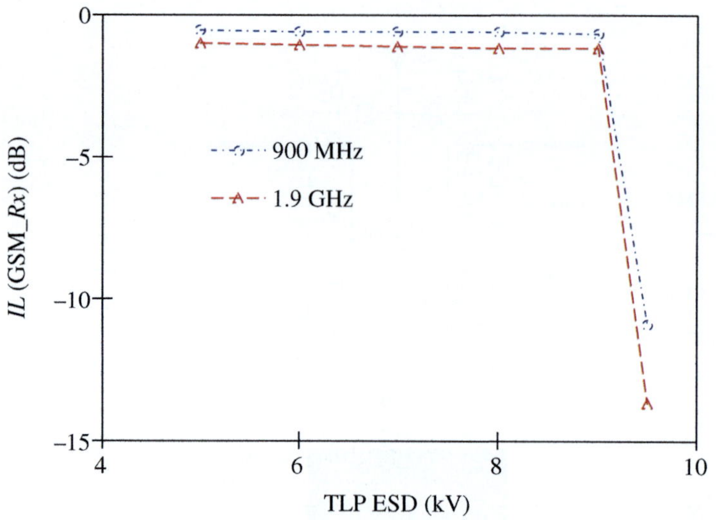

Figure 10.26 Measured IL for the ESD-protected SP10T switch in GSM R_x mode under continuous ESD stressing by TLP test shows excellent insertion loss up to 9 kV ESD zapping, then ESD failure occurs to the SP10T IC, resulting in dramatical increase of IL. The SP10T chip passes 9 kV HBM ESD zapping.

i.e., higher voltage drop in the head transistors, resulting in possible gate breakdown of the head MOSFETs. To overcome this design problem, a feed-forward capacitor (FCC) technique is applied, as depicted in Figure 10.22c, to force-even the voltage drops across the upper transistors, M1 and M2. Figure 10.25 clearly shows that, without using FFC technique, the voltage drops across MOSFETs in the circuit branch is very uneven, with the first and second MOSFETs suffering very high voltage drops, possibly causing gate breakdown. With the FFC technique in place and the FFC capacitors are optimized as C_{f1} = 60 fF and C_{f2} = 40 fF, the voltage distribution across all MOSFETs in the branch becomes fairly even. The ESD protection target for the SP10T is 8 kV for HBM ESD protection, which is provided by an anti-parallel diode string, shown in Figure 10.24c. Though the ESD diode strings are optimized by TCAD ESD simulation, the record-high 8 kV ESD protection still introduces C_{ESD} ~ 400 fF, which will affect SP10T specs. This is clearly observed in Figure 10.25, which shows that ESD protection corrupts the even voltage distribution realized by using FFC technique initially, which becomes very uneven again across the MOSFETs in the branch again. Next, careful ESD-SP10T co-design is conducted to fine-tune the FCC capacitors according to the ESD-induced C_{ESD}, i.e., the original C_{f1} = 60 fF and C_{f2} = 40 fF are modified to C_{f1} = 55 fF and C_{f2} = 33 fF. Figure 10.25 readily shows that after ESD-SP10T co-design optimization, the voltage distribution across the MOSFET stack in the branch becomes even again. The SP10T ICs are fabricated in a 180 nm SOI CMOS with a die photo shown in Figure 10.24d. Measurement shows excellent specs: $P_{-0.1dB}$ = 36.4 and 34.2 dBm, insertion loss (IL) = 0.48 and 0.81 dB in ^{T}x, and isolation of 43.1 and 40 dB, at 900 and 1.9 GHz, respectively. The SP10T achieved a record 8.5 kV HBM ESD protection for the full SP10T chip. Figure 10.26 depicts the measured IL in GSM R_x mode, maintaining excellent isolation under TLP ESD stressing to up to 9 kV ESD zapping.

10.3 Summary

This chapter discusses the complex ESD-circuit interactions in details. On one hand, the core circuit may affect ESD protection. Primarily, unwanted internal parasitic ESD-like structures may compete with intentional ESD protection structures, resulting in pre-mature ESD failure regardless how well ESD protection structures are designed. Possibly, strong and high-frequency RF signals may cause accidental ESD mis-triggering in normal IC operations. The circuit-to-ESD influences are validated in design examples. On the other hand, any ESD protection structures inevitably introduce parasitic effects, mainly the ESD-induced parasitic capacitance, leakages, noise coupling, and extra self-generated noises, which can seriously degrade RF IC performance, as validated in design examples. Good RF ESD protection solutions must be optimized to minimize the ESD-induced negative effects. Unfortunately, no matter how well an ESD protection structure may be optimized, the ESD-induced parasitic effect always exists. Therefore, ESD–RFIC co-design is required to balance the design requirements for both ESD protection and core circuit performance. The ESD–RFIC co-design flow is described in details using practical RF IC design examples, confirming both the importance and usefulness of ESD-RFIC co-design techniques for advanced RF ICs.

References

1 Feng, H.G. (2001). A mixed-mode simulation-design methodology for on-chip ESD protection design. An MS thesis. Illinois Institute of Technology.

2 Gong, K. (2001). ESD protection in copper interconnect and ESD-to-circuit performance influences. An MS thesis. Illinois Institute of Technology.

3 Gong, K., Feng, H., Zhan, R., and Wang, A. (2002). A study of parasitic effects of ESD protection on RF ICs. *IEEE Trans. Microw. Theory Tech.* 50 (1): 393–402. https://doi.org/10.1109/22.981291.

4 Wang, A., Feng, H., Zhan, R. et al. (2005). A review on RF ESD protection design. *IEEE Trans. Electron Devices* 52 (7): 1304–1311. https://doi.org/10.1109/TED.2005.850652.

5 Chen, G., Feng, H., Wang, A., and Cheng, Y. (2005). Noise analysis of ESD structures and impacts on a fully-integrated 5.5 GHz LNA in 0.18 μm SiGe BiCMOS. *Proceedings of IEEE European Conference on Wireless Technology*, pp. 261–263. https://doi.org/10.1109/ECWT.2005.1617707.

6 Guan, X., Wang, X., Lin, L. et al. (2008). ESD-RFIC co-design methodology. *Proceedings of IEEE RFIC*, pp. 467–470.

7 Wang, A., Lin, L., Wang, X., and Liu, H. (2008). Emerging challenges in ESD protection for RF ICs in CMOS. *J. Semicond.* 29 (4): 628–636.

8 Guan, X., Chen, G., Lin, L. et al. (2007). A new ESD-aware power amplifier design method. *Proceedings of IEEE ASICON*, pp. 1363–1366.

9 Wang, A.Z., Feng, H., Zhan, R. et al. (2002). ESD protection design for RF integrated circuits: new challenges. *Proceedings of IEEE Custom IC Conference (CICC)*, pp. 411–418. https://doi.org/10.1109/CICC.2002.1012860.

10 Wang, X., Tang, H., Lin, L. et al. (2011). Co-design of ESD protection and UWB RF front-end ICs. *Sci. China* 54 (10): 2209–2220. https://doi.org/10.1007/s11432-011-4416-3.

11 Lu, F., Ma, R., Dong, Z. et al. (2016). A systematic study of ESD protection co-design with high-speed and high-frequency ICs in 28 nm CMOS. *IEEE Trans. Circuits Syst. I: Regul. Pap.* 63 (10): 1746–1757. https://doi.org/10.1109/TCSI.2016.2581839.

12 Feng, H., Chen, G., Zhan, R. et al. (2003). A mixed-mode ESD protection circuit simulation-design methodology. *IEEE J. Solid-State Circuits* 38 (6): 995–1006. https://doi.org/10.1109/JSSC.2003.811978.

13 Wang, X.S., Wang, X., Lu, F. et al. (2014). Concurrent design analysis of high-linearity SP10T switch with 8.5 kV ESD protection. *IEEE J. Solid-State Circuits* 49 (9): 1927–1941. https://doi.org/10.1109/JSSC.2014.2331956.

11

ESD Layout Designs

11.1 Layout is Critical to ESD Protection

The complexity and, hence, challenge for on-chip electrostatic discharge (ESD) protection design is rooted in its multiple-coupling nature, i.e., the electro-thermal-transient-materials-process-device-circuit-layout coupling effects. To achieve the design goal of optimization and prediction of any ESD protection designs, it is very important to thoroughly understand the ESD protection mechanisms and to fully use ESD simulation design techniques, in addition to wisely excise prior design experiences, in any practical ESD protection design practices. Nevertheless, ESD design simulation still cannot guarantee ESD protection design successes in the real world. It is not unusual that a chip fails in ESD testing, even though an individual standalone ESD protection structure has been fully "verified" by very comprehensive ESD simulation. One of the common ESD design failure problems is due to inappropriate ESD layout designs, which often causes unexpected problems under ESD stressing at chip level such as parasitic ESD discharge structures or local current (heat) crowding. This chapter thoroughly discusses the importance of physical layout design in ESD protection designs, which should never be overlooked in real-world IC designs.

Let us make it crystal-clear: *Layout is Critical to ESD protection*! As discussed before, the circuit-to-ESD influences are inevitable in practical ESD protection designs. Similar to IC designs, even if full circuit simulation is completed at schematic level, there are unlimited ways to construct integrated circuit (IC) layout designs for a given circuit schematic, especially for analog and RF ICs where the "art" of designs is in play. Fortunately, there are mature and powerful CAD tools to validate IC physical designs such as design rule checking (DRC), layout versus schematic (LVS) checking, parasitic extraction, power and timing analysis, etc., in the post-simulation phase to verify IC physical designs before tape-out for Si fabrication. Unfortunately, ESD protection designs are much more complicated that are still beyond the capability of existing ESD simulation software, often leaving the designers in darkness until the Si is back for testing, and get shocked. The main ESD layout design challenges are discussed below. First, any unwanted and random parasitic device structures (i.e., *ESD-like* devices) on an ESD protected chip inherently exist, which often lead to early ESD failures. In fact, many commonly used ESD protection structures, such as ggMOS or gcMOS ESD protection units, rely entirely on parasitic bipolar junction transistor (BJT) for ESD triggering and discharging, which certainly suggests ESD protection sensitivity to any parasitic devices on a chip. For example, a required guard-ring (GR) for any ESD protection structure may bring in a parasitic BJT or SCR device, which may be turned on before triggering an intentional ESD protection, or simply compete with any designed ESD protection structure in ESD discharging. Since these parasitic conducting devices are never optimized to handle high

Practical ESD Protection Design, First Edition. Albert Wang.

ESD currents, low-level ESD failure frequently occurs in a parasitic ESD-like structure. Hence, any parasitic ESD-like devices cannot be considered as ESD protection structures. These parasitic ESD-like devices simply have the probability of conducting currents during ESD events, but not "designed" for ESD discharging. With extensive testing work, suitable ESD DRC rules can be developed, mainly by relaxing the critical dimension (CD) in ESD protection device layout, for a designer to follow in ESD protection circuit designs to avoid possible parasitic ESD conduction. However, ESD DRC has very limited checking capability at chip scale. Often, some random parasitic devices associate with ESD-circuit interfaces and inner circuit core may kick-in to compete against the intentional ESD protection devices at pads, resulting in early ESD failures. Unless a powerful CAD tool is available to automatically and accurately extract arbitrary parasitic ESD-like devices anywhere on a chip, the potential layout-induced early ESD failure cannot be rooted out. Second, because ESD events are ultrafast and energy-heavy (i.e., very large transient current surges), and semiconductors (e.g., Si) are very poor in thermal conduction, local ESD overheating (i.e., hot spots) can easily cause ESD thermal damages. In principle, local ESD hot spots are attributed to uneven heat generation; hence, lattice temperature distribution, across a chip, which is directly tied to local current crowding (i.e., heat crowding). Since current conduction on a chip is entirely decided by the device layout and metal interconnects, therefore, ESD layout plays a key role in routing large ESD currents, therefore, lattice temperature mapping. For example, it is well-known that the discontinuity in current conduction or a sharp turning in a current flow line will create unwanted local current crowding, which results in local thermal crowding effects at the corners and edges of any ESD protection structures. Such an ESD device corner/edge effect is deadly for ESD protection designs, which is often directly associated with poor ESD physical layout designs. Unfortunately, such ESD corner/edge effect is extremely hard to deal with in practical ESD protection designs. The rule of thumb for a good ESD protection layout design is to ensure ESD discharge uniformity and to avoid sharp-turning or any bottle neck in ESD discharge paths, which are associated with the ESD Si device shapes, ESD metal interconnects routing, and ESD contact and via placement. Third, an artistic mind can be very beneficial for ESD protection layout planning at chip scale. It is recognized that ESD protection physical layout is a design headache because ESD protection structures have large sizes and there are many ESD protection units needed on a chip, which makes full-chip layout floor planning very difficult. It is therefore necessary to think smartly in ESD layout designs. For example, an ESD protection structure can be placed under a bonding pad, hence, saving significant Si asset. However, putting an ESD protection device under a bonding pad requires careful reliability evaluation for the pad in order to avoid damaging the ESD protection structure during pad-bonding procedures. In summary, both simulation-based quantitative design and careful layout design are critical to ESD protection design optimization and prediction at chip level [1–4].

11.2 Basic ESD Protection Layout

This section presents general ESD layout considerations for commonly used basic ESD protection structures, including diode, BJT, MOSFET, and SCR device structures.

A wise analog circuit engineer knows that, in analog IC designs, as long as delivering the specs, a design should be "the simpler, the better". The same golden rule applies to ESD protection designs too. PN junction diodes are the simplest ESD protection units, which should always be used if even possible. An ESD PN junction diode can be made in many fashions, e.g., N^+/P^+, p-well/n-well (PW/NW), N^+/PW or P^+/NW. The design keys for making ESD protection didoes including ESD

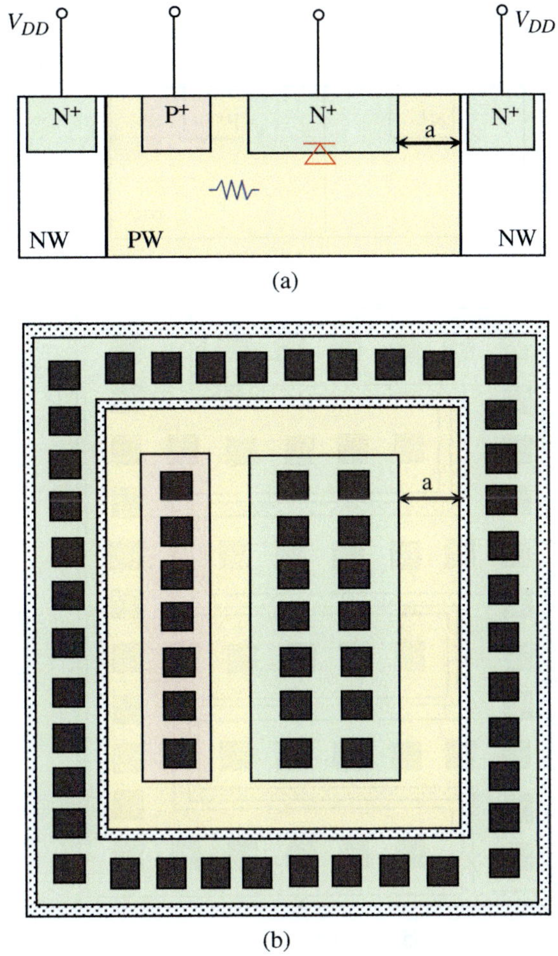

Figure 11.1 A classic vertical (bottom discharging) N+/PW diode ESD protection structure: (a) cross-section, and (b) layout.

triggering voltage (V_{t1}), ESD discharging resistance (R_{ON}), thermal dissipation, and ESD current handling capability (I_{t2}), etc., which are entirely determined by the diode structure and layout. Both vertical (conduction through a diffusion bottom area) and lateral (conduction via diffusion sidewall) diodes can be used for ESD protection. Figure 11.1 depicts a classic and conceptual vertical N^+/PW ESD protection diode in complementary metal-oxide-semiconductor (CMOS) where the ESD current is discharged vertically through the bottom of the diode PN junction. The diode bottom area determines the ESD discharge efficiency, i.e., J_{t2}, and the ESD R_{ON} is mainly defined by the lateral resistance in the P-Well that is controlled by the lateral spacing between N^+ and P^+ (pick-up) diffusion regions in the layout. Typically, an ESD protection structure is surrounded by a GR or a double-GR. Hence, the CD (a) between N^+ and its adjacent GR N-Well is critical to avoiding potential lateral parasitic NPN that may be turned on accidently during ESD stressing, leading to possible early ESD failure. A PW/NW sidewall diode is another commonly used diode ESD protection structure, as shown in Figure 11.2 where ESD discharge occurs through the deep PN junction sidewall, not the PN junction bottom. A deep PW/NW sidewall ESD diode has a key advantage for ESD discharging. Since the heat generated by an ESD transient can only be dissipated downward through the substrate bottom for an IC, a deeper PN junction is generally preferred because of its deeper hot spot location and easiness for thermal dissipation. However, the main disadvantage for a

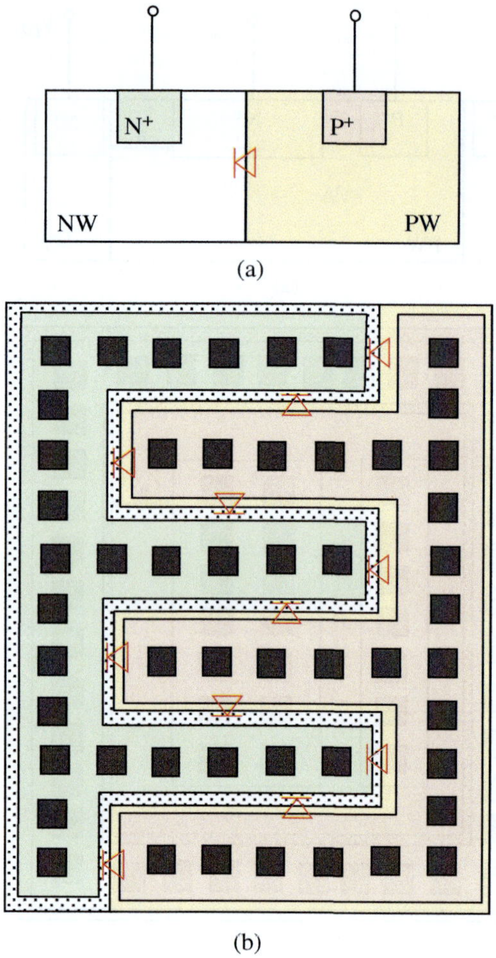

Figure 11.2 A classic lateral (sidewall discharging) NW/PW sidewall diode ESD protection structure: (a) cross-section, and (b) interdigitated layout.

sidewall PN diode is that ESD current can only be discharged through the PN sidewall, not good for shallow junctions, and the large diffusion bottom area mainly contributes to ESD parasitic effects including C_{ESD}, I_{leak} and noises. Therefore, a sidewall ESD diode is typically designed as an interdigitated structure (comb) in the layout as depicted in Figure 11.2. The total sidewall periphery in layout determines the ESD discharge efficiency. The area ratio of sidewall periphery to diffusion bottom should be maximized in layout design.

At advanced CMOS nodes, both gated and shallow trench isolation (STI) isolated diodes are used for ESD protection as depicted in Figure 11.3a. In comparison, a gated diode allows ESD discharging straight laterally through the channel between N^+ and P^+ diffusions under the isolate gate, while an STI diode suffers from a curvature ESD discharging path around the STI plug that causes ESD current and thermal crowding, as depicted in Figure 11.3b–d for a design in a foundry 28 nm CMOS technology [5]. On the other hand, a gated ESD diode introduces more parasitic C_{ESD} than a STI diode. These factors should be considered in ESD protection diode layout designs. The TCAD ESD simulation observation is validated in TLP ESD measurements, which shows that for the STI and gated ESD diode samples of same size ($W = 40\,\mu m$) fabricated in 28 nm 0.85 V core CMOS, the measured results are ESD $C_{ESD} \sim 59$ fF for a gated diode and $C_{ESD} \sim 33$ fF for a STI diode and ESD $I_{t2} \sim 2.4$ A for the gated diode and $I_{t2} \sim 1.9$ A for the STI diode, respectively.

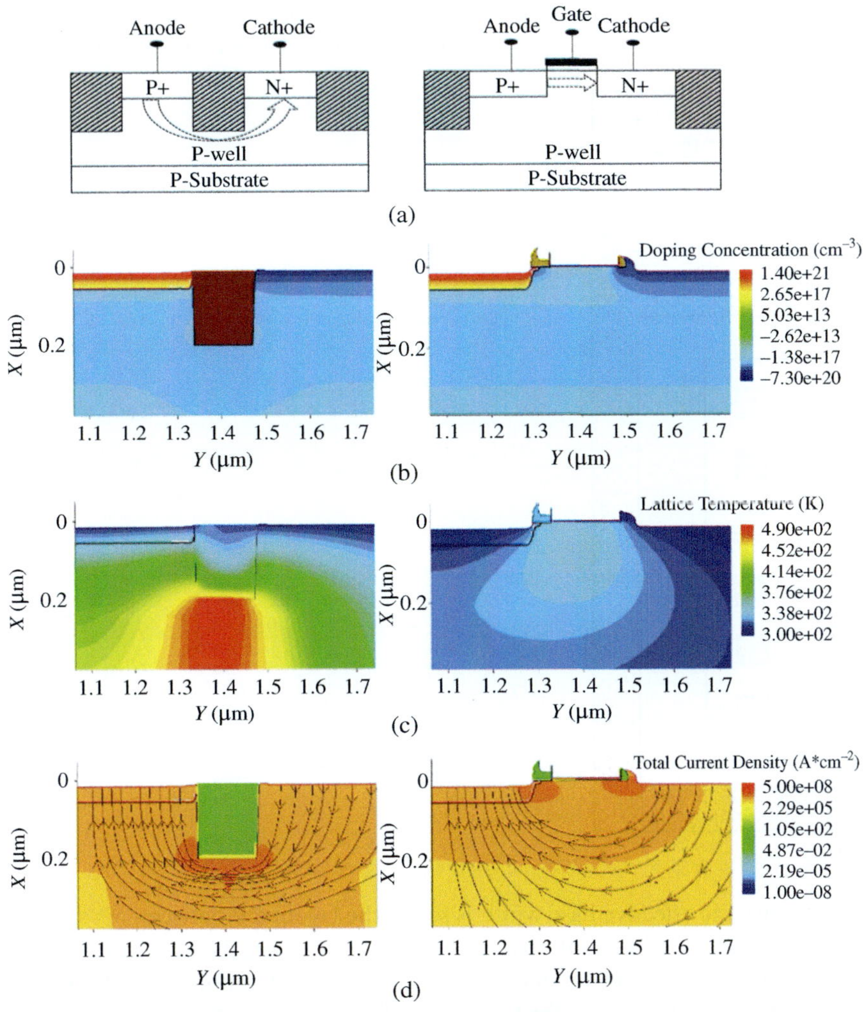

Figure 11.3 A gated ESD diode versus a STI ESD diode made in a 28 nm CMOS: (a) cross-sections, (b) simulated cross-section views, (c) lattice temperate maps, and (d) ESD discharging current contours by TCAD ESD simulation.

BJT is an efficient ESD protection structure. In fact, except for ESD protection diodes, most other active ESD protection structures, such as MOSFET, SCR, and their derivatives, rely on internal "parasitic" BJT structure(s) for active low-resistance ESD discharging. It is therefore critical to understand BJT ESD protection designs. A BJT ESD protection structure can be in a vertical or lateral BJT fashion, making its ESD discharging efficiency very different. A vertical BJT is clearly preferred for ESD protection due to its very high current gain (i.e., ultrathin base width), low conduction resistance, and vertical heat dissipation channel. However, a vertical BJT requires a dedicated base diffusion, which often only exists in a BiCMOS technology. Figure 11.4 depicts an exemplar layout for a vertical NPN ESD protection device where the N^+ to P^+ spacing is a key CD in layout design. The ESD discharging efficiency for such a vertical BJT ESD protection is largely limited by the emitter diffusion area. To minimize the ESD discharging resistance, a symmetric layout is often used, such as a CBEBC finger structure. Alternatively, in normal CMOS technologies, no dedicated base diffusion is available, and hence, a lateral BJT may be used for ESD protection, which

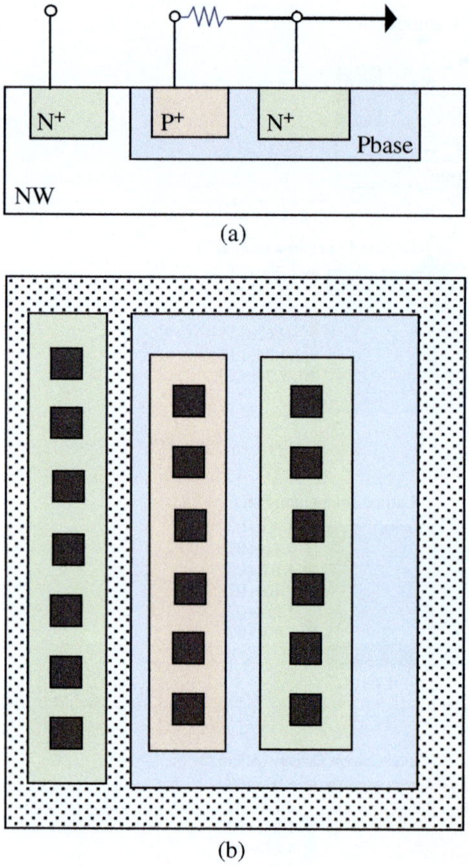

Figure 11.4 A classic vertical NPN ESD protection device: (a) cross-section, and (b) layout top view.

obviously is not ideal because the effective base width is set by the layout CD between N^+ and P^+ diffusions, which is much bigger than its counterpart in a vertical BJT where the base width is vertical diffusion difference between base and emitter. To ensure straight and uniform ESD discharge current flow, special attention must be given for the layout of diffusion fingers, metal interconnects, and contacts and vias. Further, the lateral diffusion spacing between inner diffusions and guard-rings must be carefully considered to avoid possible parasitic lateral BJT conduction.

MOSFET ESD protection has been widely used in CMOS technologies for years, typically in ggMOS and gcMOS structures. Figure 11.5 shows an exemplar ggNMOS ESD protection structure that relies on its parasitic lateral NPN for ESD discharging. Therefore, the NMOS channel length (L) becomes critical to ESD conduction efficiency because it is the equivalent NPN base width, which is limited by the CMOS lateral CD, i.e., the minimum channel length in a CMOS at a given technology node. Typically, to minimize the ESD discharging R_{ON}, the Source Contact to Gate Spacing (SCGS) is set to CD limit (minimum), while the Drain Contact to Gate Spacing (DCGS) is set to 4–5 μm (long enough) to form a given drain extension resistance to ensure multiple-finger ESD triggering uniformity. However, this DCGS design trick does not apply to all technologies such as a silicided CMOS or advanced CMOS at sub-45 nm nodes. Obviously, the ESD discharging efficiency is determined by the S/D diffusion areas, and the typical finger width (W) is bounded by the outer contacts of ggNMOS layout.

Silicon-controlled rectifier (SCR) is considered extremely area-efficient for ESD discharging, mainly due to its NPN–PNP gain amplification effect, which also leads to a very low ESD R_{ON}

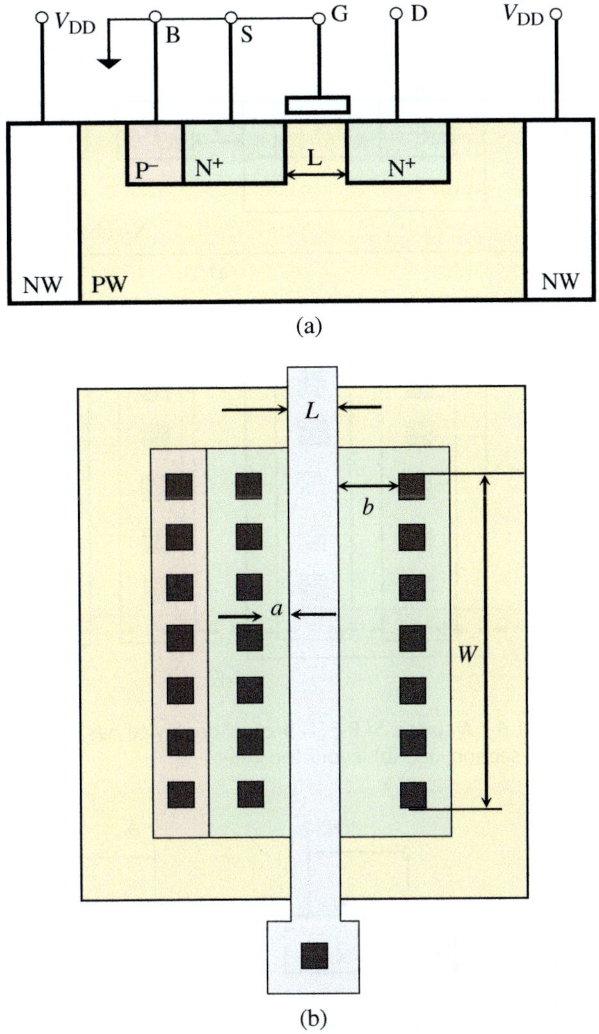

Figure 11.5 A classic ggNMOS ESD protection device: (a) cross-section, and (b) layout top view.

and ESD holding voltage V_h associated with its deep snapback I–V characteristics. Figure 11.6 depicts a classic SCR ESD protection structure in CMOS technology, which relies on a coupled vertical NPN and lateral PNP for a deep snapback ESD discharging I–V behaviors. In layout design for an SCR ESD protection structure, the key CD dimensions include the Pwell extension over the cathode N^+ diffusion (a) and the Pwell to anode P^+ diffusion distance (b). Since $a+b$ determines the total ESD R_{ON}, they should be smaller. However, too small a or b value may lead to lateral diffusion punch-through, resulting in unwanted parasitic ESD conduction that leads to early ESD failure. While an SCR ESD protection is area-efficient due to its large $\beta_{NPN} \times \beta_{PNP}$ product, an original SCR in CMOS has a rather high ESD triggering V_{t1}, determined by the Pwell/Nwell reverse breakdown, which is unsuitable for many LV CMOS ICs. Hence, low-V_{t1} SCR ESD protection structure has been a constant pursuit in SCR-type ESD protection designs. Figure 11.7 depicts an exemplar layout for middle-voltage SCR (MVSCR) ESD protection structure where an extra N^+ diffusion plug is added to the Pwell boundary, so that instead of relying on reverse breakdown of the Pwell-N-substrate junction, the N^+/Pwell junction breakdown triggers

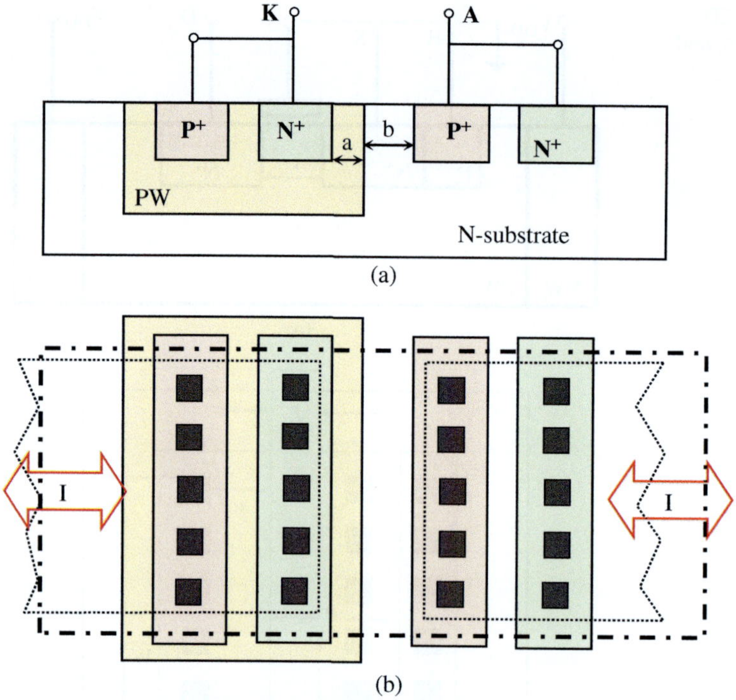

Figure 11.6 A classic SCR ESD protection device has high V_{t1} due to high Pwell/N-substrate breakdown: (a) cross-section, and (b) layout top view.

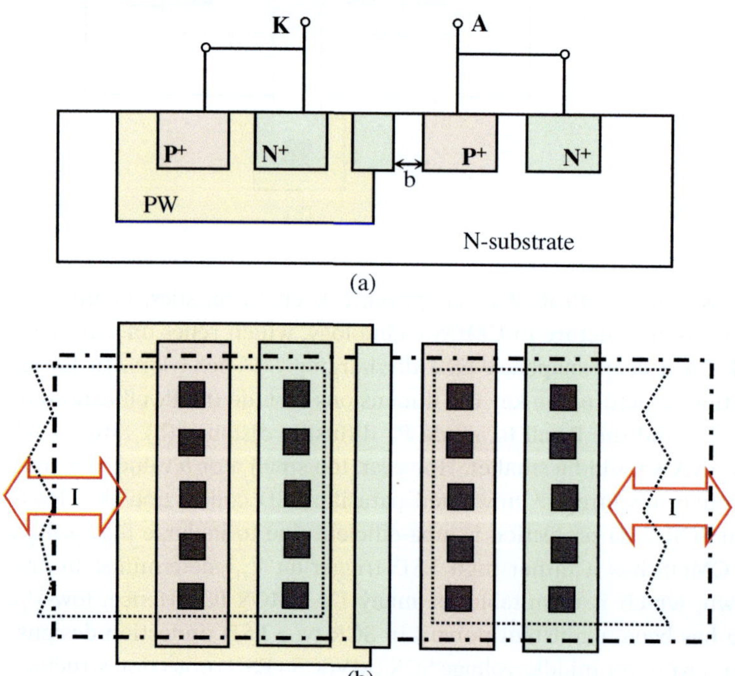

Figure 11.7 A MVSCR ESD protection device has a N^+ plug and relies on N^+/Pwell breakdown for a reduced V_{t1}: (a) cross-section, and (b) layout top view.

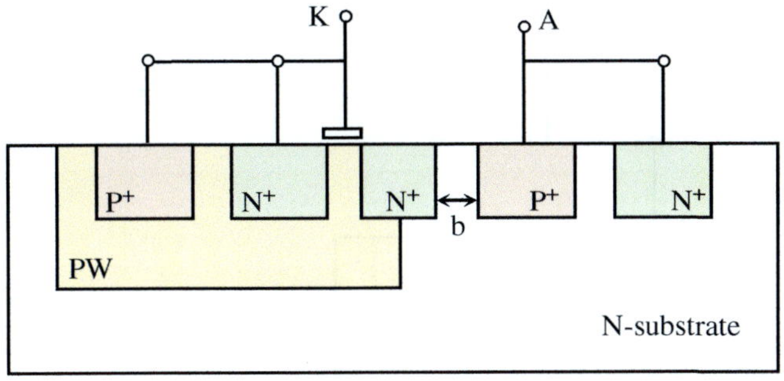

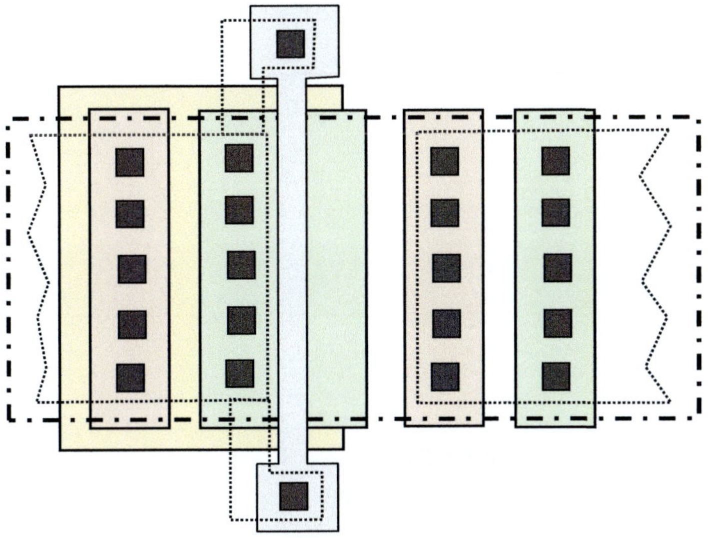

Figure 11.8 An LVSCR ESD protection device relies on an embedded NMOS to achieve low-V_{t1}: (a) cross-section, and (b) layout top view.

the SCR ESD discharge, resulting in a smaller V_{t1}. To further reduce the V_{t1}, a low-voltage SCR (LVSCR) ESD protection structure is depicted in Figure 11.8 where a MOSFET is embedded into the SCR that results in an even lower V_{t1}, while maintaining the desired deep snapback ESD discharging I–V characteristics. While SCR ESD protection mechanism is straightforward in concept, layout design plays an important role in practical designs. As an example, low-voltage SCR ESD protection structures of various layout dimensions are designed and fabricated in a foundry 28 nm CMOS technology [6]. As depicted in Figure 11.9a, the key layout dimensions are the anode Nwell to cathode N$^+$ spacing (L1) and the cathode N$^+$ to P$^+$ spacing (L2). Two SCR design splits with L2 = 2 μm and L2 = 0.15 μm (the minimum CD) are designed in this study. Figure 11.9b presents the measured SCR ESD discharging I–V characteristics by TLP testing. It is observed that the SCR of L2 = 2 μm behaves like a "normal" SCR device featuring $V_{t1} \sim 3$ V and a very low snapback holding voltage of $V_h \sim 1.75$ V as expected. However, the SCR of L2 = 0.15 μm shows a $V_{t1} \sim 4.2$ V and a fairly high holding voltage of $V_h \sim 3$ V, not expected for a "good" SCR device. This is attributed to the fact that a longer L2 of 2 μm offers a large enough R_{sub} that helps to quickly turn on the parasitic lateral NPN (Q_2) that then triggers the parasitic vertical PNP (Q_1),

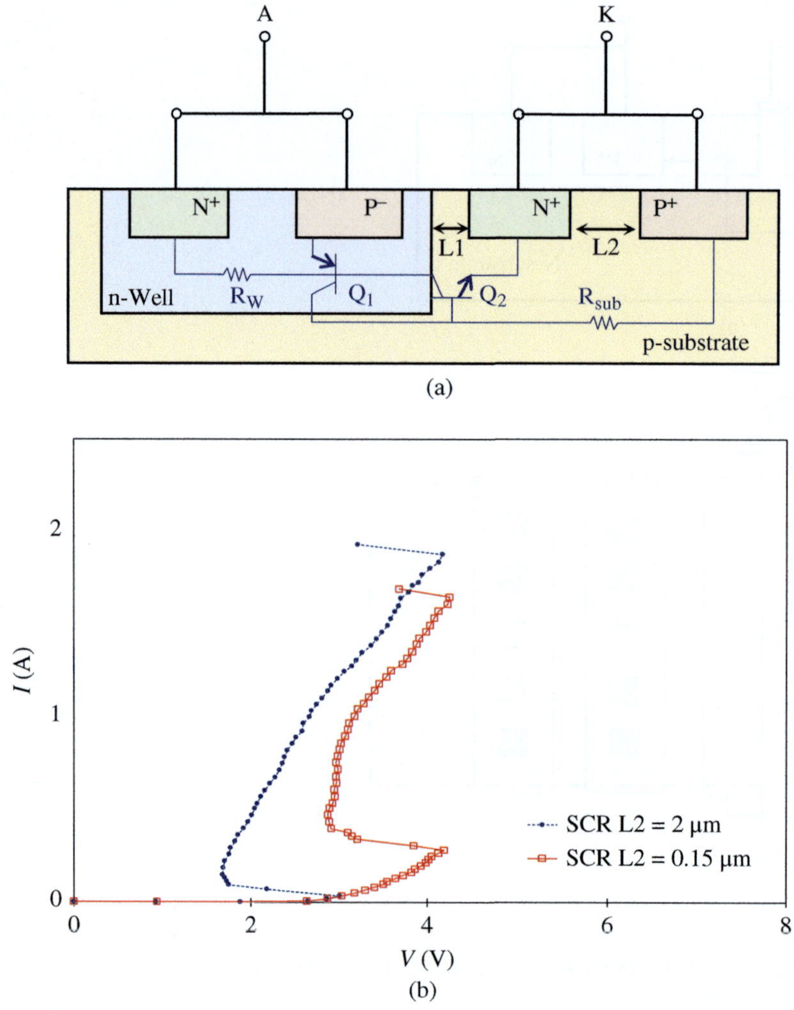

Figure 11.9 An example of LV SCR device fabricated in a 28 nm CMOS shows impact of layout dimensions (L2) on SCR triggering behaviors: (a) cross-section, and (b) measured ESD discharge I–V curves by TLP for the two design splits of L2 = 2 and 0.15 μm.

hence fires up the SCR of Q_1–Q_2, which features a deep snapback I–V curve with very low V_h. However, for the SCR of minimum L2 = 0.15 μm, the R_{sub} is too small to build up a required voltage drop to turn on Q_2. Hence, before the SCR is triggered by an ESD pulse, a parasitic BJT may be turned on to conduct the ESD transient, which features a much higher holding V_h, indicating a failed SCR design due to poor layout. Consequently, the functional SCR of L2 = 2 μm has higher ESD current handling capability of $I_{t2} \sim 1.88$ A than $I_{t2} \sim 1.65$ A for the failed SCR of L2 = 0.15 μm. The importance of layout design is clearly demonstrated in this LV SCR design example.

The next example shows how layout design can be used to realize a low-C_{ESD} SCR ESD protection in poly-Si layer, highly desired for high-frequency broadband RF ICs. As discussed previously, SCR is more area-efficient and introduces less ESD parasitic effects. Nevertheless, traditional SCR ESD protection structures are made of in-Si PN junctions, which inherently bring in parasitic ESD effects including C_{ESD}, I_{leak}, and noises associated with the whole PN junction area. To minimize such in-Si

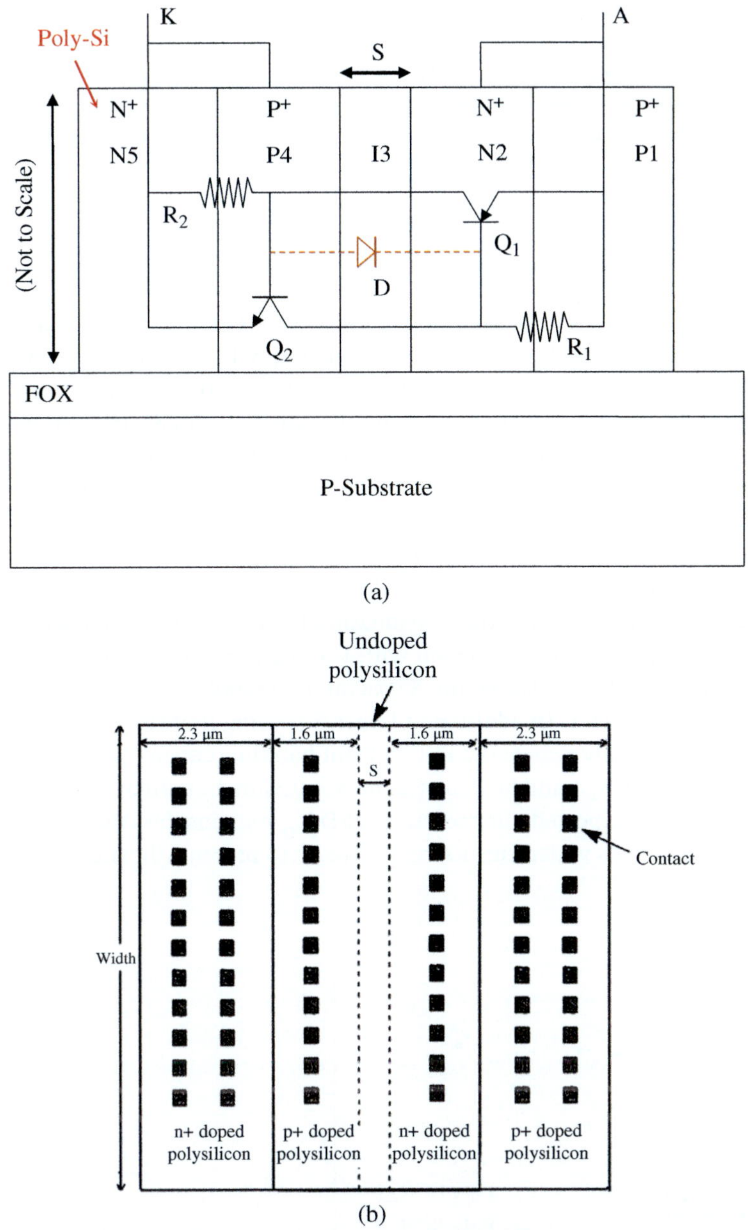

Figure 11.10 A low-capacitance Poly-Si SCR ESD protection structure implemented in BEOL of a foundry 0.35 μm SiGe BiCMOS: (a) cross-section and equivalent circuit, and (b) layout top view.

PN junction effect, a poly-Si SCR ESD protection device is devised, which is realized in the poly-Si layer in the backend of the line (BEOL) deck in CMOS [4]. As depicted in Figure 11.10, a poly-Si SCR has two major advantages over its bulk-Si counterpart. First, since a poly-Si SCR ESD device sits above the field oxide (FOX) on top of Si substrate, it only has small sidewall PN junction, which is much less than the junction interface in bulk SI, hence, substantially reduces PN-induced C_{ESD}, I_{leak}, and noises. Second, since a poly-Si SCR ESD protection device is isolated from the conducting

Table 11.1 A summary for poly-Si SCR ESD design splits.

Devices	W (μm)	S (μm)	V_{t1} (V)	I_{t2} (A)
SRC_1	75	0.8	11.0	2.1
SRC_2	75	2.4	15.0	1.4
SRC_3	30	0.8	11.0	1.5
SRC_4	75	0	9.0	2.8

Si substrate, it removes the troublesome global noise coupling problem inherent to an Si substrate. Further, the poly-Si SCR ESD protection structures can be realized simply by layout designs as shown in Figure 11.10b. This novel poly-Si SCR ESD protection device is demonstrated in a foundry 0.35 μm SiGe BiCMOS technology. To study the layout impacts on the poly-Si SCR ESD protection devices, design splits of varying layout dimensions, summarized in Table 11.1, are designed and fabricated: SCR_1 of $W = 75$ μm and $S = 0.8$ μm, SCR_2 of $W = 75$ μm and $S = 2.4$ μm, SCR_3 of $W = 30$ μm and $S = 0.8$ μm, and SCR_4 of $W = 75$ μm and $S = 0$ μm, where W is the SCR finger width and S is width of an undoped poly-Si strip used to control the ESD triggering V_{t1} of SCR ESD devices. For comparison, classic bulk-Si SCR ESD protection devices for equivalent ESD protection level are designed as a reference. Figure 11.11 depicts the TLP measured ESD discharge I–V curves for poly-Si SCR ESD protection devices of same $W = 75$ μm with varying $S = 0$, 0.8, and 2.4 μm. It is readily observed that the ESD V_{t1} is controlled by the S spacing monotonically, i.e., a wider S-gap increases the V_{t1} as listed in Table 11.1. In addition, a larger S-spacing increases the ESD discharge resistance, and hence substantially reduce the ESD current handling capability, I_{t2}, due to ESD heating. Figure 11.12 compares SCR_1 and SCR_3 of same $S = 0.8$ μm, but different finger widths of $W = 75$ μm and $W = 30$ μm. As expected, the measured ESD V_{t1} remains the same; however, I_{t2} is much higher for SCR_1 due to its wider finger size. Figure 11.13 presents the measured

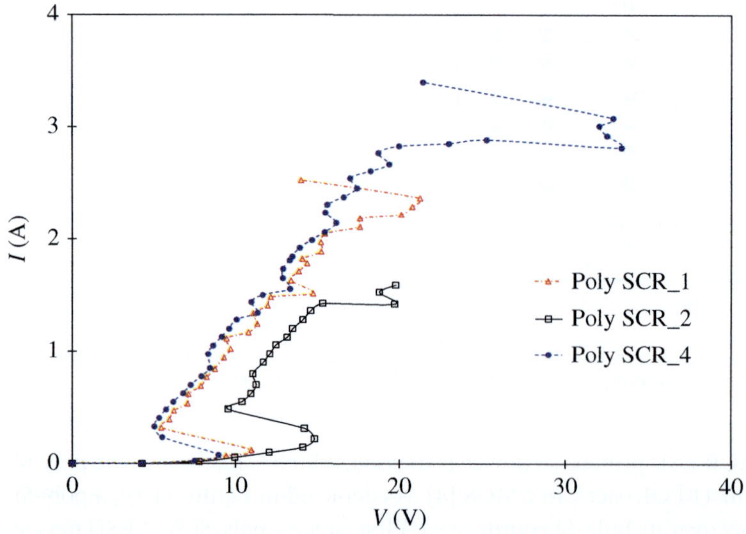

Figure 11.11 A comparison for measured ESD discharging I–V curves for Poly-Si SCR ESD device splits of $W = 75$ μm with varying undoped spacing of $S = 0$, 0.8, and 2.4 μm for SCR_4, SCR_1, and SCR_2 shows layout impact.

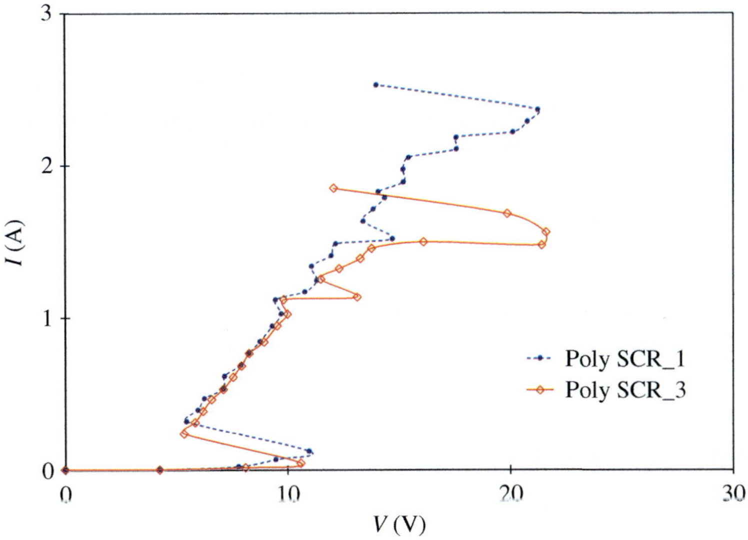

Figure 11.12 A comparison for measured ESD discharge I–V curves for Poly-Si SCR ESD device splits of $S = 0.8\,\mu m$ with different finger width $W = 30$ and $75\,\mu m$ for SCR_3 and SCR_1 shows layout impact.

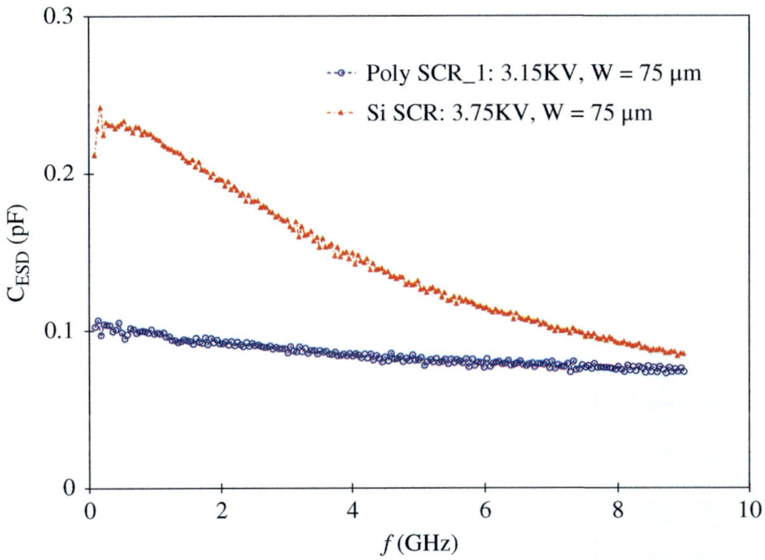

Figure 11.13 A comparison for measured parasitic ESD-induced C_{ESD} for Poly-Si and bulk Si SCR ESD device splits of same ESD protection level shows the advantage of Poly-Si SCR: lower and flatter C_{ESD} across 9 GHz frequency bandwidth.

ESD-induced C_{ESD} for a poly-Si SCR_1 device with its bulk-Si counterpart, both sized to equivalent ESD protection levels by design. It is clear that the poly-Si SCR ESD protection device induces much less parasitic C_{ESD} than its bulk-Si counterpart across a 0–9 GHz frequency bandwidth. It is also noticed that the measured C_{ESD} for poly-Si SCR ESD device is fairly flat across the 9 GHz bandwidth. Both features are very beneficial for high-frequency broadband RF ICs.

11.3 Advanced ESD Protection Layout

A good ESD protection design takes well more than basic ESD design layout skills. There are many subtle details that ought to be carefully considered in practical ESD protection physical designs. Making an ESD protection layout design "electrically" correct, as validated by DRC and LVS checking, is just a baby step in practical ESD protection designs. It is ESD layout design optimization that makes a good designer different from ordinary ESD engineers, much like in analog and RF IC designs. In ESD protection structure layout design, critical factors to consider include triggering mechanisms that are often affected by a parasitic device, holding voltage that is affected by bipolar gain, discharging resistance that determines heat generation, and current flow lines that may induce current crowding and local overheating, etc. All these are nontrivial in ESD protection layout designs.

11.3.1 Advanced ESD Layout Considerations

Figure 11.14 depicts one classic example where DCGS and SCGS spacings are carefully considered in design of MOSFET ESD protection structures. For years, a well-known golden rule for designing

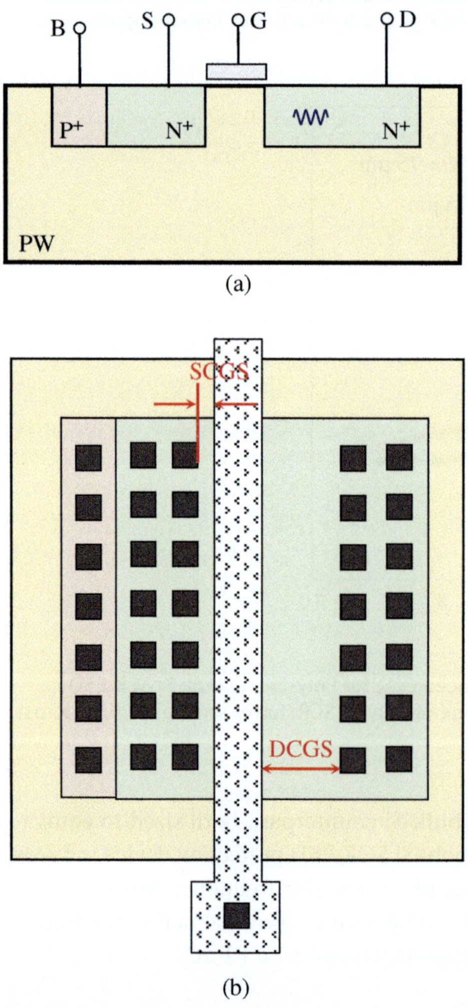

Figure 11.14 A classic ggNMOS ESD protection requires minimum SGCS and large DCGS in layout design to optimize ESDV.

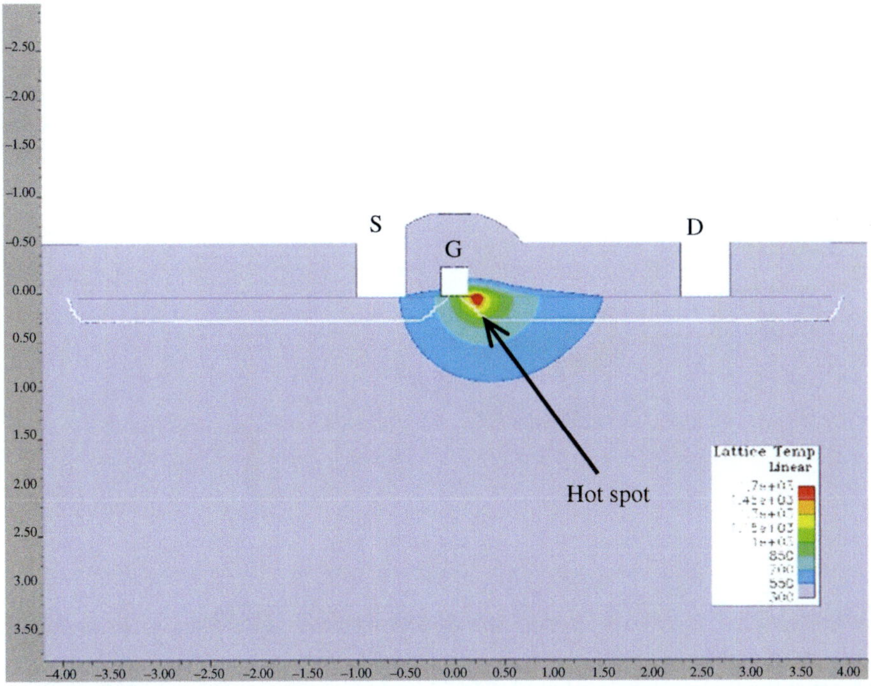

Figure 11.15 Transient TCAD ESD simulation shows an ESD-induced hot spot at the Drain corner in a bare-Si NMOS ESD protection structure made in 180 nm CMOS and the heat can spread rapidly laterally, causing melting in contacts or metal interconnects at Drain.

good MOSFET ESD protection structures in CMOS has been to minimize SCGS (i.e., minimum CD at a given technology node) and a long DCGS of 4–5 μm, which has been validated experimentally [6–8]. The underlying rationale is that direct relationship between measured ESDV and DCGS has been widely reported in experiments, i.e., ESDV increases with DCGS until saturates at DCGS ~4–5 μm in old IC technologies. Several factors may contribute to the magic ESDV ~ DCGS relationship. For example, the hot spot in MOSFET ESD protection devices is located at the Drain junction corner and a large DCGS spacing prevents heat spreading from reaching out to the Drain contacts and metals where metal melting may occur, as illustrated in Figure 11.15. In addition, a large DCGS introduces a series resistance in the Drain extension region that acts as a ballasting resistor to ensure uniform ESD triggering in a common multiple-finger MOSFET ESD protection structures. Uniform ESD triggering is critical to achieving higher ESD protection that is proportional to the finger numbers in such ESD protection structures (i.e., design scalability). However, a large DCGS certainly increases the total ESD discharge resistance R_{ON} that generates more heat. Large DCGS may also degrade CDM ESD protection. Therefore, a design balance is needed in layout design for DCGS. On the other hand, SCGS does not seem to suffer from the heat spreading effect of a hot spot at the Drain corner, hence should be minimized to reduce the total ESD discharge R_{ON}. This SCGS-DCGS golden rule apparently does not apply to silicide CMOS because the silicidation of the Drain extension effectively short-circuits the DCGS-induced series resistance. In addition, this minimum SCGS rule cannot be applied to CMOS at advanced nodes, e.g., sub-180 nm, because a hot spot at the Drain corner can readily spread the heat across the very short MOSFET channel and reaches to the Source contact and metal area, as indicated in Figure 11.16, causing metal melting. Transient TCAD ESD simulation should be used to assist ESD layout design in this regard [1]. In addition, a combined Silicon-metal ESD simulation method is developed to

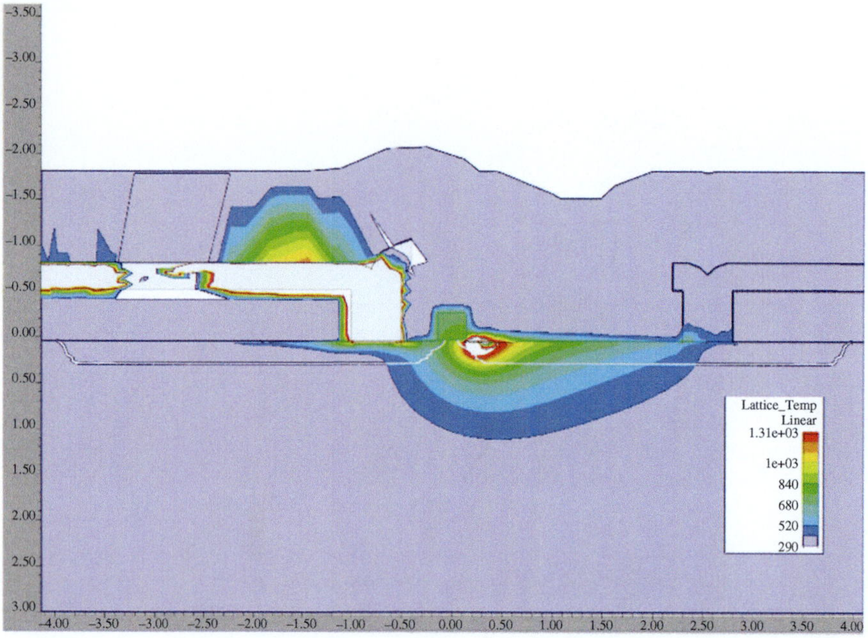

Figure 11.16 Transient TCAD ESD simulation, using a combined Si-metal ESD simulation technique, shows an ESD-induced hot spots at the Drain corner in a full NMOS ESD protection structure made in 180 nm CMOS. Two possible ESD thermal failures exist: first, overheating in metal interconnects, and second, the hot spot at the Drain spreads heat into the Source region, if the channel is ultrashort, causing melting in contacts or metal interconnects at the Source.

simulate full ESD protection structures (Si ESD devices + ESD metal interconnects) that is able to identify possible overheating in ESD metal interconnects, which may also cause ESD failures as depicted in Figure 11.16 [9]. In the next example, Figure 11.17 depicts thorough layout consideration in designing multiple-finger ggNMOS ESD protection structure. In fact, there are many different ways to layout a multiple-finger ggNMOS ESD protection, however, not every layout design will achieve high ESD protection level scalable to the finger width and numbers as expected. The key design consideration in ggNMOS layout is to ensure uniform ESD triggering across all fingers and a uniform ESD current conduction flow. At large ESD current level, it is easy to experience ESD current crowding, leading to local overheating and early ESD thermal failure. It is proven that certain finger layout patterns for the source (S), drain (D), body (B), and gate (G) in a large multiple-finger ggNMOS structure are advantageous to achieve high ESD protection. For example, a BSGD-DGSBSGD-DGSB finger pattern, as shown in Figure 11.17, achieves higher ESD protection robustness over other layout patterns such as B-DGSGD-B-DGSGD-B, BSGDGSGDGSB and BSGDGSBSGDGSB [10].

In addition to layout considerations for the Si device structure of an ESD protection unit, the contacts and metal interconnects also play a critical role in overall ESD protection physical designs. While the Si device structure determines the ESD-critical parameters (V_{t1}, I_{t1}, t_1, V_h, I_h, R_{ON}, V_{t2}, I_{t2}, β, etc.), contacts/vias and metal interconnects are often the ESD weak points that often lead to early ESD failure, even though the active Si device is designed to be very ESD-tough. For the contacts and vias in an ESD protection structure, transient ESD characterization for each contact and via must be experimentally evaluated in order to decide *how many* contacts and vias for needed for ESD device terminals to handle a given amount of ESD current, which is often overlooked

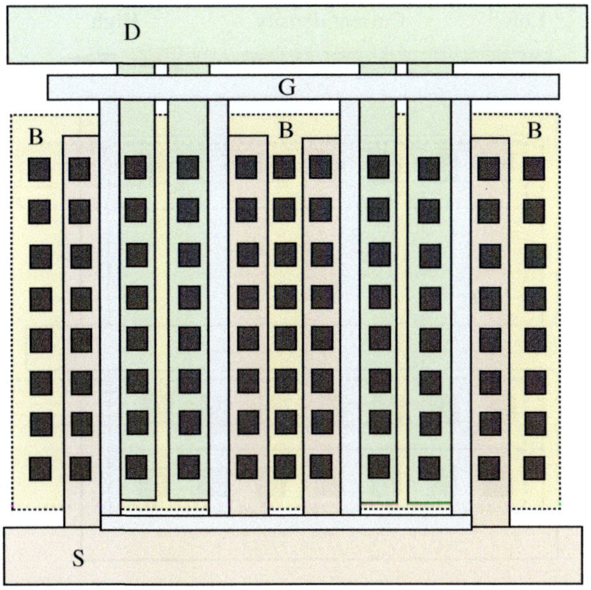

Figure 11.17 Illustration for a preferred BSGD-DGSBSGD-DGSB layout pattern for a multiple-finger ggNMOS ESD protection structure, achieving enhanced ESD robustness in testing.

by a designer. For ESD metal interconnects, two key factors must be considered. First, adequate ESD *metal width* must be used to handle the large ESD currents. This is guided by experimentally evaluating metal interconnects under transient ESD stressing, which cannot be obtained using the metal DC/AC metal stressing data normally offered in PDK. Of course, being too conservative, i.e., using excessive metal for ESD interconnects, will introduce too much metal-induced parasitic $C_{\text{ESD-metal}}$, affecting IC performance. Second, ESD *metal routing* decides the ESD discharge current pathway, which may result in unexpected ESD current crowding and local overheating if not designed properly. Such ESD contact/via and metal routing problem cannot be identified by normal DRC and VLS checking; however, it is a critical layout consideration. In principle, the rule of thumb for ESD contact/via and metal interconnects design is to ensure a *smooth* and *straight* low-*R* ESD discharging pathway in the layout, both vertically (*contacts/via effect*) and laterally (*metal routing effect*). If ever possible, one should avoid any turning in ESD metal routing and ensure absolutely no sharp-turning in the ESD conduction path. Figure 11.18 depicts two exemplar metal routing scenarios for a ggNMOS ESD protection structure, illustrating the metal routing impacts. In Figure 11.18a, a poor *anti-parallel* metal routing is used, which will cause ESD current crowding at one end of the whole ggNMOS ESD protection structure, resulting possible ESD thermal failure at the same end due to local thermal over stressing. On the other hand, Figure 11.18b depicts a better *parallel* ESD metal routing scenario where the large ESD current flows into the Drain and out of the Source in the same direction straightly, hence balances the overall ESD discharging current flow across the whole ggNMOS ESD protection structure. Figure 11.19 depicts a preferred metal routing scenario for a ggNMOS ESD protection structure that ensures a *straight* ESD conduction pathway. On the other hand, Figure 11.20 illustrates a careless, though imaginary, poor layout scenario that not only uses an anti-parallel metal routing but also unevenly distributed the contacts, which further worsen the ESD current crowding effect on one end of the ggNMOS ESD protection structure. Similarly, Figure 11.21 depicts a disastrous contact/via placement scenario where the contacts and vias are not line up straightly, causing unnecessary ESD current turning in the vertical direction, which further worsen the ESD current crowding effect. Indeed, while

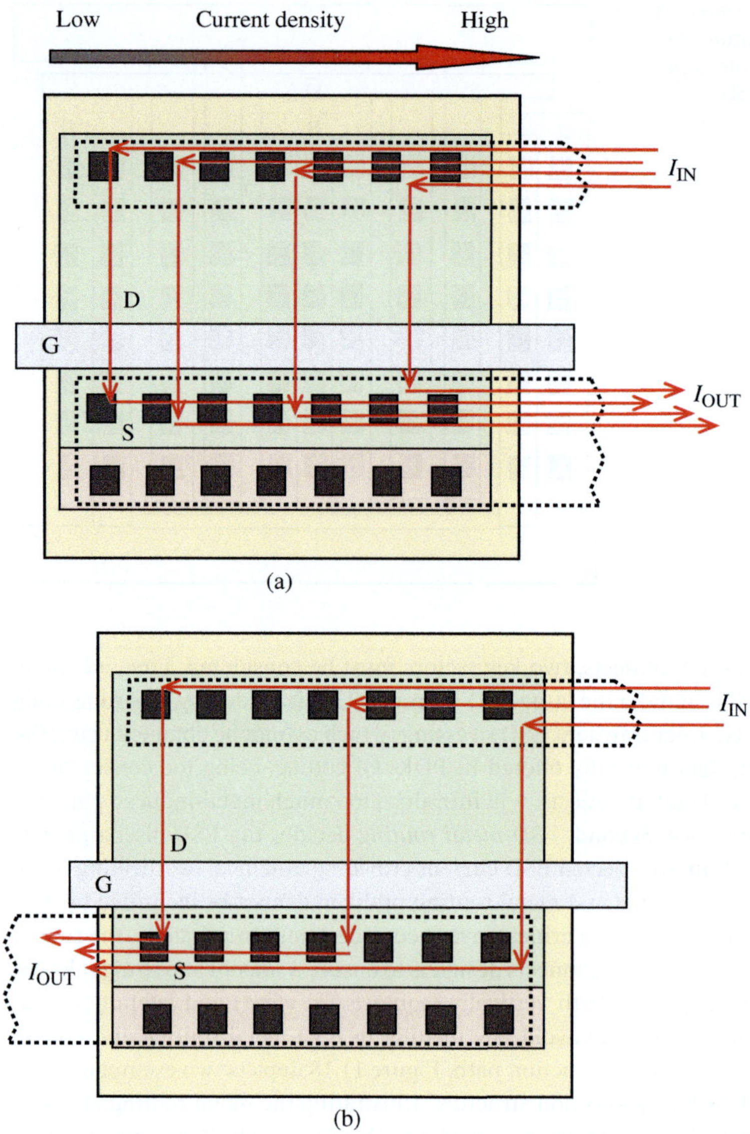

Figure 11.18 Two different ESD metal routing scenarios for a ggNMOS ESD protection may lead to different ESD protection results: (a) an anti-parallel metal routing causes unwanted ESD crowding on one end (right) of the structure, and (b) a parallel metal routing achieves even ESD thermal distribution.

it is impractical to provide absolutely golden design rules for ESD contact/via and metal interconnects layout, burning a few brain cells will really worth it in ESD metal interconnect layout designs.

11.3.2 ESD Design Layout is an Art

As stated, like analog and RF IC designs, ESD protection circuit design is artistic in nature, not just meaning that novel ESD protection structure can be created but also reflecting the fact that ESD layout design can truly be a piece of art. Other than various subtle considerations in ESD layout

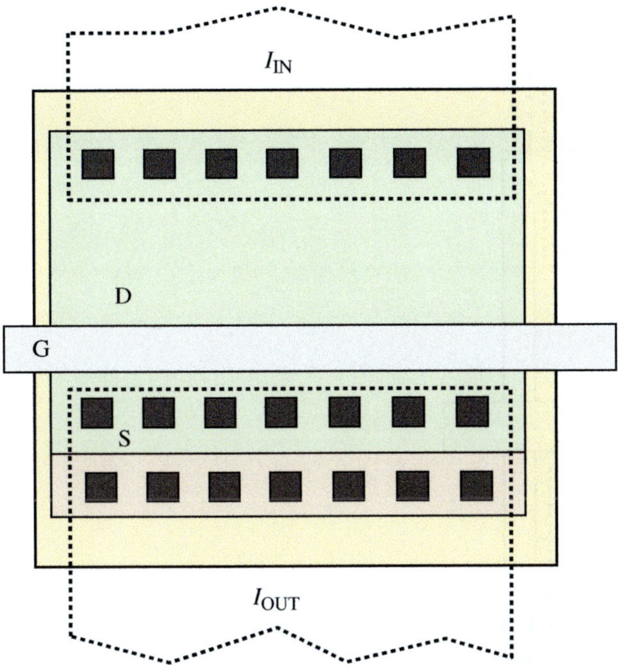

Figure 11.19 An ideal straight ggNMOS ESD metal routhing ensures smooth ESD discharge current pathway, avoiding possible current and thermal cowding.

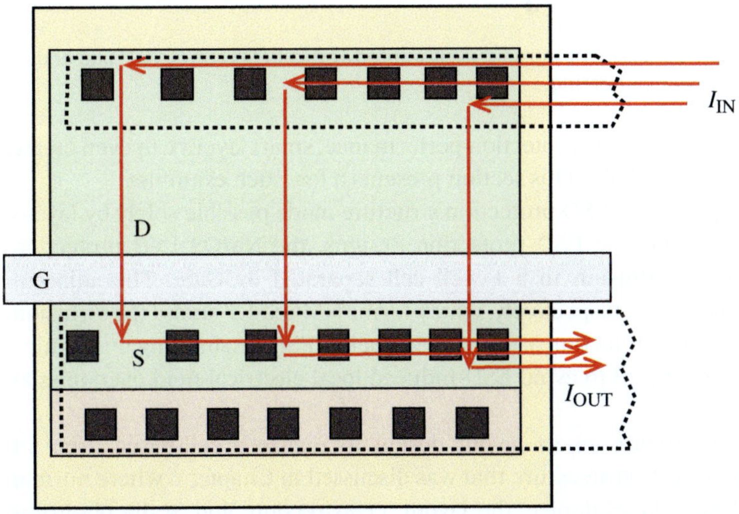

Figure 11.20 Poor contact placement in combination of an anti-parallel metal routing in ggNMOS layout can worsen local ESD current crowding and overheating effects.

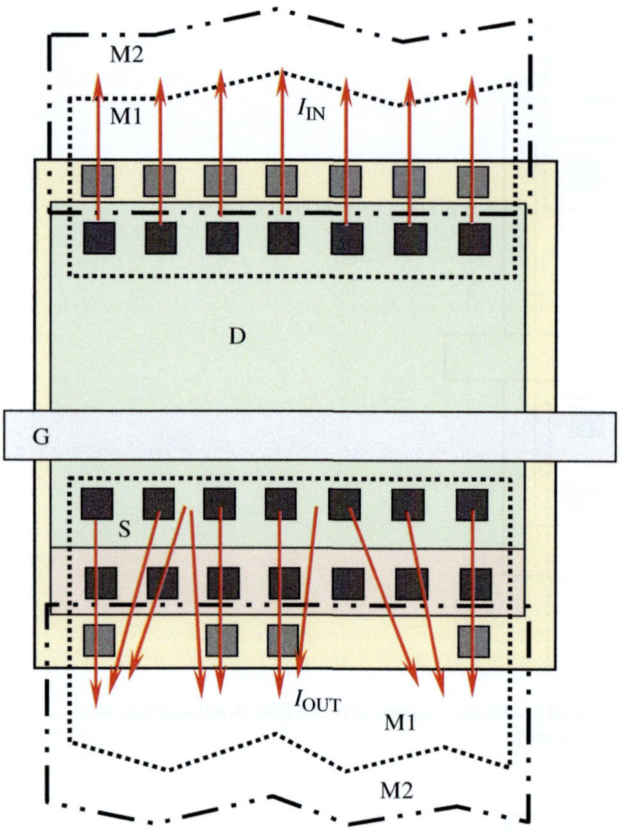

Figure 11.21 Careless contact and via placement introduces unwanted vertical ESD current turning and crowding, resulting in serious local ESD overheating.

designs that may meaningfully improve ESD protection performance, smart layout can even create unique ESD protection structure as a whole. This section presents a few such examples.

Figure 11.22 depicts waffle-type NMOS ESD protection structure made possible solely by layout design [11]. Unlike traditional finger-type ESD protection designs, the NMOS ESD protection structure contains an array of N^+-diffusion in a P-well cell separated by Gate. The adjacent N^+-diffusions serve as Source and Drain, respectively, which forms NMOSFET across the diffusion borders. This waffle-type NMOS ESD protection array achieves better ESD discharging uniformity. Further, the sharp cell corners are cut-off to avoid ESD-induced local electrical field overstress at corners.

The next example shows how to use smart layout design to realize a cell-based array of dual-polarity SCR (dSCR) ESD protection structure that was discussed in Chapter 6 where normal finger layout was presented. Figure 11.23 depicts the layout, cross-section, and Si die photo for a dSCR cell array implemented in a foundry 0.6 μm BiCMOS technology [12]. Figure 11.23a shows one cell of the dSCR array structure comprising a central P^+ diffusion as one terminal, an N^+ diffusion ring as second terminal, and an outer Nwell guard-ring as the boundary between adjacent cells. Figure 11.23b depicts the cross-section of a working dSCR ESD protection device along the A–A' cutline for two adjacent cells. The ESD discharging mechanism of dSCR ESD

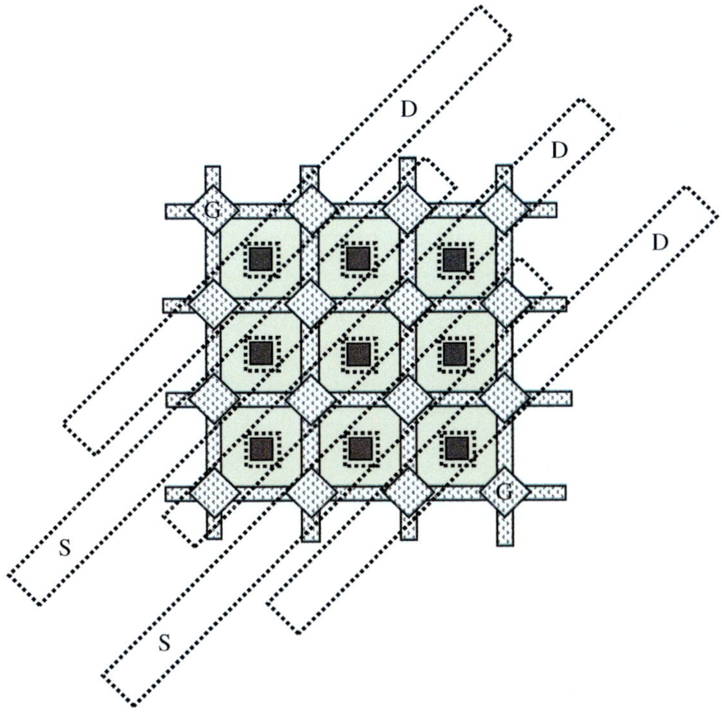

Figure 11.22 A waffle layout design of NMOS ESD protection structure contains a S/D diffusions array.

device was discussed in Chapter 6. Figure 11.23c shows a die photo for a fabricated 3 × 3 dSCR array ESD protection structure. The main benefit of this dSCR cell array is to ensure uniform ESD discharging among small cells, hence achieving high ESDV of ~1.66 V/μm^2 in ESD testing.

The three-terminal multiple-polarity SCR ESD protection structure can also be realized by artistic layout design, as depicted in Figures 11.24 and 11.25, which are pad-centric designs [13, 14]. Putting ESD protection structure surrounding or under a bonding pad has two main advantages: saving Si asset and being layout-friendly. Figure 11.24a depicts one layout design example realizing the three-terminal all-in-one (all-mode) SCR ESD protection structure with a die photo shown in Figure 11.24b for a design fabricated in a foundry BiCMOS technology featuring a deep N-isolation. The square-bad design passed 14 kV HBM ESD zapping and 15 kV IEC ESD zapping, which is very area-efficient, achieving ESDV ~ 2.8 V/μm^2, more than doubled that of ESDV ~ 1.15 V/μm^2 for a traditional finger-type SCR ESD protection structure fabricated in the same BiCMOS technology [13]. To further improve ESD discharging uniformity and eliminate ESD current crowding, a round-pad layout design for the three-terminal multiple-polarity SCR ESD protection was designed as depicted in Figure 11.25 where the SCR cross-section views for the individual embedded SCR ESD device is shown. The embedded SCR ESD devices can discharge the ESD pulses of different stressing modes, i.e., PD, ND, PS, NS, and DS ESD pulsing modes, respectively [14].

Layout art can be applied to other traditional ESD protection structures to make it surrounding or under a bonding pad too. Figure 11.26 depicts a pad-oriented complementary ggNMOS and ggPMOS ESD protection sub-net in CMOS to protect I/O pad against ESD pulses with respect to V_{DD} and GND buses [7]. The inset shows the cross-section view of the ggPMOS ESD protection device

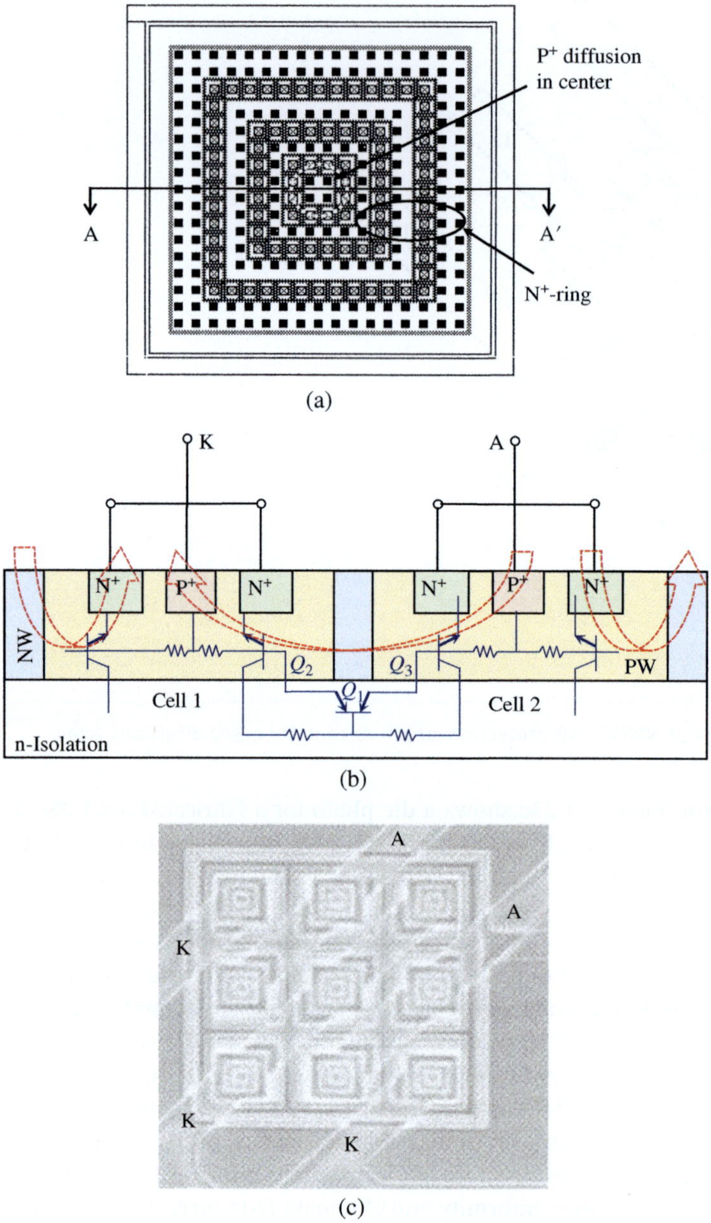

Figure 11.23 A dSCR cell array ESD proteciton structure in BiCMOS: (a) one-cell view, (b) cross-section along A–A′ cutline of two adjacent cells showing a dSCR device structure, and (c) die photo of a 3×3 array.

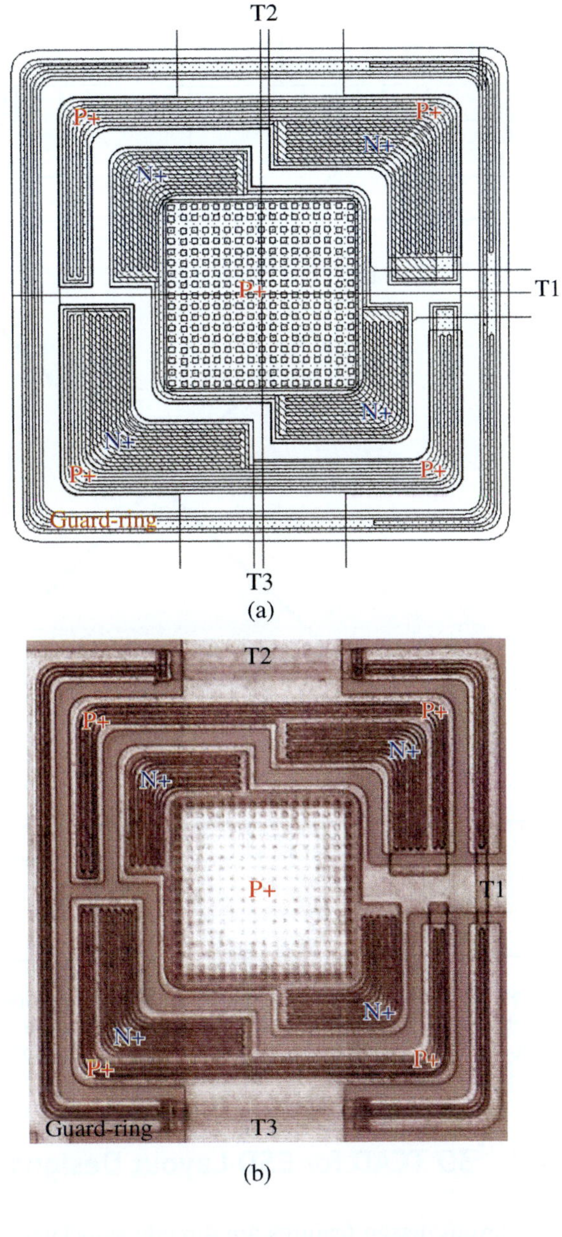

Figure 11.24 A square-pad based three-terminal mutiple-polarity SCR ESD proteciton structure designed and fabricated in a foundry BiCMOS technology: (a) layout, and (b) die photo.

where SCGS, DCGS, and guard-ring can be seen. This layout design has the ggNMOS/ggPMOS ESD protection structures surrounding one pad, making it layout-friendly and large enough to achieve high ESD protection level. Figure 11.27 presents a layout design for a ggNMOS-triggered low-V_{t1} SCR ESD protection structure surrounding a pad implemented in CMOS [7]. The inset depicts the cross-section view, showing critical device dimensions, i.e., L_1 being the ggNMOS channel length and L_2 being the N^+–Nwell spacing. These device dimensions must be carefully designed to optimize ESD protection.

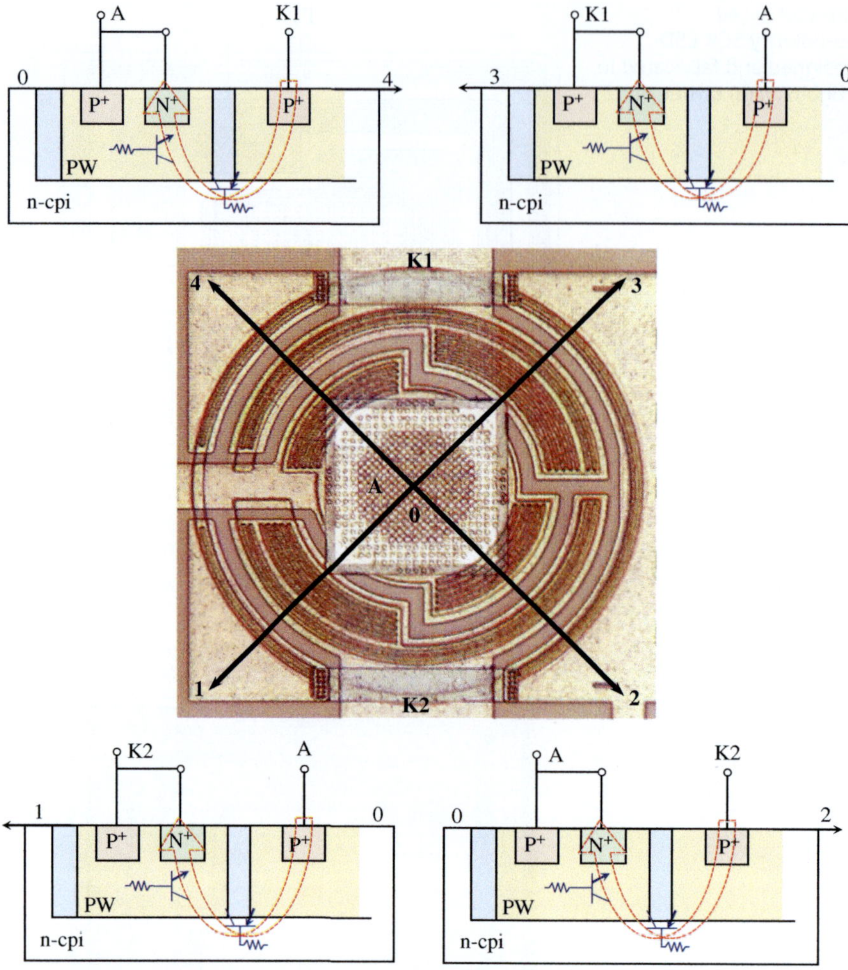

Figure 11.25 A circular-pad based three-terminal multiple-polarity SCR ESD protection structure designed and fabricated in a BiCMOS contains individual embedded SCR ESD devices to discharge PD, ND, PS NS and DS mode ESD pulses in each diagonal direction. Each embedded SCR ESD device is depicted in its cross-section view along the corresponding cutline.

11.4 3D TCAD for ESD Layout Designs

Since layout design features are directly associated with local ESD current crowding and overheating, particularly at the corners and edges of an ESD protection structure, true 3D TCAD ESD simulation can be very useful in guiding ESD layout designs with subtle details. As discussed in Chapter 8, true 3D TCAD ESD simulation is needed in studying local ESD thermal crowding effect, which cannot be revealed by 2D or 2.5D TCAD ESD simulation. This section discusses design of a scalable Sudoku-like diode-triggered low-V_{t1} SCR (DTSCR) cell array ESD protection structure with detailed layout considerations by aid of true 3D TCAD ESD simulation. This Sudoku DTSCR ESD protection array is implemented in a foundry 22 nm fully depleted silicon-on-insulator (FDSOI) CMOS technology [15]. As discussed, SCR ESD protection structure is generally ESD-robust and area-efficient. However, SCR devices in CMOS suffer from very high

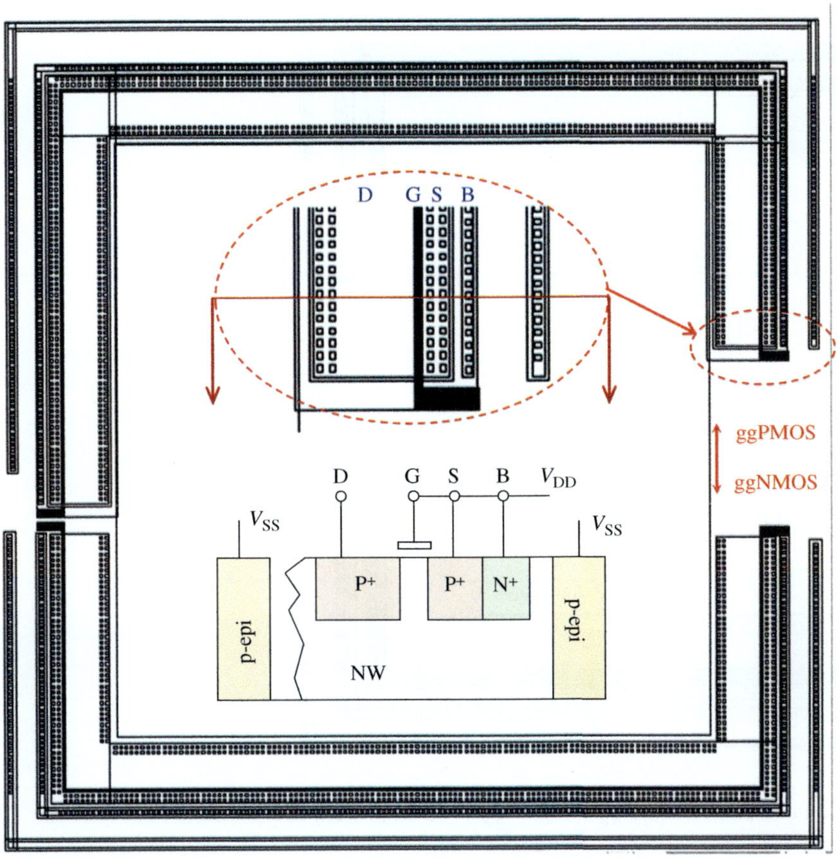

Figure 11.26 Illustration of a pad-oriented complementary ggNMOS/ggPMOS ESD protection structure in CMOS.

V_{t1}, making SCR ESD protection devices not suitable for many LV CMOS ICs. DTSCR utilizes diode(s) to assist ESD triggering, hence achieving very low ESD V_{t1}. On the other hand, like most traditional ESD protection structures, SCR and DTSCR ESD protection structures commonly use long multiple finger layout patterns, which is neither area-efficient nor layout-friendly for a chip. Cell-based array ESD protection structures are often used to improve ESD discharge uniformity, hence, improve ESD robustness and area efficiency, while being layout-friendly. With this in mind, a Sudoku-like DTSCR ESD array structure is designed in a 22 nm FDSOI CMOS. True 3D TCAD ESD simulation is used to analyze the corner/edge effect, i.e., local ESD current crowding and overheating in a real-world DTSCR ESD protection structure. The TCAD-based layout design analysis also serves to realize scalable Sudoku DTSCR ESD array design by layout design optimization.

Since a DTSCR ESD protection structure consists of a SCR ESD protection core and a trigger-assisting diode(s), it is important to study the Sudoku SCR ESD core device arrays first, which is depicted in Figure 11.28 where a 3 × 3 SCR array is shown in its 3D device structure by 3D TCAD simulation, including cross-section of SCR structure, and layout top view with embedded sub-circuit. The Sudoku SCR ESD array has two different cells, each has a central P^+ diffusion pick-up surrounded by an N^+ diffusion ring inside a Pwell and a central N^+ diffusion pick-up surrounded by a P^+ diffusion ring inside a Nwell, respectively. A working SCR ESD device

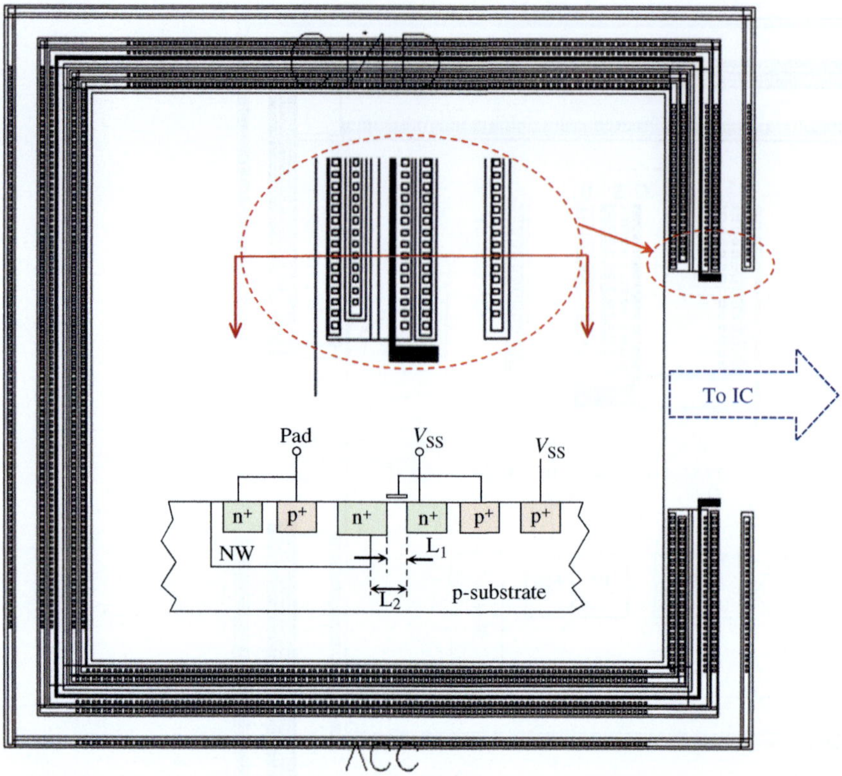

Figure 11.27 Illustration of a pad-oriented ggNMOS-triggered SCR (LVSCR) ESD protection structure in CMOS.

is formed across the boundary of two adjacent SCR cells of different types, with two electrodes being anode (A) and cathode (K). As per the PDK Design Rules, the dimensions for the central P^+/N^+ diffusions and the surrounding N^+/P^+ diffusion rings are set to 1 μm. STI separates the central P^+ and its surrounding N^+-ring (same for the central N^+ and its surrounding P^+-ring), defined as the inner-cell isolation (STI2). The inter-cell isolation is provided by STI1. Design optimization sets STI1 = STI2 = 1 μm. For a large Sudoku SCR ESD array, each inner ESD cell has four across-border SCR ESD devices contributing to ESD discharging, while the edge ESD cells only have two across-border SCR ESD devices contributing to ESD discharging. Compared to traditional long finger-type SCR ESD devices, the Sudoku SCR ESD array is more area-efficient for ESD discharging. The fabricated Sudoku SCR ESD structure is a 3 × 3 array with a dimension of 17 μm each side and an area of 289 μm^2. Transient 3D ESD simulation was conducted by stressing the Sudoku SCR ESD structure with an HBM waveform of 5 kV. Figure 11.29 presents the 3D ESD discharging current density and lattice temperature contours by transient 3D TCAD ESD simulation, which shows the critical 3D ESD discharging behaviors across the ESD cell boundaries. It is observed that, for the inner ESD cells, transient ESD discharging current and heating peak across the cell borders. However, the outer edges of the edge cells do not contribute to ESD discharging, therefore, showing lower ESD heating. The 3 × 3 sudoku SCR ESD array has 12 ESD discharging channels. It is recognized that though a Sudoku SCR ESD array is generally area-efficient in ESD discharging, the outer edges of the edge cells certainly reduce the total ESD discharging area efficiency. A Sudoku DTSCR ESD array consists of a SCR ESD core and an ESD

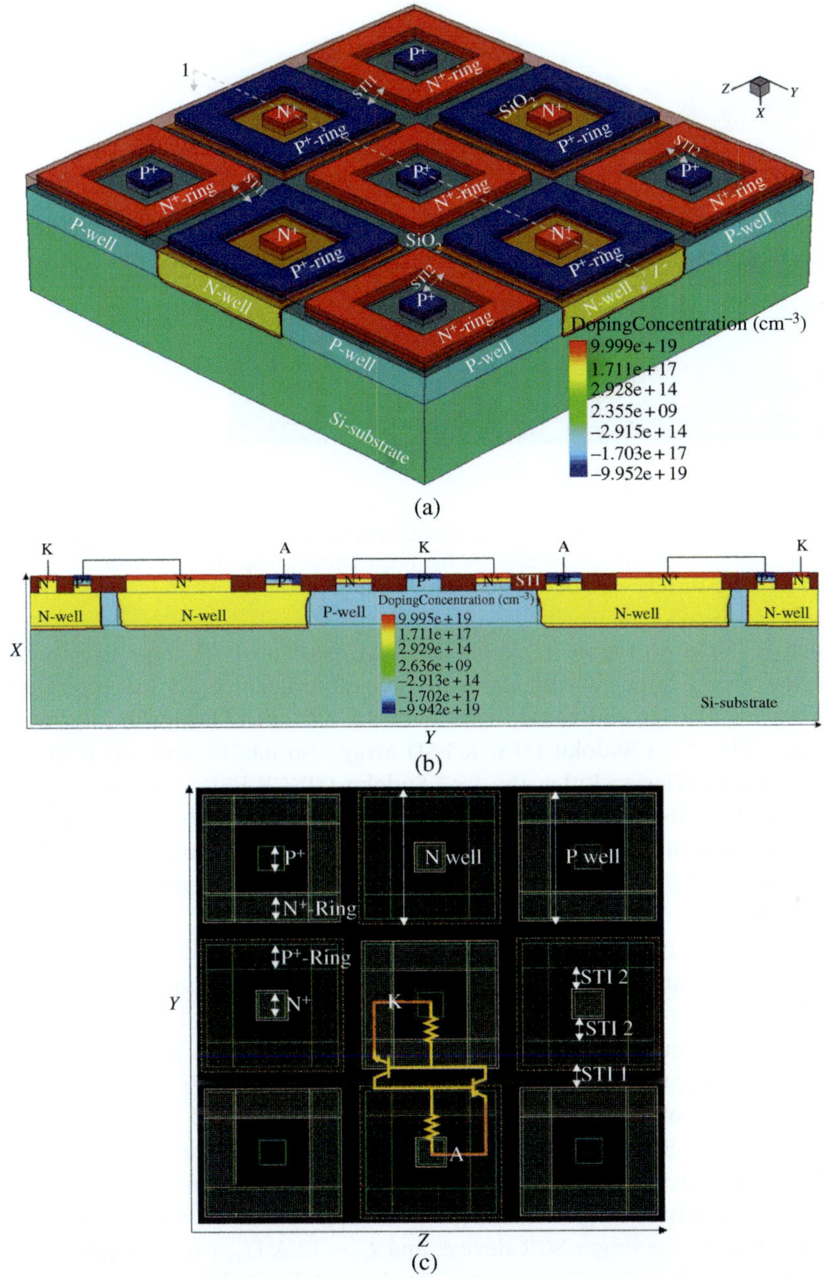

Figure 11.28 A 3 × 3 Sudoku-SCR ESD array by 3D TCAD where dual-polarity SCR ESD devices are formed across the cell borders: (a) 3D view by TCAD, (b) cross-section view on X–Y plane along the 1–1′ cutline, and (c) Y–Z plane layout view with SCR equivalent sub-circuit.

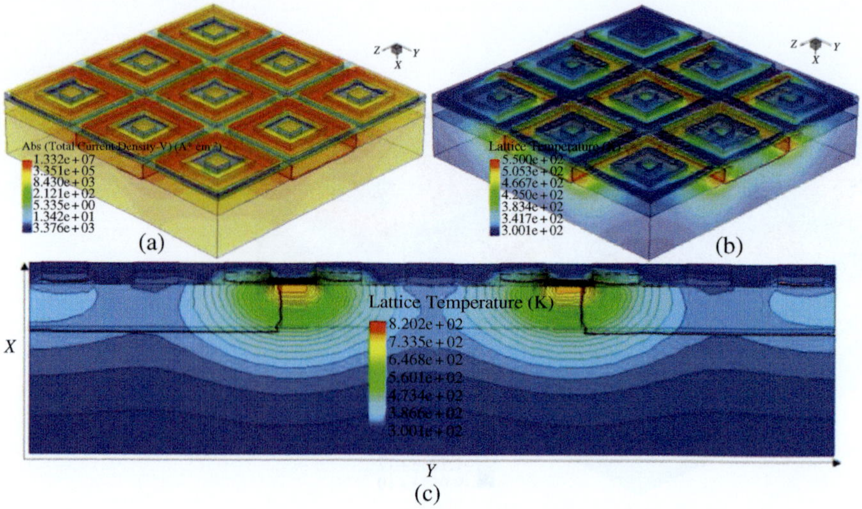

Figure 11.29 3D transient ESD simulation for the 3 × 3 Sudoku-SCR ESD array by 5 kV HBM ESD zapping shows transient 3D ESD discharging I-density map, (b) 3D lattice temperature T_{max}-map, and (c) cross-section view of transient ESD heating.

triggering diode(s), which is shown in Figure 11.30 for its TCAD-generated 3D array structure, including cross-section view and layout view with embedded sub-circuit. Unlike its SCR core, the Sudoku DTSCR ESD array has its edge and corner cells featuring an "open" layout to integrate the trigger-assisting diodes. This 3 × 3 Sudoku DTSCR ESD array also has 12 across-boundary ESD discharging paths. As per the Design Rules, the 3 × 3 Sudoku DTSCR ESD array has a total area of ~484 μm^2, larger than its Sudoku SCR counterpart for the same ESD protection level. Figure 11.31 depicts the 3D transient ESD discharging *I*-density and T-contour maps under 5 kV HBM ESD stressing. The critical 3D ESD discharging behaviors are clearly observed across the cell boundaries where transient ESD heating also peaks.

The design splits for comparison include the following: a 3 × 3 Sudoku SCR ESD array and a 3 × 3 DTSCR ESD array with the same total ESD discharging path width of 60 μm, a long finger SCR ESD device of total layout area of 420 μm^2 and a long finger DTSCR ESD device of layout size of 570 μm^2, targeting on the same ESD protection level for SCR and DTSCR ESD protection structures, respectively. The fabricated ESD structures are characterized by TLP and VFTLP testing, respectively. Figure 11.32 compares the measured ESD discharging *I*–*V* curves by TLP testing for the Sudoku and finger SCR and DTSCR ESD structures, respectively. It is clearly found that Sudoku-ESD arrays achieve much higher ESD I_{t2} and ESD protection area efficiency (J_{t2}) over their finger-ESD counterparts: $I_{t2} \sim 3.5$ A ($J_{t2} \sim 12$ mA/μm^2) for Sudoku-SCR device and $I_{t2} \sim 2.45$ A ($J_{t2} \sim 5.9$ mA/μm^2) for finger-SCR device; and $I_{t2} \sim 3.1$ A ($J_{t2} \sim 6.5$ mA/μm^2) for Sudoku-DTSCR device; and $I_{t2} \sim 2.2$ A ($J_{t2} \sim 3.3$ mA/μm^2) for finger-DTSCR device, respectively. Figure 11.33 presents the measured ESD discharging *I*–*V* curves by VFTLP testing, which also shows higher ESD area efficiency for Sudoku-ESD arrays over their finger-ESD counterparts: i.e., $J_{t2} \sim 43.9$ mA/μm^2 for Sudoku-SCR device and $J_{t2} \sim 32.6$ mA/μm^2 for finger-SCR device; and $J_{t2} \sim 27.0$ mA/μm^2 for Sudoku-DTSCR device and 19.7 mA/μm^2 for finger-DTSCR device, respectively. Measurement also shows that $V_{t1} \sim 1.92$ V for the Sudoku-DTSCR array is much lower than that for Sudoku-SCR array of $V_{t1} \sim 10.8$ V, suitable for LV CMOS ICs. In designs, the

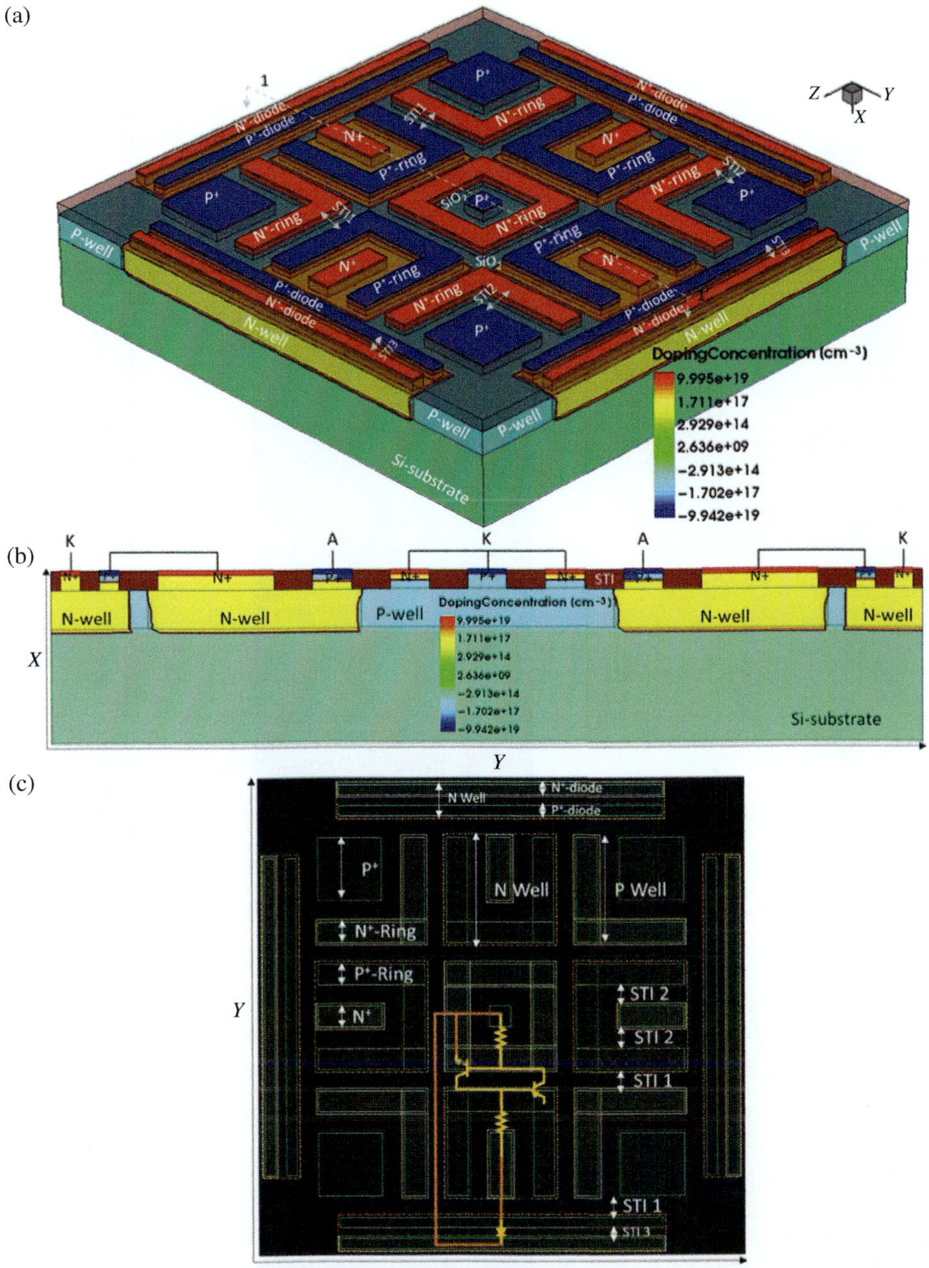

Figure 11.30 A 3 × 3 Sudoku-DTSCR ESD array by 3D TCAD where dual-polarity DTSCR ESD devices are formed across the cell borders: (a) 3D view by TCAD, (b) cross-section view on $X-Y$ plane along the 1–1′ cutline, and (c) $Y-Z$ plane layout view with DTSCR equivalent sub-circuit.

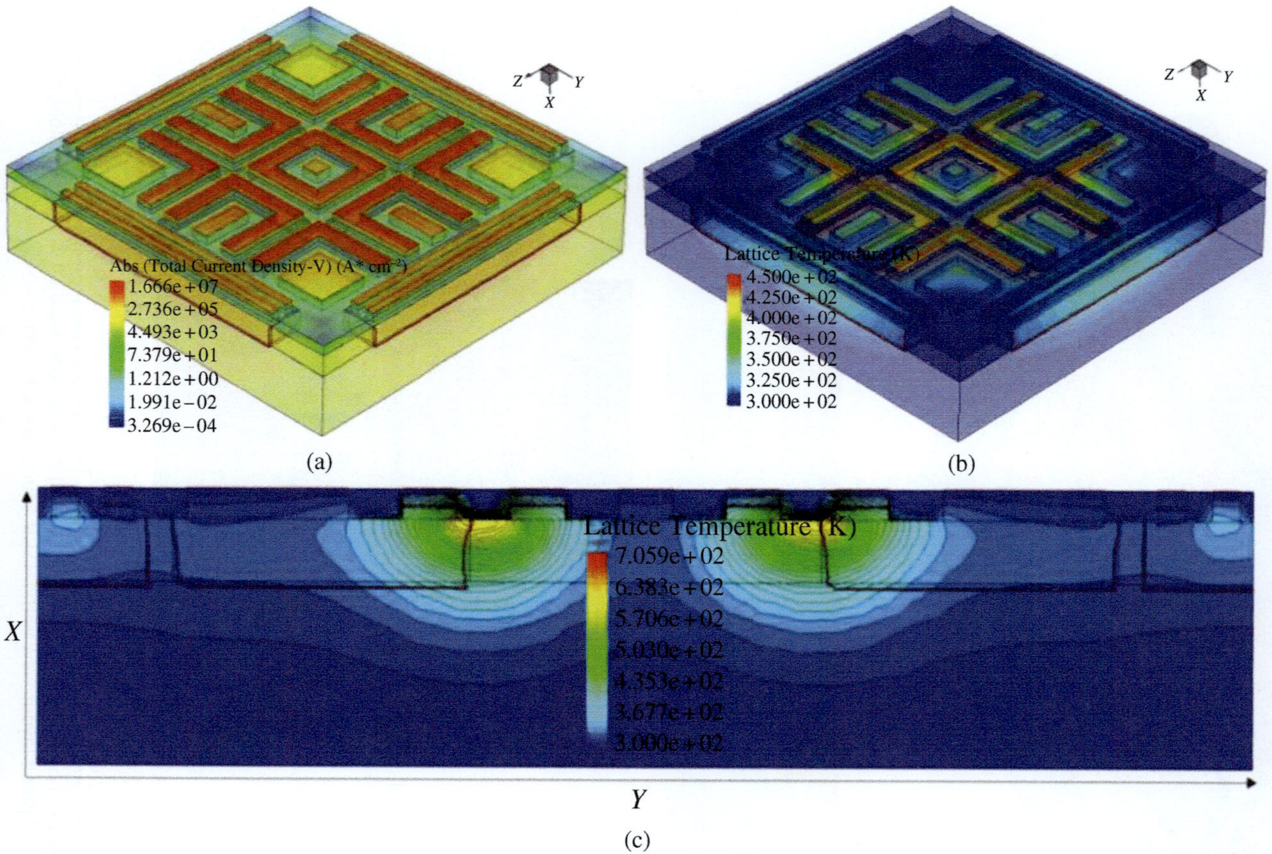

Figure 11.31 3D transient ESD simulation for the 3 × 3 Sudoku-DTSCR ESD array by 5 kV HBM ESD zapping shows transient 3D ESD discharging behaviors: (a) 3D ESD discharging I−density map, (b) 3D lattice temperature T_{max}-map, and (c) cross-section view of transient ESD heating.

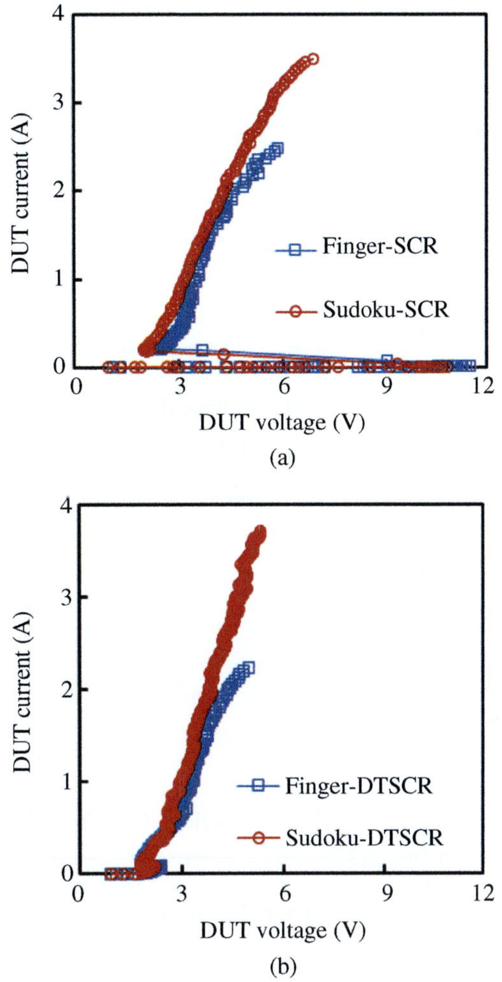

Figure 11.32 TLP-measured ESD discharging *I*–*V* curves for Sudoku and finger ESD protection structures: (a) SCR ESD structures, and (b) DTSCR ESD structures.

V_{t1} of Sudoku-DTSCR arrays can be tunable by using one or more triggering diode(s) to meet the needs of specific ICs.

Design scalability for the Sudoku DTSCR ESD arrays is desired, yet the practical layout design optimization is a rather involving design task, which must consider several design factors: While a Sudoku array is generally area-efficient because the across-border all-perimeter ESD conduction maximizes ESD discharging capacity at the cell level, which however does not apply to the edge and corner cells in a Sudoku array, resulting in reduced ESD area efficiency for smaller Sudoku arrays. Intuitively, smaller cells improve ESD discharging uniformity for a large Sudoku ESD array. However, using smaller cells suffers from relatively more corner current/heat crowding effect at the cell level. Further, one design uncertainty is how to select the cell dimensions in an array, including the line width of P^+/N^+ rings that may affect the equivalent cell ESD conduction path width per side (L_{eq}). A Sudoku ESD array design strategy is developed for $N \times N$ Sudoku SCR and DTSCR ESD arrays, as depicted in Figure 11.34 where the $N \times N$ array contains a 3×3 Sudoku device, the same as that fabricated and discussed previously. The key Sudoku ESD array dimensions include: the cell

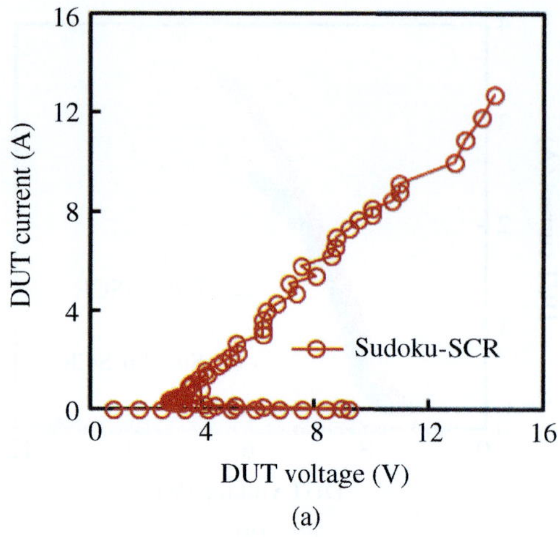

(a)

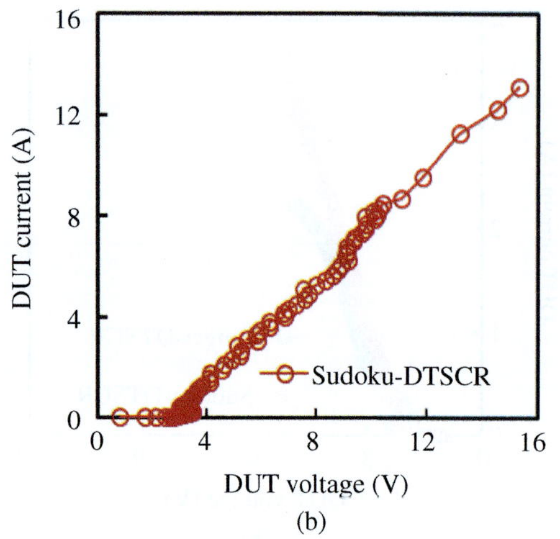

(b)

Figure 11.33 VFTLP-measured ESD discharging I–V curves for Sudoku ESD arrays: (a) SCR ESD protection structure, and (b) DTSCR ESD protection structure.

dimension (L), cell-to-cell isolation spacing (S), equivalent ESD discharging side length of P^+/N^+ rings (L_{eq}) that is varying because the two ring ends are bounded by the inner and outer edges of the P^+/N^+ diffusion rings, which will affect the cell corner effect, the total equivalent ESD discharging width ($W_{\text{eq-total}}$) for the Sudoku ESD array, and the ESD failure current density ($J_{t2\text{-}N} = I_{t2}$ per area) of a Sudoku ESD array (i.e., Sudoku ESD area efficiency). For layout designs, the following formulas hold:

$$W_{\text{eq-total}} = \frac{4L_{\text{eq}}N^2 - 4L_{\text{eq}}N}{2} = 2N(N-1)L_{\text{eq}} \tag{11.1}$$

where $I_{t2\text{-}N}$ is measured by TLP for an $N \times N$ Sudoku ESD array, which is normalized to $W_{\text{eq-total}}$ as

$$I_0 = \frac{I_{t2-N}}{W_{\text{eq-total}}} = \frac{I_{t2-N}}{2N(N-1)L_{\text{eq}}} \tag{11.2}$$

The Sudoku ESD array area efficiency is then derived as

$$J_{t2-N} = \frac{I_{t2-N}}{A} = \frac{2N(N-1)L_{\text{eq}}}{(NL+NS-S)^2} I_0 \tag{11.3}$$

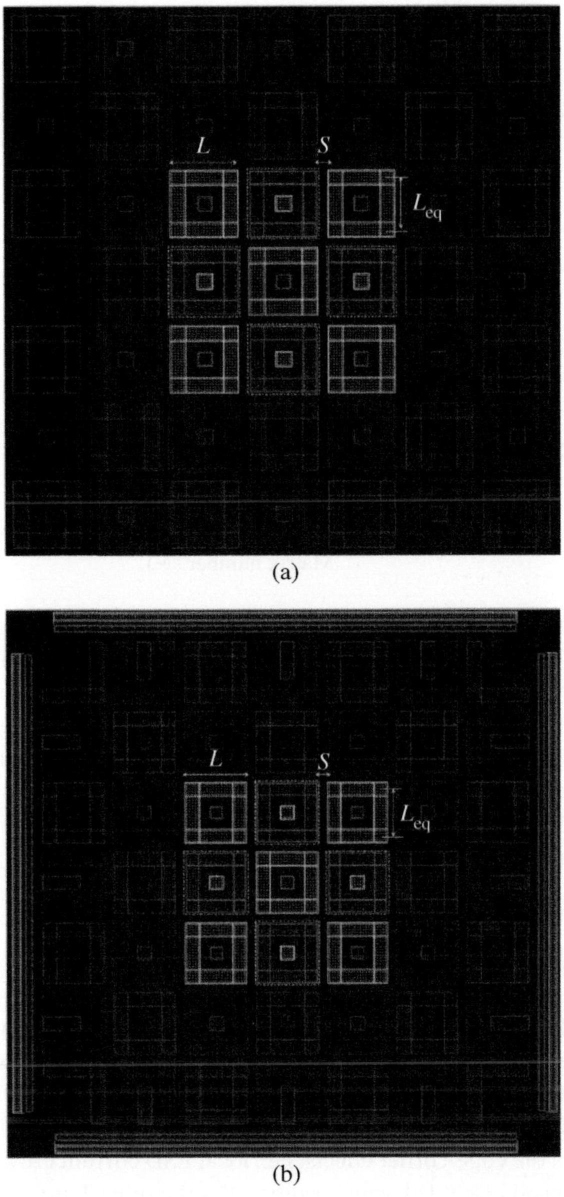

Figure 11.34 Layout of $N \times N$ scalable Sudoku ESD arrays contains a smaller 3×3 array device fabricated in 22 nm FDSOI CMOS: (a) Sudoku-SCR ESD array, and (b) Sudoku-DTSCR.

where the $N \times N$ Sudoku ESD array layout size is

$$A = (NL + NS - S)^2 \tag{11.4}$$

Equation (11.3) implies that the equivalent ESD discharging cell dimension, L_{eq}, directly affects ESD area efficiency, making designing Sudoku DTSCR ESD arrays very involving because one has to decide on not only the cell dimension but also the width and position of the P^+/N^+ diffusion rings, which leads to some uncertainties in layout designs. However, for a large Sudoku DTSCR ESD array with $N \rightarrow \infty$, it gives,

$$J_{t2}|_{N\rightarrow\infty} = \frac{2L_{eq}}{(L+S)^2} I_0 = \frac{I_{t2-N}}{N(N-1)(L+S)^2} \tag{11.5}$$

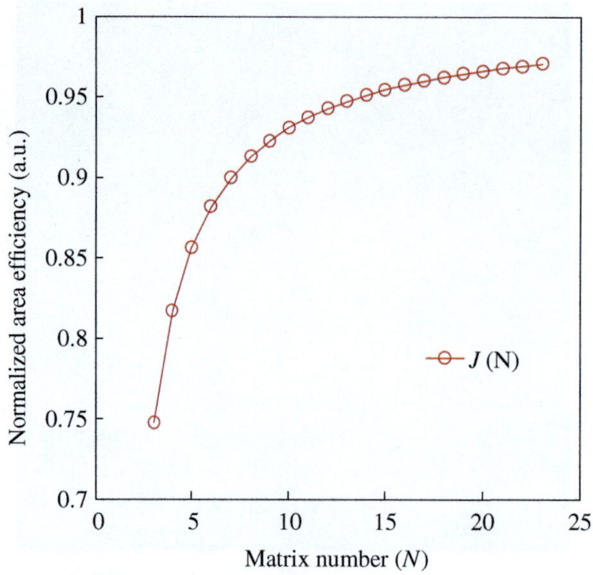

Figure 11.35 Normalized ESD area efficiency for large Sudoku DTSCR ESD arrays ($N \rightarrow \infty$) versus the array scale (N) shows design scalability and suggests array design guidelines. The Sudoku DTSCR ESD arrays are designed in a 22 nm FDSOI CMOS.

which indicates that the pain in accuracy of choosing the L_{eq} will not be a sensitive design factor for a large array. Figure 11.35 gives the normalized ESD area efficiency versus array size for the $N \times N$ Sudoku DTSCR ESD array, which clearly shows that a larger Sudoku ESD array improves its ESD area efficiency substantially, however, the degree of improvement tends to saturate for a very large Sudoku array. This analysis leads to a quantitative design strategy and design guidelines for optimizing ESD area efficiency of scalable Sudoku DTSCR ESD array structures in light of properly handling the subtle design trade-off complexity including cell size, cell layout, and array size.

11.5 Summary

This chapter discusses in details the ESD layout impacts on ESD design performance. In general, different designers will make different layout designs for the same ESD protection structures and schematics, which likely results in different ESD protection performance and Si area consumption. ESD layout design is critically important in practical ESD protection designs because of the edge/corner effects, i.e., local ESD current crowding and overheating (hot spots) are directly related to ESD layout specifics. On the other hand, smart ESD layout design can also save Si area and make an ESD protection structure more layout-friendly. Conceptual ESD layout examples for common ESD protection structures are presented. Special and subtle ESD layout design considerations are discussed in details using ESD design examples. Well-thought-out ESD layout design can improve ESD robustness. Smart and artistic layout design may lead to novel ESD protection structure designs, such as uniform cell arrays and pad-oriented ESD protection designs, both improving ESD robustness and making full-chip layout design easier. 3D TCAD ESD simulation helps to reveals the critical layout impacts on ESD protection performance, which is discussed using a Sudoku DTSCR ESD protection arrays.

References

1 Feng, H., Chen, G., Zhan, R. et al. (2003). A mixed-mode ESD protection circuit simulation-design methodology. *IEEE J. Solid-State Circuits* 38 (6): 995–1006. https://doi.org/10.1109/JSSC.2003.811978.

2 Zhan, R., Feng, H., Wu, Q. et al. (2004). ESDInspector: A new layout-level ESD protection circuitry design verification tool using a smart-parametric checking mechanism. *IEEE Trans. Comput. Aided Des. Integr. Circuits Syst.* 23 (10): 1421–1428. https://doi.org/10.1109/TCAD.2004.833613.

3 Zhan, R., Feng, H., Wu, Q., and Wang, A. (2003). ESDExtractor: A New Technology-Independent CAD tool for arbitrary ESD protection device extraction. *IEEE Trans. Comput. Aided Des. Integr. Circuits Syst.* 22 (10): 1362–1370. https://doi.org/10.1109/TCAD.2003.818140.

4 Xie, H., Feng, H., Zhan, R. et al. (2005). A new low-parasitic polysilicon SCR ESD protection structure for RF ICs. *IEEE Electron Device Lett.* 26 (2): 121–123.

5 Lu, F., Ma, R., Dong, Z. et al. (2016). A systematic study of ESD protection co-design with high-speed and high-frequency ICs in 28 nm CMOS. *IEEE Trans. Circuits Syst. I: Regul. Pap.* 63 (10): 1746–1757. https://doi.org/10.1109/TCSI.2016.2581839.

6 Rountree, R. and Hutchins, C. (1985). NMOS protection circuitry. *IEEE Trans. Electron Devices* ED-32: 910–917.

7 Research Database. Wang's Lab, University of California.

8 Feng, H. (2001). A mixed-mode simulation-design methodology for on-chip ESD protection design. An MS thesis. Illinois Institute of Technology.

9 Wang, A. (1999). ESD protection design using copper interconnects: more robustness and less parasitics. A Project Report to SRC Copper Design Contest.

10 Wang, A., Tsay, C., and Deane, P. (1998). A study of NMOS behaviors under ESD stress: simulation and characterization. *Microelectron. Reliab.* 38: 1183–1186.

11 Baker, L., Currence, R., Law, S. et al. (1989). A waffle layout technique strengthens the ESD hardness of the NMOS output transistor. *Proceedings of EOS/ESD Symposium*, pp. 175–181.

12 Wang, A.Z. and Tsay, C.H. (1999). A compact square-cell ESD structure for BiCMOS ICs. *Proceedings of IEEE BCTM*, pp. 46–49.

13 Feng, H., Zhan, R., Gong, K., and Wang, A.Z. (2001). A new pad-oriented multiple-mode ESD protection structure and layout optimization. *IEEE Electron Device Lett.* 22 (10): 493–495.

14 Feng, H.G., Zhan, R.Y., Wu, Q. et al. (2002). A circular under-pad multiple-mode ESD protection structure for ICs. *IEE Electron. Lett.* 38 (11): 511–513.

15 Li, C., Zhang, F., C. Wang, et al. (2021). Analyze Scalable Sudoku-Type DTSCR ESD Protection Array Structures in 22 nm FDSOI. *Proceedings of IEEE Journal of Electron Devices Society (J-EDS).* September 2021, doi: 10.1109/JEDS.2021.3110955.

12

ESD versus IC Technologies

12.1 IC Technologies and ESD Protection

One simply could not overlook the influences of integrated circuit (IC) technologies on electrostatic discharge (ESD) protection. Since the invention of complementary metal-oxide-semiconductor (CMOS) IC technology in 1963, CMOS scaling of various flavors has been the driving force for continuous advances in CMOS IC technologies, from above 1 μm to 2 nm in critical dimension (CD) today, which has been relentlessly increasing both IC performance and chip complexity. Since ESD events dump large transient energy upon an IC, aggressive scaling down in CMOS technologies certainly makes ICs more vulnerable to ESD stressing, leading to higher revenue losses as chips become more and more complicated and costly. On the other hand, simply shrinking an ESD protection structure following the same CMOS scaling path will certainly reduce the ESD robustness of the same ESD protection structure from generation to generation. In one early study, P^+/Nwell diode ESD protection devices made in IC technologies along the scaling path from 1.2 to 0.25 μm nodes were used as a benchmark to evaluate technology scaling impacts on ESD protection. As summarized in Table 12.1, keeping the same finger size of these ESD protection diodes while otherwise following the technology scaling features, the ESD protection level of those ESD didoes decreases monotonically, at no surprise [1]. Yet no worry, the world is not ending. Over years, active research in ESD protection designs has allowed the industry to continuously meet the ESD protection needs for advanced ICs, though not easy. Nevertheless, it is clear that IC technologies can directly affect ESD protection designs and performance, which must be considered.

12.1.1 ESD Metal Interconnects

ESD failures in metal interconnects account for a large portion of ESD damages to ICs, which has been a constant design pain to IC engineers. Often, designers devote their energy to design an ESD protection structure, i.e., a bare Si device, to make the device ESD-tough. Frequently, a tested individual ESD protection structure is applied to an IC, but fails ESD zapping at chip level due to damage to ESD metal interconnects. Simply put, the risk to ESD metal interconnects is often overlooked, but should not be overlooked in IC designs. An IC engineer often thinks that simply following the ESD metal Design Rules (DR) in the PDK is sufficient. In reality, two issues may exist in ESD metal design considerations. First, the ESD current handling capability of metals, as well as contacts and vias, is not accurately considered in ESD protection designs. In core ICs, layout versus schematic (LVS) and design rule checking (DRC) of ESD metal interconnects is used to make validate circuit schematics and CD compliance, which unfortunately is not enough for dealing with

Practical ESD Protection Design, First Edition. Albert Wang.

Table 12.1 Summary of technology scaling impact on P+/Nwell ESD diodes following the same scaling features.

Nodes (μm)	1.2	0.7	0.5	0.25
t_{ox} (Å)	235	150	135	70
Epitaxy (μm)	12	2.5	2.0	2.0
Well depth (μm)	4.0	1.4	1.2	0.9
L_{eff} (μm)	1.0	0.7	0.5	0.25
P^+ diode (μm)	4.0	2.0	2.0	1.5
Ti:Si	No	Yes	Yes	Yes
STI	No	No	Yes	Yes
ESD diode L (μm)	74	60	43	30
ESD diode W (μm)	120	120	120	120
ESDV (kV)	**9.0**	**6.5**	**4.3**	**3.0**

very high ESD currents (i.e., ESD metal current handling capability) in metal interconnects. Second, even if a designer wants to carefully design the ESD metal interconnects, it may be difficult to do so because metals, as well as contacts and vias, are often not fully characterized for their ESD current handling capability by transient ESD zapping test. Indeed, the PDK DRs offer some guidelines on current handling capability of metals; however, the PDK Specs for metals/contacts/vias are generally characterized using accelerated DC/AC stressing measurements, which does not reflect their transient ESD current handling capacity at all. Further, many PDK DRs for ESD metals set a universal specific metal width, insensitive to the material properties of metals (i.e., Al or Cu) and contacts and vias (e.g., Ti, W, alloys, etc.). For example, one may see a suggested 20 μm in width for ESD metals in a PDK for many CMOS technologies, regardless of the types of metals and ESD protection targets. This is apparently too rough a DR for designing ESD metal interconnects. On the other hand, transition from Al to Cu metal interconnects in CMOS has substantial impacts on ESD performance. Cu has a melting temperature of ~1357 K, much higher than 933 K for Al, which means using the same metal width will improve ESD protection in CMOS featuring Cu interconnects compared to that using Al metals. Otherwise, a quantitative design for ESD metal interconnects for a given ESD protection level requires much less metals in Cu interconnects than that in Al interconnects, which leads to much less ESD metal induced parasitic capacitance that is very beneficial to high-speed ICs. This is confirmed in a study of ESD protection in a foundry one-poly-six-metal (1P6M) 180 nm CMOS featuring both Cu and Al interconnects [2]. In this study, a combined technology CAD (TCAD) Si-metal ESD simulation technique is used to quantitatively evaluate the ESD current handling capability, which is validated in transmission-line-pulsing (TLP) ESD measurements. The simulated and measured ESD metals, including Al and Cu wires, are compared to PDK DR for ESD metals in terms of the maximum sustainable current density (J_C) for all six metal layers (i.e., M1 = M2, M3 = M4, and M5 = M6 in metal thickness in the 180 nm CMOS used). Figure 12.1 presents the comparison data, which readily shows that the ESD metal DR is too conservative, mainly because they are mostly based on DC/AC metal aging tests, which is very different from that under transient ESD stressing. It also shows that Cu is much tougher than Al in terms of ESD protection. It is further observed that the combined Si-metal TCAD ESD simulation technique is realizable, showing good matching between ESD simulation and testing results.

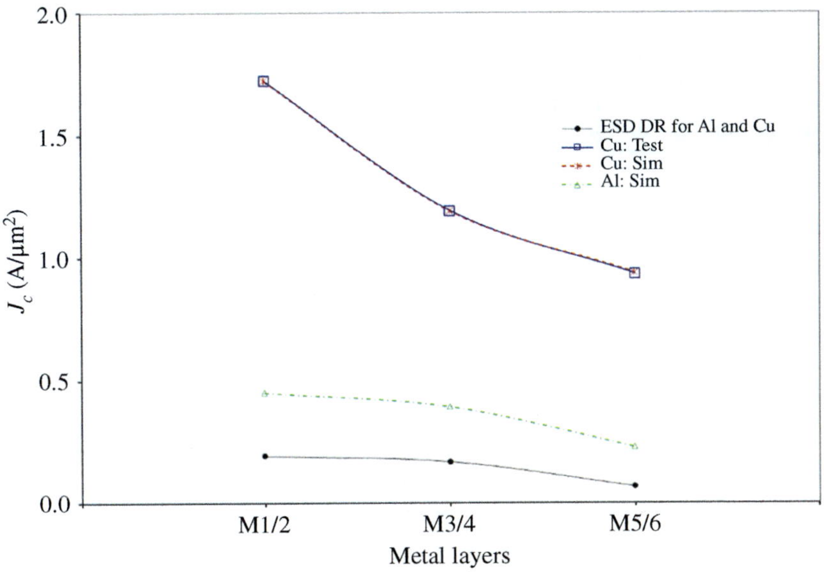

Figure 12.1 Comparison for maximum sustainable current density (J_C) for Al and Cu interconnects based on the DC and AC stressing data typically seen in PDK Design Rules, and transient ESD stressing simulation and ESD testing results by TLP stressing.

12.1.2 Technology-ESD Co-Development

Shrinking has been the driving force that continuously improves CMOS IC technologies, making many "impossible" a reality today, for example, RF CMOS operating well into 100s GHz spectrum. On the other hand, many unique technologies have been developed to dramatically enhance CMOS technologies as well, for instance, lightly-doped drain (LDD), silicidation, SiGe, and SOI CMOS, to name a few. These new process technology features may have positive or negative impacts on ESD protection because, after all, almost all new process technologies were developed to improve core CMOS transistors, not for ESD protection structures. "No free lunch" is a perfect statement here. The main goal of ESD designers is to take full advantage of the positive technology features while overcome those negative factors, in order to design better ESD protection structures or at least keep the ESD protection level as IC technologies continuously advance to smaller nodes. This section discusses various technology impacts on ESD protection designs.

Let us first discuss the LDD feature in CMOS, which is a standard process feature in advanced CMOS technologies to address the hot electron effect. Unfortunately, due to its shallow junction and light doping, LDD in metal-oxide-semiconductor field-effect transistors (MOSFET) ESD protection structures can seriously degrade ESD protection performance [3]. The LDD-induced ESD degradation may be resolved in several ways. For example, a deeper double-doped-drain (DDD) technique may be used to replace LDD to recover the LDD-induced ESD degradation. However, DDD will negatively affect CMOS device performance. Another solution is to use an extra implantation step for the ESD protection structure only, called ESD implant, which aims to make a deeper LDD junction within ESD protection structures for better ESD protection. The most popular solution is to use an extra LDD-blocking mask to cover the ESD protection structure area so that the standard CMOS LDD implant will not be applied to the ESD protection devices, as illustrated in Figure 12.2. Apparently, the extra process steps will introduce extra fabrication costs and make the whole wafer fabrication cycle longer, while recovering LDD-induced ESD protection loss.

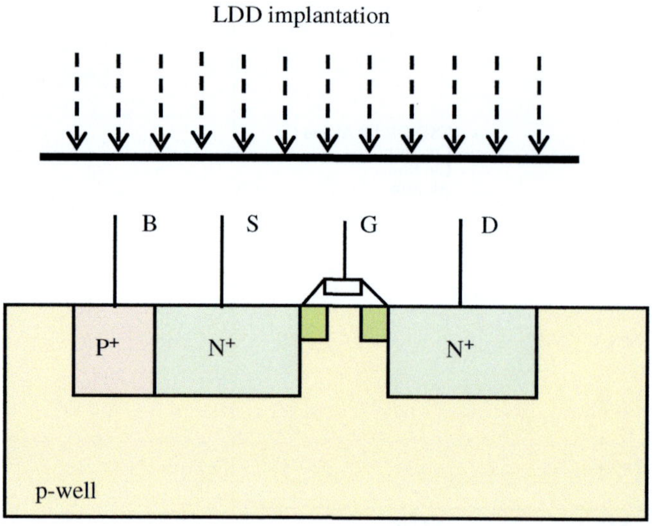

Figure 12.2 An LDD-blocking mask is commonly used to block LDD implant in the ESD protection device area, resulting in a MOSFET ESD protection structure without LDD diffusions.

Sicilidation technology is also widely used in modern CMOS to silicide the Si Drain and Source diffusions, and the poly-Si Gate layers, aiming to dramatically reduce the series resistance by converting a thin surface layer of a thickness of *d* into an ultralow resistive alloy layer as depicted in Figure 12.3a, which can significantly improve CMOS transistor performance. Unfortunately, silicidation can substantially degrade ESD protection performance of MOSFET ESD protection structures. The main reason is that, in a MOSFET ESD protection device, a large *drain-contact-to-gate spacing* (DCGS) distance is required to ensure a ballasting resistor in the Drain extension region that helps to enforce a uniform ESD triggering voltage across multiple fingers of the MOSFET ESD protection structures. However, the benefit of reducing the Drain extension resistance in a MOSFET means the ballasting resistor in a MOSFET ESD protection structure is effectively removed, which will result in nonuniform ESD triggering of the multiple fingers, leading to lower ESD protection level. One easy solution is to introduce a silicidation-blocking mask to cover the MOSFET ESD protection device area during silicidation process, so that while the Drain extension resistance in a normal CMOS transistor is removed, the same Drain extension resistance in the MOSFET ESD protection device remains there, as illustrated in Figure 12.3b. Obviously, using an extra silicidation-blocking mask increases IC fabrication costs and adds more fabrication time.

SOI CMOS has a major advantage over its bulk CMOS counterpart, which is very low parasitic capacitance. This is mainly because it eliminates the body PN junction that inherently produces substantial PN junction capacitance. SOI also reduces junction leakage and, more importantly, prevents the global noise coupling effect. SOI CMOS has been widely used for RF ICs and high-performance chips. However, SOI is inherently poor for ESD protection performance, mainly because the very thin active Si layer (e.g., 5–20 nm in a 22 nm FDSOI CMOS) is sandwiched between two oxide layers on top and underneath, which are both thermal insulators. Therefore, SOI ESD protection suffers seriously ESD protection degradation, up to ~50% [4]. Active research has been done to improve SOI ESD protection. Figure 12.4 depicts a simple solution where a MOSFET ESD protection device is made in the region where the underneath oxide BOX layer is removed; hence, the ESD-induced heat can be easily dissipated downward through the bulk Si substrate, just like in bulk CMOS [4]. Figure 12.5 shows a ggNMOS ESD protection device made in the thin active

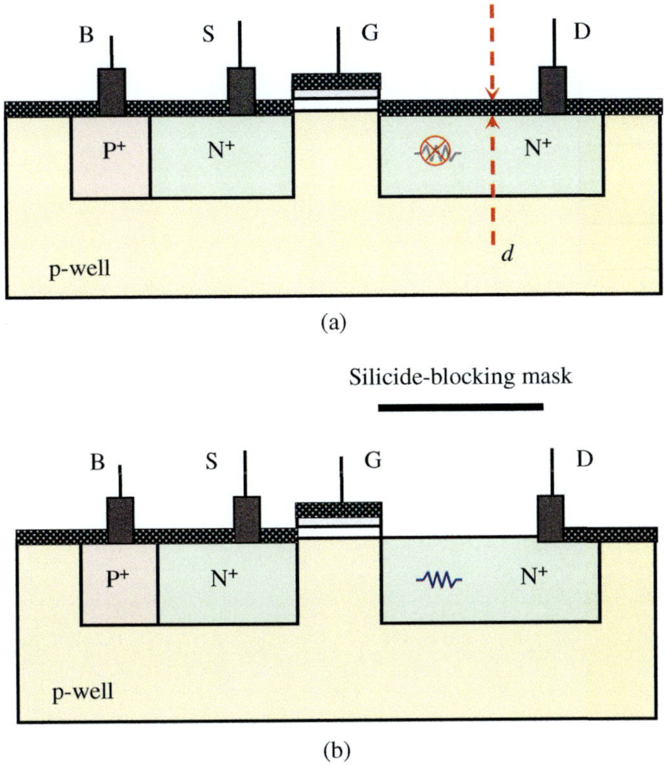

Figure 12.3 Illustration of silicidation impact on MOSFET ESD protection devices: (a) the silicided Drain reduces the Drain extension resistance needed for good ESD protection, and (b) a silicidation-blocking mask can prevent silicidation of the Drain extension in the MOSFET ESD protection area, hence keeps the Drain resistance.

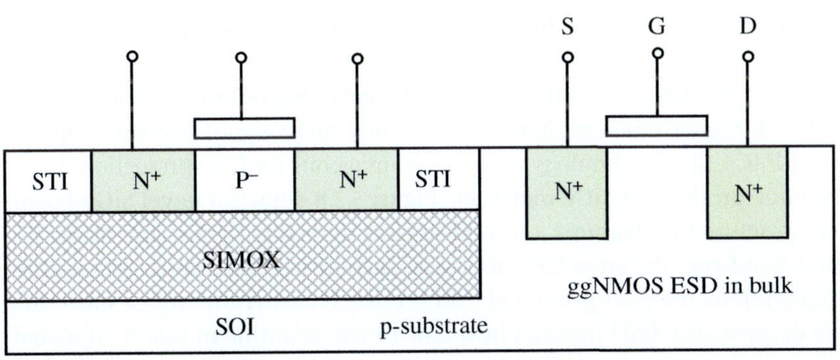

Figure 12.4 A ggNMOS ESD protection device made in SOI CMOS removes the underneath separation by implantation of oxygen (SIMOX) layer to improve its thermal conductivity.

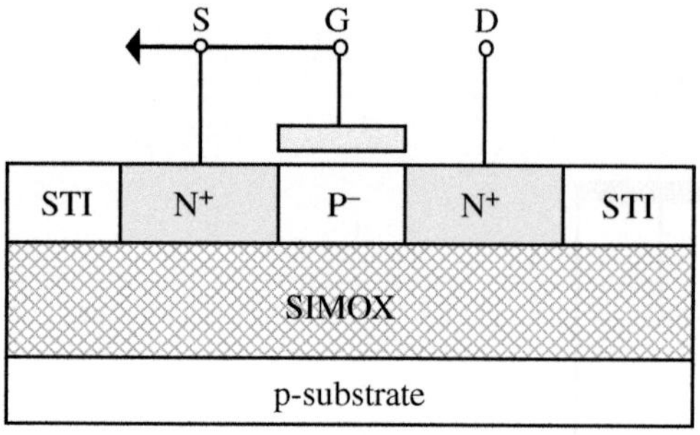

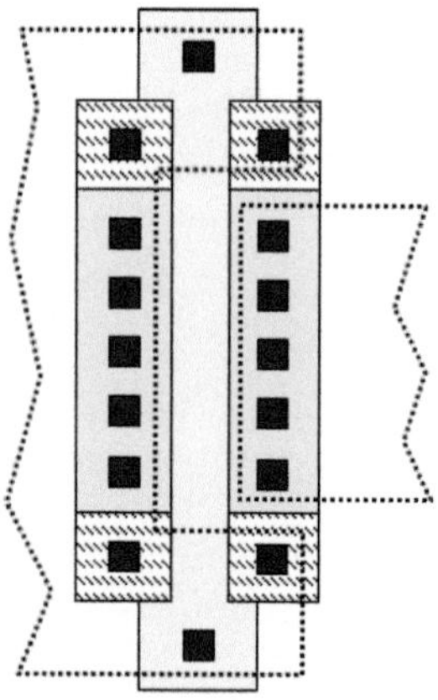

Figure 12.5 Illustration of a ggNMOE ESD protection device made in SOI CMOS: (a) cross-section, and (b) layout view.

Si layer that relies on a lateral NPN to discharge the positive ESD pulse and a body-drain diode to shunt the negative ESD transient, given that a body connection is available [5]. Figure 12.6 shows several ESD protection diode structures made in the thin active Si layers [6]. Figure 12.7 depicts a tunable-V_{t1} SCR ESD protection structure made in the above-Si poly-Si layer, which can be readily migrated into the thin active Si layout to realize high-performance SCR ESD protection devices in SOI CMOS [7].

SiGe technology has advantages over its traditional Si CMOS technologies in that SiGe features narrower band-gap and much higher mobility, hence significantly increases its transistor operation frequencies, good for RF ICs. SiGe technology may be advantageous for ESD protection if one can manage to use the unique property of SiGe materials. Figure 12.8 depicts a novel SiGe-based ggNMOS ESD protection structure that features a thin SiGe layer embedded under the MOSFET channel. Due to its narrow band-gap, the large ESD discharge current will be confined to the thin SiGe layer deep in the Si; therefore, the ESD-generated heat is pushed deeper into the Si substrate, making it much easier to dissipate the ESD-induced heat downward, resulting in much improved ESD protection [8].

From the previous discussions, it becomes obvious that IC technologies have direct impacts on ESD protection, positively or negatively, which have to be addressed in order to keep up ESD robustness for advanced ICs. The conventional approaches of developing and qualifying a process flow for core CMOS device optimization first, followed by evaluating ESD protection performance, identifying any new process induced ESD degradation, and then find ways to fix the ESD degradation problems, is no longer suitable for advanced IC technologies. As IC technology node rapidly advances

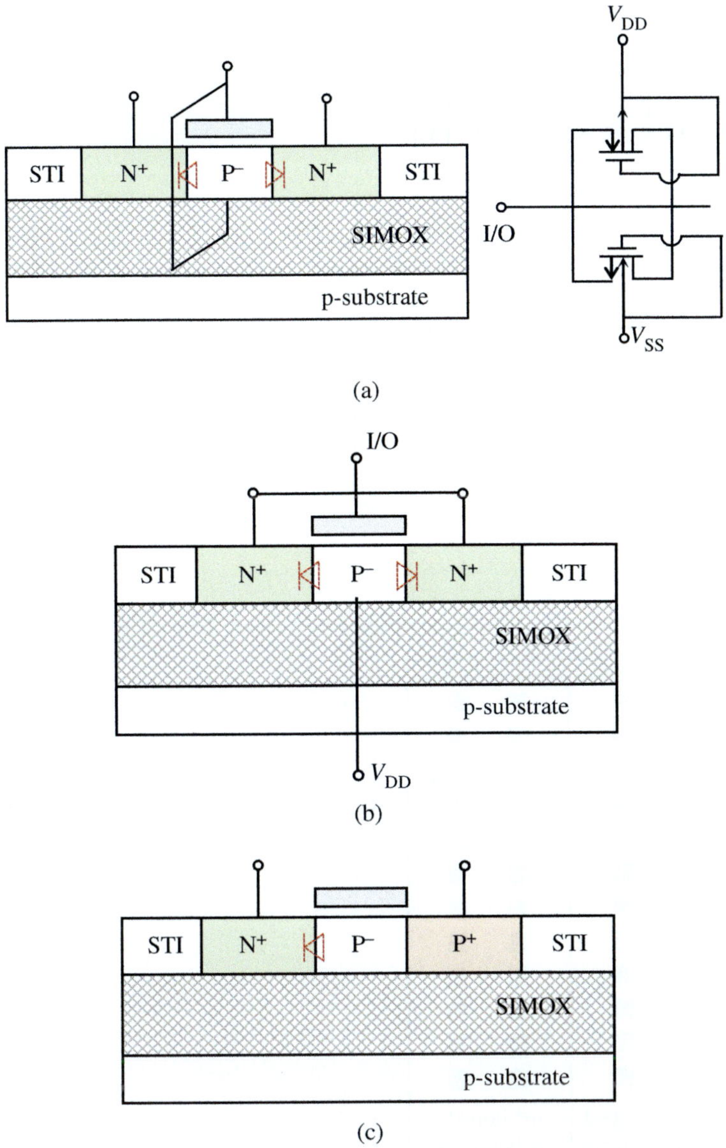

Figure 12.6 Various diode ESD protection devices in SOI CMOS: (a) a double-diode ESD protection sub-net, (b) a grounded-gate diode ESD protection device, and (c) a gated ESD protection diode.

to 2 nm and heterogeneous integration technologies become a key alternative to ensure continuous IC technology advances, the price, in terms of both time-to-market, and development and production costs, of using *the fixing-ESD-problem after process qualification approach* must be changed. The fundamental principle for new IC technology development should follow the *Technology-ESD Co-Development* approach where ESD protection must be considered in the early process development phase and any trade-offs must be well-balanced for simultaneous optimization of both core transistors performance and ESD protection [9, 10]. This is simply analogous to the ESD-RFIC co-design method for whole-chip ESD protection design as detailed in Chapter 10 [11, 12]. It is worth noting that, in new technology development, many process recipes or steps may be adjusted

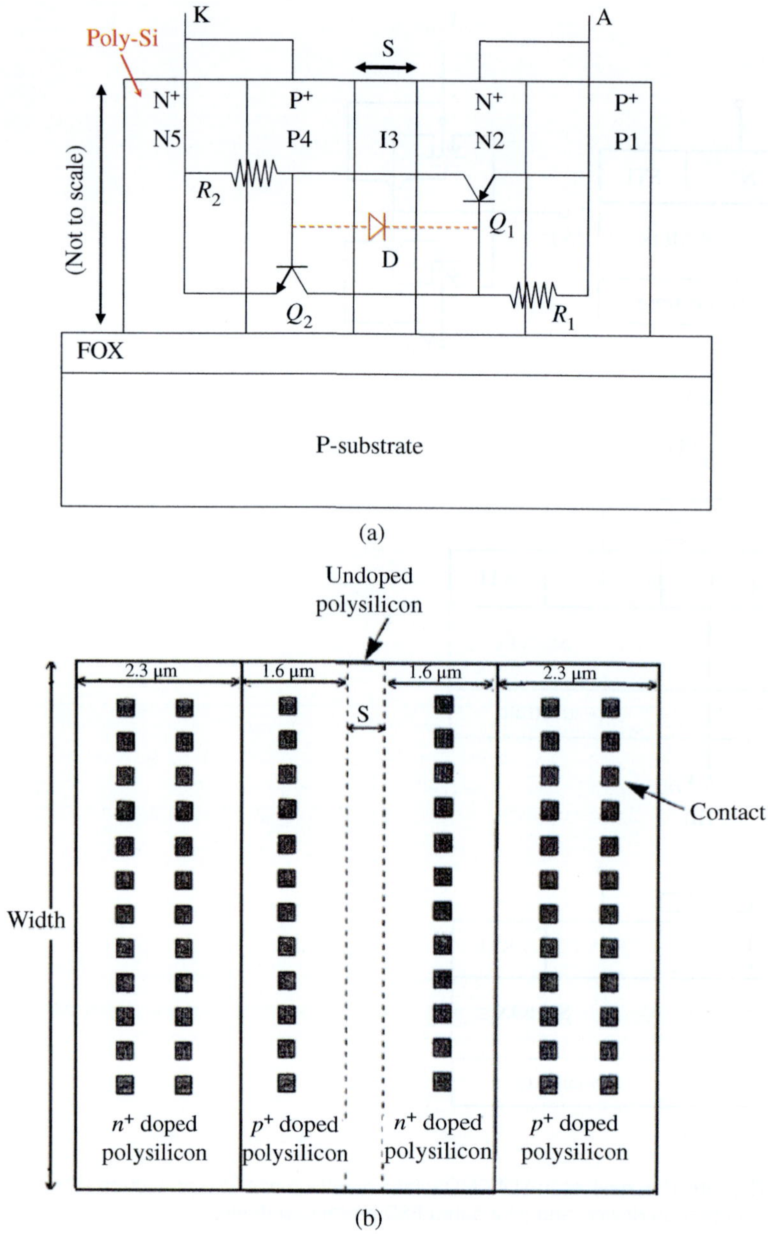

Figure 12.7 A low-capacitance Poly-Si SCR ESD protection structure implemented in BiCMOS BEOL can be readily migrated into SOI CMOS: (a) cross-section, and (b) layout view.

for both transistors and ESD protection performance considerations. For example, LDD doping and shallow junction have negative effects on ESD protection, yet should a low LDD implant dosage of 4×10^{15} cm^{-2} or 5×10^{15} cm^{-2} be chosen for a CMOS technology? Often, one may discover that a minor variation in LDD doping may improve ESD robustness without affecting transistor specs much, at least not unacceptable to ICs. In fact, it might not be a surprise to find that when asking a process engineer about how a recipe value is selected exactly what it is, the response could be more

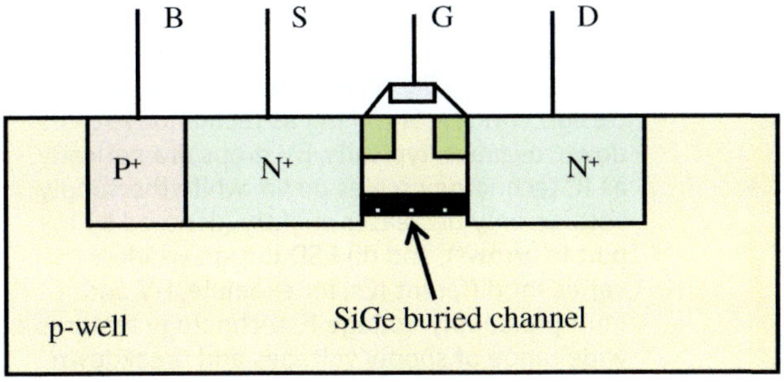

Figure 12.8 A SiGe-based ggNMOS ESD protection structure features a thin buried SiGe layer underneath the conduction channel for improved ESD heat dissipation.

qualitative, than quantitative, or even simply be referred to as a "historical" decision. This said, it must be pointed out again that quantitative design and development approach aided by TCAD simulation is critical to new IC technology development, advanced IC designs, and ESD protection designs, today more than ever before, as the technologies advance to 2 nm node and beyond, and IC chips become bigger and more complicated. Treating process development, IC design and ESD protection design "separately" will result in unbearable losses in time-to-market and production costs. The new IC era requires technology-ESD co-development, as well as ESD-IC co-design.

12.1.3 Graphene Heat Spreading

IC technologies utilize many different materials including semiconductors, metals, and dielectrics. Some new materials, not conventionally used in CMOS technologies, may be used to enhance ESD protection robustness, and potentially, can also improve core devices or even introduce new device functionalities. Exploring new materials for new devices through heterogeneous integration is considered a main pathway to future IC technologies beyond 28 nm node. For example, nanotubes have been investigated for possibly replacing Cu as metal interconnects. The excellent electrical and thermal conductivity of nanotubes may potentially be used as ESD metal interconnects, and vertical contacts and vias, to mitigate the risk of ESD failures in metal interconnects and contacts/vias. Another very promising materials for ESD protection is graphene. Graphene has been explored to make nanosheet FET devices. On the other hand, due to its unique materials properties, such as ultrahigh electron mobility, ultrahigh thermal conductivity, superior Young's Modulus and superior mechanical strength, graphene has been explored to make mechanical switch for ESD discharging and replacing traditional ESD metal interconnects, which will be discussed in details in Chapter 15 [13]. More interestingly, graphene sheet can be embedded into a CMOS device to serve as a superior thermal spreader to quickly remove the heat generated by large ESD discharge currents. It takes wisdom to explore new materials for new ESD protection mechanisms and structures where, again, ESD-technology co-development plays a critical role in discovering future ESD protection solutions.

12.2 Technology Affects ESD Design Window

From ESD protection design theory, one of the most significant, yet inevitable, technology impacts on ESD protection designs is on the ESD Design Window. As depicted in Figure 12.9, any good on-chip ESD protection solution must meet the relevant ESD Design Window established by the

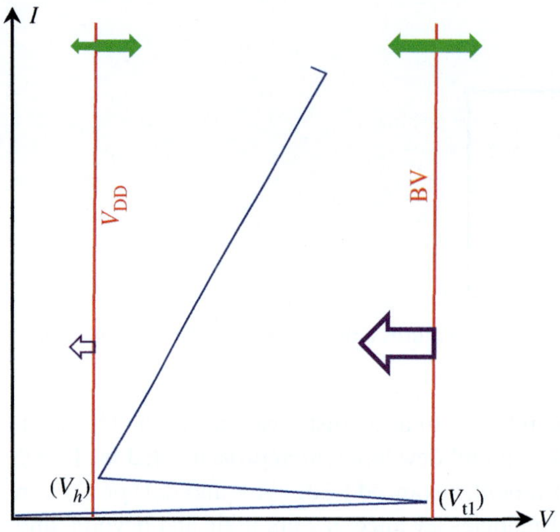

Figure 12.9 An ESD Design Window can change as per technologies and ICs, which brings in two ESD design uncertainties: (i) ESD Design Window Shrinking as technology scales down, because, typically, BV drops dramatically as IC technology scales down, while the supply voltage only decreases slightly (denoted by the purple arrows), and (ii) ESD design window varies for different ICs, for example, HV and multiple-supply-voltage IC technology features a wide range of supply voltages and breakdown voltages (denoted by the green arrows).

ESD-critical parameters. Simply speaking, the ESD discharge I–V curve must be confined within a safe voltage (e.g., BV, on the upper end) and the supply voltage (e.g., V_{DD}, on the lower end), including a safety margin [14]. Careful quantitative ESD protection design should ensure $V_{DD} < (V_{t1}, V_h, V_{t2}) <$ BV on a chip. The challenge is that, as IC technology scales down aggressively, e.g., down to 3 nm node in pilot product today, the breakdown voltage drops dramatically (e.g., $BV_{GS} \sim 0.95$ V for logic core in a 5 nm FinFET CMOS, much lower than $BV_{GS} \sim 3.2$ V for logic core in 28 nm FinFET CMOS), while the supply voltage only decrease slightly due to the CMOS technology nature (e.g., $V_{DD} \sim 0.75$ V in 5 nm CMOS, not much lower than $V_{DD} \sim 0.85$ V in 28 nm CMOS). This leads to the *ESD Design Widow Shrinking* Effect, which makes on-chip ESD protection design extremely challenging [15]. On the other hand, another main pathway to advanced chips, other than simply technology scaling, is to realize chip heterogeneity, e.g., mixed-signal and RF circuits, and using MEMS, sensors, photonic, magnetic, and bio-inspired devices. Such chip complexity can be a design nightmare for ESD protection because the ESD Design Window may vary across a chip. For example, HV and multiple-voltage mixed-signal ICs have multiple power domains, each may have its own unique ESD critical parameters, hence different local ESD design windows. As illustrated in Figure 12.9, HV and multiple-supply ICs may have a wide range for the BV specs and V_{DD} and V_{SS} values on a chip. Particularly, HV ICs require a large V_h, which is often very difficult to realize in IC designs as discussed in Chapter 7 [16]. In principle, the relationship between IC technologies, IC types, and ESD Design Window must be carefully considered in designing complex chips in advanced IC technologies.

12.3 Lowering ESD Protection for Advanced ICs?

When ICs were born in 1958 (hybrid in Ge) and 1959 (monolithic in Si) and CMOS process was created in 1963, the IC-ers probably never expected that the life could be so bumpy and sometimes very dangerous due to various ESD failures. In 1970s, ESD failure was recognized as a reliability problem to ICs, leading to R&D efforts to develop ESD protection measures. Through 1980s until mid-1990s, significant efforts have been devoted to understand the ESD failure problems and to develop various on-chip ESD protection solutions, which has continuously improved the ESD robustness for ICs in terms of ESD protection ratings. Over years, the industry commonly considers 2 kV HBM ESD protection level as a baseline ESD protection requirements for IC products, while much higher ESD Specs, e.g., 8 kV + HBM, have been achieved for many analog ICs. For CDM ESD rating, 500 V has been the baseline for most IC products. The same trend for more CDM ESD protection robustness, 500 V and higher, has been seen for ICs. However, since early 2000's, the never-ending pursuit for ever higher ESD protection Specs, typically by HBM and CDM testing, have become increasingly difficult to ESD designers as IC technologies rapidly scale down, from 250 to 3 nm in pilot today. Meanwhile, IC chips become larger (thousands of pins) and more complex (e.g., multiple cores), running faster (>100 Gbps) and operating at higher frequency (1s–100s GHz). The main technical barrier for achieving better on-chip ESD protection is not that individual/standalone ESD protection devices could not be made tougher, but that making an IC more ESD-robust at chip-level becomes very challenging. The root cause to such ESD design challenge is that any ESD protection structures, mostly in-Si PN-junction-based active devices, introduce significant ESD-induced parasitic effects, including C_{ESD}, I_{leak}, ESD self-generated noises and C_{ESD}-induced global noise coupling, which can seriously affect IC performance in terms of speed, frequencies, bandwidth, and data rates. For example, even an C_{ESD} of a few tens of *fF* may be deadly to a millimeter wave RF IC (28–68 GHz for wireless mobiles) and high-speed I/O circuits (100 Gbps and beyond). Consequently, there has been a recommendation to *lower* the ESD protection ratings for advanced IC products, i.e., 1 kV HBM and 250 V CDM. It even aggressively suggests that, with good ESD control in place, the ESD protection ratings can be dropped to 500 V for HBM and 125 V for CDM ESD qualification for future ICs [17–20]. The basis for such recommendation of lowering the ESD ratings for advanced ICs follows: First, "good" ESD control measures currently exist in IC fabrication environments, which reduce the ESD failure risks. Second, there is lack of correlation between device (IC) level ESD protection ratings and system (end products) level field return of products due to ESD failures. Third, the focus of product ESD safety should migrate from device (IC) level to system (end product) level ESD protection. Unfortunately, such analysis and recommendations fail to address the real problem, which is that the same or even higher ESD risks still exist in the real world regardless of IC technology advances (scaling), chip performance (*x*-GHz or *y*-Gbps), and complexity (*z*-cores). Further, any "good" ESD control measure is indeed important in the production environments, but mostly irrelevant to the end users. It is important to understand that ICs will be used to make end products that will mostly be used by end users who have been the enablers for today's IC prosperity. For example, consumer electronics represent a major portion of electronics system products, such as smartphones, tablets, and game consoles, where the consumers cannot be asked to be ESD-aware or ESD-controlled (e.g., wears a wrist strap) when using the products. In fact, the risk of ESD failures is actually higher for such consumer electronics because the devices are hand-handled by human beings, and hence will suffer from ESD zapping every second. Sure, for the enterprise systems, such as data centers and mainframe super computers, after production and installation, human touching becomes rare, therefore, tough ESD control can be utilized throughout the system's life cycle. On the other hand, the lack of clear correlation between IC level ESD ratings and system level ESD failures (i.e., product field returns) remains a complicated problem with many unknown factors. For example, ESD testing models (e.g., HBM, CDM) for ICs are different from ESD measurement methods for systems (e.g., IEC61000-4-2); therefore, one should be very cautious in trying to correlate (or not) these ESD failure data. Further, an increased HBM ESD rating of one specific IC may not be directly translated into better system level ESD protection, e.g., field return rates of products due to ESD-like failures because a system contains many different IC devices and much more something else. So, any trial of directly mapping the failure data of systems products with a specific IC can be over-simplified. Just because one could not see a "link" for the time being does not means the relationship does not exist in reality. The above said, it is indeed the fact that ESD protection design for more sophisticated ICs in more advanced IC technologies is becoming extremely challenging. It is therefore imperative to fully understand the various ESD failure physics and testing mechanisms at both IC and system levels, and more importantly, to explore truly novel or revolutionary ESD protection solutions. Recently efforts in this direction include developing graphene-based ESD protection structures and investigating what is fundamentally wrong with the existing pad-based CDM ESD protection solutions and CDM ESD testing methods [13, 21, 22].

12.4 Summary

This chapter discusses how IC technologies may affect ESD protection, positively and negatively. The nature of the transient-electro-thermal-materials-process-device-circuit-layout-system multiple coupling effects for ESD protection makes on-chip ESD protection designs highly depending upon IC technologies. Most likely, new process features adopted for advanced IC technologies for a better transistor, such as LDD implant and silicidation, may substantially degrade ESD protection device performance. The common fixes for such ESD degradation problems, such as using extra LDD-blocking and silicide-blocking masks, increase the product costs and the time-to-market. On the other hand, some new technology features, e.g., SiGe and Cu interconnects, may be quite beneficial to ESD protection. As IC technologies continue to scale, and IC chip complexity and performance increase rapidly, designing on-chip ESD protection structures for the same level of ESD robustness along the technology scaling pathway becomes very challenging because any ESD protection structures will negatively affect the IC performance (e.g., data rates) under protection. It is important to carefully consider all aspects of IC technology impacts on ESD protection designs. It is also important to adopt the ESD-technology co-development principle in developing new IC technologies and ESD protection solutions. For future ICs in future technologies, it is also imperative to explore nontraditional revolutionary ESD protection solutions that are fundamentally different from today's existing in-Si PN-based active ESD discharge structures that have been in use for several decades.

References

1 Voldman, S. and Gross, V. (1993). Scaling, optimization and design considerations of electrostatic discharge protection circuits in CMOS technology. *Proceedings of EOS/ESD Symposium*, pp. 251–260.

2 Wang, A. (1999). ESD protection design using copper interconnects: more robustness and less parasitics. *A Project Report to SRC Copper Design Contest.*

3 Shabde, S., Simmons, G., Baluni, A., and Back, D. (1984). Snapback induced gate dielectric breakdown in graded junction MOS structures. *Proceedings of International Reliability Physics Symposium*, pp. 165–168.

4 Chan, M., Yuen, S., Ma, Z. et al. (1994). Comparison of ESD protection capability of SOI and bulk CMOS output buffers. *Proceedings of International Reliability Physics Symposium*, pp. 292–298.

5 Verhaege, K., Groesenken, G., Colinge, J., and Maes, H. (1993). Double snapback in SOI nMOSFET's and its application for SOI ESD protection. *IEEE Electron Device Lett.* 14 (7): 326–328.

6 Voldman, S., Assaderaghi, F., Mandelman, J. et al. (1998). Dynamic threshold body- and gate-coupled SOI ESD protection networks. *J. Electrostat.* 44: 239–255.

7 Xie, H., Feng, H., Zhan, R. et al. (2005). A new low-parasitic polysilicon SCR ESD protection structure for RF ICs. *IEEE Electron Device Lett.* 26 (2): 121–123.

8 Choi, C., Park, Y., Lee, S., and Kim, K. (1996). Novel ESD protection transistor including SiGe burried layer to reduce local temperature overheating. *IEEE Trans. Electron Devices* 43 (3): 479–489.

9 Feng, H. (2001). A mixed-mode simulation-design methodology for on-chip ESD protection design. An MS thesis. Illinois Institute of Technology.

10 Feng, H., Chen, G., Zhan, R. et al. (2003). A mixed-mode ESD protection circuit simulation-design methodology. *IEEE J. Solid-State Circuits* 38 (6): 995–1006. https://doi.org/10.1109/JSSC.2003.811978.

11 Wang, A., Feng, H., Zhan, R. et al. (2005). A review on RF ESD protection design. *IEEE Trans. Electron Devices* 52 (7): 1304–1311. https://doi.org/10.1109/TED.2005.850652.

12 Gong, K., Feng, H., Zhan, R., and Wang, A. (2002). A study of parasitic effects of ESD protection on RF ICs. *IEEE Trans. Microwave Theory Tech.* 50 (1): 393–402. https://doi.org/10.1109/22.981291.

13 Chen, Q., Ma, R., Zhang, W. et al. (2016). Systematic characterization of graphene ESD interconnects for on-chip ESD protection. *IEEE Trans. Electron Devices* 63 (8): 3205–3212. https://doi.org/10.1109/TED.2016.2582140.

14 Lin, L., Wang, X., Tang, H. et al. (2009). Whole-chip ESD protection design verification by CAD. *Proceedings of EOS/ESD Symposium,* pp. 28–37.

15 Shi, Z., Wang, X., Liu, J. et al. (2012). Programmable on-chip ESD protection using nano crystal dots mechanism and structures. *IEEE Trans. Nanotechnol.* 11 (5): 884–889. https://doi.org/10.1109/TNANO.2012.2204767.

16 Wang, S., Yao, F., Wang, L. et al. (2013). Design and analysis of full-chip HV ESD protection in BCD30V for mixed-signal ICs. *IEEE International Symposium on Circuits and Systems,* pp. 1059–1062.

17 ESD Association (2020). *Electrostatic Discharge (ESD) Technology Roadmap.* ESD Association.

18 ESD Association (2018). *White Paper 1: A Case for Lowering Component Level HBM ESD Specifications and Requirements.* Industry Council on ESD Target Levels.

19 ESD Association (2009). *White Paper 2: A Case for Lowering Component Level CDM ESD Specifications and Requirements.* Industry Council on ESD Target Levels.

20 (2019). *White Paper 3: System Level: Part II: Implementation of Effective ESD Robust Designs.* ESD Association, Industry Council on ESD Target Levels.

21 Ma, R., Chen, Q., Zhang, W. et al. (2016). A dual-polarity graphene NEMS switch ESD protection structure. *IEEE Electron Device Lett.* 37 (5): 674–676. https://doi.org/10.1109/LED.2016.2544343.

22 Di, M., Li, C., Pan, Z., and Wang, A. (2020). Pad-based CDM ESD protection methods are faulty. *IEEE J. Electron Devices Soc.* 8: 1297–1304. https://doi.org/10.1109/JEDS.2020.3022743.

13

ESD Circuit Simulation by SPICE

At this time, it should be crystal-clear that serious and accountable on-chip electrostatic discharge (ESD) protection design should be quantitative and fully guided by ESD simulation in order to address the complex ESD multiple-coupling effects. As discussed before, technology CAD (TCAD)-based mixed-mode ESD simulation design method is powerful that can reveal the ESD discharge functions and behaviors in details. However, TCAD ESD simulation has many disadvantages: computing hungry, time-consuming, not suitable for large chip, and requiring solid knowledge on semiconductor device physics and integrated circuit (IC) process technologies. For ordinary IC designers, and large and complex chips, SPICE-like circuit level ESD simulation should be more practical.

13.1 ESD Device Behavior Modeling

The foundation for SPICE-type circuit simulation is device modeling for which Berkeley short-channel IGFET model (BSIM) device modeling technique is probably the most successful and widely used device modeling technique in analog and mixed-signal IC designs. The grand technical barrier to realizing full-chip ESD protection circuit simulation using transistor-level circuit simulation tool, such as SPICE, is the lack of general and accurate ESD device models. Hence, device physics based mixed-mode TCAD ESD protection circuit simulation without using any compact device models becomes the main option for ESD protection design optimization and predication [1]. This is because any meaningful and accurate ESD protection design simulation must be able to analyze *ESD discharge functions* at chip level. Without accurate ESD device models, SPICE circuit simulation can only offer some general circuit analysis, such as timing, but is incapable of revealing the critical ESD discharge functionalities, which is essential to ESD simulation accuracy at circuit level. Unfortunately, due to the complexity of ESD multiple-coupling effects, i.e., transient-electro-thermal-materials-process-device-circuit-layout-system coupling effects, at the time of this writing, accurate ESD device modeling is still rather impractical [2, 3]. Although major efforts have been devoted to developing various ESD device models, many key physical phenomenon and behaviors still cannot be accurately modeled, for example, the transient local current/thermal crowding at device corners and edges, thermal boundary conditions (i.e., thermal resistance and capacitance) around any transient hot spots inside an ESD protection device. Without fully and accurately revealing the ESD discharge functionalities of ESD protection structures, using SPICE-like circuit simulators for full-chip ESD protection circuit simulation cannot be accurate and predictive. Nevertheless, while major research effort is still on-going in exploring novel and accurate ESD device compact modeling techniques, which is a big research

Practical ESD Protection Design, First Edition. Albert Wang.

topic beyond the scope of this book, one powerful and practical alternative is to use ESD device behavior modeling technique for circuit-level ESD design simulation [4, 5].

ESD device behavior modeling accurately describes the *measured* ESD discharging I–V characteristics for any ESD protection structures fabricated. With the ESD behavior models in place for given ESD protection structures, when these same ESD protection devices are used on an IC chip, SPICE-like ESD protection circuit simulation can then be conducted at full chip level, which can reveal circuit-level ESD discharge functionalities in time domain. Therefore, circuit-level ESD protection simulation using SPICE will be able to optimize and predict full-chip ESD protection circuit designs by ordinary IC design engineers without much pain as typically associated with conducting TCAD ESD design simulation. ESD device behavior modeling technique is certainly limited in terms of simulating ESD protection circuit because it only allows ESD protection simulation of an IC using pre-fabricated and characterized ESD protection devices, for which the accurate ESD device behavior models are already developed. However, this seems to be nothing different from RF IC simulation involving using certain on-chip inductors from a validated inductor device library.

Figure 13.1 depicts the general method for a scalable ESD device behavior modeling technique, which comprising three main tasks: extraction, coding, and validation [2–5]. The ESD behavior modeling flow starts with design and fabrication of a selected list of various ESD protection structures to be used in a given IC technology. To allow scalable ESD device behavior modeling, device dimensions and layout for carefully selected ESD protection structures should be thoroughly considered and defined in ESD protection device designs. Next, the fabricated ESD protection devices will be fully characterized for their transient ESD discharge behaviors. Typically, transmission-line-pulsing (TLP) testing will be used for human body model (HBM) ESD discharge characteristics, and VFTLP will be applied for CDM ESD protection functionalities. After transient ESD characterization, the measured ESD discharging I–V curves will be carefully analyzed to extract the ESD-critical parameters, including V_{t1}, I_{t1}, t_1, V_h, I_h, R_{ON}, V_{t2}, I_{t2}, etc., for each fabricated ESD protection device [6]. A thorough analysis of the measured ESD discharge I–V characteristics and the extracted ESD-critical parameters will be used to establish mathematical relationship between the ESD-critical parameters and the ESD device design splits (i.e., sizes, dimensions, layout patterns, etc.), which will lead to various suitable *fitting equations* for the concerned ESD protection structures with the device scalability accounted for. In the next step, considering the varying and unique ESD discharging I–V characteristics obtained in actual transient ESD testing, typically associated with idealities in any real-world ESD protection structures, the measured ESD discharging I–V curve for a given ESD protection structure will be partitioned into several segments. Partitioning an ESD discharging I–V curve divides a measured nonlinear ESD discharge I–V curve into a set of piece-wise functional line segments, corresponding to different on-set points and phases of the whole ESD discharge cycle for the ESD protection device, which may or may not be related to the actual ESD device physics, analogous to BSIM SPICE device modeling. Key fitting parameters for the piece-wise functional I–V curves can then be obtained for a given ESD protection structure, which is critical to addressing the scalability of ESD protection device physical designs. Next, Verilog-A language will be used to faithfully describe the piece-wise ESD discharging I–V curves for a given ESD protection device. Lastly and most importantly, the Verilog-A coded ESD device behavior models must be validated by SPICE simulation to ensure accuracy of full-chip ESD protection circuit simulation by SPICE. As examples, Figure 13.2 shows two N^+/Pwell ESD protection diodes featuring gated and shallow trench isolation (STI) isolation in forward ESD conduction modes and Figure 13.3 depicts a two-diode triggered DTSCR ESD protection structure in CMOS. In building up a scalable ESD protection device library for IC designs, selected ESD protection structures of varying sizes

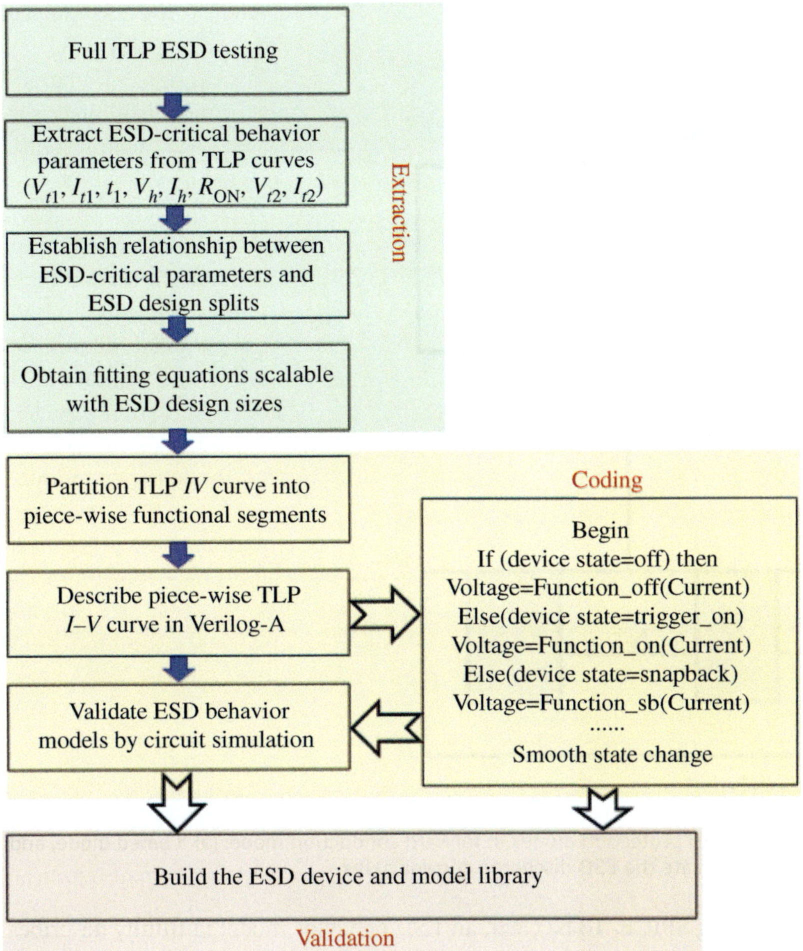

Figure 13.1 A flowchart for the scalable ESD device behavior modeling method includes extraction, coding, and validation steps.

(e.g., the finger width, W) are designed and characterized, as shown in Figure 13.4, for the TLP-measured $I_{t2} \sim W$ and $R_{ON} \sim W$ curves for exemplar N$^+$/Pwell gated ESD protection diodes fabricated in a foundry 28 nm CMOS [2]. Partitioning the TLP-measured ESD discharge I–V curves are illustrated in Figure 13.5 for a sample N$^+$/Pwell gated ESD protection diode and in Figure 13.6 for a sample two-diode DTSCR ESD protection structure fabricated in the foundry 28 nm CMOS [2]. Partitioning the measured ESD discharge I–V curves is critical to accuracy of ESD device behavior modeling because any real-world ESD protection structures will not behave exactly as theoretically expected and/or simulated due to many varying factors in fabricated devices. Hence, it is important to fully account for the actually measured ESD discharging I–V characteristics of fabricated ESD protection structures that will be used in real ICs in order to ensure the accuracy of full-chip ESD protection circuit simulation by SPICE utilizing the extracted ESD device behavior models. With the ESD device behavior models developed and validated, one can start to build up a library of various scalable ESD protection structures with the associated ESD behavior models in a given IC technology, which can then be used for full-chip ESD protection circuit

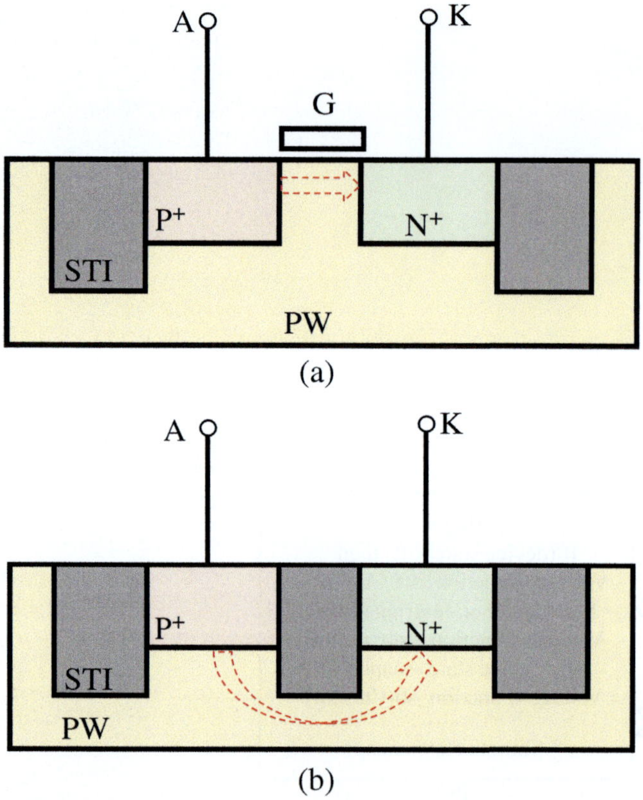

Figure 13.2 Exemplar N⁺/Pwell ESD protection diodes in forward conduction mode: (a) a gated diode, and (b) an STI diode. The red arrows indicate the ESD discharge current paths.

simulation and verification using SPICE. To be clear, an ESD behavior model faithfully describes the corresponding individual/standalone ESD protection device fabricated and measured, while SPICE ESD protection circuit simulation using the same ESD protection device(s) with ESD behavior model(s) enables accurate circuit-level ESD circuit simulation, accounting for ESD discharging functionalities at full chip level.

13.2 Full-Chip ESD Circuit Simulation by SPICE

13.2.1 Principle for ESD Circuit Simulation by SPICE

With accurate and scalable ESD device behavior models in place, an ordinary IC designer can handily select suitable ESD protection devices from the ESD protection device library in a process design kit (PDK) package for a specific chip and conduct whole-chip circuit-level ESD simulation by SPICE to both optimize and verify the IC chip with desired ESD protection before a tape-out. The full-chip ESD protection circuit simulation flow is depicted in Figure 13.7, which consists of five tasks: Core IC design, ESD protection design, ESD design library, ESD-IC co-design, and full-chip ESD testing simulation [2, 3]. The complete design flow for an ESD-protected IC certainly starts with the design of the core IC chip with required Specs for both core circuit and ESD protection. In the next design phase, various ESD protection structures are designed, fabricated and measured in

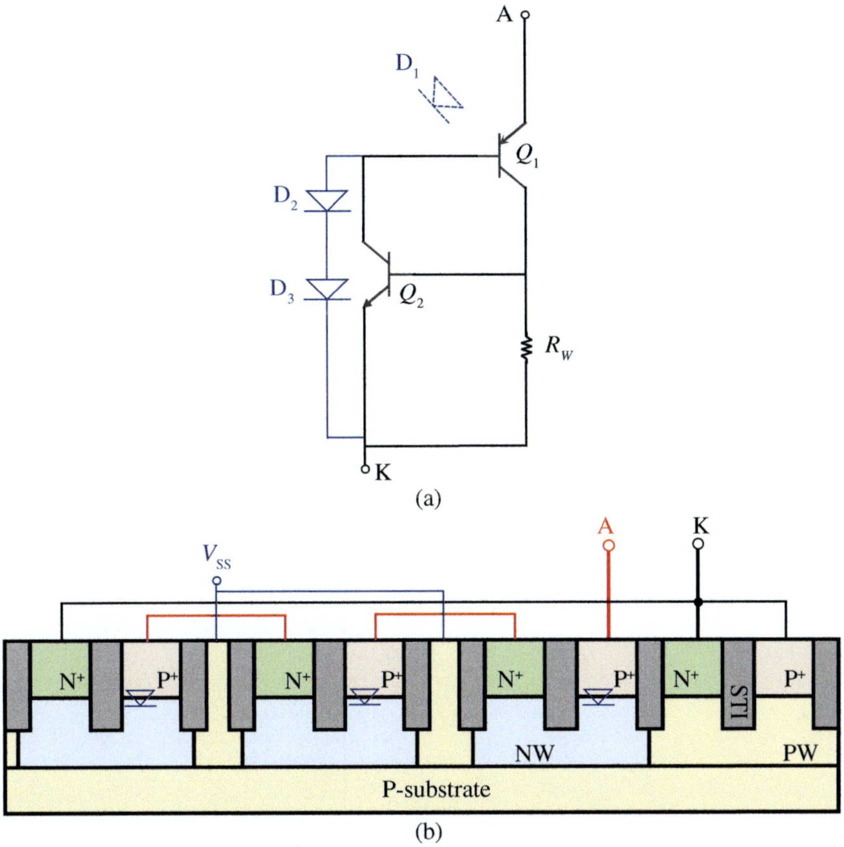

Figure 13.3 A two-diode-triggered DTSCR ESD protection structure in CMOS: (a) equivalent circuit, and (b) cross-section view.

a given IC process technology. In the third phase, a verified ESD protection device library is built up, which includes testing the ESD protection structures fabricated, extracting the ESD-critical parameters, building the ESD device behavior models and completing the ESD protection device library. Typically, in the IC industry, building up ESD protection device behavior models and device libraries is done by an IC foundry or the design support/service team in a fabless design house. An ordinary IC designer can simply select a suitable ESD protection structure in the PDK library to construct the whole chip, i.e., core IC + ESD protection circuit according to the design Specs. In the fourth design phase, ESD-IC co-design can be thoroughly conducted through full-chip ESD protection circuit simulation by SPICE using the given ESD device behavior models. It is worth noting that, for accurate ESD protection circuit simulation, the series resistance of the ESD metal interconnects must be fully included because even small metal bus resistance may play a role in ESD circuit simulation due to the very large ESD transient current involved. During ESD-IC co-design, careful design trade-offs must be considered to ensure simultaneous design optimization of both the core circuit and the ESD protection structures on a chip. After pre-simulation, layout verification and post-simulation, and before chip tape-out, one can perform thorough *ESD protection zapping testing simulation* for the whole chip by SPICE simulation using the ESD device behavior models, which is the last design phase. This full-chip ESD zapping test simulation is critical to avoiding ESD failure debugging and ESD design iterations. In addition, complete ESD zapping

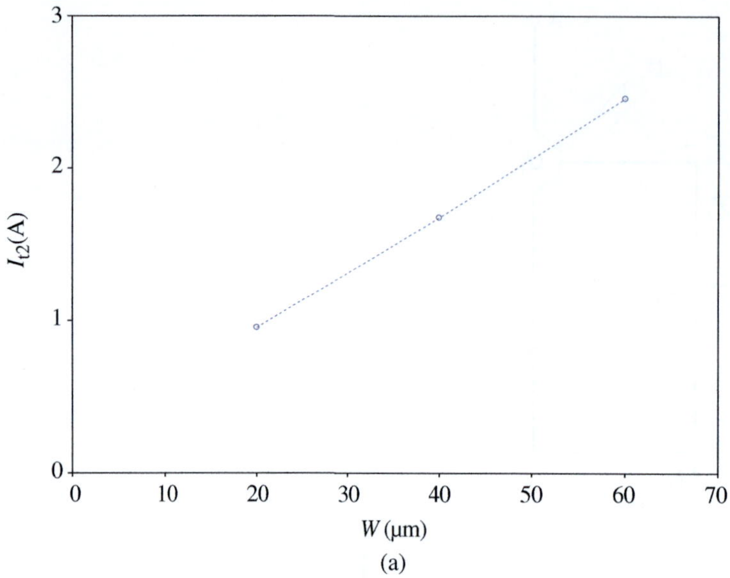

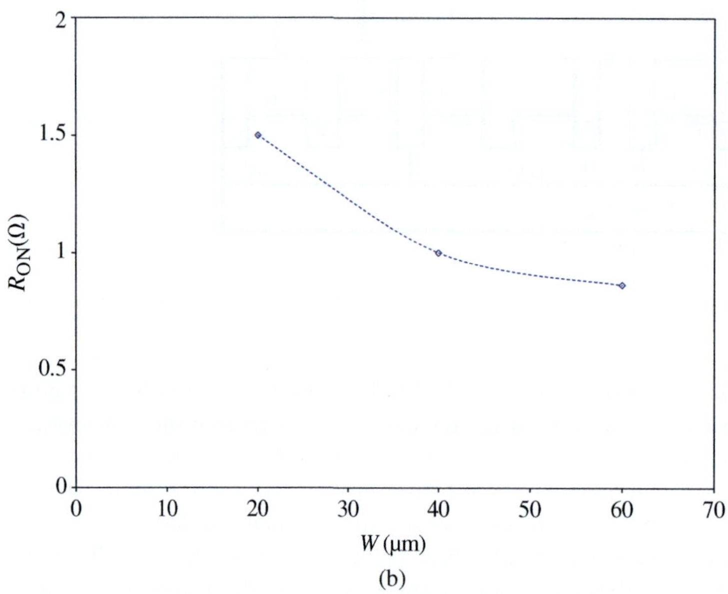

Figure 13.4 Measured I_{t2} (a) and R_{ON} (b) for a sample gated N^+/Pwell ESD protection diode of different device finger width fabricated in a foundry 28 nm CMOS shows the scalability of the diode ESD protection device.

test is very tedious, costly, and time-consuming. Further, an ordinary IC design engineer may not be in charge of ESD testing of the Si dies fabricated, making design revision difficult. SPICE-based full-chip ESD zapping test simulation offers ordinary IC designers a powerful means to check if the ESD protection design will work for a specific IC chip.

In principle, SPICE-based full-chip ESD protection circuit simulation is similar to conventional IC simulation with the following exceptions: (i) an ESD pulse will be used as the stimulus, (ii) the chip-level ESD zapping routines will be simulated, and (iii) ESD failure criteria will be used

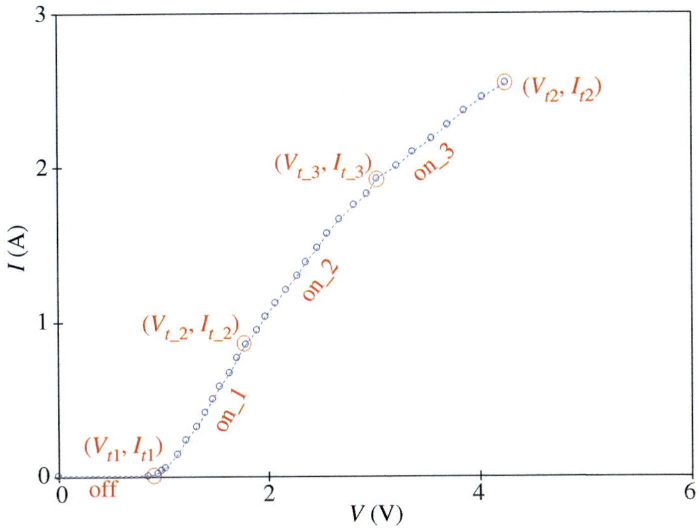

Figure 13.5 The TLP-measured ESD discharging *I–V* curve for an exemplar N^+/Pwell gated ESD protection diode of $W = 60\,\mu m$ fabricated in a 28 nm CMOS is partitioned into piece-wise functional line segments.

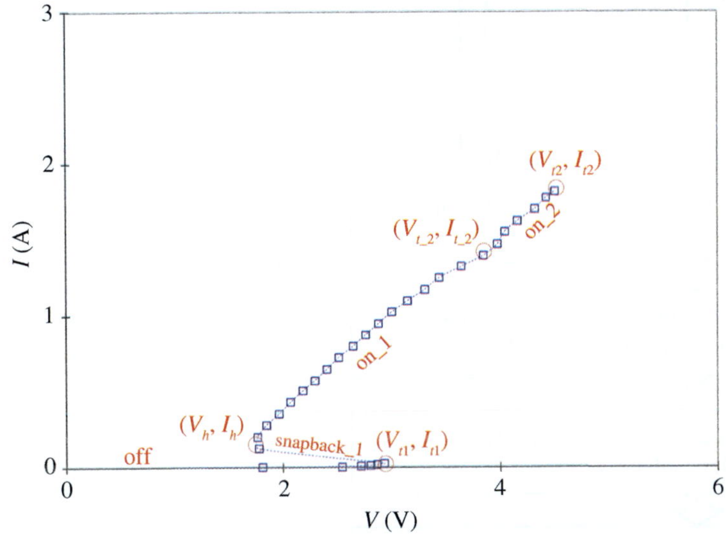

Figure 13.6 The TLP-measured snapback ESD discharging *I–V* curve for an exemplar two-diode DTSCR ESD protection device of $W = 50\,\mu m$ fabricated in a 28 nm CMOS is partitioned into piece-wise functional line segments.

to analyze ESD protection performance at chip level. Therefore, full-chip ESD protection circuit simulation must use suitable transient ESD pulses as the input "ESD signals," which can be the ESD pulse waveforms per various industrial ESD testing standards or models, such as HBM, CDM, IEC, TLP, VFTLP, etc. Next, ESD protection circuit simulation is quite different from normal circuit simulation in two aspects: First, ESD events involve extremely large currents, while normal circuit simulation only handles small signals. Second, chip-level ESD testing simulation routines are very complicated. The industrial ESD zapping test procedures require each pad (I/O, control

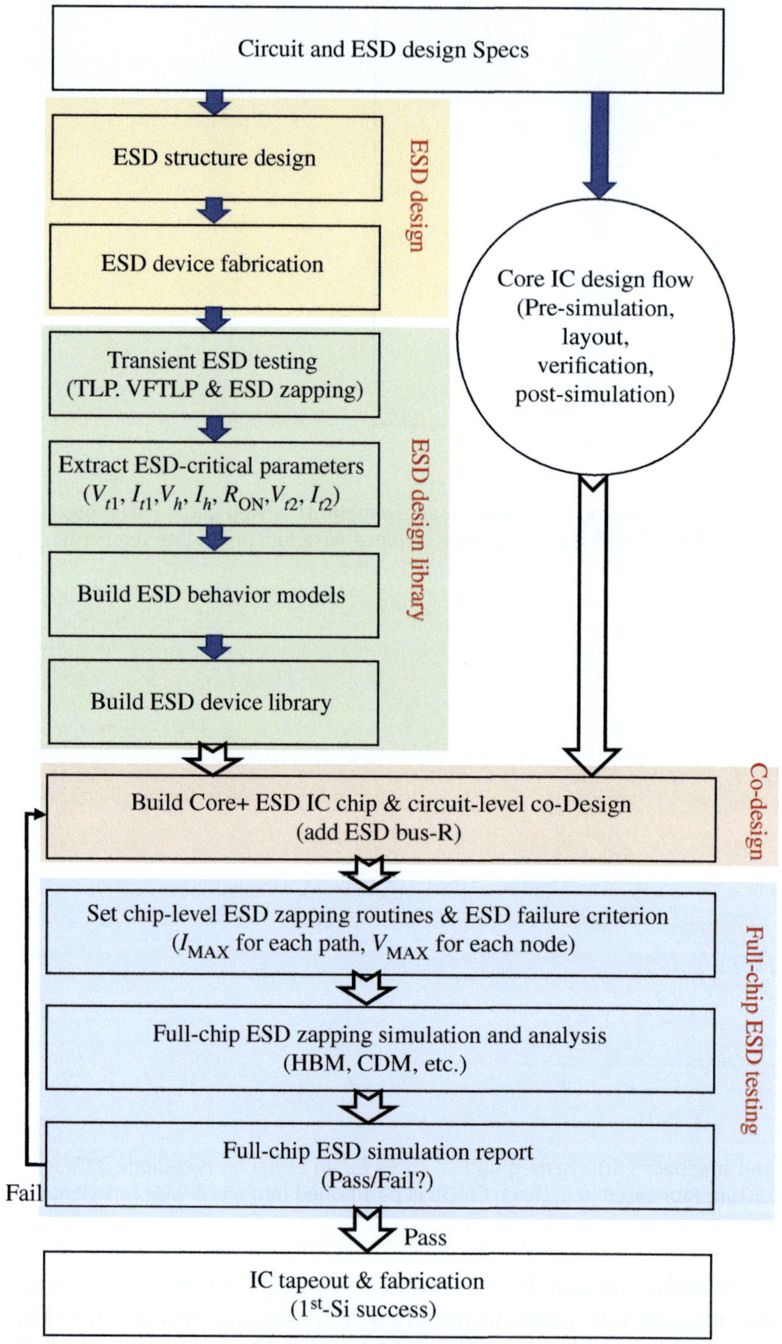

Figure 13.7 A flowchart of full-chip ESD protection circuit design simulation and verification using SPICE simulation and ESD device behavior models. The ESD protection circuit simulation routine include whole-chip ESD zapping test simulation.

and supply) be zapped by various ESD pulses with respect to a reference point, with all other pads on a chip being handled per ESD testing standards, e.g., all open or all grounded in bundles. An industrial ESD testing standard states that each pad must be zapped with reference to a positive supply pad (e.g., V_{DD}) and a negative supply pad (e.g., V_{SS}) or ground (GND) positively (i.e., PD and PS ESD modes) and negatively (i.e., ND and NS ESD modes). Each supply pad (e.g., V_{DD}) must be zapped with respect to a V_{SS} pad or GND pad positively (i.e., DS ESD mode) and negatively (i.e., SD ESD mode). An ESD test standard also requires each pad be zapped several times for each zapping mode, e.g., three times. Overall, full-chip ESD zapping procedure is extremely tedious, time-consuming, and costly, which makes the SPICE-based chip-level ESD circuit simulation method even more desirable and valuable. Therefore, a proper full-chip ESD zapping routine should be defined and programmed for complex ESD protection circuit simulation. Further, analyzing full-chip ESD circuit simulation results is more complicated than normal SPICE circuit simulation. In ESD protection circuit simulation, the ESD-critical parameters and the ESD Design Window must be included to analyze the full-chip ESD simulation results in terms of at least two ESD failure criteria: The Criterion-1 is to compare the maximum allowed voltage (V_{MAX}) at the protected nodes, typically determined by the BV at the protection nodes, with the simulated ESD clamping voltage (V_{ESD}) at the same nodes. If $V_{ESD} < V_{MAX}$ holds for all pads, then the chip passes the ESD testing. If $V_{ESD} \geq V_{MAX}$ at any node occurs, then a voltage breakdown ESD failure happens. The Criterion-2 is compared the maximum sustainable current (I_{MAX}) of each ESD discharging path against the simulated maximum ESD charging current (I_{ESD}) in the same path. Normally, I_{MAX} is same as the I_{t2} of the ESD protection structure used or the equivalent I_{t2} for a given ESD discharging path. If $I_{ESD} < I_{MAX}$ holds for all ESD discharging paths, then the chip passes the ESD testing. If $I_{ESD} \geq I_{MAX}$ occurs in any conduction channel, then the chip fails the ESD testing. In practical designs, a safety margin (e.g., 10–20%) is often defined for an IC. The results for the whole-chip ESD circuit simulation must be analyzed carefully as per the two ESD failure criteria. If an ESD failure occurs, one needs to check and revise the ESD protection designs, for example, whether an ESD device has suitable V_{t1}, R_{ON}, or I_{t2}. If the chip passes the comprehensive circuit-level ESD circuit simulation, one can proceed to tape-out the design for fabrication and expect first-Si ESD design success with confidence. This SPICE-based whole-chip ESD protection circuit simulation and verification method will be further illustrated in practical design examples in Section 13.2.2.

13.2.2 Circuit-Level ESD Design Verification by SPICE

The following real-world design example shows how to use circuit-level SPICE simulation to verify full-chip ESD protection circuit designs. In this example, a simple input buffer IC core, as shown in Figure 13.8 for its simplified functional diagram, is designed and fabricated in a foundry 28 nm CMOS technology [3]. Figure 13.9 shows the layout view of the input IC core circuit, which can be compared with the die image to pin-down the ESD hot spot later for ESD failure debugging. The foundry 28 nm CMOS technology used features V_{DD} = 0.9 V and typical I/O breakdown voltages of BV_{GS} = 8.52 V and BV_{DS} = 7.01 V, respectively, setting the ESD Design Window for this design. The targeted ESD protection level for this input buffer is 2 kV in HBM ESD mode, which utilizes an N^+/Pwell STI diode for pull-down ESD protection (PD-ESD) against GND and a P^+/Nwell STI diode for pull-up ESD protection (PU-ESD) against V_{DD}, and an active RC power clamp for rail-to-rail ESD protection, respectively. Selected from the ESD device library, these ESD protection structures have accurate ESD device behavior models validated for circuit simulation. TLP testing shows I_{t2} = 1.8 A (~2.7 kV HBM) for the PD-ESD diode and I_{t2} = 2.3 A (~3.5 kV HBM) for the PU-ESD diode in forward ESD conduction mode, respectively. The reverse ESD triggering

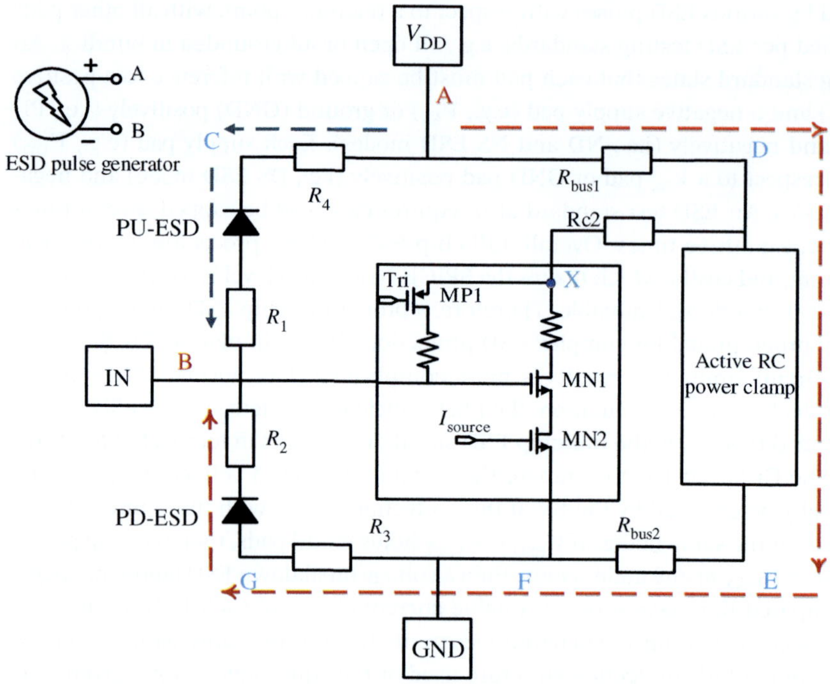

Figure 13.8 A simplified functional diagram for the input buffer IC core with full-chip ESD protection for chip-level ESD protection circuit simulation. The ESD metal interconnects resistances were extracted from its layout. The IC was designed and fabricated in a 28 nm CMOS.

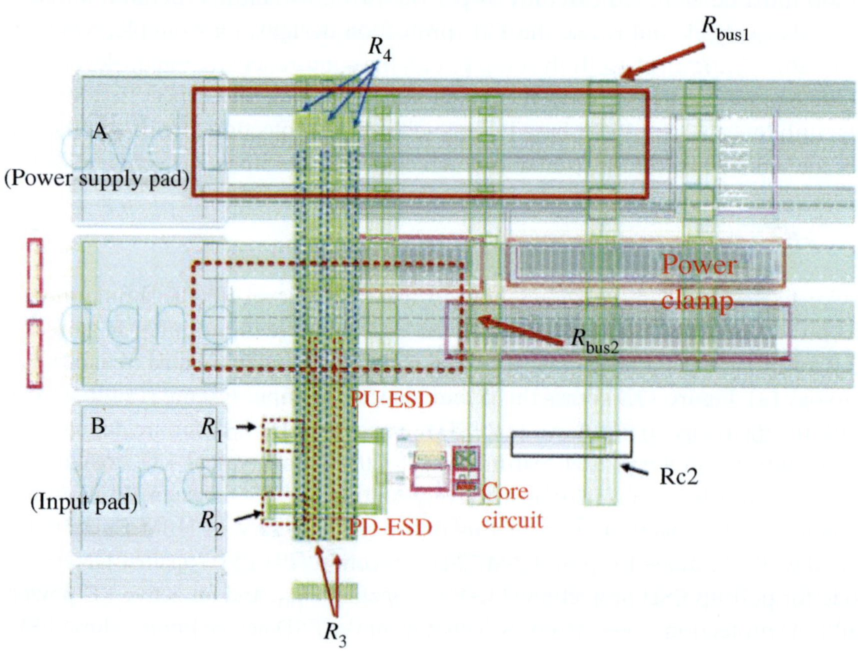

Figure 13.9 Layout of the IC core from which the ESD metal interconnects resistances can be extracted for ESD circuit simulation.

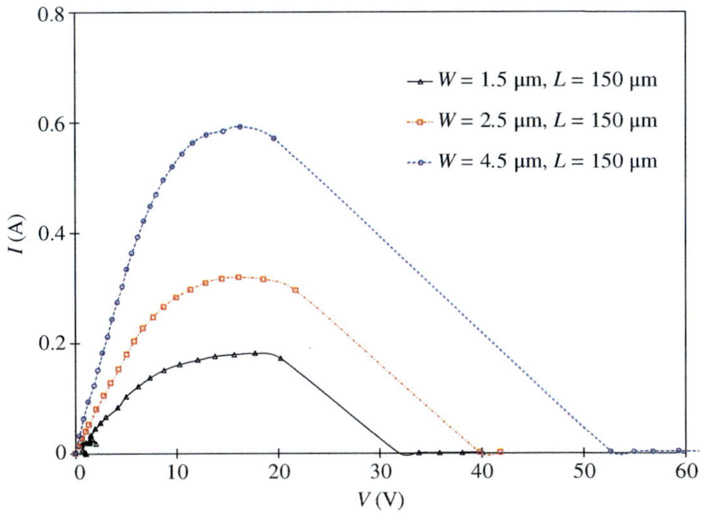

Figure 13.10 Measured ESD discharge $I-V$ curves by TLP for sample ESD metal interconnects using the Mx metal stack (M3–M7) fabricated in a foundry 28 nm CMOS.

voltage is $V_{t1} \sim 7.98\,\text{V}$ for the PU-ESD diode discharging in the reserve direction. Whether an ESD design will discharge in forward or reverse mode depends entirely on the full-chip ESD protection schematics, their ESD-critical parameters, and the actual ESD zapping modes. To account for the ESD metal interconnects resistance that cannot be ignored due to the large ESD transient currents, the metal wires in the 28 nm CMOS is characterized by TLP testing for different metal stacks. Figure 13.10 depicts the transient ESD discharge $I-V$ curves by TLP testing for a sample Mx metal stack (i.e., M2–M5 metal layers), resulting in sheet-resistance of around $R_{\square} = 1\,\Omega/\square$ for Mx stack and $R_{\square} = 0.3\,\Omega/\square$ for the My stack (i.e., M8–M10 metal stack), both used for ESD metal interconnects in this design. Based upon the TLP-measured metal sheet resistance and the circuit layout, the key bus resistances under ESD stressing are extracted as $R_{\text{bus1}} = 1.5\,\Omega$ between V_{DD} and power clamp anode (using M10, $L = 183\,\mu\text{m}$ and $W = 36\,\mu\text{m}$), $R_{\text{bus2}} = 0.3\,\Omega$ from the power clamp cathode to GND, $R_1 = 1\,\Omega$ between Input pad and cathode of PU-ESD, $R_2 = 1\,\Omega$ between Input and PD-ESD anode, $R_3 = 1.15\,\Omega$ from PD-ESD cathode to GND and $R_4 = 0.6\,\Omega$ from V_{DD} to PU-ESD anode, respectively. These ESD metal bus resistors are sizable that must be included in full-chip ESD schematic test bench for ESD zapping simulation. Next, comprehensive full-chip ESD protection circuit simulation is performed by SPICE ESD protection circuit simulation. The input stimuli for ESD simulation are 2 kV HBM ESD pulses following the required ESD zapping routines and stressing polarities. The simulated node voltages and branch currents during ESD stressing are thoroughly examined for the input buffer IC. For example, in a zapping case that applies a negative HBM ESD pulse to the Input pad with reference to the V_{DD} pad, i.e., ND ESD mode, the incident ESD pulse occurs at the node B with the node A grounded. Figure 13.11 presents the simulated transient ESD voltages at various key circuit nodes under 2 kB HBM ESD zapping. Figure 13.12 gives the simulated transient ESD conduction currents in key ESD discharge paths and transient voltage for the core circuit transistors. From the full-chip ESD protection circuit schematic representation, the expected main ESD discharging path is the route-ADEFGB (Red line, discharging through the ESD power clamp and PD-ESD diode in forward mode), which is readily confirmed by ESD circuit simulation. From Figure 13.11, it is clearly observed that ESD power clamp has a voltage drop of $V_{\text{DE}} < 2.5\,\text{V}$, the PD-ESD diode has a voltage drop less than

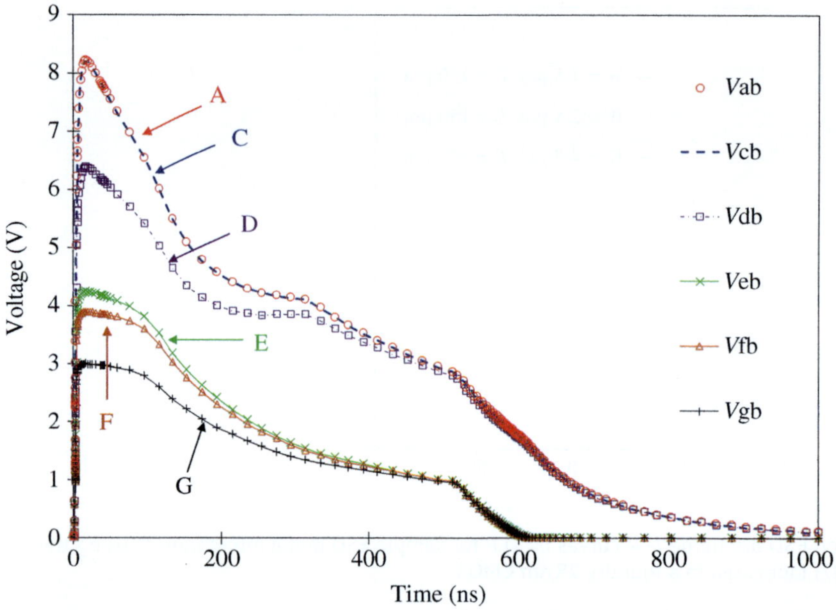

Figure 13.11 Simulated transient node voltages for the ESD-protected input buffer IC using the new ESD simulation method under an ND mode 2 kV HBM ESD zapping to the input pad (B) against V_{DD} (A). Note that all voltages are referred to node-B.

2.7 V, and V_{DD}-GND voltage drop V_{DF} is less than BV_{DS} = 7.01 V of the core circuit transistor. This states that the PD-ESD diode and the ESD power clamp are designed successfully to protect the input buffer IC. Looking into the details further, it is observed that a transient current of ~40 mA in the route-ACB (Marked in Blue, discharging through PU-ESD diode in reverse mode) exists, attributed to the voltage at Node-C that exceeds the reverse BV ~ 7.98 V of the PU-ESD diode (i.e., its reverse V_{t1}) during ESD stressing. Figure 13.12 also shows $V_{DS} < BV_{DS}$ ~ 7.01 V for MP1 and $V_{GS} < BV_{GS}$ ~ 8.52 V for MN1 of the core circuit during ESD stressing. Overall, the SPICE ESD circuit simulation and analysis confirm that the ESD protection design can pass 2 kV HBM zapping level. However, it finds that the PU-ESD diode (in reverse mode) seems to be an ESD weak point, which is confirmed by the emission microscopy (EMMI) thermal image depicted in Figure 13.13 where a hot spot is clearly observed at the PU-ESD diode location under HBM zapping, matching the chip layout very well. Further analysis reveals that this hot spot may be associated with the higher-than-expected ESD metal resistance that leads to a total voltage drop reaching to 4 V. With this observation from SPICE ESD simulation, one can optimize the ESD protection circuit design by re-designing the ESD metal interconnects for a reduced ESD discharging R_{ON}.

The next design example is a relatively larger 7-bits pseudorandom binary sequence (PRBS) generator circuit designed and fabricated in commercial 28 nm CMOS featuring V_{DD} = 0.9 V, BV_{DS} ~ 5.6 V, and BV_{GS} ~ 5.2 V that define the ESD Design Window [3]. Figure 13.14 depicts the functional diagram for the PRBS IC core circuit, including D flip-flop and XOR gate, where full-chip ESD protection is provided by anti-parallel gated ESD diodes at I/O pads and a low-V_{t1} DTSCR ESD power clamp. The ESD protection diodes and DTSCR devices are characterized by TLP, and the ESD device behavior models are extracted and validated for SPICE ESD protection circuit simulation. To fit into the ESD Design Window, the ESD triggering voltages are designed to be V_{t1} ~ 3.19 V for the DTSCR and V_{t1} ~ 1.03 V at 10 mA for the diodes in forward mode,

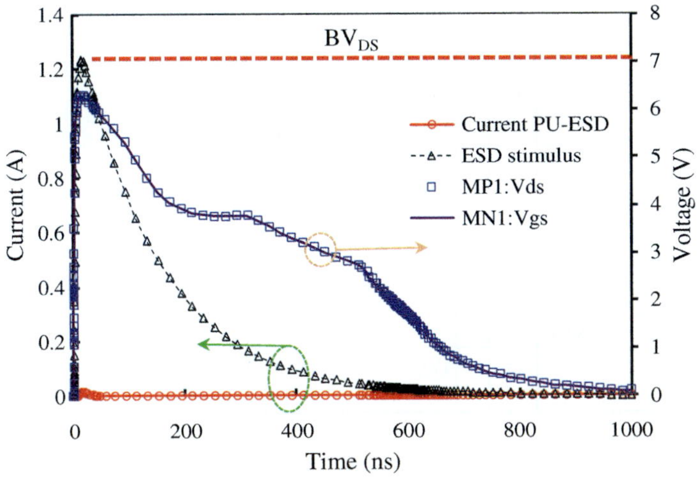

Figure 13.12 Simulated branch currents (Left axis) and core circuit transistor voltages (Right axis) under the ND mode ESD zapping from Input pad (B) to V_{DD} (A).

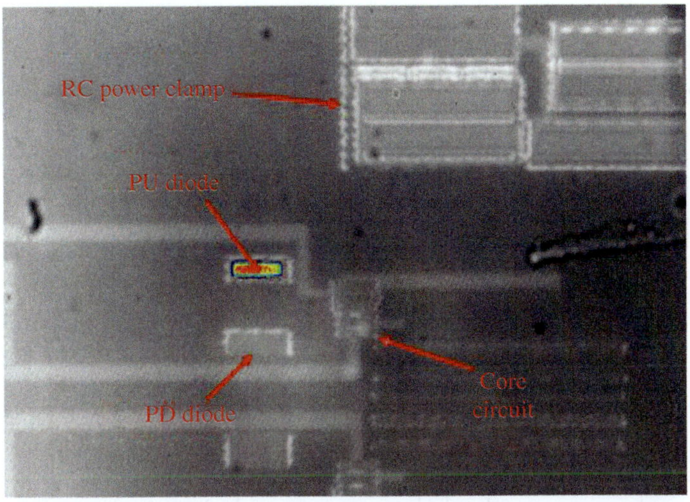

Figure 13.13 EMMI image under HBM ESD zapping shows a hot spot at the PU-ESD diode when comparing with its layout, indicating an ESD weak point in the design that can be modified.

extracted from TLP testing. Complete full-chip transient ESD circuit simulation is conducted by SPICE ESD protection circuit simulation using the ESD behavior models and HBM ESD pulses as stimuli for all ESD zapping routines. Figure 13.15 depicts two exemplar ESD circuit simulation cases for Input-to-V_{SS} and V_{DD}-to-V_{SS} ESD zapping. Shown in Figure 13.15a, during positive Input-to-V_{SS} zapping (PS mode), D_8 is the intended ESD discharging path by design. However, two unintentional ESD discharging paths, D_5 + DTSCR and $D_5 + D_2 + D_4$, may be possible during PS ESD zapping. Figure 13.16a presents the simulated ESD discharging currents for all possible paths, which clearly shows that the ESD pulse is mostly discharged through the input ESD diode (D_8) in forward conduction mode as designed, with negligible current leaking through the other unwanted paths. In the positive V_{DD}-to-V_{SS} zapping case (DS mode) shown in Figure 13.15b, the DTSCR power clamp is the intended ESD discharging channel as confirmed by ESD simulation

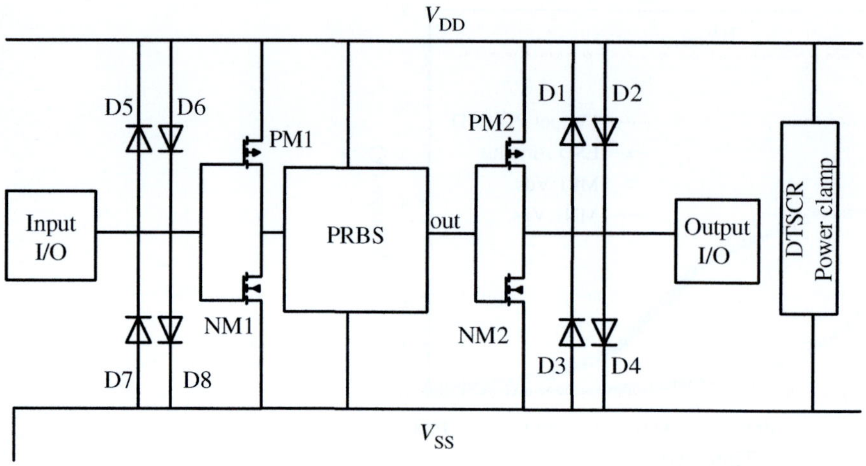

Figure 13.14 A functional schematic for the PRBS IC core using anti-parallel gated diodes (D1–D8) for I/O ESD protection and a low-V_{t1} DTSCR device for power rail ESD protection.

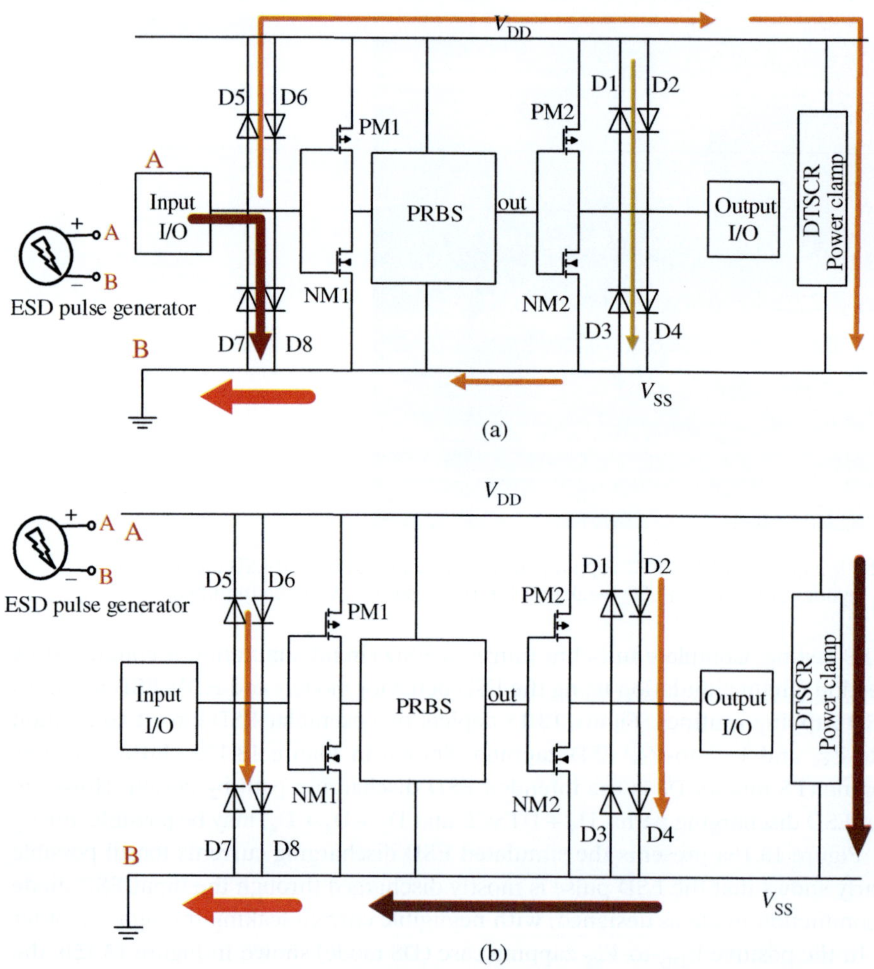

Figure 13.15 Two exemplar whole-chip ESD zapping simulation cases: (a) positive Input-to-V_{SS} PS ESD zapping, and (b) positive V_{DD}-to-V_{SS} DS ESD zapping. The arrowed lines represent the possible ESD discharging channels with the line width suggesting the relative amount of ESD discharging currents.

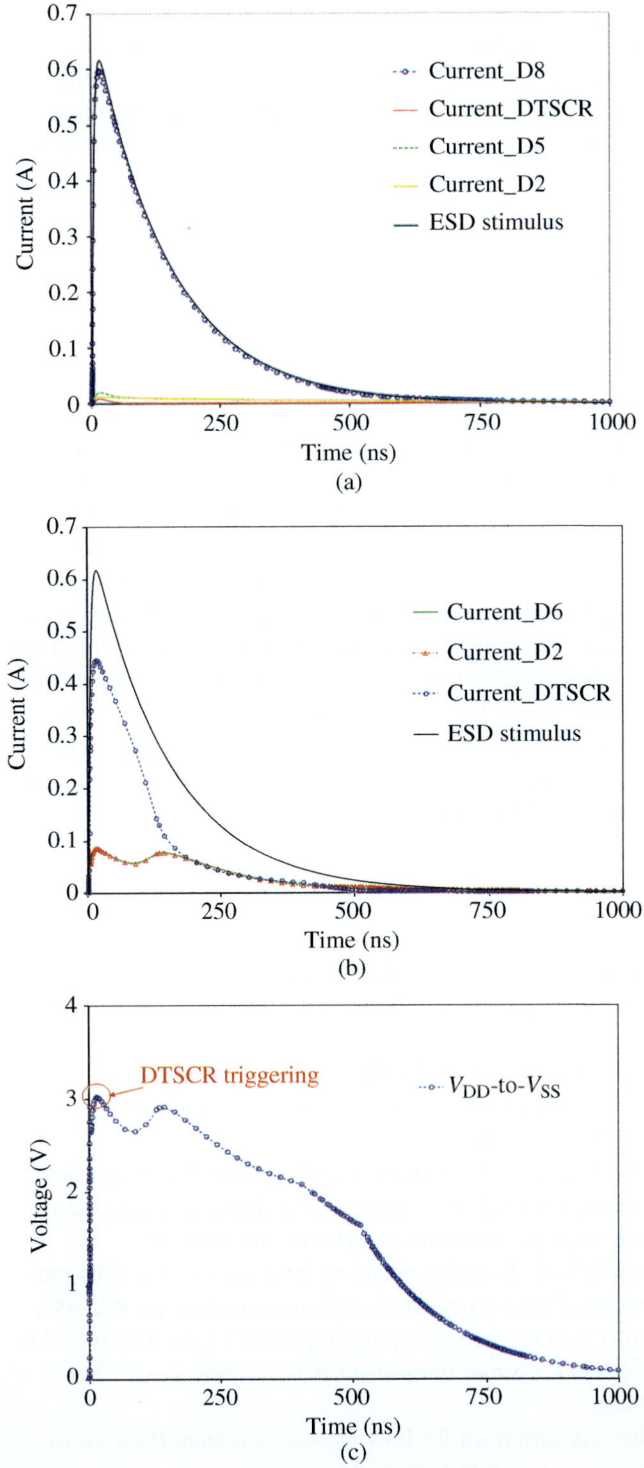

Figure 13.16 Simulated transient ESD discharging behaviors for two ESD zapping cases of Figure 13.15: (a) ESD current distribution during Input-to-V_{SS} PS ESD zapping, (b) ESD current distribution under V_{DD}-to-V_{SS} DS ESD zapping, and (c) transient rail-to-rail voltage in t-domain under V_{DD}-to-V_{SS} DS ESD zapping.

depicted in Figure 13.16b, i.e., >70% of the ESD stimulus conducting through DTSCR due to its low R_{ON}. Meanwhile, the other two possible conduction channels, i.e., $D_2 + D_4$ and $D_6 + D_8$, also discharge a small amount of the ESD current as expected. Figure 13.16c shows the transient V_{DD}-to-V_{SS} voltage where the peak voltage corresponds to the V_{t1} of DTSCR and is much lower than the breakdown voltage during HBM zapping. In summary, the two design examples prove that full-chip SPICE-based ESD protection circuit simulation, including complex ESD zapping test simulation, can be used to verify whole-chip ESD protection design where the critical ESD discharge functionalities are included in the ESD device behavior models.

13.3 Summary

Though TCAD-based mixed-mode ESD protection design simulation is very powerful for conducting circuit-level ESD-function-based simulation, it is not user-friendly to ordinary IC circuit designers. Alternatively, SPICE-based ESD protection circuit simulation can be used for whole-chip circuit-level ESD protection simulation to optimize and validate ESD protection design before Si fabrication. ESD device behavior modeling is critical to accuracy of ESD protection circuit simulation by SPICE because ESD device behavior models can precisely account for all ESD discharge function details. The limitation for using ESD device behavior modeling technique is that the ESD behavior models are only accurate for the given ESD protection structures that are fabricated and characterized in advance. The scalability of ESD device behavior models depends entirely on the design splits for ESD protection devices in an ESD device library. Through comprehensive whole-chip ESD zapping test simulation, the SPICE ESD protection circuit simulation serves to verify ESD protection designs at chip level before tape-out, hence, enhances the chance of achieving first-Si ESD protection design success in real-world IC designs.

References

1 Feng, H., Chen, G., Zhan, R. et al. (2003). A mixed-mode ESD protection circuit simulation-design methodology. *IEEE J. Solid-State Circuits* 38 (6): 995–1006. https://doi.org/10.1109/JSSC.2003.811978.

2 Lu, F., Ma, R., Dong, Z. et al. (2016). A systematic study of ESD protection co-design with high-speed and high-frequency ICs in 28 nm CMOS. *IEEE Trans. Circuits Syst. I* 63 (10): 1746–1757. https://doi.org/10.1109/TCSI.2016.2581839.

3 Zhang, F., Wang, C., Lu, F. et al. (2019). A full-chip ESD protection circuit simulation and fast dynamic checking method using SPICE and ESD behavior models. *IEEE Trans. Comput. Aided Des. Integr. Circuits Syst.* 38 (3): 489–498. https://doi.org/10.1109/TCAD.2018.2818707.

4 Wang, L., Wang, X., Shi, Z. et al. (2013). Scalable behavior modeling for nano crossbar ESD protection structures by Verilog-A. *Proceedings of IEEE Conference on Nanotechnology*, pp. 452–455.

5 Wang, L., Wang, X., Shi, Z.T. et al. (2013). Scalable behavior modeling for 3D field-programmable ESD protection structures. *Proceedings of IEEE Custom Integrated Circuits Conference (CICC)*. https://doi.org/10.1109/CICC.2013.6658492.

6 Wang, A., Feng, H., Zhan, R. et al. (2005). A review on RF ESD protection design. *IEEE Trans. Electron Devices* 52 (7): 1304–1311. https://doi.org/10.1109/TED.2005.850652.

14

Emerging ESD Protection

14.1 Emerging ESD Protection Challenges

For decades since the birth of integrated circuits (ICs), significant R&D efforts have been devoted to developing various on-chip electrostatic discharge (ESD) protection solutions for ICs. The principles for ESD protection designs are generally accepted as ESD protection has complex transient-electro-thermal-materials-process-device-circuit-layout-system coupling effects, ESD protection design is a circuit-level design task, ESD protection is not universal, ESD-critical parameters must be quantitatively designed to fit the ESD design window, there exist complex ESD-circuit interactions, ESD-induced design overhead (e.g., C_{ESD}) has negative impacts on IC performance, ESD-IC co-design is critical for advanced ICs, and ESD simulation plays a key role in ESD protection design optimization and predication, etc. Following such ESD design principles, generation by generation, substantial improvements have been achieved in on-chip ESD protection designs, making the prosperity of semiconductor industry a reality today. However, as IC technologies rapidly advance into sub-28 nm nodes, chip size and complexity continuously increase, and the demands for IC performance become higher and higher, major challenges emerge in ESD protection designs for advanced ICs that call for novel, disruptive, and even revolutionary on-chip ESD protection solutions, from ESD protection mechanisms to ESD design methodologies and tools. Many emerging ESD design challenges cannot be addressed through traditional ESD protection design thinking. For example, for high-performance ICs with data rates more than 10 Gbps or of frequencies higher than 10 GHz, even a tiny C_{ESD} of a few femto Farad (*fF*) will be unbearable. Unfortunately, nobody can make $C_{ESD} = 0\,°F$ using any traditional ESD protection structures. ESD-IC co-design is very helpful, but only to certain performance limits. This is why some in the industry proposes to lower the ESD protection level for advanced ICs, e.g., human body model (HBM) < 1 kV and charged device model (CDM) < 250 V. However, this is not acceptable to the consumer market since the real-world ESD danger has never shrunk following the Moore's Law, unlike the IC scaling trend. In another consideration, electronic system products are being built using different IC chips from different vendors, but the designed ESD protection Specs, e.g., ESD V_{t1}, for various chips may not sync on the system board. For instance, when replacing a specific IC chip from one vendor to another vendor, it possibly results in unexpected ESD failures at board level. The same concern is true for chiplet-based heterogeneous integration systems, which is considered an important pathway to future ICs beyond the Moore's Law. Since different functional chiplets, e.g., digital core, analog amplifiers, RF wireless transceivers, and sensor and micro-electromechanical systems (MEMS) dies, have different ESD-critical parameters designed by different designers or companies, therefore, when putting these chiplets together heterogeneously in a microsystem, the system module may suffer from ESD failures even though

Practical ESD Protection Design, First Edition. Albert Wang.

each chiplet has its own ESD protection on a die. The problem is that these different ESD Specs on different chiplets just do not sync with each other. Looking even further, future chips contain diversified functional devices and circuits well more than Si complementary metal-oxide-semiconductor (CMOS) ICs, so, everything could be integrated together heterogeneously, less likely in the traditional monolithic fashion, which will make on-chip/in-package/on-board ESD protection very challenging. The conventional in-Si PN-junction based active ESD protection devices might not be suitable at all. Therefore, incremental improvement in traditional ESD protection designs will not be suitable for next-generation ICs and future chips. It is important to explore novel and disruptive ESD protection solutions, and further, truly revolutionary future ESD protection concepts and structures. This chapter discusses a few such emerging ESD protection design examples. Some potentially revolutionary future ESD protection concepts will be discussed in Chapter 17.

14.2 Dispensable ESD Protection

It is well known that ESD-induced parasitic C_{ESD} will negatively affect IC performance, particularly for high-speed ICs, and high-frequency and broadband RF ICs [1–5]. A nicely designed ESD protection structure may have very little C_{ESD} of ~30 fF, which can be handled by careful ESD–IC co-design to balance the performance of both IC core circuit and ESD protection structures [6, 7]. However, IC performance has been increasing rapidly to accommodate the endless demands for applications such as super-computing, wireless communications, big-data centers, and autonomous driving. For example, as data rates go beyond 20 Gbps and frequencies move into millimeter wave spectrum (>28 GHz), even $C_{ESD} \sim 30$ fF will be too much to handle by ICs. Unfortunately, it is impossible to design any ESD protection structure with $C_{ESD} = 0$ °F, and ESD-IC co-design will not be able to solve the problem. It requires new thinking of novel ESD protection solutions for such high-performance ICs. Toward this end, let us consider some electronic systems such as high-performance mainstream systems, the data rates can go beyond 100 Gbps. Such enterprise infrastructure backbones do require robust ESD protection to prevent high ESD failure costs that often occur during manufacturing, transportation, and installation. Uniquely, after installation, such mainstream systems are typically isolated and ESD risk will then no longer be a “live” threat as in consumer electronics. Therefore, it is possible that the ESD protection structures can be physically removed from the enterprise systems (i.e., making $C_{ESD} = 0$) to entirely eliminate any ESD-induced IC performance degradation at chip and system level. This idea of making $C_{ESD} = 0$ ESD protection is demonstrated in practical IC design using a novel *field-dispensable ESD protection* concept [8].

Figure 14.1 shows the functional diagram for a whole-chip field-dispensable ESD protection circuit including a high-speed IC core and fuse-based *dispensable ESD protection* devices [8]. This ESD-protected IC is designed and fabricated in a foundry 28 nm CMOS technology. The basic idea is to use fuse-based ESD protection structures to protect a high-speed IC circuit that is very sensitive to ESD-induced C_{ESD}. The designed ESD protection stays in place until a system using the IC is installed in an ESD-safe location, then the ESD protection structures will be *physically* removed by blowing out the fuses; therefore, the IC will resume its originally designed high data rates without any impact from the ESD-induced C_{ESD}. To realize the unique field-dispensable ESD protection function, as shown in Figure 14.1, a fuse is inserted between an ESD protection device and the supply rails. In the design, ESD protection diodes are used for 1 kV + HBM ESD protection, which are first optimized for minimum C_{ESD} by TCAD ESD simulation. In field applications, when the IC is in a relatively “ESD-safe” position, the fuse will be blown out by field programming

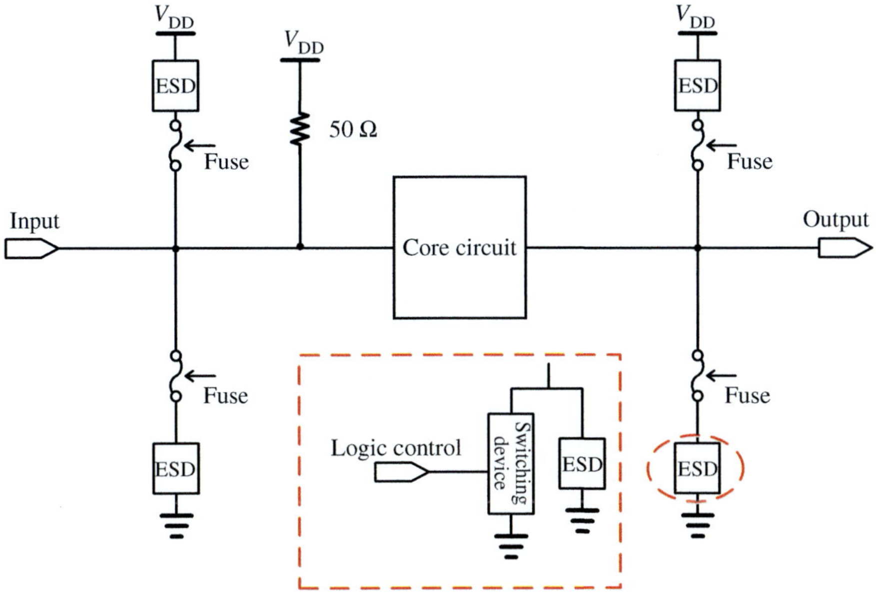

Figure 14.1 A functional diagram for ultrahigh-speed IC using fuse-based field-dispensable ESD protection where the fuses can be blown out as controlled by a switch-logic block for large pin-count chips.

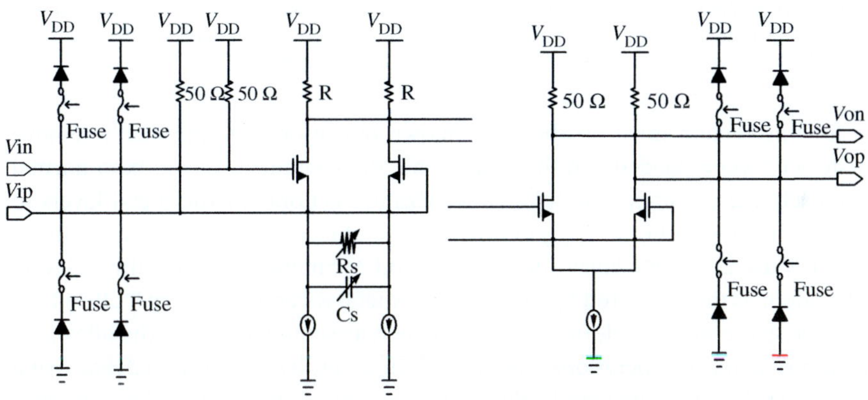

Figure 14.2 Simplified schematics for the input and output circuit blocks of the high-speed transceiver I/O link IC made in 28 nm CMOS show fuses between ESD protection devices and pads.

to physically cut off the ESD protection structures, i.e., achieving $C_{ESD} = 0\,°F$. Afterward, the IC core will resume its original high data rate mode designed for high system performance. The IC core designed is a transceiver I/O link circuit including equalizers at input and current mode logic (CML) buffers at output, supporting a high data rate of 22 Gbps. Figure 14.2 depicts the schematics for the input and output circuit blocks. Figure 14.3 show the layout for the IC designed. To make the field-dispensable ESD protection scheme more suitable for large chips with hundreds of pads, the ESD fuses can be controlled by switches through on-chip logics for field ESD programming as illustrated in Figure 14.1.

Two types of fuses are used in this design: a lateral notched metal wire and a vertical via specially designed in back-end-of-line (BEOL) in 28 nm CMOS, as shown in Figure 14.3. The fuse used to

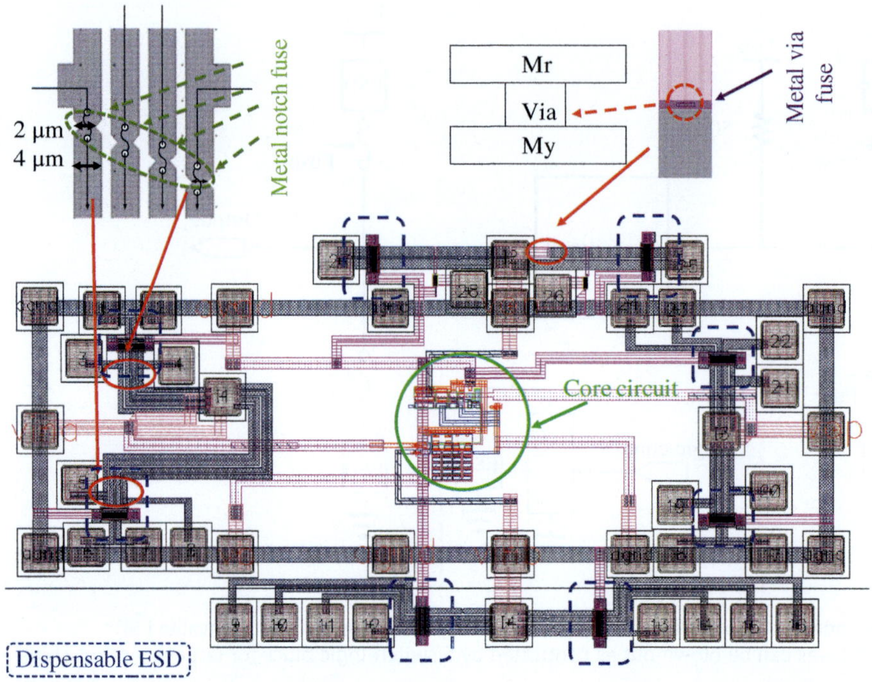

Figure 14.3 Layout of the high-speed transceiver link IC using fuse-based dispensable ESD protection diodes. Different fuse made of lateral metal lines of varying widths in different metal layers (with and without an embedded notch) and vertical vias between metal layers are designed.

enable dispensable ESD protection must be able to be blown out in order to physically remove the ESD protection devices when needed. On the other hand, the fuse must sustain both normal operations and ESD discharge currents. Careful characterization is done for the metal layers to find out the maximum current capability (I_{max}) of different metal layers of varying line width in the 28 nm 1P10M CMOS used, i.e., M1 layer and Mx, My, and Mr metal stacks (multiple layers combined). Figure 14.4 depicts the measured I_{max} for different metal wires by transient TLP and DC stressing, planned to assess the metal melting threshold, which are compared with the allowable DC and AC current ratings in normal operations given in PDK design rules. It is observed that, using a suitable metal wire, a metal fuse can be readily blown out when needed; however, the metal fuse stays safe under normal DC/AC operation currents and the targeted ESD transients. In this design, M7 and M8 metals are used to make the ESD fuses. For comparison, this design has four metal fuse splits: (1) a four-line metal line cluster of 4 µm wide, each with a 1 µm notch on both sides, (2) the same as in (1), but without a notch in metal wires, (3) a two-line cluster with each metal line of 8 µm wide with notches, and (4) a pair of 8 µm wide metal line using M8 and M9 without a notch where the vias between M8 and M9 serve as the fuse. A notch is designed to initiate a hot spot in the metal wire that leads to DC current "crowding" at the notch for easy fuse blow-out. The measured DC melting current for the metal splits 1–4 are 0.19, 0.46, 0.55, and 0.59 A, respectively.

Comprehensive ESD and IC tests are conducted including TLP, DC, and S-parameter characterization. TLP testing confirms 1500 V HBM ESD protection level at the IC output. The IC data rates are obtained by measuring the return loss at I/O pads with reference to the return loss mask per the CEI-28G-SR standard [9]. Figure 14.5 compares the measured output buffer circuit data rates in reference with the ESD protection structures. It is readily observed that the originally

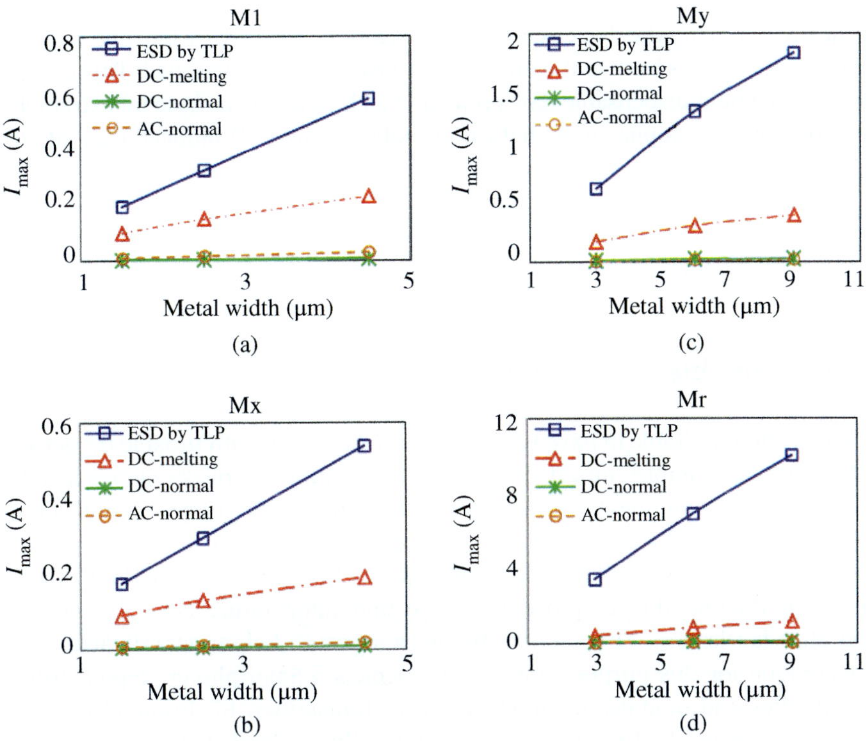

Figure 14.4 BEOL metal wires fabricated in a 28 nm 1P10M CMOS are measured by TLP and DC melting testing, and are compared with the DC and AC current ratings given in the PDK Design Rules for: (a) M1 layer, (b) Mx (M2-M6) stack, (c) My (M7 and M8) stack, and (c) Mr (M9 and M10) stack.

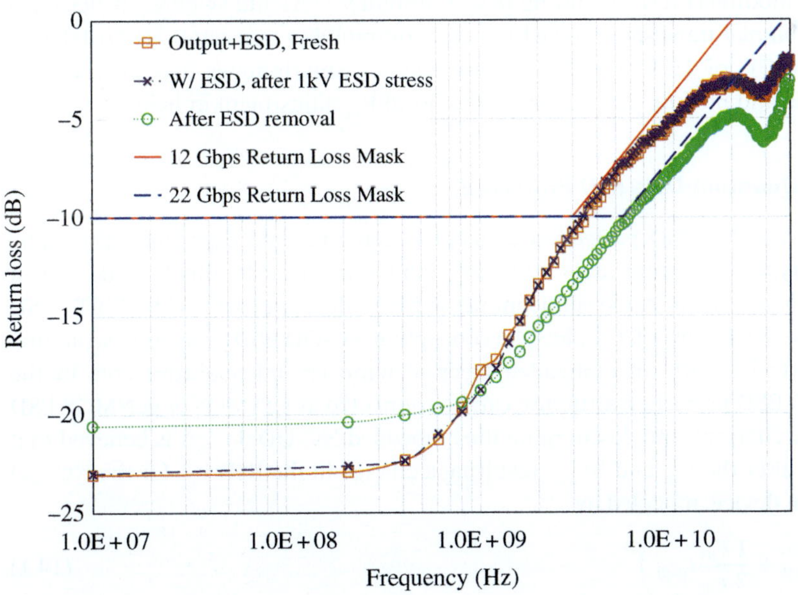

Figure 14.5 Measured output return loss for the high-speed IC fabricated in 28 nm CMOS shows that ESD-induced C_{ESD} greatly reduces the data rate from 22 to 12 Gbps, which is recovered back to originally designed 22 Gbps by physically removing the dispensable ESD protection diodes.

designed data rate of 22 Gbps dramatically decreases to 12 Gbps due to the ESD-induced C_{ESD}. However, after physically removing the dispensable ESD protection structure by blowing out the fuse, the measured data rate returns back to the original 22 Gbps as initially designed. This design example demonstrates that a field-dispensable ESD protection technique can be used for extremely high-speed ICs for certain system applications that cannot tolerate any ESD-induced C_{ESD} at all, i.e., ultrahigh-throughput enterprise backbone infrastructures, before any future ESD protection solution of $C_{ESD} = 0\,°F$ might be discovered. Of course, this is not going to be a universal ESD protection solution for high-performance consumer electronics, such as 5G smartphones that have to face ESD dangers every second during the whole life span.

14.3 Field-Programmable ESD Protection

As discussed earlier, it is often desirable to have ESD protection structures with tunable triggering voltage V_{t1} for several reasons: complex mixed-signal ICs with multiple supplies requires locally tuned V_{t1} in different supply/functional domains for whole-chip ESD design optimization; multiple-chip modules (MCM) and system boards using various dies fabricated in different IC processes may replace chips during the life cycle that requires fine tune of ESD V_{t1} in field; V_{t1} fluctuation in IC production due to process, voltage and temperature (PVT) variations need to be adjusted after fabrication. This section discusses a post-Si field-programmable ESD protection design technique for this purpose. Figure 14.6 depicts a Smartphone chipset utilizing multiple-supply IC dies and modules including receiver, transmitter, RF frontend module (FEM), data converters, baseband and interface circuits, typically implemented in different IC technologies. For example, the baseband ICs use 28 nm logic CMOS of 0.8 V, the receiver (Rx) IC including low-noise amplifier (LNA) and mixer use 90 nm radio-frequency CMOS (RFCMOS) of 1.2 V, analog-to-digital conversion/digital-to-analog conversion (ADC/DAC) use I/O CMOS of 2.5 V, and RF front-end module (FEM) including power amplifier (PA) and switches is designed in GaAs heterojunction bipolar transistor/p-type high-electron-mobility transistor (HBT/pHEMT) or silicon-on-insulator (SOI) of 3.3 V. The ESD robustness for Smartphones depends on ESD protection on each IC die and module that may need fine local V_{t1} adjustment in field.

14.3.1 Nano-Crystal Quantum-Dots ESD Protection

The post-Si field-programmable ESD protection concept is demonstrated experimentally using nontraditional nano crystal quantum dots (NC-QD) and silicon-oxide-nitride-oxide-silicon (SONOS) floating gate based metal-oxide-semiconductor field-effect transistor (MOSFET) ESD protection structures [10, 11]. Figure 14.7 depicts the conceptual NC-QD ESD protection structure, which is an MOSFET device with an embedded layer of nano crystal quantum dots in the floating gate. An NC-QD ESD protection structure can be connected as ggNMOS or gcNMOS ESD protection structures. By charging and de-charging the embedded NC-QD dots, it is believed that the tunneling effect can alter the V_{th} and BV_{DS}, resulting a change in the ESD V_{t1} for the NC-QD MOSFET ESD protection device, modeled by,

$$\Delta V_{th} \approx \frac{qn_{well}}{\varepsilon_{ox}}\left(t_{ctl} + \frac{1}{2}\frac{\varepsilon_{ox}}{\varepsilon_{Si}}t_{well}\right) \tag{14.1}$$

where t_{well} is the nano crystal well dimension, n_{well} is the nano crystal dot density, t_{ctl} is the control gate oxide thickness, and ε is the dielectric constant. Figure 14.8 illustrates possible application

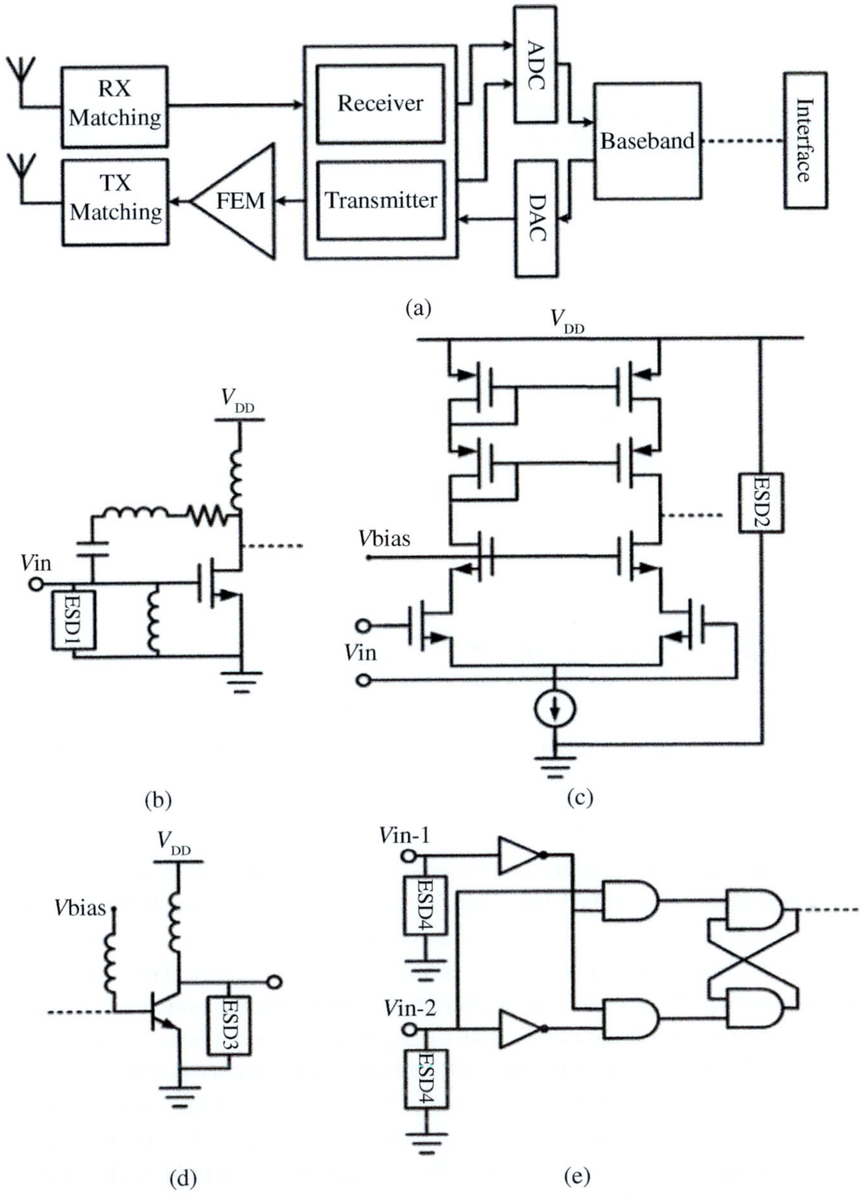

Figure 14.6 A Smartphone multiple-chip platform contains ESD-protected IC dies/modules using different supplies: (a) conceptual system diagram, (b) LNA with ESD protection at input, (c) a differential amplifier input for ADC with power rail ESD protection, (d) RF FEM with ESD protection at output, and (e) baseband IC with ESD protection at all I/O ports.

schemes for NC-QD ESD protection in ICs. Figure 14.8a depicts that charging/de-charging the nano crystal dots can be realized by filed-programming controlled by the F-program terminal. Figure 14.8b shows that an on-chip gate control logic block can be used for large ICs with many different I/O pads. Unlike a fixed-V_{t1} MOSFET ESD protection device, the NC-QD ESD protection allows fine-tuning of its V_{t1} in field by V_{t1}-programming. In the feasibility demonstration, NC-QD ESD protection structures are fabricated in a CMOS-compatible process, featuring a tunneling

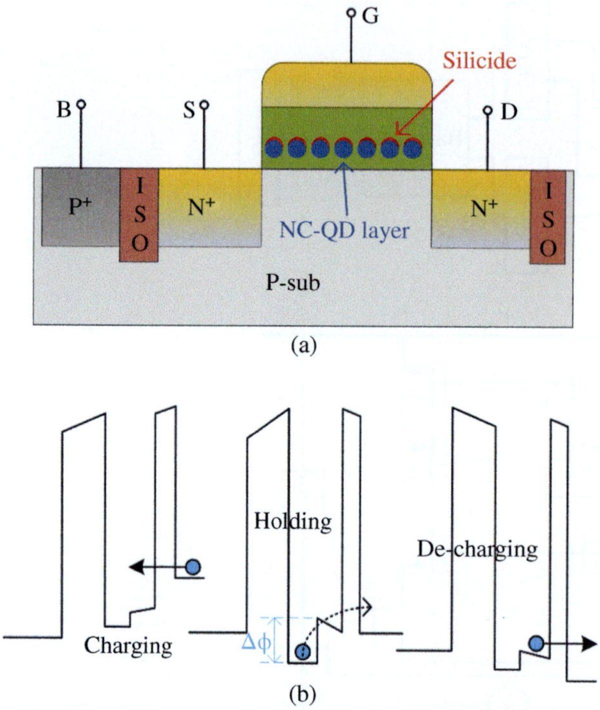

Figure 14.7 Concept of programmable NC-QD ESD protection structure uses a tunneling mechanism to adjust the ESD V_{t1}: (a) cross-section, and (b) energy band structures depicting the charging/de-charging procedures.

oxide layer of 5 nm grown on Si at 850 °C, a layer of Si nano crystal quantum dots deposited by low pressure chemical vapor deposition (LPCVD) at 600 °C, and coated by 1 nm cobalt film and annealed to form $CoSi_2$. The resulting $CoSi_2$-skinned nano crystal dots, with a dot size of 10 nm and dot density of 4×10^{11} cm^{-2}, are covered by a 20 nm control oxide layer on top. Transient ESD TLP testing ($t_r \sim 10$ nS) is conducted before and after charging/de-charging the nano crystal dots. The NC-QD programming requires moderate biasing of $V_G \cong 20$ V for 5s and a 50% duty cycle to charge the nano crystal dots. Figure 14.9 presents the TLP-measured ESD discharging I–V curves for a sample NC-QD ESD protection device before and after the NC-DQ dots charging action, which readily shows that the ESD triggering V_{t1} can be modified, as much as $\Delta V_{t1} = 2$ V for the sample devices of $L = 2\,\mu$m/$W = 100\,\mu$m, which is large enough for V_{t1} adjustment in field applications. VF-TLP testing confirms that the NC-QD ESD protection devices can respond to extremely fast ESD transients of $t_r \sim 100$ pS and $t_d \sim 1$ nS. The measured ESD leakage current is merely $I_{leak} \sim 14.6$ pA at a bias of 0.5 V.

14.3.2 SONOS ESD Protection

Similarly, a SONOS programmable MOSFET ESD protection is shown in Figure 14.10 where the energy bandgap structures under neutral, programming, and erasing conditions depict the charge variation in the float gate during programming and erasing phases, which modifies the MOSFET V_{th}, resulting in the desired ΔV_{t1} for ESD protection. The V_{t1} *versus* charges relationship is modeled by [12],

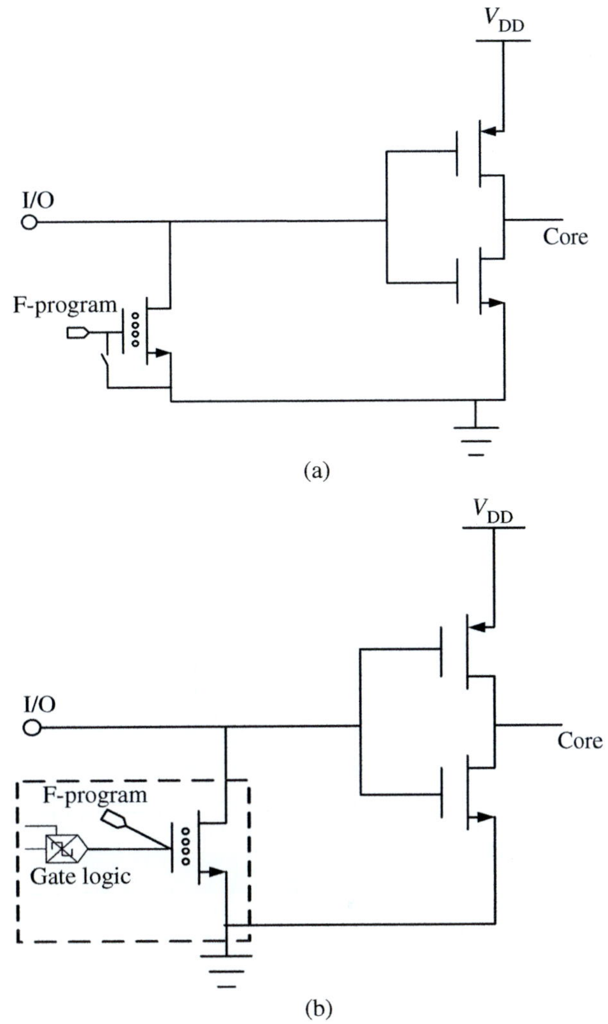

Figure 14.8 NC-QD ESD triggering programming and on-chip ESD protection schemes: (a) Scheme-1, direct gate programming by F-program control where a simple anti-fuse switch separates V_{t1}-programming function from ggNMOS ESD protection, and (b) an embedded gate logic control field-programming for large-pin-count complex chips.

$$\Delta V_{\text{th}} \approx -\frac{Q_N}{A}\left(\frac{t_{\text{top}}}{\varepsilon_{\text{ox}}} + \frac{X_C}{\varepsilon_N}\right) \tag{14.2}$$

where Q_N is the trapped charges by programming/erasing, X_C is the centroid of trapped charge measured from the top oxide and nitride interface, A is the device area, t_{top} is the top oxide thickness, ε_{ox} and ε_N are permittivity of SiO_2 and Si_3N_4, respectively. Both channel hot electron (CHE) and Fowler-Nordheim (FN) programming mechanisms may be applied. Prototype SONOS ESD protection structures are fabricated in a foundry 130 nm 1P4M logic CMOS process, featuring an oxide-nitride-oxide (ONO) thickness of 14 nm. Figure 14.11 depicts the TLP-measured ESD discharging I–V curves for a sample ggNMOS SONOS ESD protection device ($W = 10\,\mu\text{m}, L = 0.15\,\mu\text{m}$) under a programming sequence of Fresh, Erasing, and multiple-steps Programming, which clearly shows the expected modification in ESD triggering voltage with a sizable ΔV_{t1}. These prototype NC-QD and SONOS ESD protection structures show the feasibility of realizing field-programmable ESD protection structures to fine-tune the ESD V_{t1} to best fit the designed ESD design windows in complex ICs, MCM modules and systems in field applications.

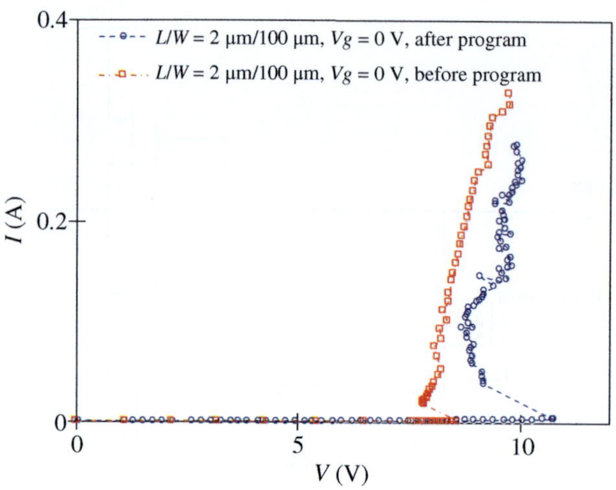

Figure 14.9 TLP-measured ESD discharge $I-V$ curves for sample NC-QD ESD protection device of $W/L = 100/2$ μm shows a change in ESD V_{t1} before and after programming, up to $\Delta V_{t1} \sim 2$ V.

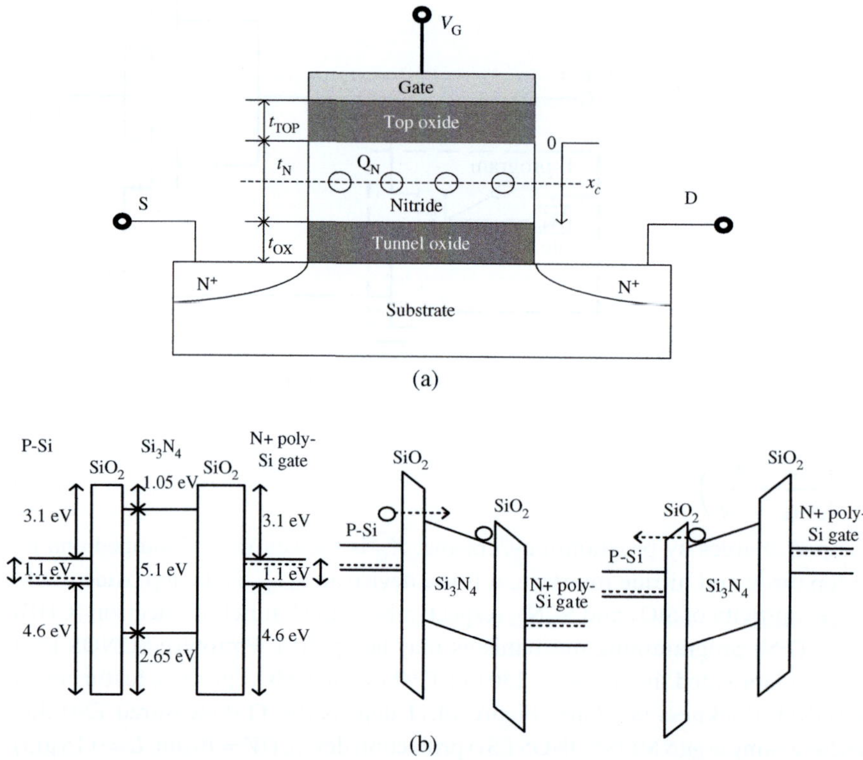

Figure 14.10 Conceptual SONOS ESD protection device shows V_{t1} programming mechanism: (a) cross-section, and (b) bandgap structures under fresh, programming and erasing conditions.

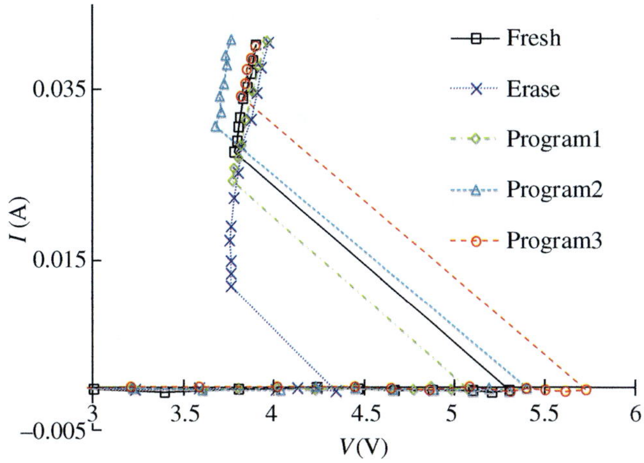

Figure 14.11 TLP-measured ESD discharge *I–V* curves for a sample SONOS ESD protection device shows adjustable V_{t1} under a programming sequence of Fresh, one-step Erasing and multiple-step Programming actions.

14.4 Interposer/TSV-Based ESD Protection

As discussed, the inevitable ESD-induced design overhead is multiple-folded, including parasitic parameters (C_{ESD}, I_{leak}, noises, etc.), die area consumption and layout floor-planning problems, all of which are becoming unacceptable to advanced ICs. These emerging ESD design challenges are much more serious for heterogeneous integration, such as 2.5D/3D chips and modules. On the other hand, some advanced IC technologies may be enablers for advanced ESD protection designs. For example, interposer and through-Silicon via (TSV) technologies. Interposer was initially developed as an electrical routing solution for MCM where several IC dies can be interconnected in 2.5D/3D fashions into a sub-system chip using a separate Si substrate dedicated to complex global metal wiring. TSV is originally proposed as a vertical electrical interconnects technique using metal pillars (i.e., vias) running through or partially through (deeply) a Si wafer to electrically connect multiple IC dies together into a 3D sub-system module or chip stack. Both interposer and TSV are demonstrated as emerging and unique high-performance ESD protection solutions for advanced and 3D ICs [13–15].

Figure 14.12 illustrates the traditional *in-die/in-plane* ESD protection scheme where various ESD protection structures are connected to pads on a chip to protect the IC (i.e., ESD protection devices

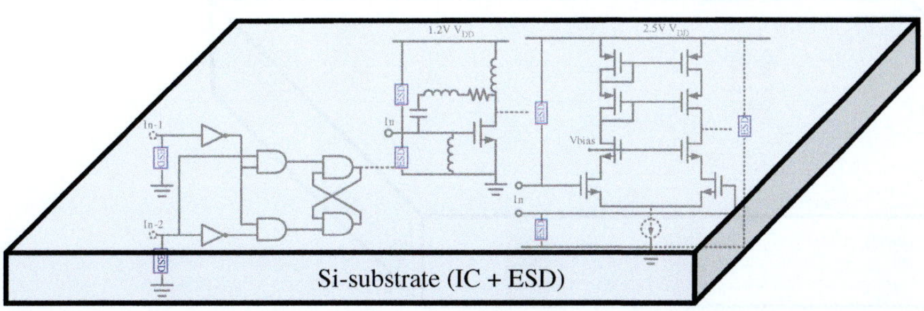

Figure 14.12 Traditional in-die/in-plane on-chip ESD protection integrates various ESD protection devices into the IC core in a side-by-side layout fashion at the same layer/level, i.e., in a lateral IC + ESD format.

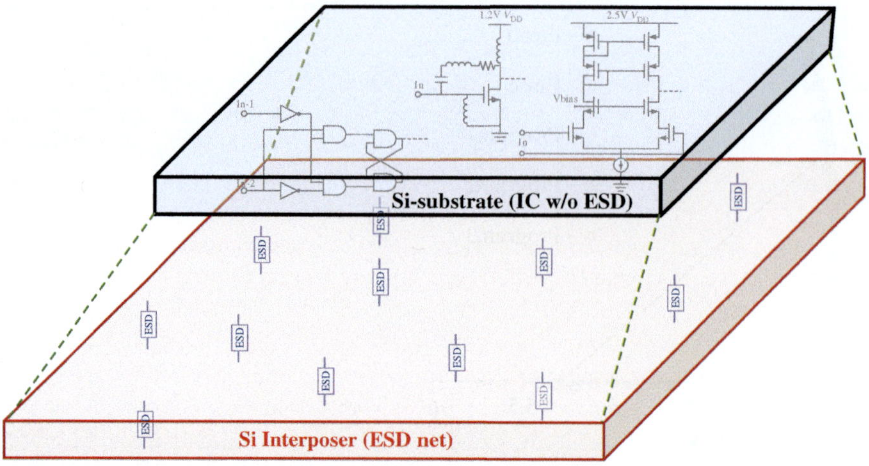

Figure 14.13 Illustration of interposer-based ESD protection scheme where the ESD protection circuit sub-net is implemented in a Si interposer that is 3D-stacked to the IC core die that has no in-plane ESD protection devices embedded.

are at the same layer/level with the core IC circuit). As discussed before, for complex ICs with large pin count (up to several thousand pins in a multiple-core GPU), not only that the ESD-induced parasitic effects will seriously affect IC performance, the large numbers of on-chip ESD protection devices also consume significant Si asset while making the whole-chip layout floor planning very difficult. Figure 14.13 depicts an interposer-based full-chip ESD protection technique that comprises two Si dies: one core die for the IC core circuit and one Si interposer die contains all ESD protection structures needed for the IC core [13]. In principle, all ESD protection structures, originally directly integrated with the IC core circuit in a side-by-side fashion, are removed from the IC core die. Instead, the ESD protection circuit network is implemented in a separate Si interposer wafer, which is then 3D-stacked with the IC core die by vertical interconnects, e.g., micro bumps and TSV pillars. The dedicated *interposer ESD* wafer is then thinned for better thermal dissipation. Apparently, an interposer ESD protection solution has many advantages: no consumption of the IC core die area, being layout-friendly, and mostly importantly, ease of thermal dissipation to avoid ESD-induced overheating.

TSV-based ESD protection scheme is depicted in Figure 14.14 where, instead of placing multiple ESD protection devices side-by-side to bonding pads on a chip, a vertical ESD protection device is

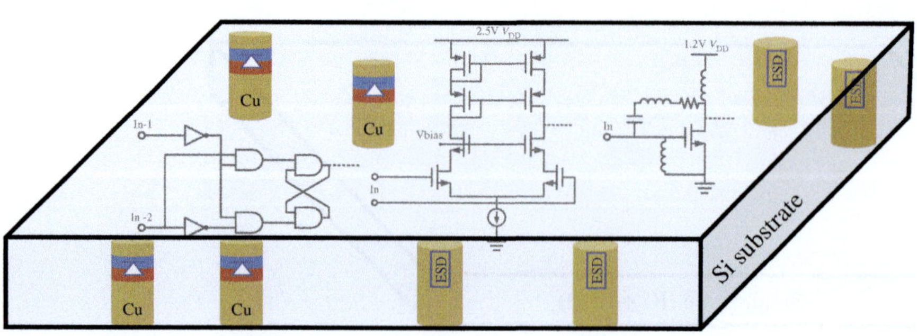

Figure 14.14 Illustration of TSV-based ESD protection scheme that places various ESD protection devices, e.g., diodes, in vertical TSV holes directly underneath the bonding pads.

placed inside a TSV hole that is under a pad [14, 15]. The *in-TSV* ESD protection device can be a simple PN diode as demonstrated in the design example below or any other active ESD protection devices, e.g., a bipolar junction transistor (BJT) or silicon controlled rectifier (SCR) ESD protection device. As in an interposer ESD protection design, since ESD protection devices are now sitting inside a vertical TSV shaft underneath a pad, an ESD protection device does not take extra Si area, which is a huge advantage over its in-plane ESD protection counterpart, especially considering the large lateral layout CD rules for ESD protection structures. This vertical in-TSV ESD protection device is entirely different from a traditional in-plane ESD protection design because the latter requires lateral routing an ESD device through lateral diffusion extension in Silicon for connecting the surface terminals. More importantly, the in-TSV ESD protection device can not only push the

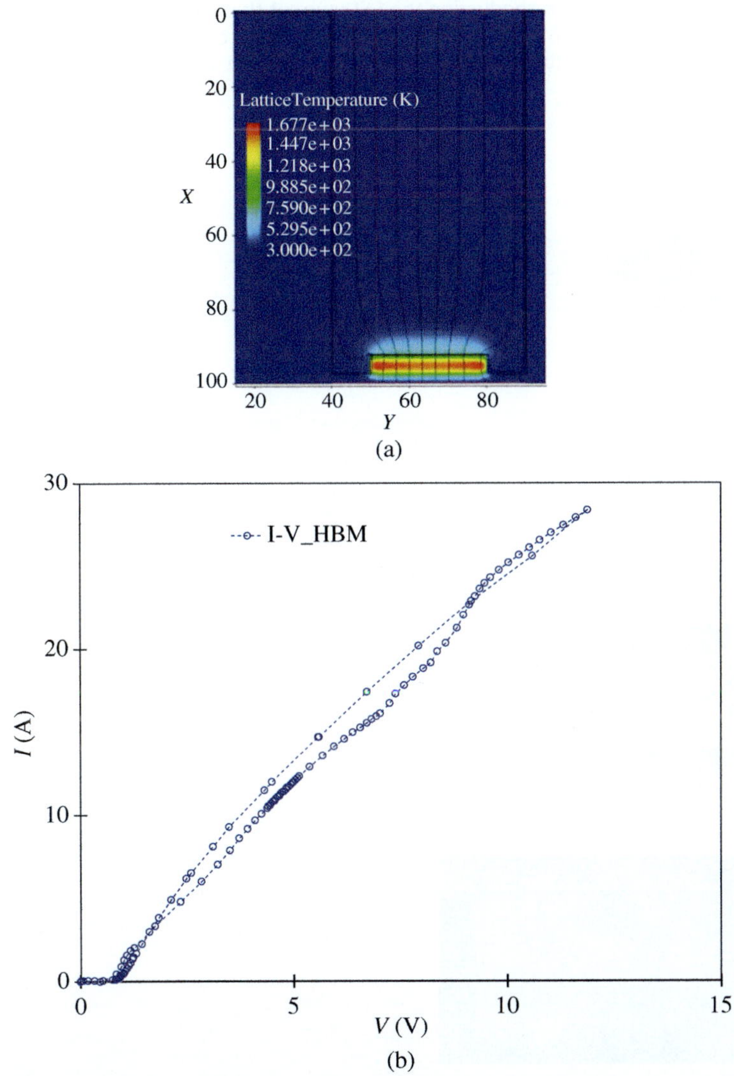

Figure 14.15 TCAD ESD simulation confirms ESD discharging function of a prototype in-TSV poly-Si/Si ESD protection diode: (a) vertical ESD discharging current conducted uniformly across the PN junction area, (b) transient ESD discharge *I*–*V* curve under HBM ESD zapping.

ESD-induced heating deep inside the Si substrate but also use a Cu pillar inside the TSV to dissipate the ESD generated heat, hence reducing the ESD thermal failure risk. Further, an Si wafer can be thinned, similar to in SOI technologies, which makes dissipating the ESD-induced transient heat much easier. Altogether, an in-TSV ESD protection solution can not only save the precious Si asset but also be able to handle the ESD overheating effect efficiently, hence dramatically enhances the ESD protection robustness without affecting the IC cores. An in-TSV diode ESD protection design is

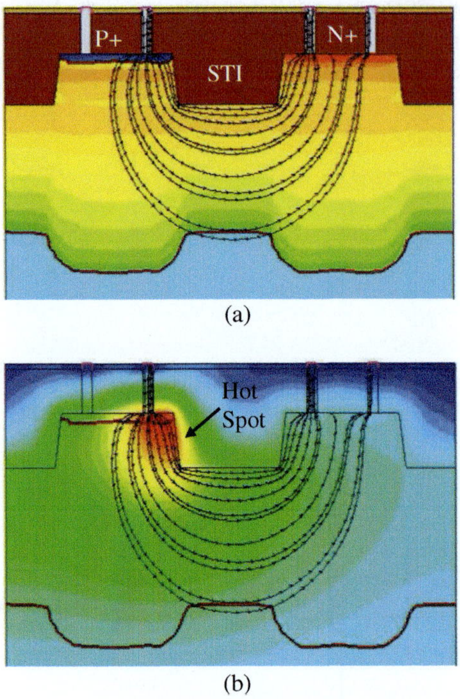

Figure 14.16 In a traditional in-plane STI ESD protection diode, though the PN junction is a "vertical" structure, the device requires "lateral" routing through Si diffusion extension to connect to the surface terminals, hence has to follow an unwanted curvature in conducting the ESD current that creates hot spots at the STI corners. (a) ESD discharge current flow, and (b) heat distribution showing a hot spot.

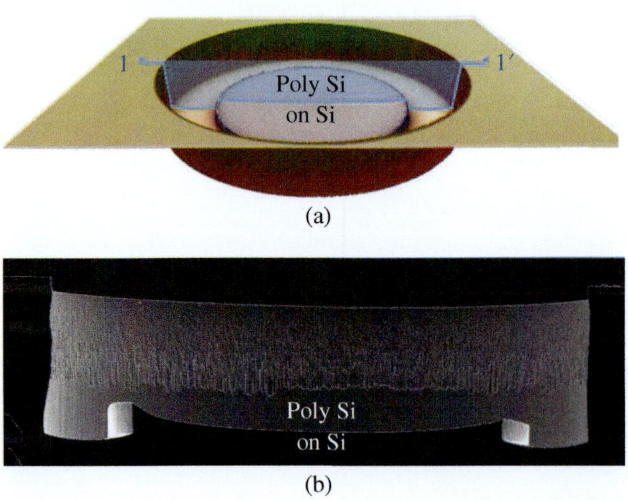

Figure 14.17 A prototype in-TSV poly-Si/Si ESD protection diode fabricated in a CMOS-compatible process: (a) a top conformal view image, and (b) a cross-section view by SEM image.

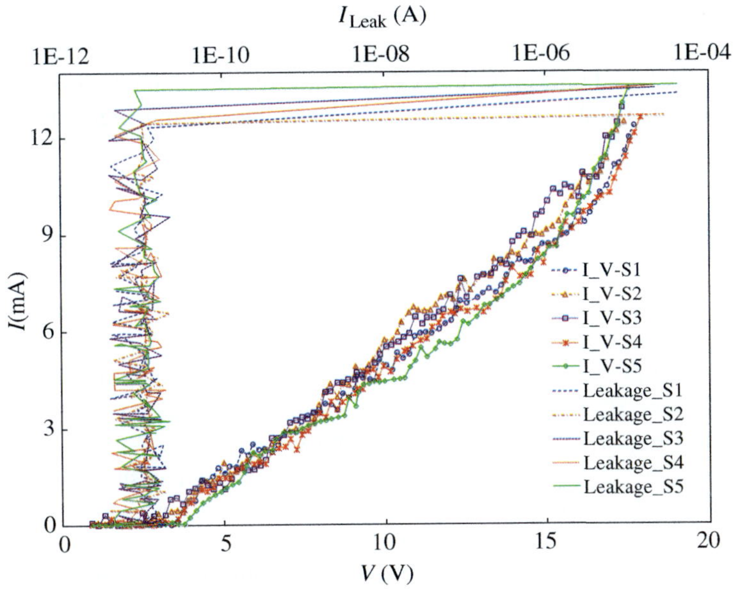

Figure 14.18 TLP-measured ESD discharging *I–V* curves for the prototype in-TSV ESD protection diodes show expected ESD discharge function.

demonstrated in a CMOS-compatible process where a simple ESD diode is made in a vertical TSV hole [15]. Figure 14.15 depicts the TCAD simulated vertical in-TSV ESD protection diode using a poly-Si/Si PN junction diode as a prototype. Figure 14.15a clearly shows the vertical ESD discharge current flow lines, and Figure 14.15b presents the transient ESD discharge *I–V* curve under HBM ESD zapping. Compared with a traditional in-plane lateral-routing shallow trench isolation (STI) ESD diode shown in Figure 14.16, the advantage of a vertical in-TSV ESD diode is obvious: the ESD discharge current runs vertically in an in-TSV diode where the ESD current conducts uniformly across the lateral PN junction area without any hot spots. In contrast, the ESD discharge current has to follow a curvature in an STI diode that inevitably generates a hot spot at the STI corner. Figure 14.17 shows images for a fabricated prototype in-TSV ESD protection diode structure. Figure 14.18 depicts the TLP-measured ESD discharging *I–V* behaviors for the prototype in-TSV poly-Si/Si ESD protection diode, demonstrating the desired vertical ESD discharging function.

Together, an interposer-based ESD protection and an in-TSV ESD protection can be combined to offer emerging ESD protection solutions for advanced ICs, including 3D ICs.

14.5 Summary

This chapter discusses the emerging challenges for ESD protection for advanced ICs. It is obvious that as IC chips become bigger and more complicated, e.g., thousands of pins on a multiple-core chip, 3D MCM, and monolithic heterogeneous integration, the traditional in-plane side-by-side ESD protection solutions become increasingly unacceptable due to the unwanted ESD-induced design overhead effects. Emerging, yet still practical, ESD protection methods are in demands. Interposer-based ESD protection and in-TSV ESD protection emerge as interesting alternative ESD protection solutions to advanced ICs, especially 3D MCM and monolithic chips. Interposer-based

ESD protection scheme uses a dedicated Si interposer substrate to house the complex ESD protection networks. TSV-based ESD protection places ESD protection diodes inside TSV holes directly underneath pads. Interposer and TSV based ESD protection can not only save the precious Si assets but also be able to handle the ESD-generated heat more efficiently, leading to possible higher ESD protection robustness.

References

1 Wang, A., Feng, H., Zhan, R. et al. (2005). A review on RF ESD protection design. *IEEE Trans. Electron Devices* 52 (7): 1304–1311. https://doi.org/10.1109/TED.2005.850652.

2 Wang, C., Lu, F., Chen, Q. et al. (2018). A study of impacts of ESD protection on 28/38 GHz RF switches in 45 nm SOI CMOS for 5G mobile applications. *Proceedings of IEEE Radio and Wireless Symposium*, pp. 157–160. https://doi.org/10.1109/RWS.2018.8304973.

3 Liu, J., Wang, X., Zhao, H. et al. (2011). Design and analysis of low-voltage low-parasitic ESD protection for RF ICs in CMOS. *IEEE J. Solid-State Circuits* 46 (5): 1100–1110. https://doi.org/10.1109/JSSC.2011.2118290.

4 Wang, X., Fan, S., Tang, H. et al. (2011). A whole-chip ESD-protected 0.14pJ/p-mV 3.1-10.6 GHz impulse-radio UWB transmitter in 0.18 μm CMOS. *IEEE Trans. Microwave Theory Tech.* 59 (4): 1109–1116. https://doi.org/10.1109/TMTT.2011.2114170.

5 Zhang, F., Li, C., Di, M. et al. (2020). Design and analysis of a 28 GHz 9KV ESD-protected distributed travelling-wave TRx switch in 22 nm FDSOI. *IEEE J. Electron Devices Soc.* 8: 655–661. https://doi.org/10.1109/JEDS.2020.2975598.

6 Lu, F., Ma, R., Dong, Z. et al. (2016). A systematic study of ESD protection co-design with high-speed and high-frequency ICs in 28 nm CMOS. *IEEE Trans. Circuits Systems I: Regul. Pap.* 63 (10): 1746–1757. https://doi.org/10.1109/TCSI.2016.2581839.

7 Wang, X.S., Wang, X., Lu, F. et al. (2014). Concurrent design analysis of high-linearity SP10T switch with 8.5 kV ESD protection. *IEEE J. Solid-State Circuits* 49 (9): 1927–1941. https://doi.org/10.1109/JSSC.2014.2331956.

8 Zhang, C., Dong, Z., Lu, F. et al. (2014). Fuse-based field-dispensable ESD protection for ultra-high-speed ICs. *IEEE Electron Device Lett.* 35 (3): 381–383. https://doi.org/10.1109/LED.2014.2300496.

9 CEI-28G-SR Standard (2010). The Physical and Link Layer Working Group of the Optical Internetworking Forum.

10 Shi, Z., Wang, X., Liu, J. et al. (2012). Programmable on-chip ESD protection using nano crystal dots mechanism and structures. *IEEE Trans. Nanotechnol.* 11 (5): 884–889. https://doi.org/10.1109/TNANO.2012.2204767.

11 Wang, X., Shi, Z., Liu, J. et al. (2013). Post-Si programmable ESD protection circuit design: mechanisms and analysis. *IEEE J. Solid-State Circuits* 48 (5): 1237–1249. https://doi.org/10.1109/JSSC.2013.2255192.

12 Arreghini, A., Driussi, F., and Vianello, E. (2008). Experimental characterization of the vertical position of the trapped charge in Si nitride-based nonvolatile memory cells. *IEEE Trans. Electron Devices* 55 (5): 1211–1219.

13 Wang, A. (2016). Interposer-based ESD protection structures. UC patent disclosure, US Provisional Patent Application, No. 62/412,105.

14 Wang, A. (2016). ESD protection structures using TSV in ICs. US Provisional Application, No. 62/385,770.

15 C. Li, M. Di, Z., Pan, H., Wu and A., Wang (2021). Vertical TSV-Like Diode ESD Protection, *Proc. IEEE Electron Devices Technology and Manufacturing Conference (EDTM)*, pp. 676–678, DOI: 10.1109/EDTM50988.2021.9421011.

15

ESD CAD for Full-Chip Design Verification

15.1 Full-Chip ESD Design Verification

As discussed, one of the biggest challenges in electrostatic discharge (ESD) protection designs is to predict ESD protection performance at chip level before Si fabrication. Indeed, in integrated circuit (IC) design practices, nobody would tape-out any design before fully validating the IC from simulation to physical verification in order to enhance the chance of first-Si success. Figure 15.1 depicts the typical IC design flow. Starting with the desired IC circuit specs, a designer needs to select suitable circuit topology and construct the circuit schematics. Pre-simulation will be thoroughly conducted to achieve the circuit specs at the schematic level. Careful physical layout will then be completed, followed by a set of layout verification routines, including design rule checking (DRC), layout versus schematic (LVS), ERC, and extraction, etc. After correcting any layout errors, post-simulation will be performed, which includes extracting the parasitic parameters from the layout data, such as bus resistance, capacitance, leakage, and noise, as well as checking the process, voltage, and temperature (PVT) corner variations, to validate that the IC will meet the Specs in a given layout. Only after the full-chip physical design verification, then can the IC be taped out for Si fabrication. Yet, there is still no guaranty for the real Si functions even after the comprehensive IC physical design verification. ESD protection design should follow the same design flow. Unfortunately and ironically, the common ESD protection designs have been following a much simpler design flow, as shown in Figure 15.1: starting from the desired ESD protection Specs, then choosing an ESD protection structure, which will be simply dropped into the IC core schematics. Typically, very limited ESD simulation, e.g., TCAD or SPICE simulation, may be applied in designing an ESD protection structure. At chip layout level, some limited ESD-flavored DRC and LVS checking may be conducted. The ESD DRC is often just a relaxed layout spacing design rules to avoid unwanted effects, such as latch-up. The ESD LVS is used to check if an ESD sub-circuit is electrically constructed correctly, such as connecting an NMOSFET into a ggNMOS or gcNMOS ESD protection structures. Sometimes, simple SPICE simulation is conducted for timing analysis, such as in a gcNMOS ESD protection sub-circuit to determine the values for the coupling resistor and capacitor. The comprehensive TCAD-based mixed-mode ESD protection design simulation method (Chapter 8) and SPICE-based ESD protection circuit-level simulation (Chapter 13) using ESD device behavior models discussed before are certainly very useful, however, still incomparable to the desired full-scale ESD protection physical design (i.e., layout level) verification procedures similar to the routines in normal IC designs. It is again important to understand that ESD protection involves multiple-coupling effects, i.e., transient-electro-thermal-materials-process-device-circuit-layout coupling effects and interacts actively with the core circuitry at full-chip level. TCAD mixed-mode ESD protection simulation

Practical ESD Protection Design, First Edition. Albert Wang.

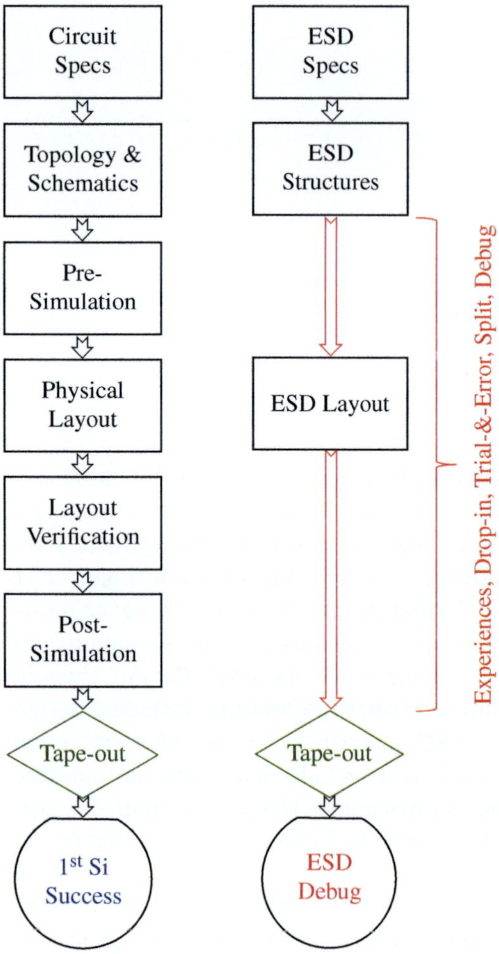

Figure 15.1 Comparing common IC design flow with typical ESD protection design procedures reveals the incompetence of traditional ESD protection design methods.

can evaluate ESD discharging physics and functions, however, TCAD ESD simulation is not suitable for large complex chips. Using 3D TCAD ESD simulation to study the ESD layout effects, such as ESD device edge and corner effects, is particularly impractical for a large circuit [1, 2]. On the other hand, SPICE ESD protection circuit simulation using ESD device behavior models is only useful at the schematic level, which cannot validate ESD physical layout designs [3–5]. The ESD-IC codesign technique is very useful to balance the ESD and IC core performance trade-offs at schematic level, however, again, cannot address the ESD layout design problems [6, 7]. Therefore, complete and meaningful ESD protection physical design verification must be able to validate all ESD discharging functionalities at both schematic and layout levels for the whole chip, which is an *ESD-function-based design verification methodology* that is far more than simple DRC and LVS checking with ESD flavors. Only after full-chip ESD-function-based design verification, then there may be a better chance to achieve first-Si ESD protection design success, instead of resorting to the traditional ESD failure debugging routines that are tedious, time-consuming, and costly. To achieve this goal, novel *ESD-function-based CAD algorithms* and software are critical, which are well more than TCAD and ECAD ESD simulation [8–12]. The ESD-function-based ESD design verification relies on smart ESD parametric checking to deal with irregular ESD protection structures to extract arbitrary ESD protection and parasitic ESD-like devices inside a given layout, to extract

ESD-critical parameters of any extracted ESD-like devices, to generate ESD netlist from layout data, and to identify any potential ESD discharging paths, both designed and unexpected, on a chip.

15.2 CAD Algorithms for ESD Design Verification

Comprehensive full-chip ESD protection circuit design verification must be conducted at physical layout level, i.e., taking the layout data (e.g., GDSII file) as the input and then performing ESD-function-based smart ESD parametric checking by CAD. The ultimate goal for ESD physical design verification is to ensure the first-Si design success, which facilitates ESD protection design optimization and prediction. The right ESD protection design principle is to make sure the design will work on the computer screen before sending it out for Si fabrication. While a design working on the screen does not guarantee it working in Si, no Si should be fabricated without working on the screen first. To enable accurate function-based chip-scale ESD physical design verification, the *smart ESD parametric checking* should have the following critical features:

- Extract all the designed ESD protection structures of arbitrary shapes (a.k.a, *Intentional* ESD devices) from the layout;
- Extract all the possible parasitic *ESD-like* structures (i.e., parasitic ESD devices) from the layout (Most ESD-like devices will not be functional under ESD stressing, though);
- Extract the *ESD-critical parameters* for the intentional and parasitic ESD devices extracted from the layout;
- Generate the *ESD netlist* consisting of all extracted ESD-like devices from the layout including their electrical connection (The initial ESD netlist contains all possible ESD-like devices, non-functional parasitic ESD devices will be eliminated, resulting in the final ESD netlist),
- Check intentional ESD devices (Any missing ESD devices in layout? Any wrong connection? Any noncompliance in the ESD-critical parameters? etc.),
- Identify all the possible *ESD discharging paths* between any two pads on a chip (Some may not be functional under ESD stressing),
- Estimate the ESD-critical parameters for each possible ESD discharging path identified (i.e., *equivalent ESD-critical parameters* for a specific ESD discharging path),
- Identify the actual ESD discharging path(s) between any two pads by removing nonfunctional parasitic ESD discharge paths through analyzing its equivalent ESD-critical parameters,
- Establish the chip-level *ESD failure criteria*, e.g., voltage breakdown, thermal breakdown (i.e., maximum current), etc.,
- Conduct *smart parametric ESD checking* at chip scale (*Quasi-static checking* by analyzing the ESD devices, ESD discharging paths and their ESD-critical parameters),
- Conduct full-chip ESD circuit simulation and ESD zapping test routines per any industrial ESD test models and standards (*Transient ESD circuit simulation* using ESD device models and ESD input stimuli).

Figure 15.2 depicts the ESD-function-based smart parametric checking mechanism and ESD CAD design verification block diagram [8–12]. The ESD CAD design verification platform comprises four functional CAD modules acting upon four input data files: the technology file provides key technology data for a given IC process including doping, diffusion, and materials; the layout file has the full physical layout data, typically in GDSII format; the ESD-critical parameters are defined in the definition file; and the reduction criteria file contains specific rules to compare and eliminate not-to-be-concerned ESD-like devices extracted. The ESDExtractor module analyzes the

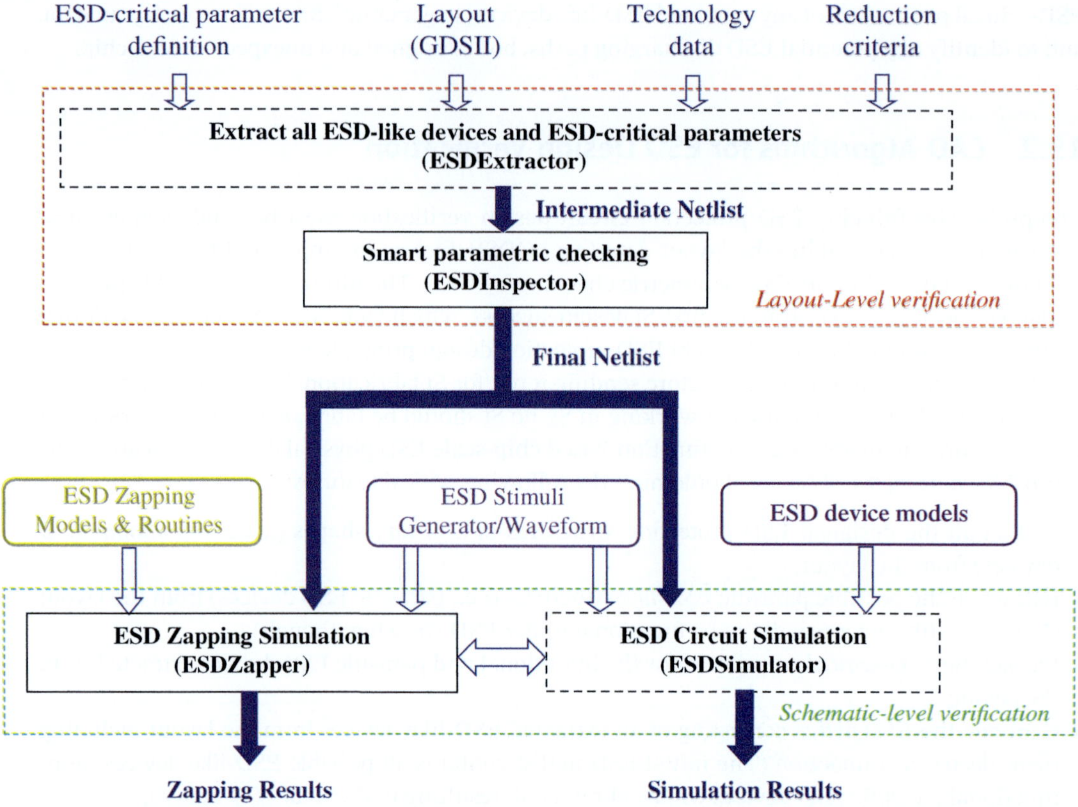

Figure 15.2 A diagram for ESD-function-based smart parametric ESD checking mechanism and ESD CAD physical design verification flow.

input layout data and extract all possible ESD devices including the intentional ESD protection structures and any unwanted parasitic ESD-like devices of arbitrary shapes, such as a BJT originated from a guard ring. Further, based on the technology data, the ESD-critical parameters for the extracted ESD-like devices will be estimated, such as equivalent ESD V_{t1}, I_{t1}, V_h, I_h, R_{ON}, V_{t2}, I_{t2}. In addition, key device parameters for some unique ESD-like devices extracted will also be calculated, such as the effective current gain, β, for a BJT and the β-product ($\beta_{NPN} \times \beta_{PNP}$) of an SCR ESD device. This extraction module will deliver a preliminary ESD netlist (intermediate level) which contains a large number of various parasitic ESD-like devices most of them will be eliminated in the device reduction phase. Extracting arbitrary ESD-like devices from a layout is very challenging for several reasons: First, unlike regular IC devices, e.g., MOSFET, ESD protection devices are irregular. Second, parasitic ESD-like devices can appear anywhere in a layout, which often can be life-threatening to ICs under ESD stressing. IC layout is a geometrical description of masks. Regular devices, e.g., MOSFET, are presented by a unique set of masks stacked on top of each other in a Manhattan style. However, ESD protection devices are irregular structures featuring complicated layout patterns and electrical connections, e.g., SCR and LVSCR structures. Regular IC devices can be recognized by Boolean logic operations of layout masks, which in general cannot recognize abnormal ESD devices. ESDExtractor module is an extraction engine that uses a subgraph isomorphism technique to recognize arbitrary ESD-like vertical or lateral devices. In recognizing ESD devices, each common ESD protection device type (e.g., ggNMOS, gcNMOS, BJT, STI diode, gated diode, SCR,

MVSCR, LVSCR, dSCR) is described by a unique device *model graph* (*MG*), while the whole IC layout data is presented as a *target graph* (*TG*). In principle, an ESD device is identified by matching TGs with MGs. In addition to the large number of ESD protection device *MGs*, another unique feature of ESD-like devices is that different ESD protection structures may share common layer elements (device redundancy), for example LVSCR, MVSCR, and ggNMOS share common elements (i.e., ggNMOS). Hence, a *decomposition* approach to improve ESD device recognition efficiency is used. The algorithm first explores the relationship between the ESD device *MGs* by a decomposition procedure by which the common part of any two or more *MGs* will only be matched for once. The model containment relationship is explicitly expressed, hence, not to extract any redundant device, i.e., a smaller device contained in a larger device structure. Each layout region is a simple polygon (non-self-intersecting polygon) consisting of a set of stacked masks described by a mask value code. Simple geometrical adjacency relationship/operation is used for ESD device definition, which is represented by a *MG* in the form of *MG (N, R)*, where the attribute *N* is a set of vertices with each *vertex* corresponding to a non-overlapping region with a specific mask value code, and the attribute *R* is a set of relationships between two different vertices of the device model, which, for ESD devices, includes geometrical adjacency relationship *R*1 and electrical connectivity relationship *R*2. Hence, a complete set of vertices forms an ESD device model, *MG*. Figure 15.3 depicts a *MG* for an exemplar ggNMOS ESD protection device, where each circle represents a vertex of the *MG*, the solid lines represent geometrical adjacency relationship *R*1 and the dashed lines represent electrical connectivity *R*2. Intentional ESD devices follow strict layout design rules and are isolated from the core circuit. Figure 15.4 shows an intentional SCR ESD device with guard-ring and pick-up diffusion plugs. Since a parasitic SCR ESD device is geometrically irregular in layout without any guard-ring and pick-ups, in order not to miss such a parasitic SCR ESD-like device in extraction, the generic *MG* of a SCR ESD device is defined by the blue dashed box. While such a SCR MG definition will not miss any parasitic ESD-type SCR devices, redundant parasitic ESD-type devices will be extracted, which may be removed by the device count reduction module later. In summary, the ESDExtractor module can extract arbitrary ESD devices, including intentional ESD device and any parasitic ESD-like devices with redundancy from the layout file.

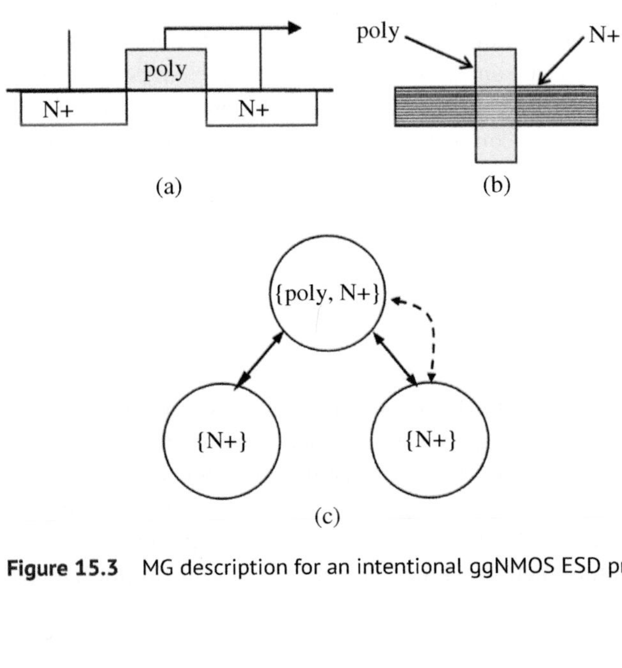

Figure 15.3 MG description for an intentional ggNMOS ESD protection device.

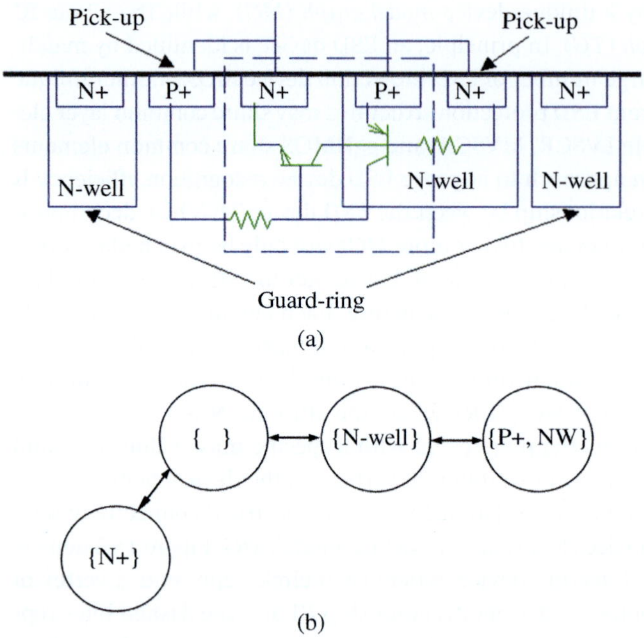

Figure 15.4 MG description for an intentional SCR ESD protection structure: (a) cross-section, and (b) MG definition.

The ESDInspector module is an ESD function checking engine that utilizes a unique smart parametric ESD checking algorithm to conduct full-chip ESD design verification at layout level by analyzing the ESD-critical parameters of the extracted ESD-like devices based on the ESD protection functions (static in nature, no transient ESD discharging simulation). In the meantime, by using the device reduction criteria, the intermediate ESD netlist will be reduced to the final ESD netlist by eliminating unconcerned parasitic ESD-like devices extracted [12]. The layout-level ESD protection design verification is to pin-down any layout design problem and to generate the final ESD netlist for schematic-level verification and ESD testing simulation. The smart parametric ESD checking algorithm can be better understood in exemplar case analysis. In Case 1, the extracted ESD intermediate netlist is examined to see if all intentional ESD protection devices are designed correctly. If any ESD device is missing or incorrect in layout and/or connection, the layout has an error (equivalent to ESD DRC and LVS checking). In Case 2, the extracted ESD-critical parameters are checked for their compliance against the ESD design Specs, e.g., V_{t1}, V_h, R_{ON}, and I_{t2} (i.e., ESD-function-based checking). If any incompliance is identified quantitatively, a design error is reported for ESD malfunction (not just spacing and connection checking). In Case 3, assume two ESD devices, A and B, are extracted that are connected in parallel, if V_{t1} of $A > V_{t1}$ of B, A will be removed because it will never be turned on (ESD device redundancy). If A and B are in parallel and have comparable V_{t1}, then neither A nor B could be removed from the netlist yet. Next, R_{ON} will be checked, and if R_{ON} of A is larger than R_{ON} of B, A will be removed since ESD current will mainly discharge into device B (i.e., analyzing ESD current share proportional to R_{ON}). Then, in Case 5, assume A and B have similar V_{t1} and R_{ON}, so both can conduct ESD currents, then V_h will be checked. Since the lower V_h will determine the pad clamping voltage, if the lower V_h is higher than the maximum allowed pad voltage, the design is faulty. Next, assume that the ESD devices A and B have comparable V_{t1}, R_{ON}, and V_h, if the I_{t2} is lower than a preset value, then the design is faulty because it cannot handle the target ESD current. If I_{t2} of device A is lower than that of device B, then A may be damaged first under ESD stressing. In Case 6, assume two extracted ESD-like devices, A and B, are in series connection, the total series V_{t1} (equivalent) will be estimated first, then compared with any existing parallel ESD structures. Other special check criteria can be preset and checked as well. For example, in Case 7 on a β-criterion, assume two ESD devices, A and B, are in parallel, if device A is an SCR ESD device with β-product < 1, then A cannot be turned on and will be removed. In Case 8 on triggering criterion, a maximum V_{t1} criterion of $V_{t1\text{-max}}$ is preset, if V_{t1} of $A \gg V_{t1\text{-max}}$, then ESD device A cannot be turned on and will be removed as a competing parasitic ESD-like device. In Case 9 of Holding criteria, a maximum V_h is preset as $V_{h\text{-max}}$, if ESD device A has a $V_h > V_{h\text{-max}}$, then device A fails in pad clamping checking and will be removed as a competing parallel ESD-like device. In Case 10 of R_{ON} criterion, a maximum R_{ON} criterion of $R_{ON\text{-max}}$ is preset. If ESD device A has $R_{ON} > R_{ON\text{-max}}$, then early ESD failure in device-A is expected, hence, an ESD design error is reported. The smart parametric ESD checking routines will go on and on until all extracted ESD-like devices are checked against specific checking criteria within the ESD netlist on a chip (not individual standalone ESD protection device). The ESDInspector performs ESD-function-based static checking and the checking outcomes are the following: First, it delivers a *final ESD netlist* for ESD schematic simulation and ESD zapping test simulation. Second, it completes simple layout-level ESD design verification without involving any transient ESD zapping routines.

The ESDSimulator module and the ESDZapper module are two closely related CAD engines for transient schematic-level ESD protection circuit simulation and ESD zapping test simulation, respectively. With a final full-chip ESD netlist generated by the ESDExtractor and ESDInspector modules, as well as accurate ESD protection device models (e.g., behavior models from TLP testing within the scope of this book) in place, the ESDSimulator engine will apply an input ESD pulse as a stimulus from an embedded ESD pulse generator to the final ESD netlist to perform transient ESD discharging simulation at the schematic level to verify whole-chip ESD protection circuit design by analyzing the dynamic ESD discharging functionalities. The ESD stimulus can be generated by an equivalent ESD generation circuit per any industrial ESD test standard or model (e.g., HBM, CDM, IEC, and TLP) that will produce the required ESD pulses, or simply an analytical ESD waveform described by mathematical formula(s) per the ESD testing requirement. The ESDSimulator must be able to handle ultrahigh ESD pulses unlike SPICE circuit simulation that typically deals with small signals only. More usefully, the ESDZapper engine can perform complex full-chip ESD zapping test simulation per any industrial ESD testing standards and models. Importantly, the ESDZapper module can simulate the complex real-world ESD testing procedures for any pads or pad combinations, by following any industrial ESD zapping test routines, including all ESD stressing modes (i.e., PD, ND, PS, ND, DS, and SD) at any pad with respect to any other pad with any pad combinations, and using any stressing sequences (e.g., typically three positive ESD pulses and three negative ESD pulses with certain time lapse). This allows ordinary IC designers to perform full-chip ESD-function-based ESD zapping test verification for an ESD-protected IC before tape-out and Si fabrication, which can dramatically reduce the costs and time-to-market of new ICs. As depicted in Figure 15.2, the inputs to the ESDZapper engine include the final ESD netlist extracted, the ESD device models, the embedded ESD stimulus generator, and the ESD zapping model and routine set. Figure 15.5 illustrates that multiple actual ESD discharging path(s) may exist between two pads on a chip under ESD stressing, including the intentional ESD discharging path designed and possibly parasitic ESD conduction channels. ESDZapper utilizes a unique *Weighted Graph Method* to identify which ESD discharging path(s) between any pair of pads on a chip will dominate during an ESD event by ESD simulation (i.e., identify the actual ESD discharge path). In a nutshell, using Figure 15.6 as an example that contains complementary ggNMOS and

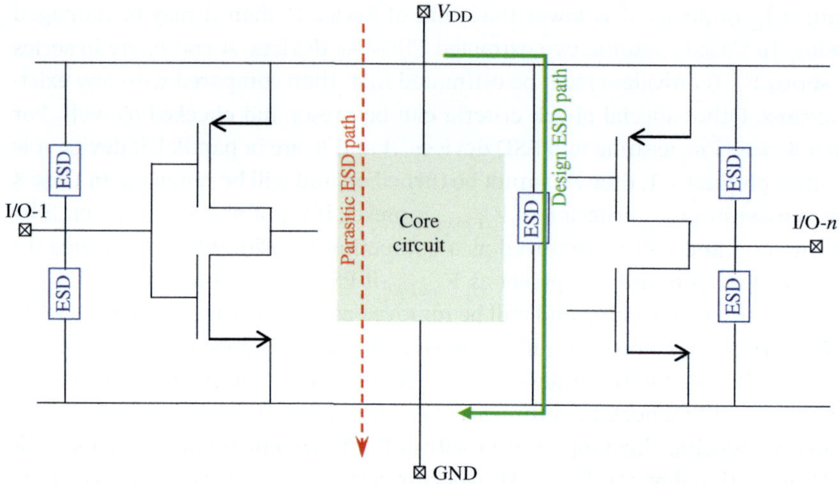

Figure 15.5 A conceptual on-chip ESD protection circuit shows multiple ESD discharging paths under power rail zapping for a chip, including an intentional ESD discharging path (solid green) through the ESD power clamp designed and a possible unwanted ESD conduction channel (dashed red) associated with any parasitic ESD-like devices inside the core circuit.

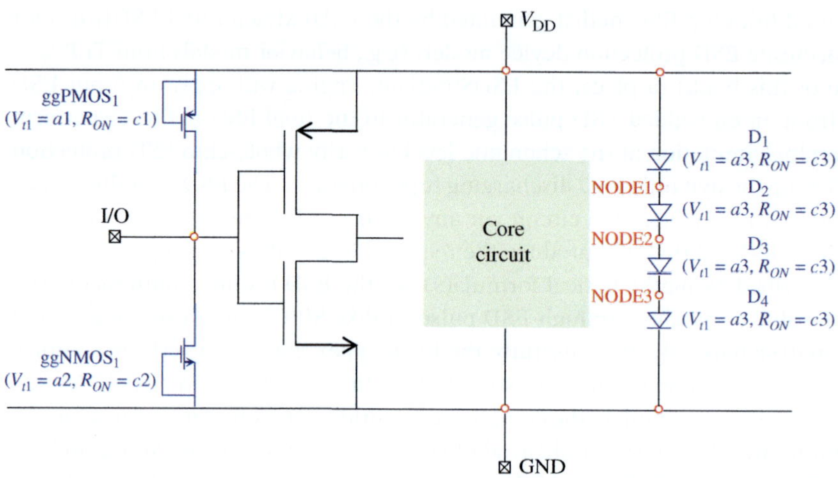

Figure 15.6 A exemplar on-chip ESD protection scheme uses a complementary ggNMOS and ggPMOS ESD protection devices at I/O pad and a diode-string power clamp, which is described by a weighted graph in the ESDZapper engine.

ggPMOS ESD protection devices at I/O pads and a diode-string ESD protection structure as the power clamp, a weighted graph is used to describe the ESD protection network where each electrical node is represented by a *Vertex* and each ESD protection device (intentional or parasitic) is an *Edge*, weighted by its ESD-critical parameters, of the weighted graph. Figure 15.7 depicts the weighted graph of the exemplar ESD protection network in Figure 15.6 [13]. Using the weighted graph for the full-chip ESD netlist, ESDZapper can analyze the weighted graph of the ESD protection circuit and identify the actual "working" ESD discharging path, which is the shortest ESD conduction path (i.e., the path with the smallest weight determined by the ESD-critical parameters)

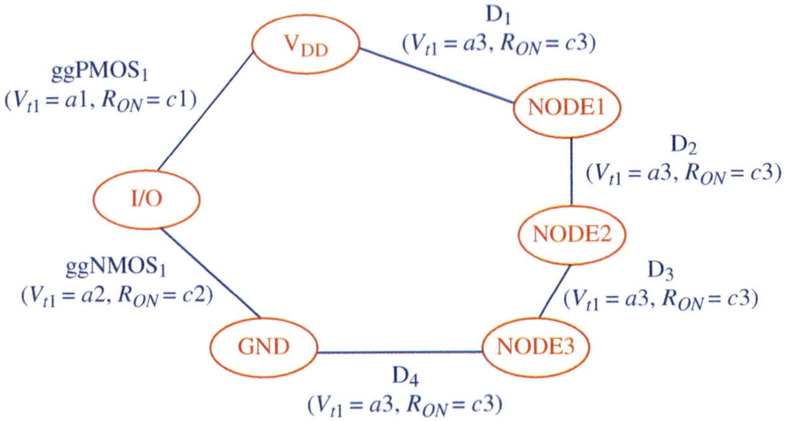

Figure 15.7 Illustration of a weighted graph for the ESD protection circuit shown in Figure 15.6 shows the vertices (electrical nodes) and the edges, being the ESD protection devices weighted by the corresponding ESD-critical parameters.

between any two vertices on the weighted graph. Therefore, the dominating ESD discharging channels between any two pads on a chip will be identified based on the ESD discharge functionalities. In summary, using the novel smart parametric ESD checking algorithm, ESD-function-based ESD physical design verification can be realized through both layout (static checking) and schematic (dynamic checking) level analyses, including ESD zapping test simulation, which allows full-chip ESD protection circuit physical design optimization, verification, and prediction.

15.3 Full-Chip ESD Design Verification Examples

This section presents design examples of full-chip ESD protection circuit physical design verification by CAD based upon ESD-function-based analysis. In the first example, a simplified logic core circuit with a CMOS buffer designed in a foundry 0.35 μm BiCMOS technology is protected by several ESD protection structures: a low-V_{t1} LVSCR ESD protection structure (LVSCR1) for one input pad (In1), an ESD protection diode in reverse mode at another input pad (In2), a complementary ggNMOS and ggPMOS ESD protection pair for the output pad (Out), and an SCR ESD protection device (SCR1) as the power clamp, as shown in Figure 15.8 [13]. For equivalent output ESD protection, the ggNMOS has one finger (GGNMOS1), while the ggPMOS has two fingers (GGPMOS1 and GGPMOS2). The ESD CAD tool takes the GDSII layout file for the chip and extracts all ESD-like devices, as shown graphically in Figure 15.9, including the intentional ESD protection devices, i.e., LVSCR1 at In1, lateral P-well/N-well diode at In2, power clamp device SCR1, and GGNMOS1, GGPMOS1, and GGPMOS 2 at Out. It is noted that the output GGPMOS ESD protection structure is extracted as two devices in parallel, GGPMOS1 and GGPMOS2 for the two fingers. In addition, several parasitic SCR ESD-like devices associated with the guard rings are extracted. The ESD-critical parameters are also calculated for the ESD-like devices using the IC technology and layout data. This forms an initial ESD netlist containing both intentional and parasitic ESD devices. The ESD-function-based smart ESD parametric checking routine is then performed, which eliminates nonfunctional parasitic ESD-like devices and leaves only one functional parasitic SCR ESD-like devices (SCR2) associated with the guard ring of the output buffer. A final ESD

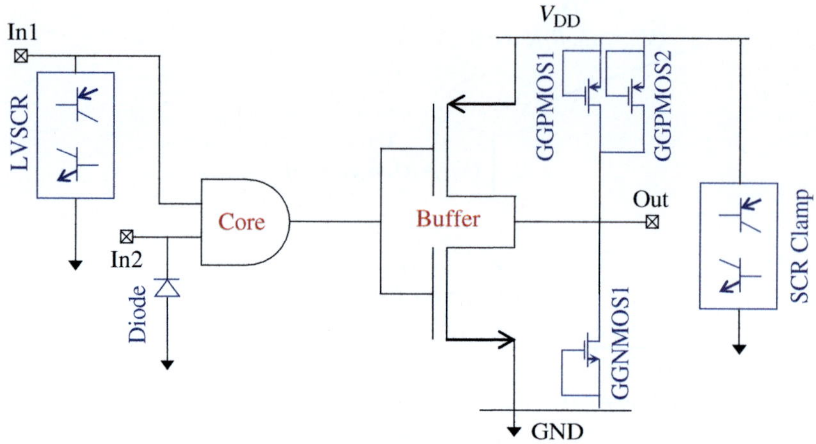

Figure 15.8 An IC circuit designed in a 0.35 μm BiCMOS contains LVSCR ESD protection at In1, diode ESD protection at In2, ggNMOS and ggPMOS ESD protection at Out, and SCR power clamp between V_{DD} and GND.

netlist is hence generated for ESD-function-based full-chip ESD protection circuit verification by simulation.

In the second example, a commercial ESD physical design verification CAD tool (SmartESD) is used to illustrate the smart ESD zapping routines at chip level [14]. SmartESD utilizes the ESD CAD algorithms described before and mimics the full IC design flow as shown in Figure 15.1. Figure 15.10 depicts the complete SmartESD CAD flow including the following key functions: define the ESD Specs, choose core Circuit Topology and ESD Protection Structures, construct Circuit + ESD Schematics, conduct Pre-Simulation, complete full-chip Layout, perform ESD layout Extraction to extract all ESD-like devices, their ESD-critical parameters and generate the Final ESD Netlist, conduct simple static layout-level ESD Physical Design Verification (e.g., ESD DCR, ESDLVS, ESDERC, and ESDxRC), then perform ESD-function-based schematic-level transient ESD Protection Circuit Simulation and comprehensive ESD Zapping Test Simulation per selected industrial ESD testing standards and models. The input data include IC Technology data, ESD-critical parameter definition file, Layout (GDSII) file, Device Reduction Criteria file, ESD Device Models, ESD Stimuli, and ESD Zapping Models and Routines. The full-chip ESD protection circuit physical design verification is conducted at both layout level (static and simplified quantitative checking) and schematic level (comprehensive transient ESD simulation), both based upon ESD discharging functionalities. The outcomes include quantitative data (textual) and graphical (via graphical user interface, GUI) illustration similar to normal IC design verification procedures, including an ESD discharge current density map (J-map), a node voltage map (V-map), and an ESD thermal temperature map (T-map) for the whole chip. Therefore, an IC designer can readily identify any ESD weak points graphically that may result in ESD failure on a chip (identified by ESD error flags), which is based on analyzing the circuit-level ESD discharging functions including the ESD layout effects (i.e., corner/edge effect), instead of the over-simplified ESD checking of layout spacings (DRC) and electrical connection (LVS) only.

The SmartESD CAD method has the following key ESD protection checking functions:

- Extract all intentional ESD protection structures and any arbitrary parasitic ESD-like devices from the layout (GDSII, graphic design system II format),
- Calculate the ESD-critical parameters for all ESD-like devices extracted,

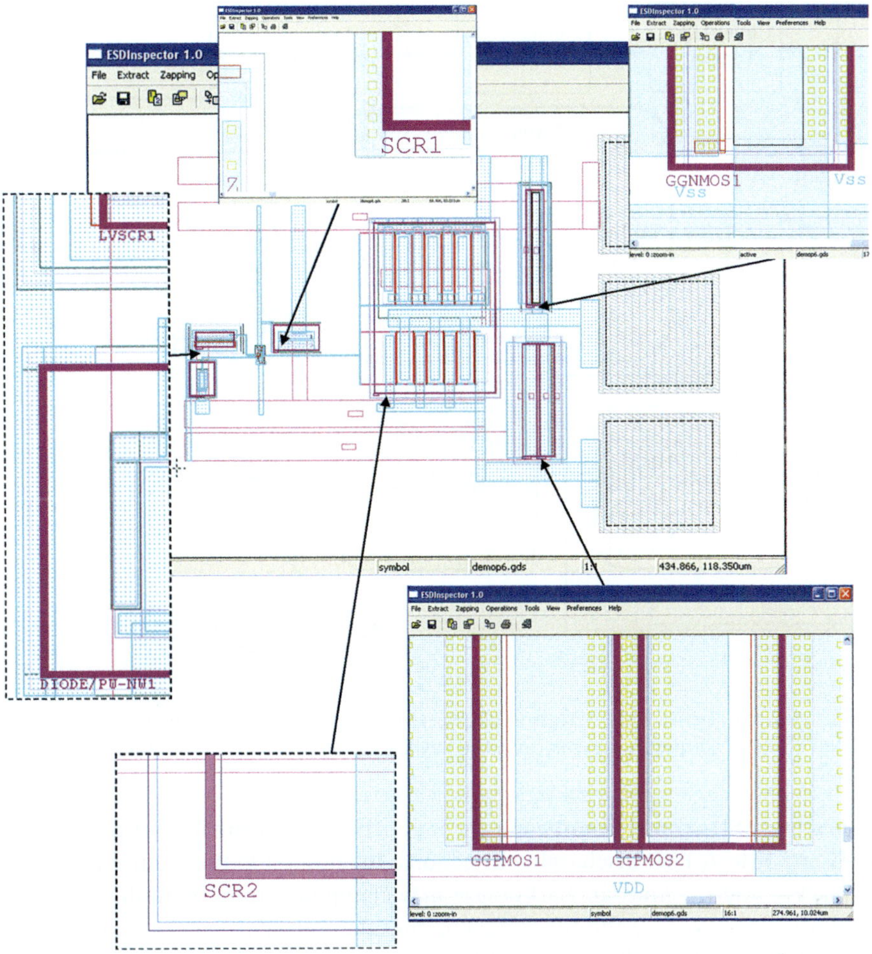

Figure 15.9 Layout view of ESD extraction outcomes for the IC shown in Figure 15.8 shows LVSCR1 at In1, Diode at In2, SCR1 as power clamp, GGNMOS1, GGPMOS1, and GGPMOS2 at Out. In addition, a functional parasitic SCR2 ESD-like device associated with the output buffer GR remains after ESD device reduction, which may compete against the intentional ESD protection devices, possibly resulting in early ESD failure.

- Generate the ESD netlist for the whole chip,
- Extract all possible ESD discharge paths between any two pads on a chip,
- Calculate equivalent ESD-critical parameters for any ESD discharge paths between two pads,
- Identify the actual ESD discharge path(s) between any two pads under specific ESD zapping model and stressing mode,
- Set up the chip-level ESD failure criteria,
- Quasi-static ESD design checking: all intentional ESD devices designed properly? (e.g., ESD-critical parameters, ESD circuit connection),
- Quasi-static ESD design checking: any ESD failures (e.g., voltage or thermal breakdown) in any ESD discharge paths? (Equivalent ESD-critical parameters versus ESD failure criteria),
- Transient ESD design checking: full-chip ESD circuit simulation and ESD zapping test simulation (using transient ESD stimuli, ESD device models, and any industrial ESD test standards/models and routines).

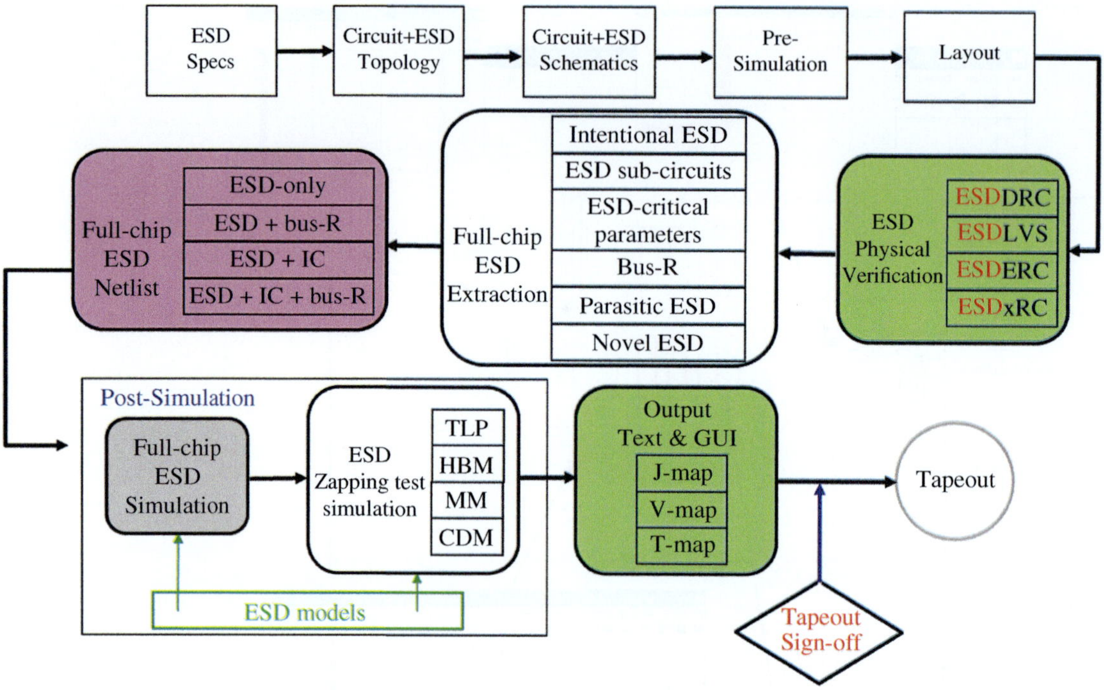

Figure 15.10 SmartESD allows full-chip ESD protection circuit design physical verification by analyzing comprehensive chip-level ESD discharging functions, with the key CAD modules and flow similar to the common IC design verification flow.

Figure 15.11 defines the ESD-critical parameters for a generic two-terminal ESD protection device in both A-to-K and K-to-A ESD discharging directions. In addition to the regular parameters, e.g., V_{t1}, V_h, R_{ON}, V_{t2}, and I_{t2}, since an ESD device may clamp at certain point during ESD discharge, the clamping voltage (V_{clamp}) and the corresponding clamping current (I_{clamp} at V_{clamp}) are also ESD-critical parameters. Typically, the peak ESD discharge current for an ESD protection device during an ESD event is I_{clamp} (V_{clamp}), i.e.,

$$I_{peak} = I_{clamp} \tag{15.1}$$

Note that while $I_{peak} = I_{t2}$ holds for an individual/standalone ESD protection device, at the chip level, the ESD protection device is typically clamped at lower maximum ESD discharge current level, i.e., $I_{clamp} < I_{t2}$. The peak voltage for the ESD protection device is defined as

$$V_{peak} \equiv \max(V_{t1}, V_h, V_{clamp}) \tag{15.2}$$

Similarly, equivalent ESD-critical parameters can be defined for any ESD discharge path on a chip that consists of multiple ESD protection devices (denoted as ESD device $i = 1, 2, \ldots n$) and ESD interconnects metal buses (denoted as ESD wire $j = 1, 2, \ldots m$) between those ESD protection devices. Therefore, the following formulas are given

$$V_{t1-path} = \sum\nolimits_i^n V_{t1-i} + \sum\nolimits_j^m (I_{bus-t1} R_{bus-j}) \tag{15.3}$$

$$V_{h-path} = \sum\nolimits_i^n (V_{h-i}, V_{clamp-i}) \tag{15.4}$$

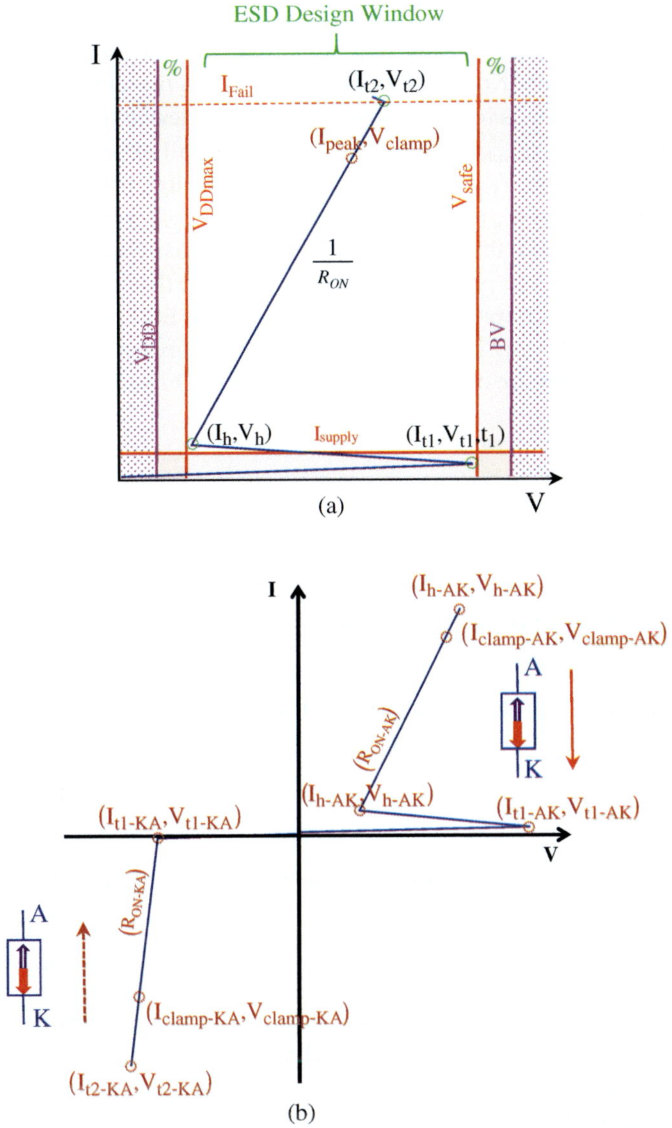

Figure 15.11 (a) Shown in an ESD Design Window, an ESD protection device may be clamped at certain point at chip level, which is lower than its ESD thermal breakdown threshold as a standalone ESD protection device, and (b) definitions of ESD-critical parameters in two possible ESD discharge directions for an ESD protection device.

$$R_{\text{ON-path}} = \sum\nolimits_i^n R_{\text{ON}-i} + \sum\nolimits_j^m R_{\text{bus}-j} \tag{15.5}$$

$$V_{t2\text{-path}} = \sum\nolimits_i^n (V_{t2-i}, V_{\text{clamp}-i}) + \sum\nolimits_j^m (I_{\text{ESD}} R_{\text{bus}-j}) \tag{15.6}$$

and

$$I_{t2\text{-path}} = Min(I_{t2-i}) \tag{15.7}$$

where R_{bus} is the series resistance of an ESD interconnects metal wire, $I_{bus\text{-}t1}$ is the bus current at ESD triggering threshold (typically, being a leakage), and I_{ESD} is the transient ESD discharge current running through an ESD conduction path. Note that, in estimating the equivalent $V_{t1\text{-path}}$ for a given ESD conduction path, $I_{bus\text{-}t1}$ corresponds to a low current at the ESD device-triggering threshold, while in calculating the equivalent $V_{t2\text{-}path}$, I_{ESD} is the full-conduction ESD discharge current in the path. In estimating the equivalent $V_{h\text{-path}}$, all individual ESD devices in the ESD discharge path may not reach to their ESD device V_h simultaneously (i.e., some ESD devices may be clamped at $V_{clamp\text{-}i}$). It is important to understand that while I_{t2} represents the maximum current handling capability of an individual/standalone ESD protection device (i.e., its ESD thermal failure threshold), the equivalent $I_{t2\text{-}path}$ for an ESD discharge path is set by the weakest ESD protection device in the conduction channel, which sets the ESD thermal failure current threshold for the given ESD discharge path on a chip. The ESD failure criteria typically include the followings:

(1) Voltage breakdown ESD failure criterion,
(2) Current/thermal ESD failure criterion,
(3) Other ESD failure criteria, e.g., soft ESD errors.

The ESD failure criteria at ESD device level (i.e., individual/standalone ESD protection structure) and full-chip level (i.e., full-chip ESD netlist) are very different, which are defined below.

Device-level ESD failure criteria based on the ESD Design Window:

1) Voltage breakdown ESD failure criterion,
   Per the ESD Design Window, a designed ESD protection device fails the voltage breakdown ESD failure criterion if (8) holds for the ESD protection structure,

$$V_{peak} \geq V_{safe} \tag{15.8}$$

2) Current/thermal ESD failure criterion,
   Per the ESD Design Window, a designed ESD protection device fails the thermal ESD failure criterion if (9) holds for the ESD protection structure,

$$I_{peak} \geq I_{t2} \tag{15.9}$$

Chip-level ESD failure criteria are more involving since any ESD failure depends on not only the individual intentional ESD protection structures but also the full-chip ESD netlist and various possible ESD discharge paths associated with large number of parasitic ESD-like devices, as well as ESD zapping routines and stressing modes.

1) Voltage breakdown ESD failure criterion,
   At chip-level, voltage breakdown-induced ESD failure occurs if (10) holds at any circuit nodes for the core devices connected to the node,

$$V_{peak} \geq V_{safe} \tag{15.10}$$

   where V_{safe} represents the breakdown voltage (BV) of the core transistor connected to the node (e.g., gate or source/drain diffusion breakdown).
2) Current/thermal ESD failure criterion,
   At chip-level, ESD current-induced ESD thermal failure occurs if (11) holds for any possible ESD discharge paths on the chip,

$$I_{peak} \geq I_{t2} - \text{path} \tag{15.11}$$

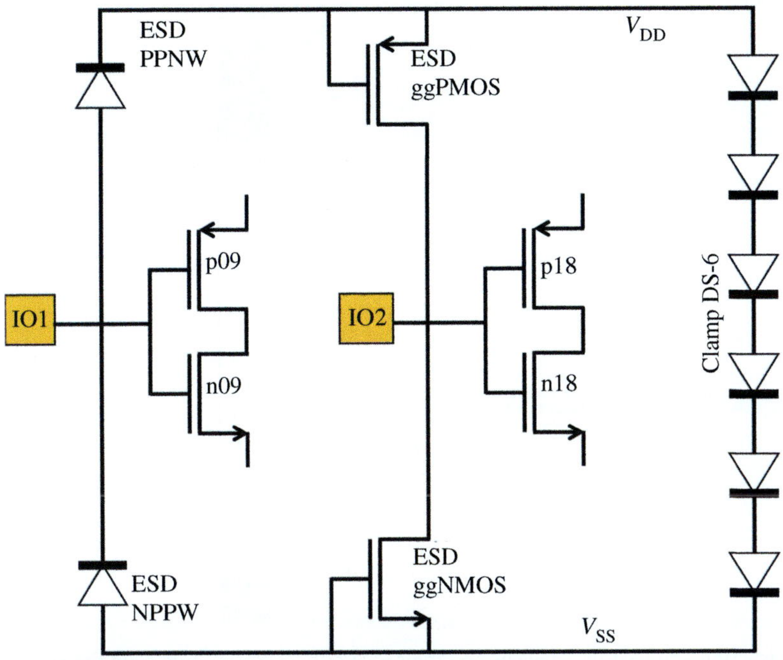

Figure 15.12 Schematic for an IC chip example designed in 28 nm CMOS features full-chip protection utilizing ESD protection diodes at input pad (IO1), ggNMOS, and ggPMOS ESD protection at output pad (IO2), and a six-diode diode-string power clamp.

It is important to understand that the weakest intentional ESD protection structure designed or, more often, a parasitic ESD-like device is typically the ESD thermal failure point on a chip. It is also worth of note that chip-level ESD failure analysis is entirely dependent upon the combination of ESD netlist and the ESD zapping routines and stressing modes (i.e., ESD stressing directions, pin combinations, and ESD test standards and models selected). Therefore, at chip level, all ESD-critical parameters for an individual ESD-like device and equivalent ESD-critical parameters for a specific ESD discharge path must be estimated in both directions, i.e., A-to-K or K-to-A.

Figure 15.12 depicts a simplified IC designed in a foundry 28 nm CMOS technology with full-chip ESD protection, including P-$^+$/N-well and N-$^+$/P-well diodes for pull-up and pull-down ESD protection at input (IO1), respectively; complementary ggNMOS and ggPMOS ESD protection at output (IO2); and a 6-diode diode-string clamp for power rail ESD protection [14]. SmartESD CAD takes the layout file (GDSII) as the input to extract all ESD-like devices and their ESD-critical parameters with the results shown in Figure 15.13. Table 15.1 lists the ESD-critical parameters for all extracted ESD-like devices under both A-to-K and K-to-A ESD stressing modes. For ESD-function-based smart parametric ESD design checking, the required ESD Design Window can be established using the voltage breakdown data listed in Table 15.2 for core CMOS transistors. The full-chip ESD protection circuit netlist is then generated for smart parametric ESD checking at layout level and transient ESD simulation at chip level. Next, ESD functionalities are analyzed to identify all possible ESD discharge paths on the chip, which is directly related to the ESD zapping routines, including the pad combinations and ESD stressing directions. In Case-1 for positive IO2-to-IO1 ESD zapping, an ESD stimulus is applied to the IO2 pad with the IO1 being the reference ground, as shown in Figure 15.14. Four possible ESD conduction paths are identified from IO2 to IO1, and their equivalent ESD-critical parameters for these ESD conduction paths are calculated based on the

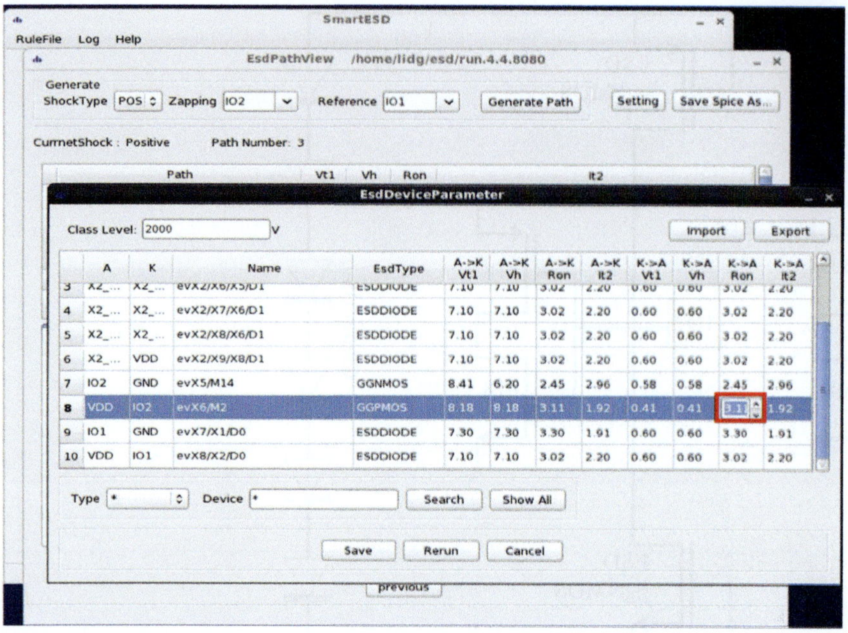

Figure 15.13 Example of extracting any ESD-like devices and their ESD-critical parameters from the chip layout file for the IC shown in Figure 15.12. The extraction accuracy depends on input data and calibration, and the values can be adjusted for the extracted ESD parameters according to measurement data (Red box).

Table 15.1 Extracted ESD devices and ESD-critical parameters.

ESD devices	A → K zapping				K → A zapping			
	$V_{t1\text{-AK}}$ (V)	$V_{h\text{-AK}}$ (V)	$R_{\text{ON-AK}}$ (Ω)	$I_{t2\text{-AK}}$ (A)	$V_{t1\text{-KA}}$ (V)	$V_{h\text{-KA}}$ (V)	$R_{\text{ON-KA}}$ (Ω)	$I_{t2\text{-KA}}$ (A)
PPNW (IO1)	7.1	7.1	3.02	2.2	0.6	0.6	3.02	2.2
NPPW (IO1)	7.3	7.3	3.3	1.91	0.6	0.6	3.3	1.91
ggPMOS (IO2)	8.18	8.18	3.11	1.92	0.41	0.41	3.11	1.92
ggNMOS (IO2)	8.41	6.2	2.45	2.96	0.58	0.58	2.45	2.96
D-String (Clamp)	42.6	42.6	18.12	2.2	3.6	3.6	18.12	2.2

ESD-critical parameters for individual ESD devices shown in Table 15.1. Further analysis of these possible ESD conduction paths are performed using their ESD-critical parameters, e.g., the equivalent ESD path triggering voltage, $V_{t1\text{-path}}$,

$$\text{Path-1}: V_{t1-\text{path1}} = 0.41 + 3.6 + 0.6 = 4.61\ \text{V} \tag{15.12}$$

$$\text{Path-2}: V_{t1-\text{path2}} = 8.41 + 0.6 = 9.01\ \text{V} \tag{15.13}$$

$$\text{Path-3}: V_{t1-\text{path3}} = 0.41 + 7.1 = 7.51\ \text{V} \tag{15.14}$$

$$\text{Path-4}: V_{t1-\text{path4}} = 8.41 + 42.6 + 7.1 = 58.11\ \text{V} \tag{15.15}$$

Table 15.2 CMOS core BV.

Core MOSFET	BV_D (V)	BV_G (V)	BV_S (V)
n09	4.26	4.26	4.26
p09	4.41	4.41	4.41
n18	9.15	9.15	9.15
p18	9.36	9.36	9.36

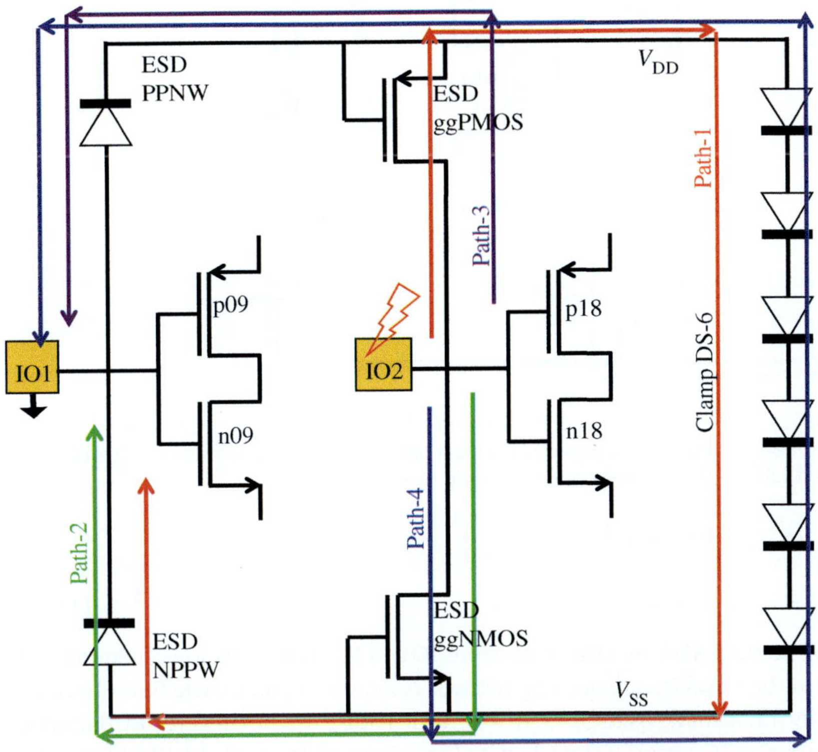

Figure 15.14 Case-1 ESD zapping scenario: in a positive IO2-to-IO1 ESD stressing event, four possible ESD conduction paths are identified and their equivalent ESD-critical parameters are estimated, which will be used to smart-check the actual ESD discharge behaviors at chip level.

It is found that the Path-1 will be the actual ESD discharge path from IO2 to IO1 because it has the lowest equivalent ESD triggering voltage of $V_{t1\text{-path1}} \sim 4.61$ V. Similarly, Figure 15.15 depicts the Case-2 for positive IO1-to-IO2 ESD stressing where an ESD stimulus is applied to the IO1 pad with respect to IO2 pad being grounded. Four possible ESD conduction paths are identified from IO1 to IO2, and their equivalent ESD-critical parameters for these ESD conduction paths are calculated, e.g., $V_{t1\text{-path}}$, listed below,

$$\text{Path-1}: V_{t1-\text{path1}} = 0.6 + 3.6 + 0.58 = 4.78\ \text{V} \tag{15.16}$$

$$\text{Path-2}: V_{t1-\text{path2}} = 0.6 + 8.18 = 8.78\ \text{V} \tag{15.17}$$

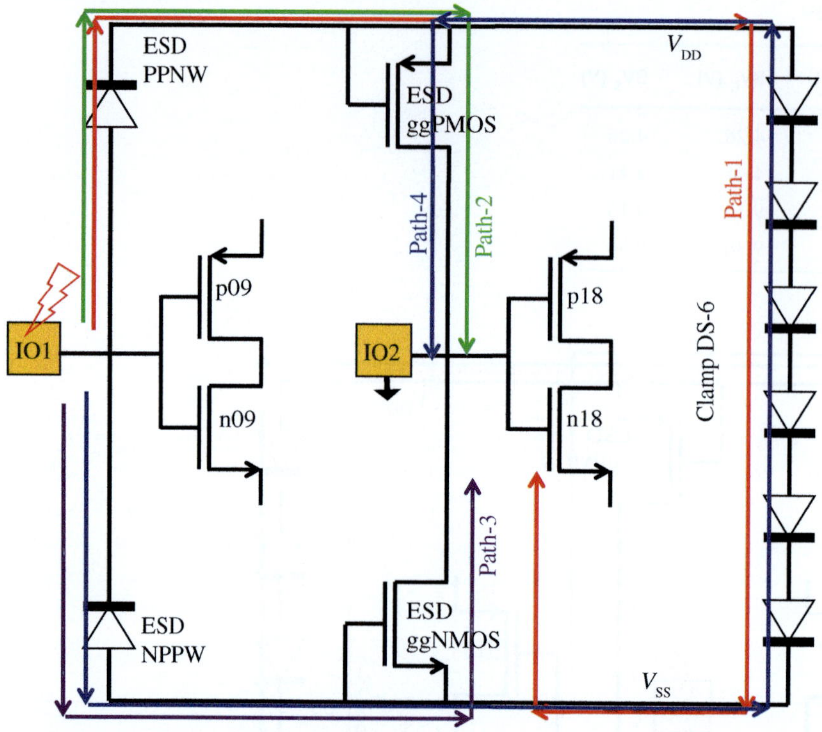

Figure 15.15 Case-2 ESD zapping scenario: in a positive IO1-to-IO2 ESD stressing event, four possible ESD conduction paths are identified and their equivalent ESD-critical parameters are estimated, which will be used to smart-check the actual ESD discharge behaviors at chip level.

$$\text{Path-3}: V_{t1\text{-path3}} = 7.3 + 0.58 = 7.88\,\text{V} \tag{15.18}$$

$$\text{Path-4}: V_{t1\text{-path4}} = 7.3 + 42.6 + 8.18 = 58.08\,\text{V} \tag{15.19}$$

The Path-1 will be the actual ESD discharge path from IO1 to IO2 due to its lowest equivalent $V_{t1\text{-path1}} \sim 4.78\,\text{V}$. Quasi-static ESD circuit checking follows. Next, smart parametric ESD checking is performed for Case-1 by transient ESD simulation, where the Path-1 will be triggered to discharge the ESD pulse that occurred at IO2 w.r.t. IO1 as depicted in Figure 15.16. A 2 kV HBM ESD pulse ($I_{max} \sim 1.33$ A) is applied to IO2 pad and the chip is checked for possible ESD failures based on both voltage breakdown and current/thermal ESD failure criteria. In comparison to the core CMOS BV data in Table 15.2 and the estimated ESD path equivalent ESD-critical parameters, since the IO2 pad is directly connected to the gates of core PMOSFET (p18) and NMOSFET (n18), it is found that $V_{t1\text{-path1}} = 4.61\,\text{V} < BV_{G\text{-}n18} = 9.15\,\text{V}$ and $BV_{G\text{-p18}} = 9.36$, hence, the breakdown voltage ESD failure criterion passed, meaning the gates of the output buffer CMOS transistors are safe under 2 kV HBM ESD zapping. Next, on ESD thermal failure checking, since $I_{peak} = I_{max} = 1.33\,\text{A} < I_{t12\text{-path1}} = \text{Min}(1.92, 2.2, 1.91\,\text{A}) = 1.91\,\text{A}$, the peak 2 kV HBM ESD pulse can safely pass through the Path-1 without overheating, hence, the positive IO2-to-IO1 zapping also passes the thermal ESD failure criterion. Following this smart parametric ESD checking algorithm, more ESD failure criteria

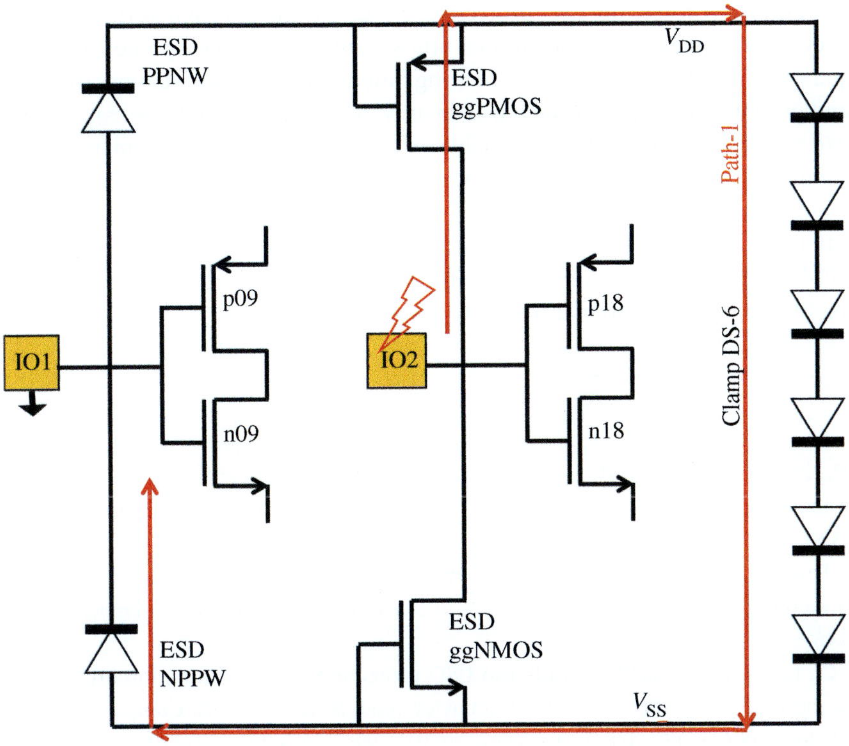

Figure 15.16 Case-1 ESD zapping scenario: in a positive IO2-to-IO1 ESD stressing event, analyzing the equivalent ESD triggering voltage suggests that Path-1 will be the actual ESD discharge path to conduct the 2 kV HBM pulse from IO2 pad to GND via IO1 pad.

associated with other concerned ESD-critical parameters can be conducted for any ESD zapping routines for the whole chip, therefore, to verify the ESD protection physical design, based on ESD discharge functionalities, before design tape-out and Si fabrication. It is worth of note that only quasi-static ESD checking routines are discussed in the examples, which seems to be simple, but the ESD checking is still rather comprehensive because it is based on ESD discharge function analysis, not just using over-simplified DRC/LVS checking of ESD flavor. More comprehensive and transient ESD discharging analysis can be conducted using ESD device models and ESD testing models for full-chip ESD protection circuit simulation and ESD zapping simulation at schematic level for more accurate ESD design analysis and verification.

15.4 Summary

This chapter discusses a powerful ESD CAD algorithm and CAD tool allowing ESD-function-based full-chip ESD protection physical design verification at both layout level and schematic level. The inputs to the ESD CAD platform include IC technology data, ESD-critical parameter definition file, Layout (GDSII) file, device reduction criteria file, ESD device models, ESD stimuli, and ESD zapping models and routines. Whole-chip ESD protection circuit physical design verification is conducted at both layout level (quasi-static and simple) and schematic level (comprehensive and transient), both are quantitative based upon ESD discharging functionalities. The outcomes include data in text format and graphical illustration, such as J-map, V-map, and T-map for a chip. Since the ESD CAD platform utilizes accurate ESD device behavior models and a smart parametric ESD checking algorithm, and analyzes transient ESD discharge functionalities for the whole chip, it allows ordinary IC designers to perform full-chip level ESD protection physical design verification in a way similar to the normal IC design verification flow. It is important to conduct full-chip ESD-function-based ESD protection circuit physical design verification before any design tape-out in order to realize first-Si design success in real-world IC designs.

References

1 Feng, H., Chen, G., Zhan, R. et al. (2003). A mixed-mode ESD protection circuit simulation-design methodology. *IEEE J. Solid-State Circuits* 38 (6): 995–1006. https://doi.org/10.1109/JSSC.2003.811978.

2 Xie, H., Zhan, R., Feng, H. et al. (2004). A 3D mixed-mode ESD protection circuit simulation-design methodology. *Proceedings of IEEE Custom Integrated Circuits Conference (CICC)*, pp. 243–246.

3 Zhang, F., Wang, C., Lu, F. et al. (2019). A full-chip ESD protection circuit simulation and fast dynamic checking method using SPICE and ESD behavior models. *IEEE Trans. Comput. Aided Des. Integr. Circuits Syst.* 38 (3): 489–498. https://doi.org/10.1109/TCAD.2018.2818707.

4 Wang, C., Lu, F., Ma, R. et al. (2016). A study of accurate extraction of ESD parasitic capacitance. *Proceedings of IEEE International Conference on Solid-State and Integrated Circuit Technology (ICSICT)*, pp. 437–440. https://doi.org/10.1109/ICSICT.2016.7998944.

5 Wang, L., Ma, R., Zhang, C. et al. (2014). Behavior modeling for whole-chip HV ESD protection circuits. *Proceedings of IEEE International Symposium on Power Semiconductor Devices and ICs (ISPSD)*, pp. 182–184. https://doi.org/10.1109/ISPSD.2014.6856006.

6 Lu, F., Ma, R., Dong, Z. et al. (2016). A systematic study of ESD protection Co-design with high-speed and high-frequency ICs in 28 nm CMOS. *IEEE Trans. Circuits Syst. I* 63 (10): 1746–1757. https://doi.org/10.1109/TCSI.2016.2581839.

7 Wang, X.S., Wang, X., Lu, F. et al. (2014). Concurrent design analysis of high-linearity SP10T switch with 8.5 kV ESD protection. *IEEE J. Solid-State Circuits* 49 (9): 1927–1941. https://doi.org/10.1109/JSSC.2014.2331956.

8 Zhan, R., Feng, H., Wu, Q., and Wang, A. (2003). ESDExtractor: A new technology-independent CAD tool for arbitrary ESD protection device extraction. *IEEE Trans. Comput. Aided Des. Integr. Circuits Syst.* 22 (10): 1362–1370. https://doi.org/10.1109/TCAD.2003.818140.

9 Zhan, R., Feng, H., Wu, Q. et al. (2004). ESDInspector: A new layout-level ESD protection circuitry design verification tool using a smart-parametric checking mechanism. *IEEE Trans. Comput. Aided Des. Integr. Circuits Syst.* 23 (10): 1421–1428. https://doi.org/10.1109/TCAD.2004.833613.

10 Zhan, R., Xie, H., Feng, H., and Wang, A. (2005). ESDZapper: A new layout-level verification tool for finding critical discharging path under ESD stress. *IEEE Proc. Asia South Pacific Design Automation Conference (ASP-DAC)*, pp. 79–82.

11 Lin, L., Wang, X., Tang, H. et al. (2009). Whole-chip ESD protection design verification by CAD. *Proceedings of EOS/ESD Symposium*, pp. 28–37.

12 Wang, A. and Zhan, R. (2007). A parameter checking method for on-chip ESD protection circuit physical design layout verification. US Patent No. 7, 243, 317B2.

13 R. Zhan (2005) "ESDcat: A New CAD Software Package for Full-Chip ESD Protection Circuit Design Verification", *A PhD Dissertation*, Illinois Institute of Technology.

14 SmartESD by TrustChip Technology Inc.

16

New CDM ESD Protection

Different electrostatic discharge (ESD) phenomena have different origins, which have been described by different industrial standards and models for ESD zapping tests, including human body model (HBM), charged device model (CDM), and International Electrotechnical Commission (IEC) [1–4]. Understanding various ESD phenomena and developing accurate models to faithfully characterize the real-world ESD discharging events have been an on-going research effort, which continuously evolves over years. Currently, HBM ESD events have been well understood, as such design and characterization of HBM ESD protection solutions have been done routinely with confidence and accuracy. As integrated circuit (IC) technologies continuously advance, and chip complexity and scale increase rapidly, CDM ESD protection emerges as a major IC reliability design challenge. Unfortunately, while many CDM test standards and models have been proposed, CDM ESD protection design and characterization remain a big headache to IC designers in design practices. Indeed, CDM ESD protection design still seems to be a black magic full of uncertainties today. Facing the unbearable losses of CDM ESD caused field returns, IC designers and product developers have many unanswered questions on CDM ESD protection: How could an IC suffer from CDM ESD failures in a field, though the same chip passed the CDM ESD testing in a lab? Why is CDM ESD testing unreliable, not reproducible over time in the same testing setting or not repeatable across different test settings built per the same CDM test standards? Why are CDM ESD testing results so random, seeming to be sensitive to everything, such as testers, operators, and environments? Why is there no quantitative relationship between CDM and HBM testing results? Are current CDM ESD testing models trustable? In summary, do we really understand CDM ESD protection?! The short answer is, NO, at least not much. The existing CDM ESD protection solutions and testing models are highly questionable [5, 6].

16.1 Misconception in CDM ESD Protection

The mystery in CDM ESD protection is rooted in its nature, which is entirely different from the origin for HBM ESD events. In principle, HBM ESD event is an *external-oriented* ESD phenomenon. When a charged human body contacts an IC, the electrostatic charges stored inside the human body will discharge into the IC chip via bonding pads. The transient energy associated with an HBM ESD pulse, as depicted in Figure 16.1a, may cause ESD damages to the IC core. The HBM ESD phenomena are well understood and HBM ESD testing standards are well developed [1, 2]. It is clear that the danger of an HBM ESD event is that a large amount of external static charges (the "alien intruders") will flow into an IC chip (the "victim") through a bonding pad (e.g., I/O, V_{DD}, V_{SS}). Therefore, the obvious HBM ESD protection strategy is to block the external electrostatic

Practical ESD Protection Design, First Edition. Albert Wang.

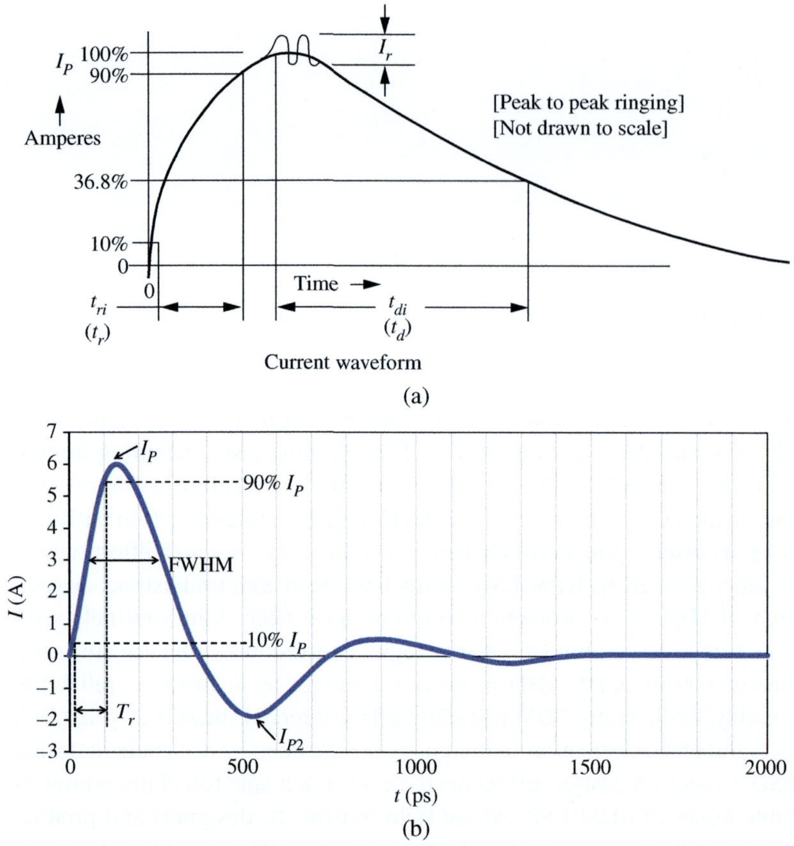

Figure 16.1 Standard ESD pulse waveforms: (a) HBM ESD model, and (b) CDM ESD model.

charges (the "devil") at the pad (i.e., the "gate") so that no electrostatic charge will flow into the IC chip. Accordingly, the classic *pad-based ESD protection solution*, as depicted in Figure 16.2, is widely used, in which on-chip ESD protection structures, with properly designed ESD-critical parameters (e.g., V_{t1}, t_1, R_{ON}, V_h, V_{t2}, I_{t2}), are connected at the pads [5]. As an incident HBM ESD pulse comes to a pad, the ESD protection device will be turned on to form a low-R conduction path to discharge the alien charges directly into the ground, i.e., blocking the incoming static charges from getting into the IC core, hence protects the IC against any HBM ESD transients. Simply speaking, the classic pad-based ESD protection method follows a *from-External-to-Internal* protocol by using an ESD protection device (the "guard") at the pad to keep any "alien" charges away from the IC core right at the "gate." Clearly, such a pad-based ESD protection solution only works if an ESD event is a "from-External-to-Internal" event, such as, HBM, MM, and IEC ESD events. Obviously, if there is no such a "gate" existing between the "intruders" and the "victim," the classic pad-based ESD protection method will not function, which seems to be the case for CDM ESD event.

On the other hand, a CDM ESD event is entirely different in nature, which is an *Internal-oriented* ESD phenomenon. Basically, CDM ESD event is a self-charging/discharging ESD event. In the real world, an IC will be charged inevitably and *slowly* in many ways, e.g., triboelectrically or field induction, during its lifetime. The induced electrostatic charges will be stored inside the IC randomly and arbitrarily in a *distributed* manner. When a self-charged IC is grounded, the electrostatic charges accumulated inside an IC will be discharged through the grounding pad (GND) into the ground,

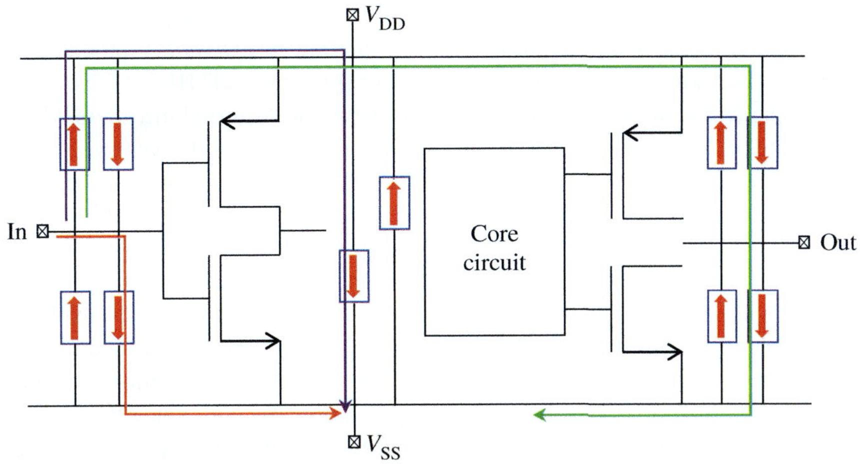

Figure 16.2 Classic pad-based ESD protection method relies on discharging the incident external charges into the ground via low-R ESD conduction path(s), which works for "external-oriented" "from-External-to-Internal" HBM ESD protection. An ESD protection device at a pad serves as a "guard" to prevent "intruders" from getting into the "door" of an IC. Positive In-to-V_{SS} zapping as an example. Arrow box = ESD protection structure.

which produces a strong and ultrafast CDM ESD transient flowing from inside the IC core outward to the ground, likely resulting in CDM ESD failures in the IC core [6]. Figure 16.1b depicts a typical CDM ESD pulse waveform [3]. In addition to the strikingly different natures of the origins of CDM and HBM ESD phenomena, it is well known that an CDM transient is extremely fast ($t_r \sim 100$ ps) and very short ($t_d \sim 2$ ns) compared to its HBM counterpart ($t_r \sim 10$ ns, $t_d \sim 150$ ns). Hence, the main R&D effort in designing on-chip CDM ESD protection structure has been to make it "faster," i.e., reducing the device triggering time (t_1) under CDM ESD zapping. Unfortunately, simply making a CDM ESD protection structure "fast" does not guarantee CDM ESD protection on a chip. Many industrial CDM ESD test standards and models were developed for various CDM ESD test set-ups [3]. For example, the latest CDM ESD standard provides details for building a field-induction charged device model (FICDM) ESD tester, as illustrated in Figure 16.3, which has two main steps: (i) to charge an IC by field induction, and (ii) to discharge the charged device

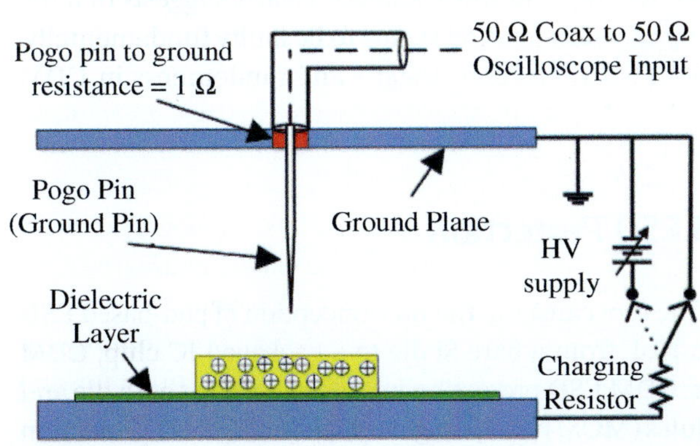

Figure 16.3 The industrial standard FICDM ESD testing set-up starts with an accelerated "short" charge induction procedure that cannot faithfully model the "long" lifetime CDM charging procedure in the real world.

(*device under test,* DUT) by applying the pogo pin to touch the DUT for grounding [3]. Obviously, the "internal-oriented" CDM ESD phenomenon follows a "*from-Internal-to-External*" protocol that is completely different from the "external-oriented," "from-External-to-Internal" HBM ESD event. The critical difference resides in the fact that the danger comes from the "external intruders" that want to get into an IC in an HBM ESD event, while the "devils" in an CDM ESD event are "trojans" that are already embedded inside an IC who want to rush out. Therefore, it is hard to believe that the same on-chip protection method will work for both HBM and CDM ESD protection. It is recently reported that the classic pad-based CDM ESD protection method, though widely accepted by the IC industry, is fundamentally faulty, which may contribute to the uncertainty and randomness of CDM ESD protection design failures commonly observed today. Several factors may explain the CDM ESD problem. First, since the "bad" electrostatic charges are "trojans" embedded inside an IC, an ESD protection device at pad loses its main protection role of being a "guard" at the "gate" to fend off any external "intruders" as in HBM ESD protection. Second, since the trojan charges inside an IC have to run through the IC core before reaching to a grounding pad and being discharged into the ground, one can expect that internal CDM damages may occur somewhere inside the IC due to high voltage or current anyway while the internal charges run from inside to outside of an IC core. Third, CDM ESD failure is closely related to the internal current route during CDM discharging. Two players will affect the internal CDM discharging routing: the amount of electrostatic charge stored inside the IC and, more importantly, their *internal distribution* (i.e., their "GPS" addresses) within a chip. In general, the internal "trojans" may rest anywhere, hence, the internal electrostatic charge distribution is random and time-variant during the lifespan of an IC. No existing model can precisely model the internal charge distribution during a CDM ESD event. Fourth, the industrial FICDM test standard utilizes an overly simplified accelerated charge induction procedure, which is very *short* compared with the real-world *lifetime* CDM charging course, hence, cannot faithfully represent the true CDM charging procedures, particularly the internal distribution of charges that is time-variant. Obviously, the charges induced into the DUT IC during the charging phase using an FICDM-based CDM tester will not have enough time to fully distribute throughout the IC chip as a true CDM ESD event does in the real world, likely leading to significant variations in CDM ESD failures. This means that the charges induced by FICDM mostly stays locally (e.g., "lumped" in the Si substrate or a package frame) as opposed to the real-world internal charge distribution inside an IC chip that is truly distributed anywhere throughout the IC die, randomly and time-variant. Therefore, any industrial FICDM CDM ESD tester is over-simplified, likely contributing to uncertainty in CDM ESD testing. The above analysis clearly suggests that the commonly used classic pad-based CDM ESD protection method seems to be faulty fundamentally, which certainly causes uncertainties in CDM ESD protection designs and randomness in CDM ESD failures in the field.

16.2 Analyzing Pad-Based CDM ESD Protection

In this section, examples are used to analyze the root causes to the misconception of pad-based ESD protection methods. As illustrated in Figure 16.4, from a bare Si die to a packaged IC chip, CDM ESD failure can occur anywhere in any phase. CDM ESD protection is hence needed at both die and package levels, including multiple-chip modules (MCM) [7, 8]. In general, full-chip ESD protection relies on a global ESD protection network on a chip for ESD protection, which contains a number of ESD protection devices connected to the pads that can be turned on by incoming ESD pulses to form low-R ESD conduction paths to discharge the incident ESD transients [6]. Figure 16.5 depicts

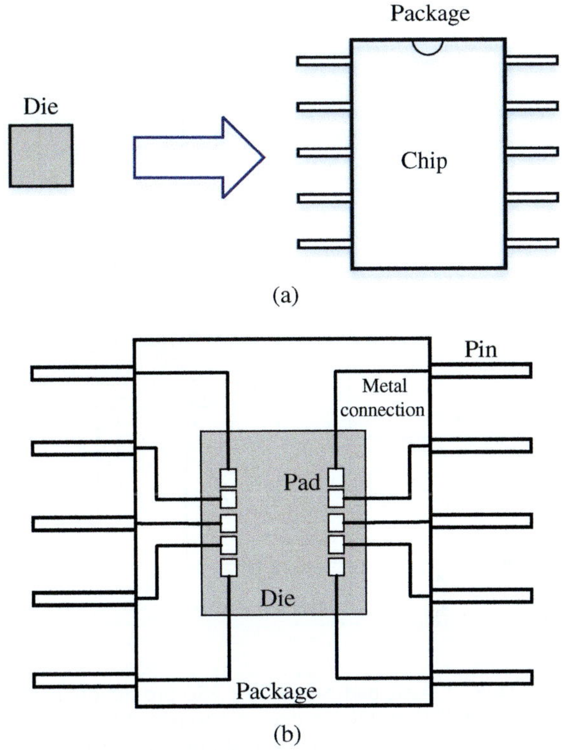

Figure 16.4 CDM ESD failure can occur at both die and package levels: (a) a die and a packaged IC, and (b) a die in a package frame where the pads are bounded to the pins. CDM-induced electrostatic charges can be stored anywhere randomly.

a CDM ESD-charged *packaged* IC with a traditional pad-based ESD protection network on chip. It is generally believed that the induced electrostatic charges are stored on the package frame and/or on the supply buses in a lumped way [9]. If this is true, when a package pin is grounded during CDM ESD stressing, the stored electrostatic charges can be charged into the ground either directly via the package frame metal buses nearby or indirectly through the on-chip ESD protection network. From the IC die view point, the package frame is "external" to the IC die and the electrostatic charges stored in the package frame are "external intruders" to the IC die, similar to HBM ESD events. Therefore, a CDM ESD event to a packaged IC is essentially a "from-External-to-Internal" ESD event to the IC die. As such, it seems that the classic pad-based ESD protection method should work for CDM ESD protection of packaged ICs. However, in real-world designs, an IC may often have an incomplete global ESD protection network on chip. For example, as shown in Figure 16.5, an ESD protection structure may not be optimized for its ESD-critical parameters (i.e., V_{t1}, t_1, R_{ON}) in both conduction directions. Consequently, under a given CDM ESD stressing condition, one specific ESD protection device may not be turned on in one direction (forward or reverse). Since the CDM-induced static charges can be stored anywhere and CDM ESD events are random in nature, there will always be cases where certain ESD protection devices cannot be turned on and the global ESD protection network is temporarily broken during certain CDM ESD stressing events. As a result, some electrostatic charges randomly stored in the package frame will inevitably flow "into" the IC die en route to be discharged into the ground through the package pins. Hence, it is likely that CDM ESD damages may occur to the internal IC core circuit even though the CDM ESD discharging process is completed through the pad-based CDM ESD protection network. In another case, also illustrated in Figure 16.5, an ESD protection device may be missing for a pad, e.g., due to layout error, which will also make the global ESD protection network incomplete on a

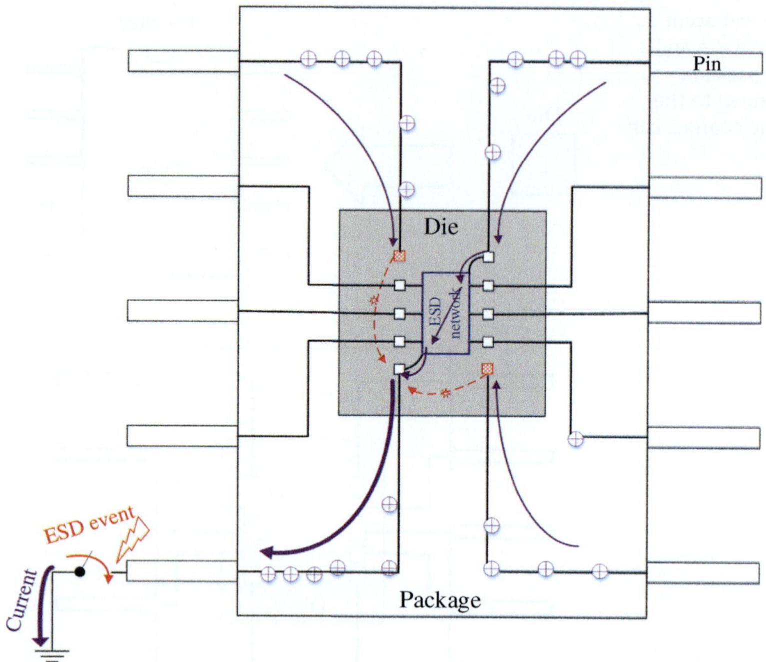

Figure 16.5 Classic pad-based CDM ESD protection method relies on the global ESD protection network on a die with ESD protection devices connected to all pads. Ideally, charges stored on the package frame and power rails can be discharged into GND without running into the IC core. However, ESD design errors due to triggering (upper-left red pad) or layout (lower-right red pad) mistakes break the global ESD protection network, causing CDM discharging current running through the internal core, resulting in CDM ESD damages to the IC die.

chip, and the electrostatic charges stored in the package may run into the internal IC core during an CDM ESD discharging event, resulting in internal CDM ESD failure to the IC die. A third case for incomplete global ESD protection network on chip is depicted in Figure 16.6 for high-speed or RF ICs. It is common in current IC design practices that some I/O pads for very high frequency or very high data rate are not ESD-protected because these I/O ports are very sensitive to any ESD-induced parasitic capacitance, hence, the design priority has to be given to IC performance, instead of ESD protection. Unfortunately, the electrostatic charges stored in the package frame will again have to run through the internal IC core, possibly causing CDM ESD damages to the IC die, even though the CDM ESD discharging is completed through the ESD protection structures at the pads. In summary, for packaged ICs, a CDM ESD event seems to be still "external-oriented" to the core IC die and the classic pad-based ESD protection method *may* or *may not* work for CDM ESD protection at chip level, entirely depending upon your luck.

Die-level CDM ESD failure becomes a major challenge for advanced ICs recently. Aggressive technology scaling down makes the gate layers and PN junctions more vulnerable CDM ESD induced voltage breakdown, while large and complex chips substantially worsens the internal accumulation and distribution of electrostatic charges from CDM induction, both factors dramatically lower the CDM ESD failure threshold. In the meantime, costs and loss of time-to-market associated with any CDM ESD failures are becoming more and more unbearable. Robust CDM ESD protection at die level hence becomes imperative. Figure 16.7 depicts a case where induced electrostatic charges are randomly distributed throughout a bare die of mixed-signal IC, anywhere and everywhere.

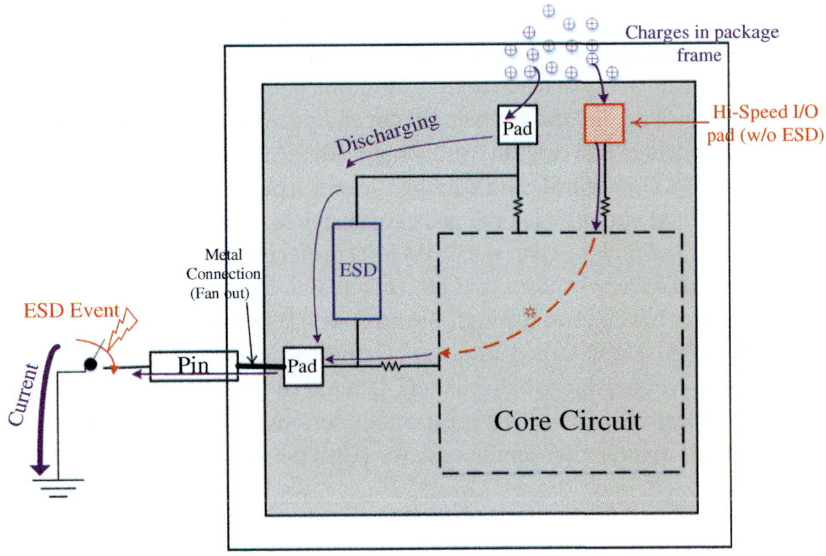

Figure 16.6 Certain high-speed and high-frequency RF signal I/O pads do not have ESD protection for packaged high-performance ICs, hence, the electrostatic charges on the package frame will have to flow into the internal IC core, possibly causing CDM ESD damages to the IC die during CDM ESD events.

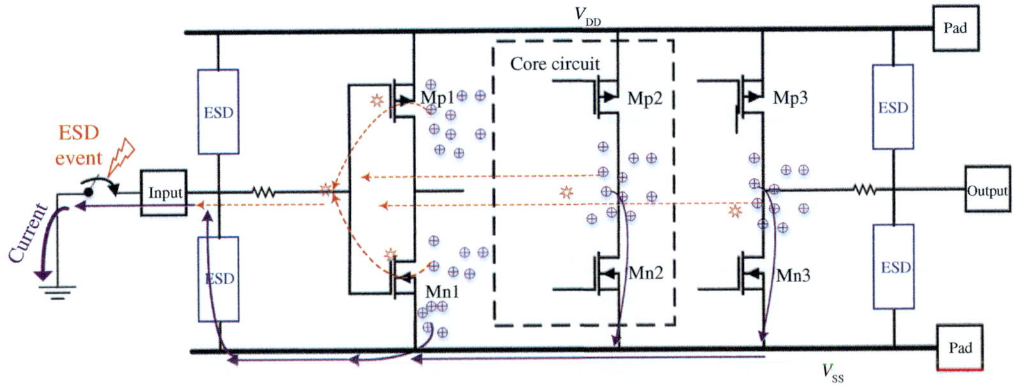

Figure 16.7 Illustration of a mixed-signal IC chip may have large amount of electrostatic charges induced over time that are distributed anywhere on the bare die. Some charges have to run through the internal circuit before reaching to the discharging pad, likely causing CDM ESD damages locally, internally, and randomly.

This IC chip does have a complete on-chip pad-based ESD protection network connected to all bounding pads. It is important to point out that CDM ESD failures are directly influenced by not only the amount of charges stored inside, but more critically, their internal distribution, which unfortunately is both unpredictable and time-variant. This is believed to be one key factor contributed to the mystery CDM ESD failures and protection designs. Figure 16.7 shows that the induced charges may be stored in the Si substrate, along the main metal buses, and local to transistors, particularly the large buffer transistors. For those electrostatic charges stored near a power rail and/or close to the GND pad, it is reasonable to expect that they can be safely discharge into a local grounding node through the pad-based global ESD protection network. Unfortunately, such ideal cases would only exist in a designer's dream. In a real world, the electrostatic charges are stored everywhere. For those charges staying far away from power rails and a GND pad, they must run through the *internal* core circuit to find a way to reach to the grounding pad before being discharged into the ground. Therefore, it is intuitive to believe that significant charge movement, en route to the discharging ground, will build up voltage and current locally inside the chip and likely cause internal CDM ESD damages within the IC die, regardless of whether there is a pad-based ESD protection network, how well the ESD protection structures at the pads are designed, if those ESD protection structures are validated by testing, and how fast can the CDM ESD protection devices be triggered. Figure 16.8 depicts other likely cases where substantial electrostatic charges are stored within a large transistor, where some charges have to run through the reverse S/D junctions or capacitively coupled through the gate before being discharged into the ground through the ESD protection structures at the pad. Therefore, it is reasonable to believe that CDM ESD damages may occur inside the MOSFET under CDM ESD stressing anyway. This is especially serious for silicon-on-insulator (SOI) and FinFET complementary metal-oxide-semiconductor (CMOS) technologies because of

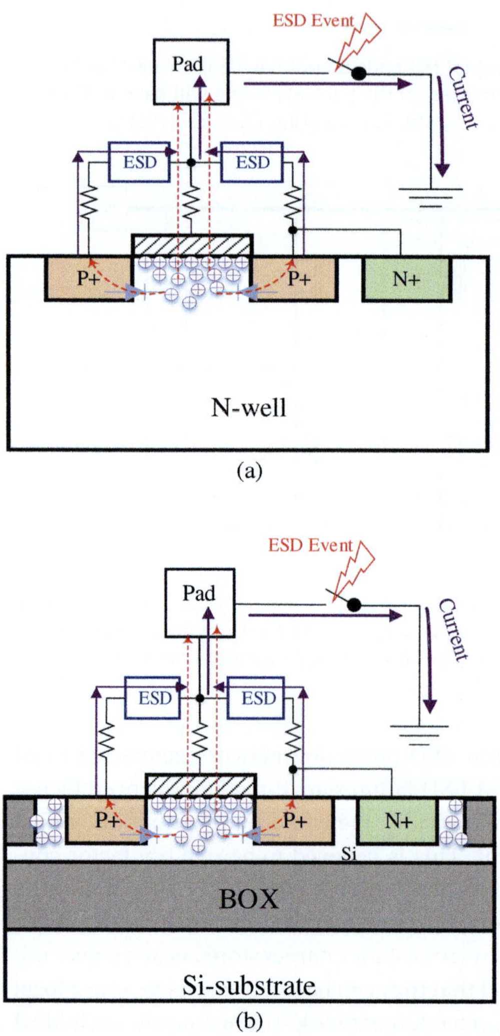

Figure 16.8 Induced electrostatic charges may be stored locally within a large transistor, which may have to run though the S/D junctions and the gate before being discharged into the ground via the ESD protection devices at the pad during a CDM ESD event: (a) bulk CMOS, and (b) SOI CMOS.

its excellent isolation nature. Very low CDM ESD failure threshold of <100 V was reported for gate damages in 14 nm FinFET technology [10]. From the above case analysis, it is clear that the classic pad-based CDM ESD protection method may not be functional for the "Internal-oriented," "from-Internal-to-External" CDM ESD events, at least not in theory.

Validation is carried out using a three-stage oscillator IC die designed in a foundry 45 nm SOI CMOS technology depicted in Figure 16.9 [11, 12]. Figure 16.9a shows that a full-chip CDM ESD protection scheme using the classic pad-based ESD protection network where the output port (CKO) features anti-parallel diode ESD protection subnets (D3–D8) and a power clamp unit (D0–D2) is used to protect the power rails. Standard ESD protection diodes from the PDK library are selected to protect the IC die. Standard FICDM test model is used in full-chip CDM ESD protection circuit simulation by SPICE. Figure 16.10 shows a conventional lumped FICDM test circuit model where the induced electrostatic charges are stored in the three capacitors (the charge reservoirs inside the IC die): i.e., the capacitance between the DUT and the induction Field Plate (C_{DUT}); the capacitance between the DUT and the discharge Ground Plate (C_{DG});

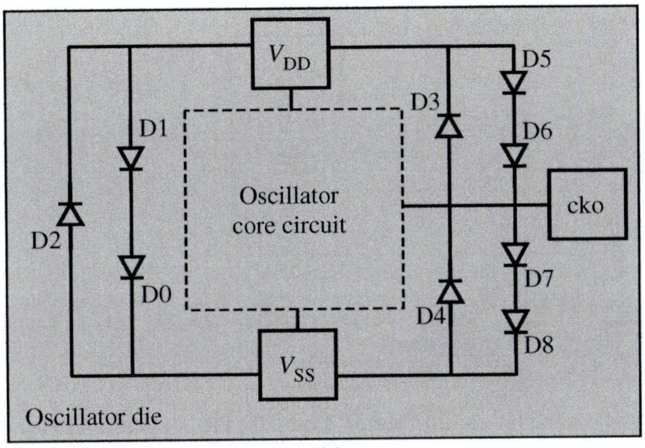

(a)

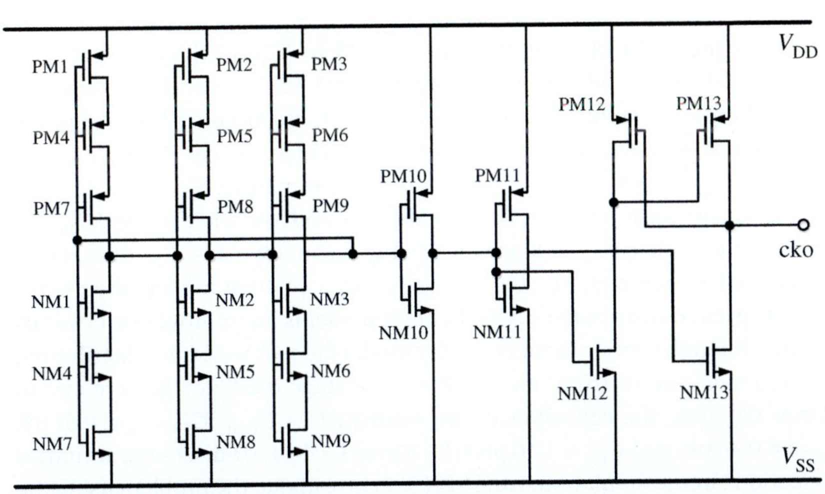

(b)

Figure 16.9 A three-stage oscillator IC designed in 45 nm SOI CMOS: (a) containing pad-based, full-chip ESD protection network, and (b) schematic for the oscillator core circuit.

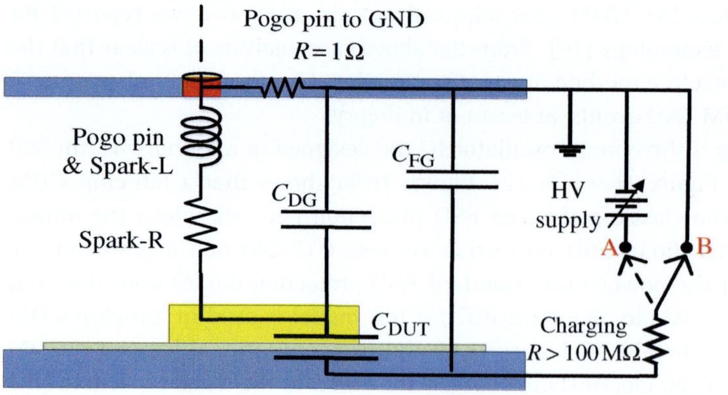

Figure 16.10 Lumped circuit model for standard FICDM tester where CDM-induced electrostatic charges are lumped into three capacitors, C_{DUT}, C_{DG}, and C_{FG}.

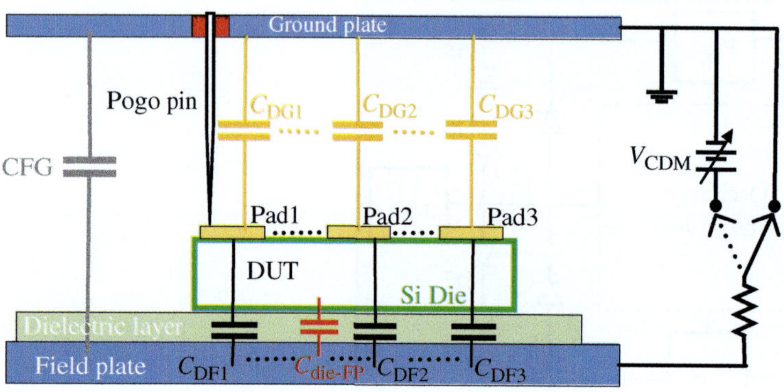

Figure 16.11 Improved pseudo-distributed charge distribution model for FICDM where CDM-induced electrostatic charges are stored in a set of capacitors, $C_{DG1,2\ldots m}$, $C_{DF1,2\ldots n}$, $C_{die\text{-}FP}$, and C_{FG}.

and the capacitance between the Field Plate and the Ground Plate (C_{FG}) [3]. Obviously, this *lumped* FICDM circuit model does not reflect the real-world CDM ESD situation because the CDM-induced electrostatic charges are actually stored inside a bare IC die in a distributed manner. Figure 16.11 depicts a *pseudo-distributed* FICDM circuit model, which breaks the C_{DG} into a set of parallel C_{DG1}, C_{DG2}, … C_{DGm} to reflect the charge distribution nature. Similarly, C_{DUT} is broken into two parts: the capacitance between Si die substrate and the Field Plate ($C_{die\text{-}FP}$), and capacitance from pad to Field Plate (C_{DF}). Since an IC chip has many pads, C_{DF} is replaced by a set of parallel capacitors for each pad, i.e., C_{DF1}, C_{DF2}, … C_{DFn}, which models the charge distribution nature too. This pseudo-distributed FICDM circuit model is an improvement over its lumped model counterpart, although it still cannot faithfully model the real-world full distribution nature of CDM charge induction and distribution. For the three-stage oscillator IC core circuit occupying a 2 mm × 2 mm die area, the capacitances are estimated as $C_{DG1,}$ $C_{DG2\ldots m} \approx 0.61$ pF, $C_{DF1,2\ldots n} \approx 2.12$ pF, $C_{die\text{-}FP} \approx 0.44$ pF, and $C_{FG} \approx 17.0$ pF [12]. Three CDM ESD discharge scenarios are studied by CDM protection circuit simulation using SPICE to investigate the problem using the classic pad-based CDM ESD protection method. Scenario-1 models using the traditional pad-based CDM ESD protection network, same as ESD protection design for the "from-External-to-Internal" HBM ESD stressing. Figure 16.12 depicts this "external-oriented" CDM ESD protection scheme

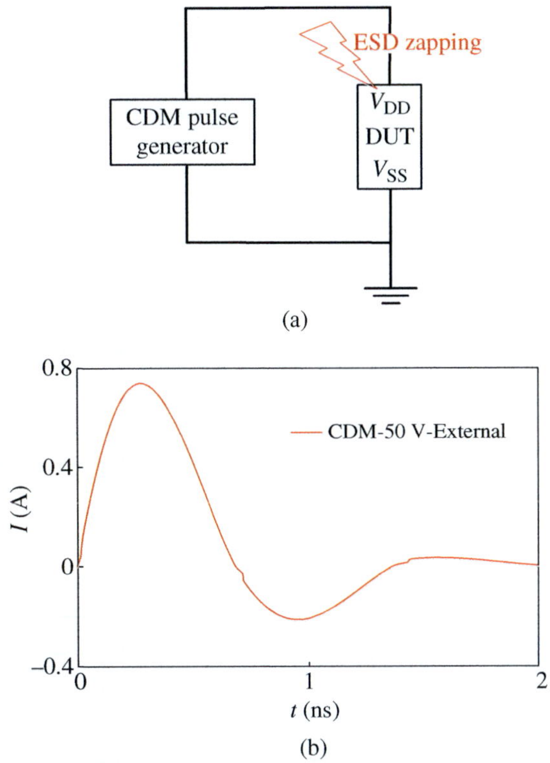

Figure 16.12 Scenario-1: Illustration of "from-External-to-Internal" discharge method to test classic pad-based CDM ESD protection: (a) CDM zapping set-up, and (b) CDM stressing waveform.

where a transient CDM ESD pulse is generated that is applied to a given pad to "externally" zap the IC die. Figure 16.13 shows the positive V_{DD}-to-V_{SS} CDM zapping mode where the V_{DD} pad is zapped with the V_{SS} pad grounded. It is observed that there will be two parallel ESD discharging paths, i.e., D0–D1 path and D2 path, each conducting in the positive and negative cycle during oscillatory CDM ESD stressing, respectively. Transient CDM ESD zapping from 50 to 500 V is simulated. Simple gate voltage breakdown CDM ESD failure criterion is used, i.e., $|BV_{GS}|$ and $|BV_{GD}|$, to examine any MOSFETs on a die. The 45 nm SOI CMOS features $BV_{OX} \approx 6.5$ V. Figure 16.14 depicts the simulated transient V_{GS} and V_{GD} for one PMOSFET (PM1) under 50 and 500 V CDM ESD zapping, which is always much lower than BV_{OX}, meaning the IC die passes 500 V CDM ESD protection test. It is also observed that V_{GS} and V_{GD} for PM1 have little change for CDM ESD stressing from 50 to 500 V, which is intuitively not reasonable. Figure 16.15 presents the simulated detailed CDM ESD discharging behaviors under 50 V V_{DD}-to-V_{SS} CDM zapping, which clearly shows that almost all transient CDM current is charged through the D0–D1 path and the D2 path during the positive and negative CDM cycle, respectively, with negligible current running through any internal MOSFETs, which further confirms "good" CDM ESD protection for the IC die. In short, Scenatrio-1 suggests that the classic pad-based CDM ESD protection method works in this case when an ESD pulse comes to the pad externally, similar to HBM.

Scenario-2 uses the "from-Internal-to-External" ESD procedure to model a real-world CDM ESD event, using the pseudo-distributed FICDM test circuit model as shown in Figure 16.11. Figure 16.16 depicts the CDM ESD protection simulation deck where "GND" and "Field" represent the CDM discharge Ground Plate and CDM induction (charging) Field Plate, respectively. V_{CDM} is the HV supply used to induce electrostatic charges into the IC die across the Field Plate, i.e., a required CDM ESD voltage rating (e.g., $V_{CDM} = 500$ V) is applied to the Field Plate and the

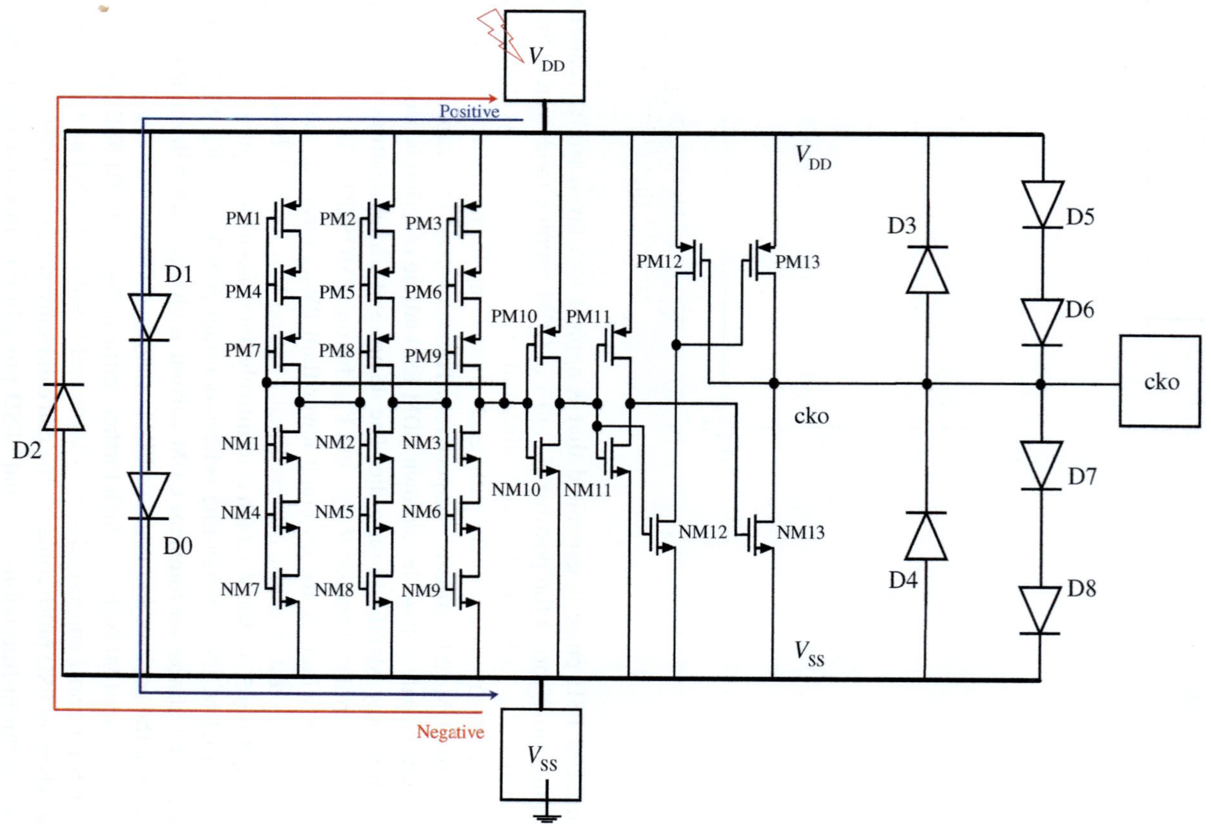

Figure 16.13 Scenario-1: V_{DD}-to-V_{SS} CDM ESD zapping case with a CDM ESD pulse applied to the V_{DD}-pad with respect to the grounded V_{SS} pad. Two CDM ESD conduction paths, D0–D1 and D2, will discharge the CDM ESD current in the positive and negative stressing cycles, respectively.

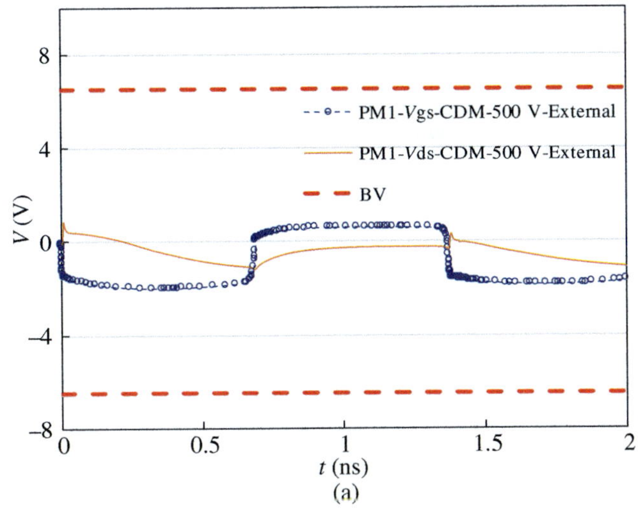

Figure 16.14 Scenario-1: Transient V_{GS} and V_{GD} behaviors for PM1 under V_{DD}-to-V_{SS} CDM ESD zapping in Scenario-1: (a) CDM 500 V zapping, and (b) CDM 50 V stressing. No voltage breakdown occurs during CDM ESD stressing.

induced charges will be stored in the C_{FG}, $C_{DG1,2,\ldots m,}$ $C_{DF1,2,\ldots n}$, and $C_{die\text{-}FP}$, according to their capacitance values in a pseudo-distributed way. Since the IC is designed in an SOI CMOS, the active Si layer and the Si substrate are separated and denoted by "sub" and "X," respectively, as shown in Figure 16.16. Figure 16.17 depicts an exemplar V_{DD}-pad CDM ESD zapping case, i.e., the V_{DD}-pad is touched by the pogo pin (GND) of the FICDM tester, which means that V_{DD} is grounded to discharge the induced charges stored inside the IC die. It is worth noting that Scenario-2 is a "from-Internal-to-External" CDM ESD stressing case, so when the V_{DD}-pad is grounded (i.e., CDM ESD zapped), other pads (i.e., V_{SS} and CKO) are floating. Since $C_{DF1,2,\ldots m}$ are much higher than other capacitors, it is reasonable to believe that most induced charges are stored at the pads. Therefore, it is readily found out that there are four possible CDM discharging paths under the V_{DD}-pad CDM zapping condition, i.e., D0–D1 and D5–D6, and D2 and D3, for

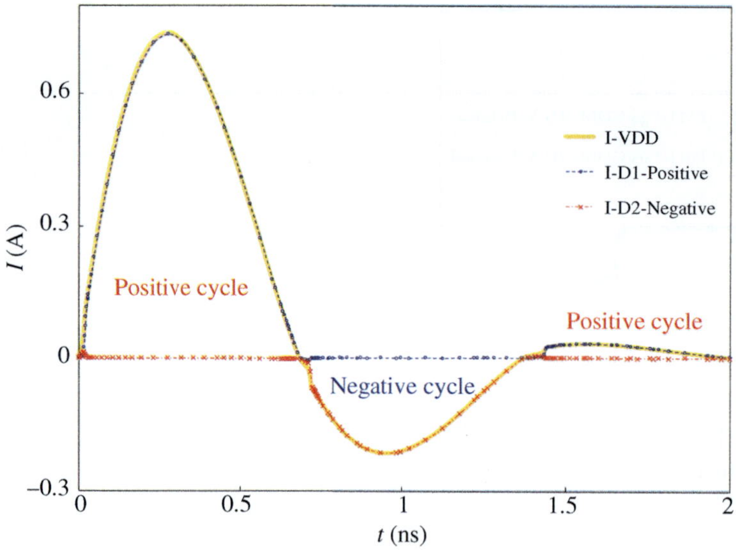

Figure 16.15 Scenario-1: transient CDM ESD discharging behaviors under 50 V V_{DD}-to-V_{SS} CDM ESD zapping shows that the CDM ESD current is discharged through the D0–D1 and D2 ESD conduction paths during the positive and negative CDM ESD zapping cycles, respectively.

the positive and negative CDM zapping cycles, respectively. Figure 16.18 presents the simulated transient voltages for PM1 transistor under 50 V CDM ESD zapping, which clearly shows possible Gate-to-Source breakdown, hence, failed the 50 V CDM ESD zapping. Figure 16.19 depicts more details of the CDM ESD discharging behaviors that shows that the D0–D1 and D5–D6 channels take almost all CDM ESD current in the positive stressing cycle, and the D2 and D3 paths conducts all CDM discharge current during the negative CDM stressing cycle, respectively. In summary, Scenario-2 states that when treating CDM ESD zapping as the "from-Internal-to-External" ESD event, the pad-based ESD protection network can still discharge the CDM-induced electrostatic charges pseudo-distributed inside an IC die. However, it fails V_{DD}-pad CDM zapping at a very low level of 50 V due to gate breakdown at PM1. This is in sharp contrast with Scenario-1 that treats CDM ESD as a "from-External-to-Internal" ESD discharge event, which can readily pass 500 V CDM ESD zapping. Both Scenario-1 and Scenario-2 cannot model the real-world CDM ESD events faithfully.

Scenario-3 is used to faithfully model the real-world CDM ESD phenomena where the CDM-induced electrostatic charges can be distributed anywhere and randomly inside an IC die that is also time-variant. For simplicity, three Splits are used, as depicted in Figure 16.20, assume that a large amount of electrostatic charges are stored around specific MOSFET transistors, i.e., NM1 for Split-1, PM9 for Split-2, and NM5 for Split-3, respectively. In each case, CDM ESD zapping occurs at the V_{DD}-pad, i.e., the FICDM pogo pin touches V_{DD}-pad for CDM discharging, during which the charges stored near the MOSFET will find their way to discharge into GND through the V_{DD}-pad. However, the internal CDM discharge currents may flow anyway they prefer according to circuit theory during the CDM ESD discharge event. This model reflects the real-world "internal-oriented" CDM ESD phenomenon by way of the "from-Internal-to-External" ESD discharge protocol. It is found that the true "from-Internal-to-External" CDM ESD discharging originated from the induced electrostatic charges randomly distributed inside an IC

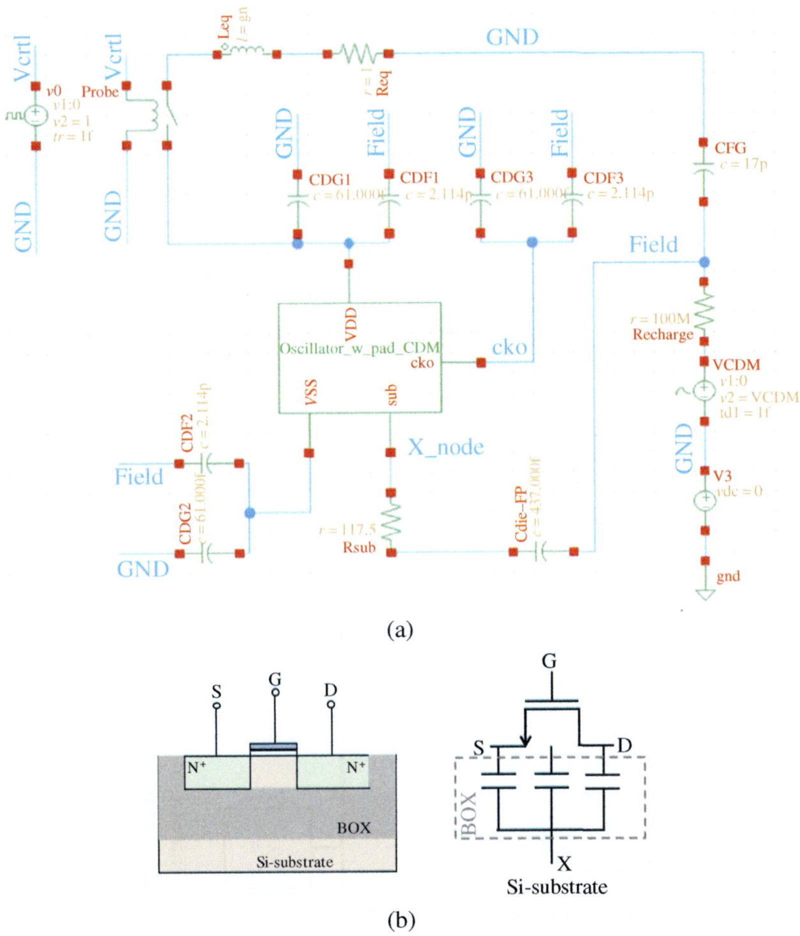

Figure 16.16 Scenario-2: The CDM ESD protection simulation deck using the pseudo-distributed FICDM testing circuit model: (a) testing diagram, and (b) MOSFET cross-section and capacitance network in SOI CMOS.

die can easily cause internal CDM ESD damages. Figure 16.21 depicts the simulated CDM ESD discharging behaviors with a small CDM ESD stimulus that is only 1% of that in the Scenario-2 case. Nevertheless, possible V_{GS} gate breakdown in PM1 (Split-1 and Split-2) and NM9 (Split-2) are observed, while PM1 can pass Split-3 and NM9 can pass Split-1 and Split-3. Scenario-3 study clearly states that the true "internal-oriented" "from-Internal-to-External" CDM ESD events may easily cause internal CDM ESD damages regardless of any traditional ESD protection at pads. The above circuit analysis confirms that using the classic pad-based CDM ESD protection method is fundamentally faulty. It further suggests that the existing industrial CDM ESD test standards and models, e.g., FICDM model, is over-simplified because the short-time charging procedure (induction) does not allow thorough distribution of the induced charges throughout the internal IC core, and the "wrong" internal charge distribution will for sure produce incorrect CDM ESD zapping test results, being random, unreproducible, and unreliable. This may be the root cause to the black magic of CDM ESD design failures today.

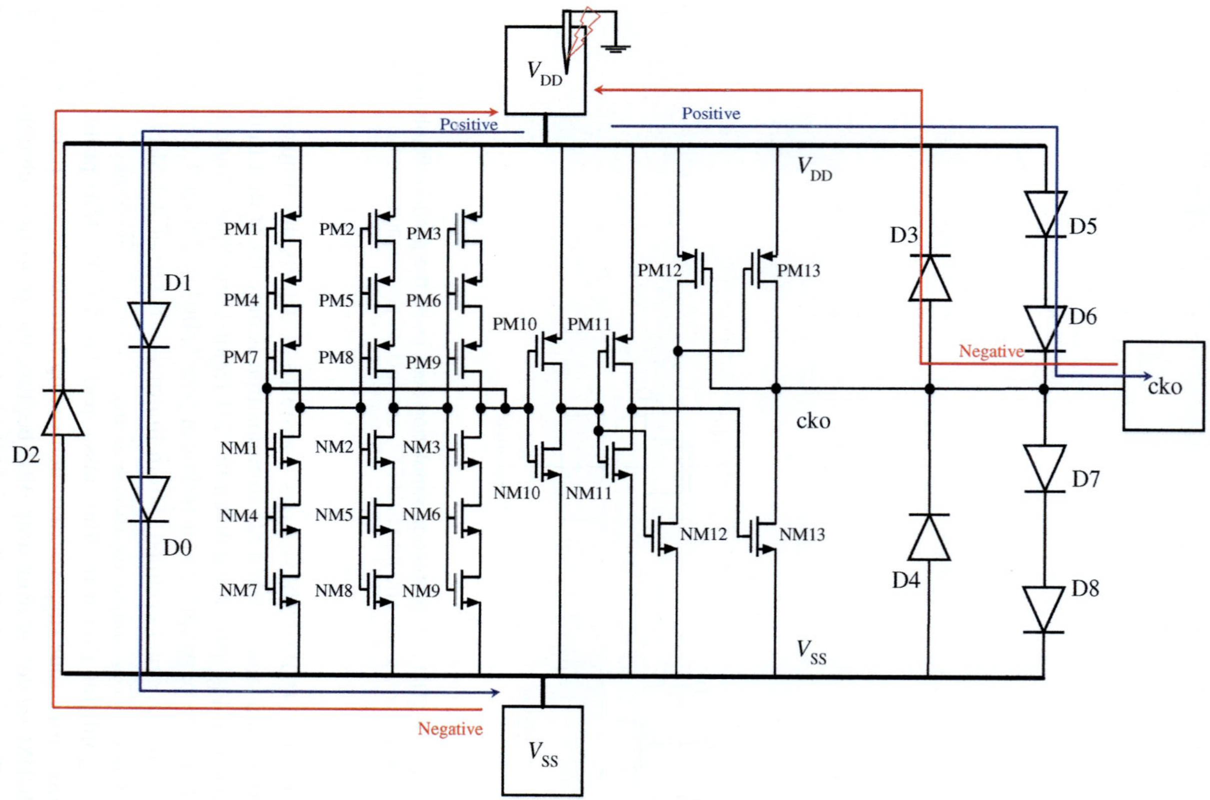

Figure 16.17 Scenario-2: V_{DD}-pad CDM ESD zapping case for a "from-Internal-to-External" CDM ESD event where the FICDM pogo pin touches the V_{DD}-pad to discharge the CDM-induced electrostatic charges stored inside the IC die. The pad-based ESD protection network seems still functional, containing four CDM ESD conduction paths, D0–D1 and D5–D6, and D2 and D3 channels to discharge the CDM ESD current in the positive and negative CDM zapping cycles, respectively. However, internal CDM ESD failures occur.

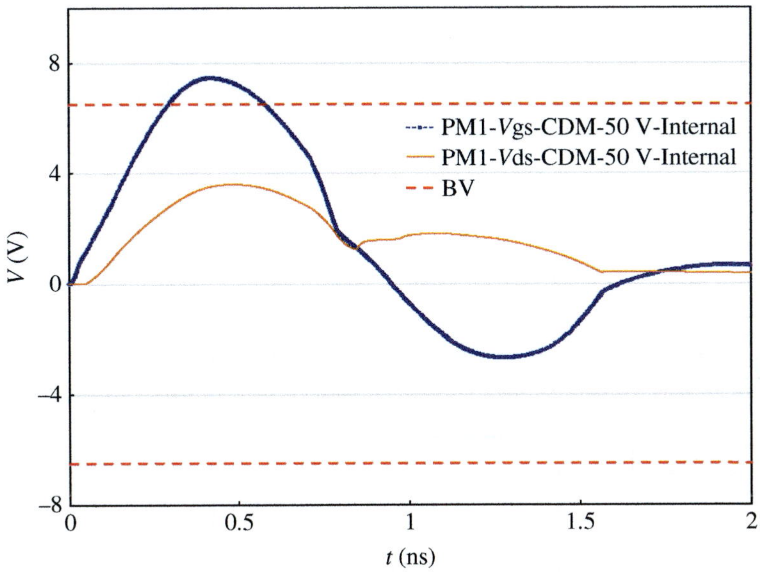

Figure 16.18 Scenario-2: Transient V_{GS} and V_{GD} behaviors for PM1 when zapping V_{DD}-pad by a 50 V CDM ESD pulse. BV_{GS} breakdown in PM1 is observed during CDM ESD stressing.

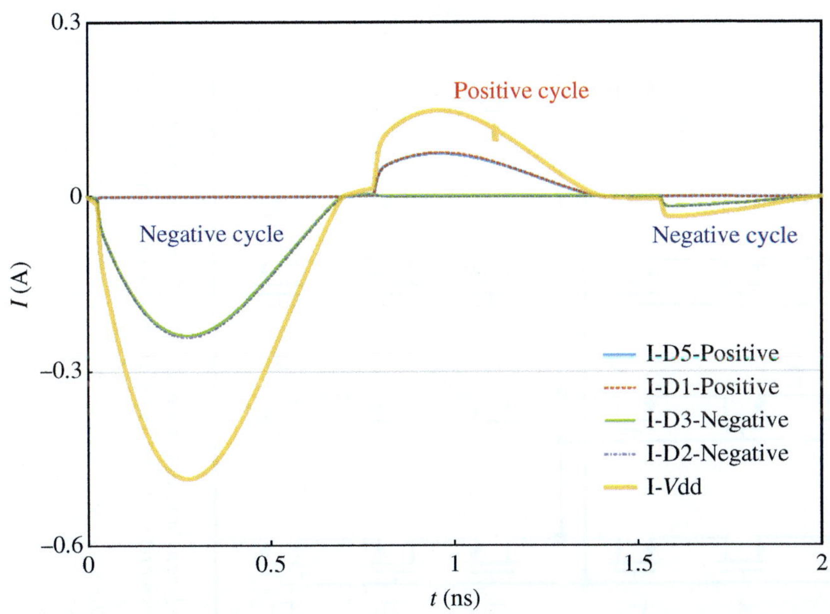

Figure 16.19 Scenario-2: Transient CDM ESD discharge behaviors under 50 V CDM ESD zapping to V_{DD}-pad shows that the CDM ESD current is discharged through the D0–D1 and D5–D6, and D2 and D3 channels to discharge the CDM ESD current in the positive and negative CDM zapping cycles, respectively.

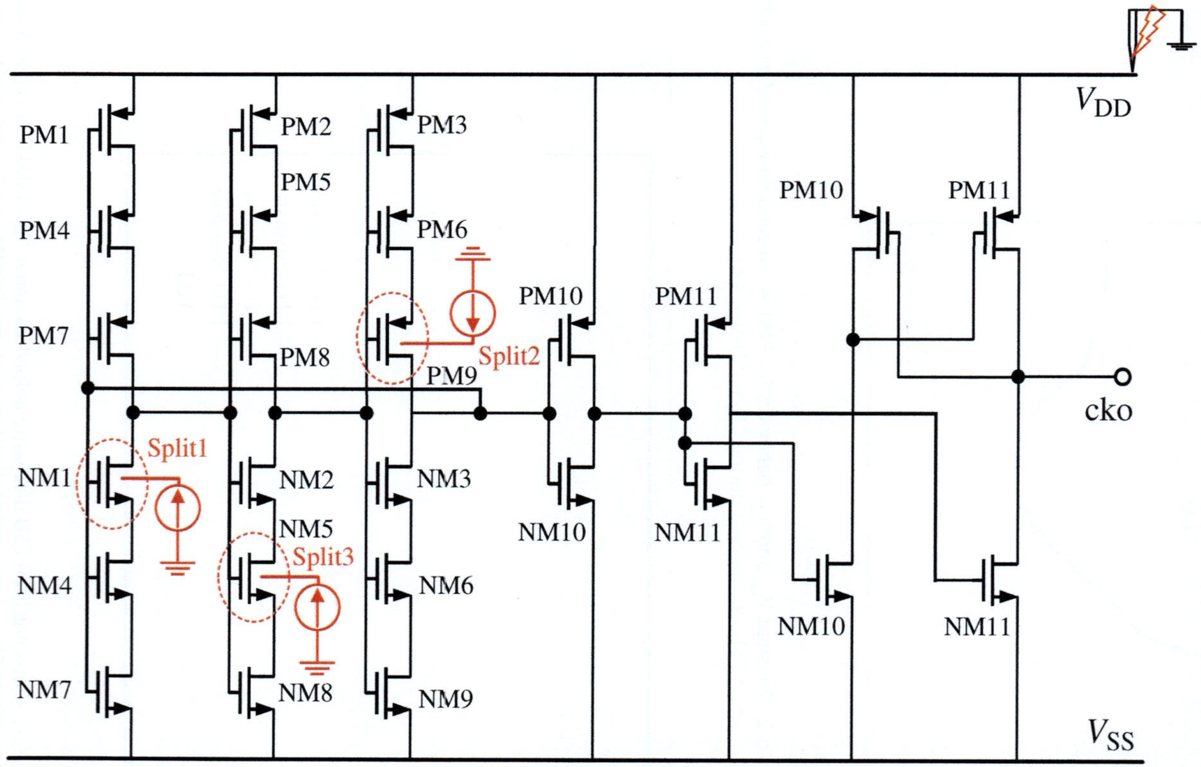

Figure 16.20 Scenario-3: true "Internal-oriented" CDM ESD zapping is a "from-Internal-to-External" CDM ESD event. Three Splits are studied modeling-induced electrostatic charges concentrated at NM1 (Split-1), PM9 (Split-2), and NM5 (Split-3).

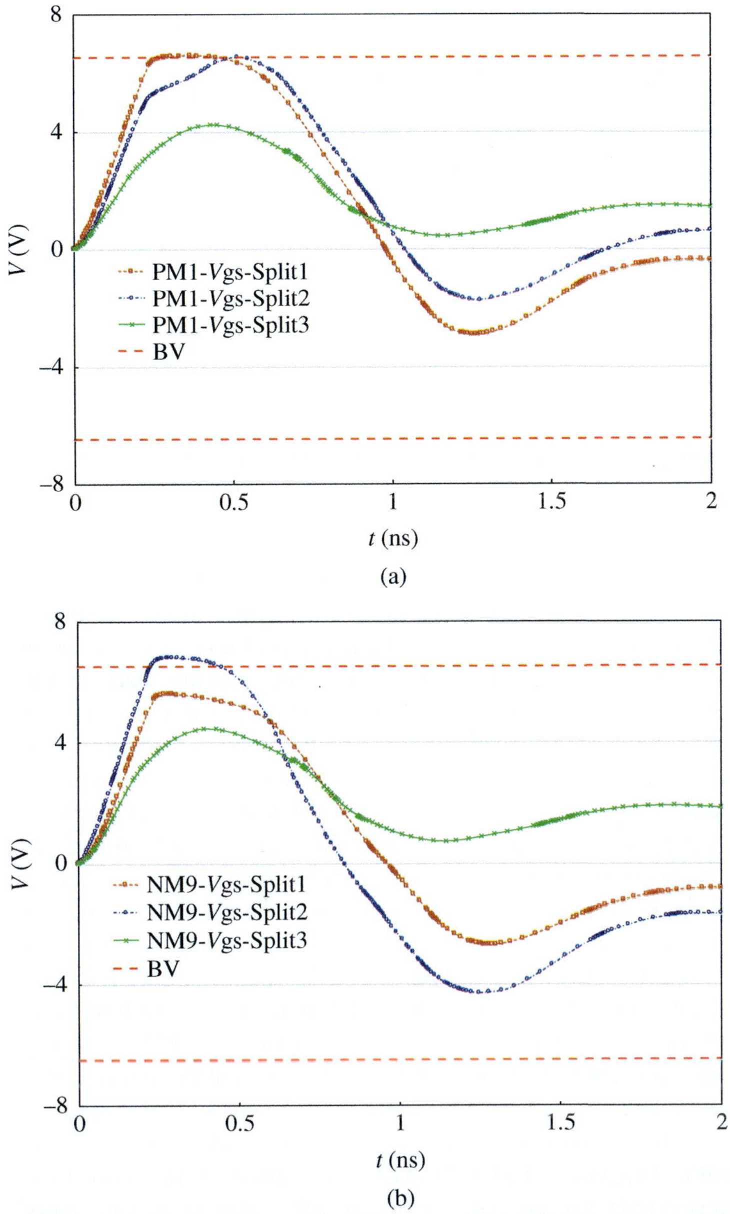

Figure 16.21 Scenario-3: Transient V_{GS} behaviors for (a) PM1 and (b) NM9 when zapping the V_{DD}-pad under a "from-Internal-to-External" CDM ESD event for Split-1, Split-2, and Split-3 show possible voltage breakdown CDM ESD failures: BV_{GS} breakdown in PM1 in Split-1 and Split-2, and BV_{GS} breakdown in NM9 in Split-2.

16.3 Internally Distributed CDM ESD Protection

As stated, since the CDM induction occurs anytime anywhere and the induced electrostatic charges can be distributed throughout an IC chip randomly, the classic pad-based ESD protection method is not suitable for CDM ESD protection. An alternative technique is a *non-pad-based*

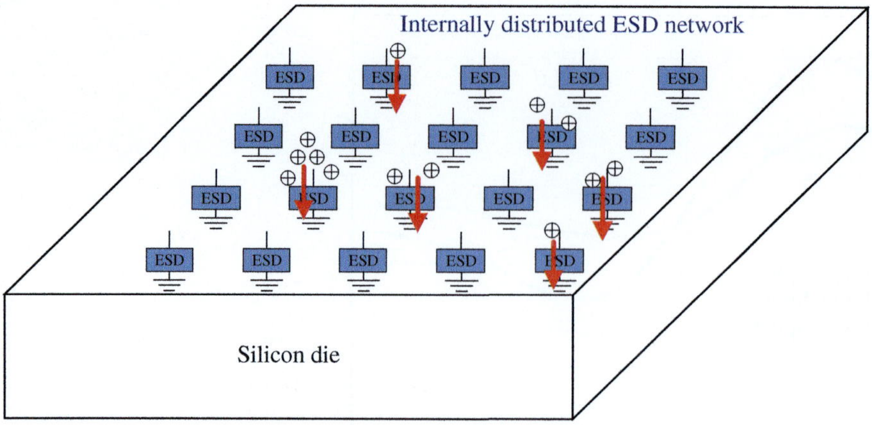

Figure 16.22 Conceptual illustration of the nonpad-based internally distributed CDM ESD protection method, utilizing an internal ESD protection device mesh for local CDM ESD discharging internally.

internally distributed CDM ESD protection method that can efficiently discharge the internal electrostatic charges as depicted in Figure 16.22 [13, 14]. First, considering the distributed nature of the CDM-induced electrostatic charges on a chip, an IC die is smartly partitioned according to the likelihood of charge distribution across a chip. The *smart chip partition* can be done by considering the functional circuit domains, layout floor plan, device properties and sizes, and IC technologies. Second, a mesh of small ESD protection devices can be designed and connected to certain internal circuit nodes per the smart portioning of the IC chip. Considering these special internal circuit nodes as internal "pads," the internal ESD protection device will function just like regular ESD protection structures at the pads. As the electrostatic charges gradually build up internally and locally, when they reach to a given local potential threshold at a circuit node, it will trigger the internal ESD protection device locally at the node to form a low-R local ESD conduction path locally to discharge the local charges around the node directly into the local ground (e.g., through a TSV to GND) without routing round the internal circuit to find an external pad to discharge. Therefore, the new *internally distributed ESD protection mesh network* can provide adequate whole-chip CDM ESD protection without using any pad-based ESD protection structures. Another advantage of using the internally distributed CDM ESD protection method is that, for the same full-chip CDM ESD protection target, the sizes of the internal ESD protection devices can be *tiny*, much smaller than their counterparts at the pads, hence reducing the unwanted ESD-induced C_{ESD}. To the bonding pads, an internally distributed CDM ESD protection follows "from-Internal-to-External" CDM ESD discharging protocol. However, to an internal circuit node with an internal ESD protection device, it can still be viewed as "pad"-based ESD protection internally and locally. From this view point, VFTLP testing can still be used to characterize the internal ESD protection devices. In practical designs, the internally distributed CDM ESD protection mesh can be realized in various ways. For example, Figure 16.23 depicts the realization of internally distributed CDM ESD protection using an in-TSV ESD protection diode mesh or an interposer CDM ESD protection layer as discussed in Chapter 14.

The concept of internally distributed CDM ESD protection is validated using a three-stage oscillator IC designed in a foundry 45 nm SOI CMOS technology [14]. This design has two splits for comparison: Split-1, shown in Figure 16.24, uses the traditional pad-based CDM ESD protection

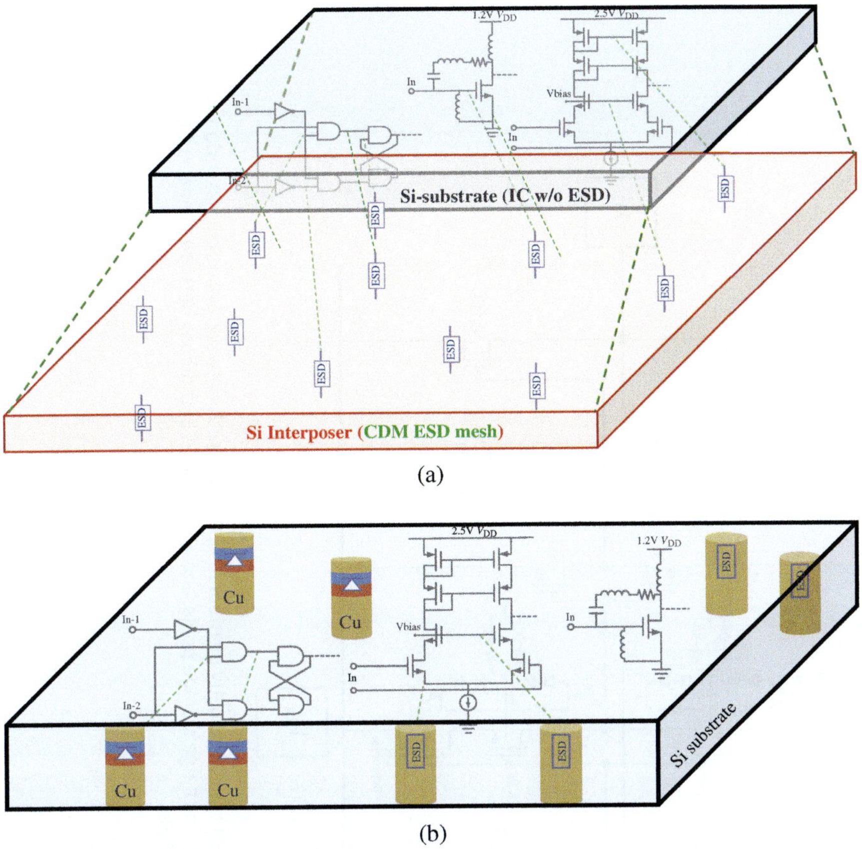

Figure 16.23 The internally distributed CDM ESD protection can be realized using (a) interposer CDM ESD mesh network, or (b) in-TSV CDM ESD mesh network.

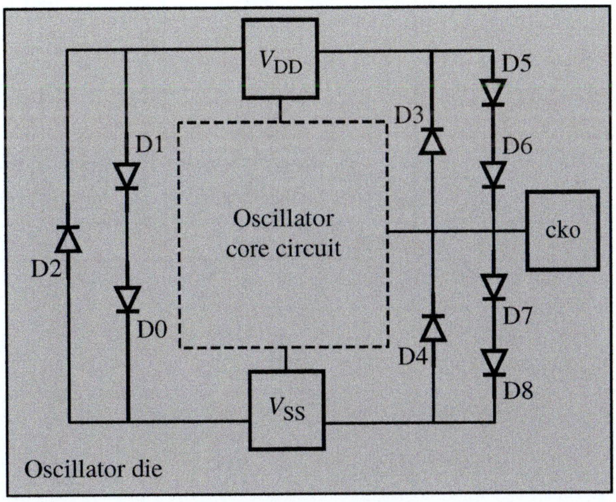

Figure 16.24 Split-1: Functional diagram for a three-stage oscillator IC core protected by traditional pad-based CDM ESD protection network.

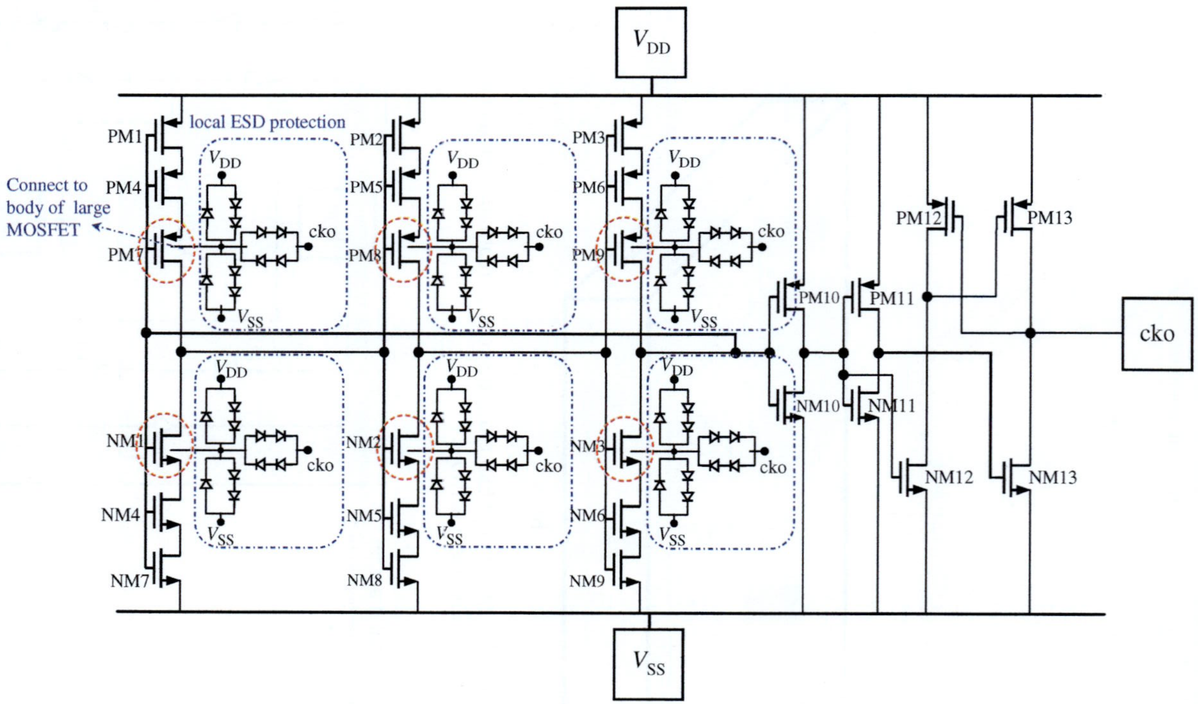

Figure 16.25 Split-2: The same three-stage oscillator IC core is protected by an internally distributed CDM ESD mesh network. Smart chip partitioning considers large MOSFETs (PM7, PM8, PM9, NM1, NM2, and NM3) as the main charge storage pools, and anti-parallel ESD didoes nets are connected to the internal circuit nodes to achieve local CDM ESD discharging inside the IC die.

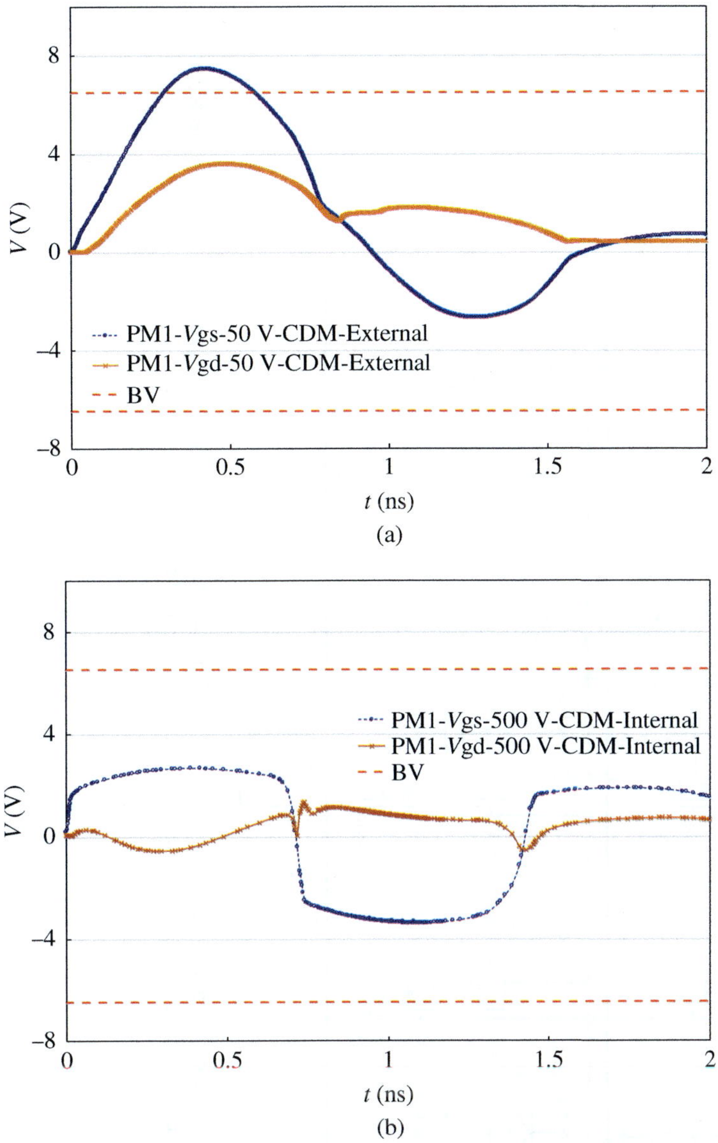

Figure 16.26 Transient V_{GS} and V_{GD} behaviors for PM1 comparison: (a) Split-1: pad-based CDM ESD protection fails gate breakdown under low level of 50 V CDM ESD stressing, and (b) Split-2: internally distributed CDM ESD protection successfully passes a high level of 500 V CDM ESD zapping for the same IC core.

method where the output pad (CKO) is protected by anti-parallel ESD diodes (D3–D8) and the power rails are protected by D0–D2. Split-2, depicted in Figure 16.25, utilizes the new internally distributed CDM ESD protection method that has an internal mesh of small ESD protection diodes. As a simple example, large MOSFETs (PM7, PM8, PM9, NM1, NM2, and NM3) is considered the main pools to store the CDM-induced electrostatic charges, which can be locally discharged into

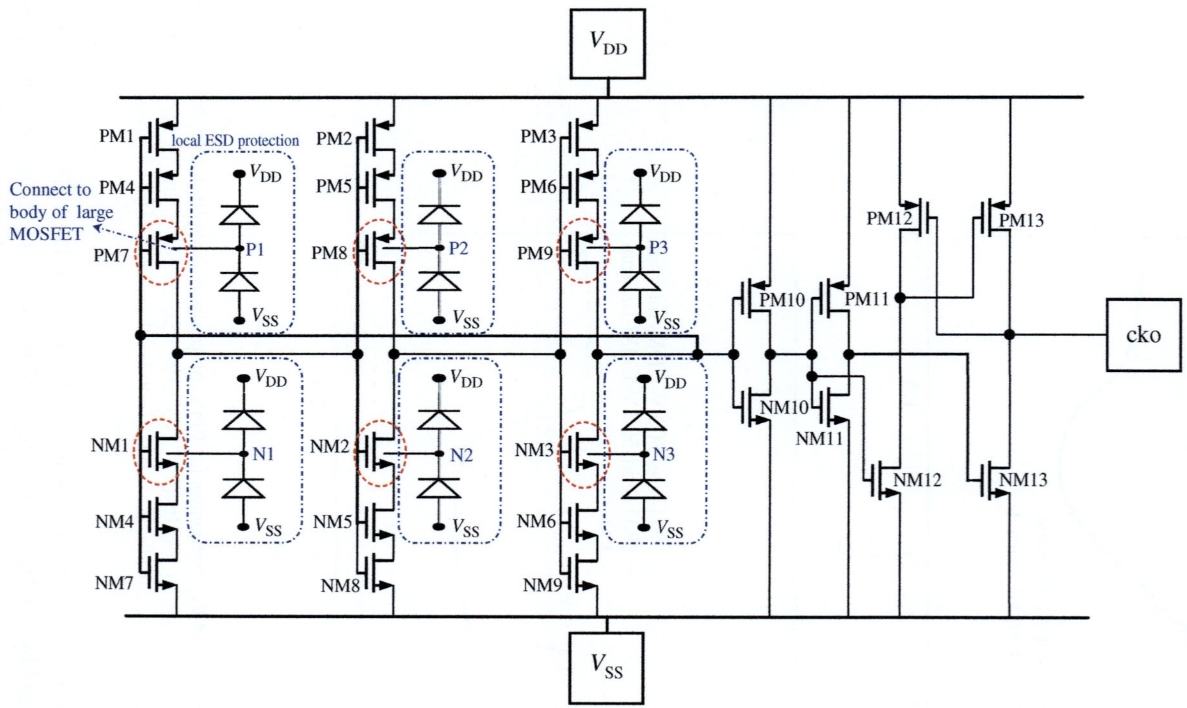

Figure 16.27 The same three-stage oscillator IC core with internally distributed CDM ESD mesh network using simple ESD protection diodes, connected at internal circuit nodes (P1, P2, P3, N1, N2, and N3) was fabricated in a 45 nm SOI CMOS and measured by VFTLP.

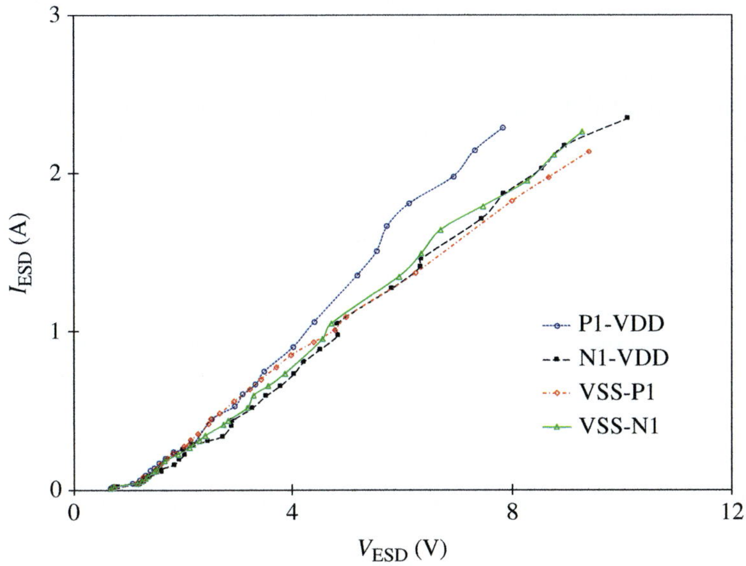

Figure 16.28 VFTLP testing for the internal CDM ESD protection didoes inside the three-stage oscillator IC core fabricated in a 45 nm SOI CMOS confirms CDM ESD discharging internally and locally in both directions.

local GND when the node potential builds up to a preset threshold level. The CDM ESD protection target is 500 V, which requires a finger width of 360 μm for an ESD diode at the bonding pad, but a much smaller size of 60 μm is used for internal ESD protection devices in this design. Figure 16.26 presents the simulated transient V_{GD} and V_{GS} for PM1 under 50 V CDM ESD zapping for Split-1 and 500 V CDM ESD stressing in Split-2. This example shows clearly that the pad-based CDM ESD protection fails due to gate breakdown at a much lower level of 50 V CDM zapping in Split-1. However, the internally distributed CDM ESD protection in Split-2 passes a much higher level of 500 V CDM ESD stressing even using much smaller internal ESD protection devices. In another design example for the same three-stage oscillator IC, simple ESD diodes are used to protection large MOSFETs at the internal circuit nodes (P1, P2, P3, N1, N2, and N3), as shown in Figure 16.27 [14]. This IC was fabricated in 45 nm SOI CMOS and characterized using VFTLP for CDM ESD evaluation. Figure 16.28 depicts the VFTLP-measured ESD discharging I–V behaviors for ESD conduction from P1 and N1 nodes to V_{DD}, as well as from V_{SS} to P1 and N1, respectively. The VFTLP testing clearly shows that the internal ESD protection diodes can be triggered by ultrafast CDM ESD pulses, hence offering CDM ESD protection to the internal circuit nodes locally.

16.4 Summary

This chapter discusses in details the misconception of traditional pad-based CDM ESD protection methods. By intuitive analysis and through circuit design examples, it argues that, while the classic pad-based ESD protection method works nicely for the “External-oriented” "from-External-to-Internal" ESD discharge events (e.g., HBM, MM, and IEC), this approach is fundamentally wrong for the "Internal-oriented" "from-Internal-to-External" CDM ESD protection. This is attributed to the fact that real-world CDM induction occurs anytime anywhere and the induced electrostatic charges are distributed throughout an IC chip in a random and time-variant way during its lifetime. Therefore, using pad-based CDM ESD protection method, even if the stored charges may eventually find a way to a grounding pad to discharge into the GND, these electrostatic charges, while routing through the internal circuit, may build up internal voltage and current, and cause internal and local CDM ESD damages inside the IC die regardless of any pad-based ESD protection structures used. An alternative non-pad-based internally distributed CDM ESD protection method can be used to efficiently handle the "from-Internal-to-External" CDM ESD discharging challenge, which can be realized in various ways in practical IC designs.

References

1 MIL-STD-883E, Method 3015.7 (1989). *Electrostatic Discharge Sensitivity Classification*. Dept. of Defense, Test Method Standard, Microcircuits.

2 ANSI/ESDA/JEDEC JS-001-2017 (2017). *For Electrostatic Discharge Sensitivity Testing – Human Body Model (HBM) – Component Level*. An American National Standard jointly developed by ESD Association and JEDEC, May 12, 2017.

3 ANSI/ESDA/JEDEC JS-002-2018 (2018). *For Electrostatic Discharge Sensitivity Testing – Charged Device Model (CDM) – Device Level*. An American National Standard jointly developed by ESD Association and JEDEC, December 17, 2018.

4 IEC 61000-4-2 (2008). *Electromagnetic Compatibility, Part 4: Testing and Measurement Techniques, Section 2: Electrostatic Discharge Immunity Test*. The International Electrotechnical Commission (IEC).

5 Wang, A., Feng, H., Zhan, R. et al. (2005). A review on RF ESD protection design. *IEEE Trans. Electron Devices* 52 (7): 1304–1311. https://doi.org/10.1109/TED.2005.850652.

6 Wang, A. (2002 Edition). *On-Chip ESD Protection for Integrated Circuits: An IC Design Perspective*. Springer. ISBN-13: 978-0792376477.

7 Voldman, S. (2017). *ESD Testing: From Components to Systems*. Wiley.

8 Jack, N. (2012). Charged device model ESD protection and test methods for ICs. A PhD dissertation. University of Illinois at Urbana-Champaign.

9 Wang, H., Zhang, F., Li, C. et al. (2018). Chip-level CDM circuit modeling and simulation for ESD protection design in 28 nm CMOS. *Proceedings of IEEE International Conference on Solid-State and Integrated Circuit Technology (ICSICT)*. https://doi.org/10.1109/ICSICT.2018.8564936.

10 Karp, J., Hart, M., Fakhruddin, M. et al. (2016). FinFET MPSoC 32 Gb/s transceivers: custom ESD protection and verification. *Proceedings of IEEE MWSCAS*. https://doi.org/10.1109/MWSCAS.2016.7869952.

11 Di, M., Wang, H., Zhang, F. et al. (2019). Does CDM ESD protection really work? *Proceedings of IEEE Workshop on Microelectronics and Electron Devices (WMED)*. https://doi.org/10.1109/WMED.2019.8714145.

12 Di, M., Li, C., Pan, Z., and Wang, A. (2020). Pad-based CDM ESD protection methods are faulty. *IEEE J. Electron Devices Soc. (J-EDS)* 8: 1297–1304. https://doi.org/10.1109/JEDS.2020.3022743.

13 Wang, A. (2019). Internally distributed CDM ESD protection. US Patent App. #62936355.

14 Mengfu Di, Cheng Li, Zijin Pan and Albert Wang (2021). "Non-Pad-Based in Situ in-Operando CDM ESD Protection Using Internally Distributed Network", *IEEE J. of Electron Devices Society (J-EDS).* DOI: 10.1109/JEDS.2021.3112736

17

Future ESD Protection Outlook

17.1 The Fundamental ESD Protection Problem

So far the electrostatic discharge (ESD) failures and on-chip ESD protection mechanisms are well discussed. It is also agreed upon that on-chip ESD protection design remains a big design challenge. One of the major on-chip ESD protection design problems is that any ESD protection structure come up with ESD-induced design overhead, which includes parasitic parameters, such as, C_{ESD}, I_{leak}, noises, and C_{ESD}-induced global noise coupling, as well as Si asset consumption and integrated circuit (IC) layout difficulty. Such ESD-induced design overhead problem rapidly becomes more and more unacceptable to large and complex ICs designed in advanced IC technologies. One fundamental ESD protection problem is that almost all ESD protection solutions utilize *in-Si PN-junction-based* active devices to discharge ESD transients. As depicted in Figure 17.1, commonly used ESD protection devices, such as diodes, bipolar junction transistors (BJTs), metal-oxide-semiconductor field-effect transistors (MOSFETs), and silicon-controlled rectifiers (SCRs), and their derivatives of all kinds, all reside inside a semiconductor (i.e., Si) substrate and contains multiple PN junctions. Unfortunately, PN junctions are inherently noisy, leaky, and has parasitic junction capacitance. More robust ESD protection for consumer electronics generally requires a larger device size, if using same kind of ESD protection structures, which will introduce more parasitic effects that can seriously degrade IC performance, particularly for multiple-GHz and wide bandwidth RF ICs and high-throughput ICs of beyond 10 Gbps. This inherent ESD parasitic problem becomes even worse for large-pin-count complex chips where full-chip ESD protection requires large numbers of ESD protection devices on a chip, which directly translate more ESD parasitic effects and Si consumption, as depicted in Figure 17.2. Obviously, one could not completely eliminate this fundamental ESD protection problem that is the "DNA" of almost any existing in-Si PN-based ESD protection structures, which have been widely used for decades. Revolution in on-chip ESD protection is hence needed in order to completely root out this fundamental ESD parasitic problem.

Theoretically, the best on-chip ESD protection structure can be any *ideal ESD switch*, as depicted in Figure 17.3, which stays OFF during normal IC operations, but can be swiftly turned ON by any incident ESD transient to create a low-R ESD conduction path to clamp pad voltage to a safe level while discharge large ESD current without overheating. While ESD discharging I–V curves are typically characterized for an ESD protection structure, ESD protection devices can be non-electronic in nature. For example, a mechanical switch may be a preferable ESD protection device due to potentially "zero" C_{ESD}, I_{leak}, noises, and C_{ESD}-induced global noise coupling. However, a mechanical ESD switch must be ultrafast (~100 ps) and super tiny (i.e., μm-scale) that can be integrated into complementary metal-oxide-semiconductor (CMOS) ICs. These are huge research challenges

Practical ESD Protection Design, First Edition. Albert Wang.

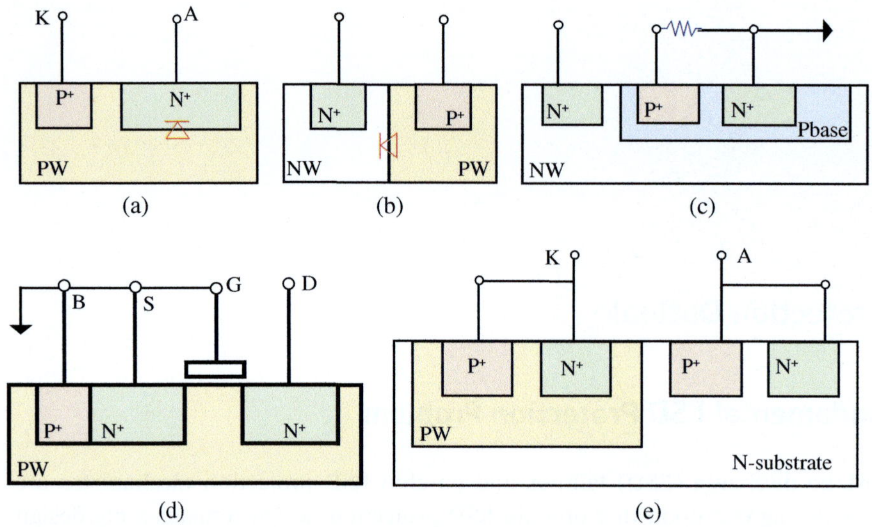

Figure 17.1 Commonly used in-Si PN-based ESD protection structures include (a) vertical diode, (b) lateral diode, (c) BJT, (d) MOSFET, (e) SCR, and their derivatives, which not only inevitably have significant PN-induced parasitic C_{ESD}, I_{leak}, and noises, but also consume large Si area and make IC chip layout difficult.

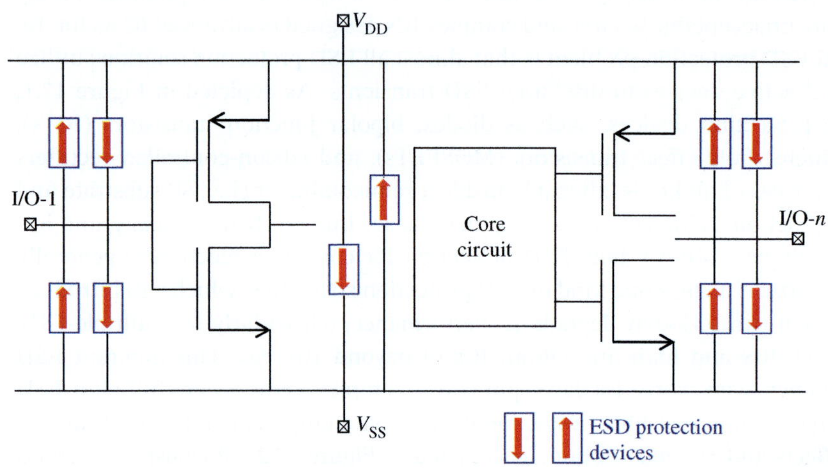

Figure 17.2 Classic pad-based full-chip ESD protection method requires large numbers of ESD protection devices on a chip, which significantly worsens the ESD-induced design overhead problem to ICs.

to overcome. Sections 17.2, 17.3, and 17.4 discuss a few disruptive ESD protection device concepts toward this direction.

17.2 Above-IC Nano-Crossbar Array ESD Switch

Imagine to pull an ESD protection structure out of the Si substrate and place it in the *back-end-of-line* (BEOL) of CMOS IC, and to replace the PN junctions by some *phase-changing* materials to make a magic ESD switch, here comes an *above-IC* nano-crossbar array ESD protection device as depicted in Figure 17.4 [1–3]. The switch array structure contains a set

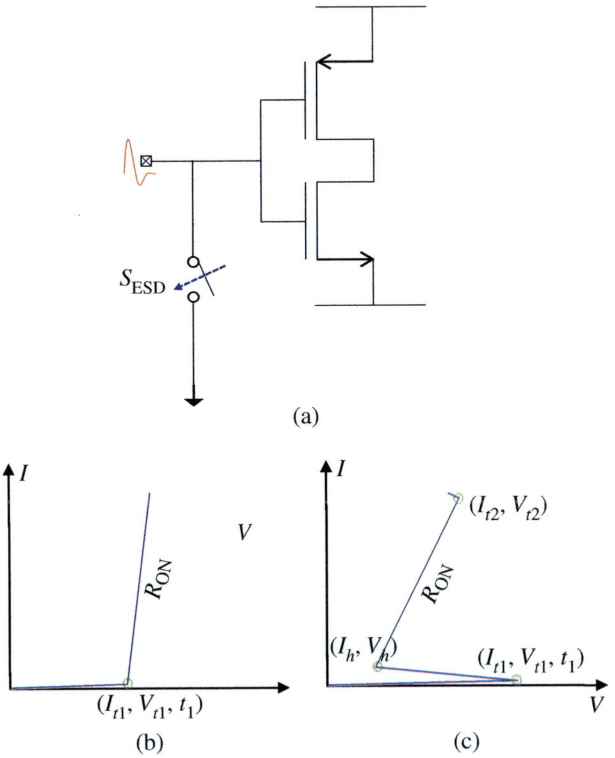

Figure 17.3 An ideal ESD switch of any kinds may serve as a good ESD protection device: (a) ideal ESD switch, (b) classic simple-turn-on ESD discharging $I-V$ curve, and (c) typical deep snapback ESD discharging $I-V$ behavior.

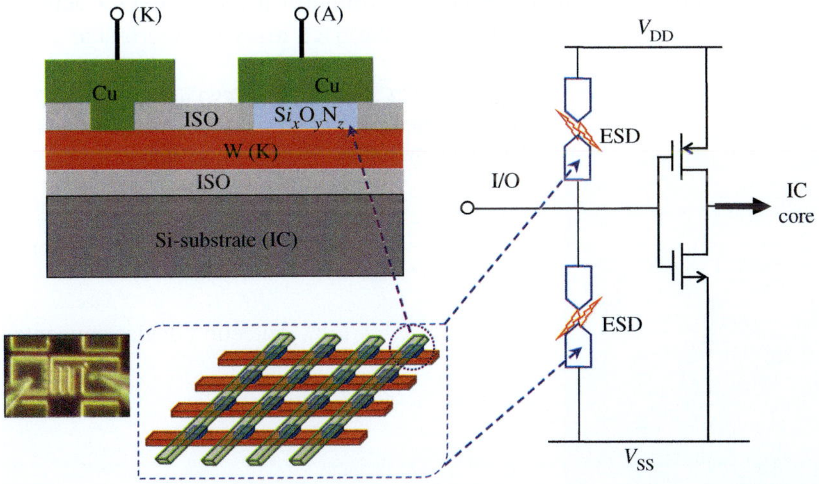

Figure 17.4 Illustration of the phase-changing nano-crossbar array ESD protection switch: its cross-section, array structure, its on-chip ESD protection scheme, and an image of a fabricated sample device.

of nano-crossbar ESD switching devices. Each nano-crossbar node ESD protection device is a two-terminal phase-changing switch containing a Cu electrode (anode, A) and a W/Cu electrode (cathode, K) with a special phase-changing insulator film in between. Ideally, this nano-crossbar ESD switch has very low parasitic capacitance, leakage, and noises. In normal IC operations, the nano-crossbar ESD switch stays OFF. Under ESD stressing, the insulating layer rapidly changes its phase to form a low-R conduction channel, hence, to turn ON the switch to discharge ESD pulses for ESD protection. In prototype demonstration, the $Si_xO_yN_z$ family nanophase-transition dielectric was used to fabricate the nano-crossbar ESD protection devices. Figure 17.5 shows a new *dispersed local ESD tunneling model* (ESD-DLT) that explains the ultrafast ESD switching effect: The annealing process pre-distributes the Cu ions throughout the $Si_xO_yN_z$ insulating materials. Under ESD stressing, the strong electrostatic field will trigger local tunneling actions, driving free carriers to hop among the predispersed Cu ions, hence, changing the $Si_xO_yN_z$ film from insulating (OFF) to conducting (ON) at a very fast speed for ESD triggering. After the ESD zapping is over, the local tunneling action will stop immediately, changing the $Si_xO_yN_z$ layer back to insulating phase, hence, turning the nano-crossbar device from ON to OFF. Unlike physical filament (i.e., conducting stings)-based switching mechanism, the local tunneling mechanism enables superfast switching that is critical to ESD protection. To handle large ESD transients, a nano-crossbar node-array ESD switch structure can be used to protect ICs. A CMOS-compatible process module was used to fabricate the nano-crossbar array ESD switches in the BEOL deck of CMOS ICs, which confirms the unique ESD switching mechanism in prototype devices. Figure 17.6 depicts the measured I–V curve for a prototype single-node nano-crossbar ESD switch (1 μm × 1 μm in crossbar node size) by transmission-line-pulsing (TLP) testing (t_r = 10 ns), showing the desired ESD discharge characteristic. Figure 17.7 presents the measured ESD discharging I–V behavior for a sample single-node nano-crossbar ESD switch (30 μm × 30 μm in node size) by VFTLP zapping (t_r = 100 ps), confirming that the ESD switch can respond to ultrafast CDM ESD transients. Figure 17.8 is the measured ESD discharging I–V curve for a sample single-node, nano-crossbar ESD switch (5 μm × 5 μm) by TLP stressing in both directions, which shows that this nano-crossbar ESD switch can provide ESD protection in dual directions, symmetrically. Figure 17.9 depicts the measured ESD discharging I–V characteristic for a sample 5 × 5 array nano-crossbar ESD

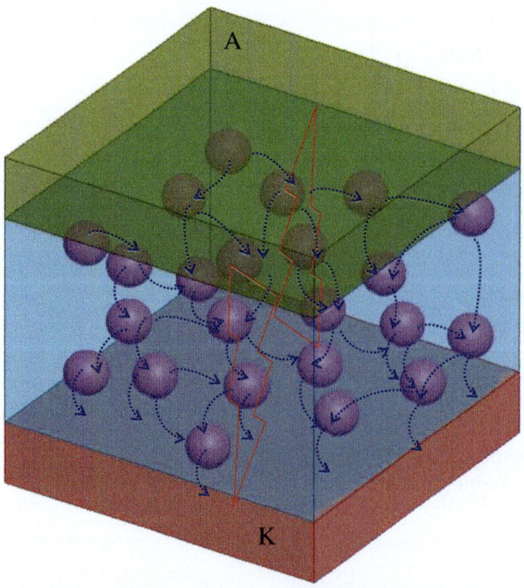

Figure 17.5 A *dispersed local ESD tunneling model* explains the ultrafast ESD switching mechanism as confirmed by 100 ps VFTLP ESD testing.

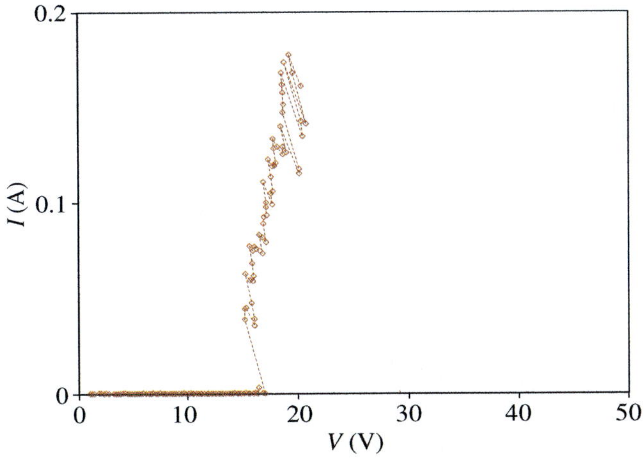

Figure 17.6 TLP-measured ESD discharging I–V curve for a sample single-node nano-crossbar ESD switch (1 μm × 1 μm) confirms the phase-switching ESD protection mechanism (t_r = 10 ns).

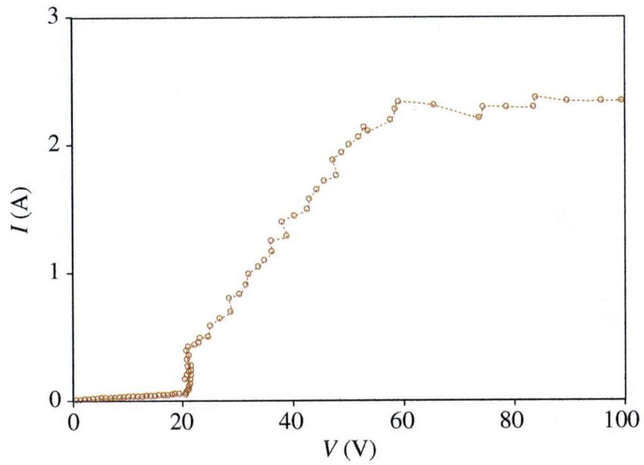

Figure 17.7 VFTLP-measured ESD discharging I–V curve for a sample single-node nano-crossbar ESD switch (30 μm × 30 μm) suggests that the phase-changing switch is a potential solution for ultrafast CDM ESD protection (t_r = 100 ps).

switch structure (crossbar node size of 5 μm × 5 μm) by TLP stressing, which not only confirms the dual-polarity ESD discharging function but also reveals multiple ESD triggering behaviors. The observed multiple ESD triggering phenomenon is attributed to nonuniform turn-on of all nano-crossbar nodes within the array as expected. Figure 17.10 presents the measured leakage currents of large numbers of nano-crossbar ESD switch devices under normal IC operation voltage, showing extremely low leakage of I_{leak} ~ 1 s of pA only. It is found that the ESD triggering voltage can be changed by design variations, i.e., device dimensions, insulators, and metal ions, etc., as given in Figure 17.11, where the measured V_{t1} varies from 1 to 28 V. Figure 17.12 compares two sample single-node nano-crossbar ESD switches (20 μm × 20 μm in node size) using Cu and Pt as the metal ions sources, showing different ESD triggering and holding behaviors, which offers another way to optimize the nano-crossbar ESD protection switches. In addition to the unique advantage

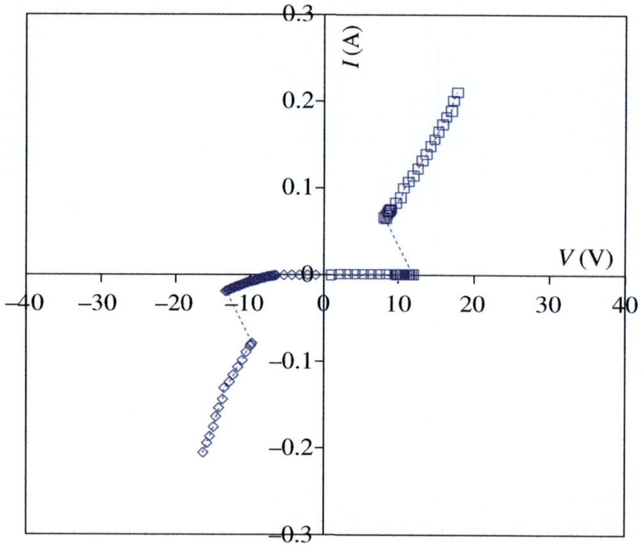

Figure 17.8 TLP-measured ESD discharging *I–V* curve for a sample single-node nano-crossbar ESD switch (5 μm × 5 μm) shows perfectly symmetric dual-polarity ESD discharging *I–V* characteristics.

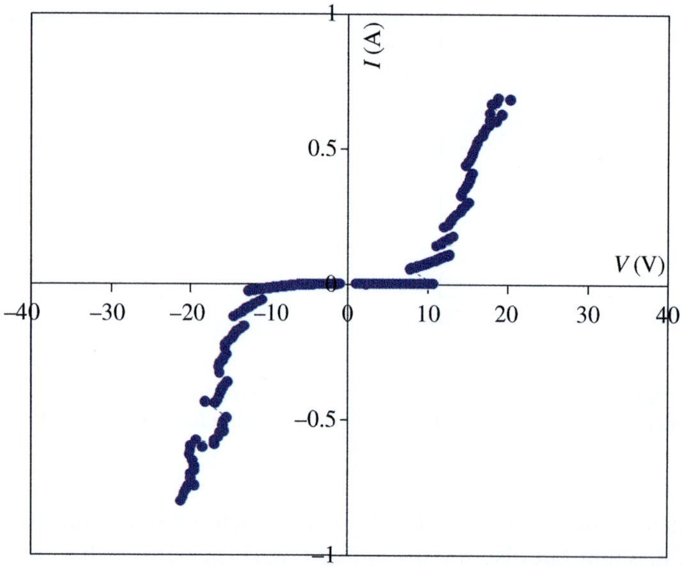

Figure 17.9 TLP-measured ESD discharging *I–V* curve for a sample 5 × 5 nano-crossbar ESD switch array (5 μm × 5 μm in node size) shows symmetric ESD discharging *I–V* and multiple-triggering characteristics.

of not using any PN junction in silicon and being in the BEOL deck of CMOS IC, the symmetric dual-polarity ESD discharge feature means that the total head counts of ESD protection devices may be reduced by up to 50% on a chip, as depicted in Figure 17.13, which is a major benefit not only inducing lower ESD parasitic effects but also consuming less Si area and making complex chip layout easier.

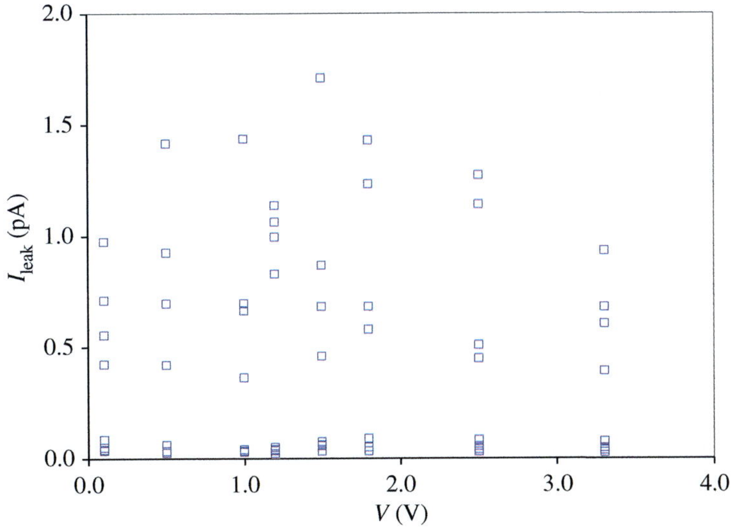

Figure 17.10 Measurement shows ultralow leakage currents of the nano-crossbar ESD switch devices (5 μm × 5 μm in node size) of merely a few pA, orders of magnitude lower than any traditional in-Si PN-based ESD protection structures.

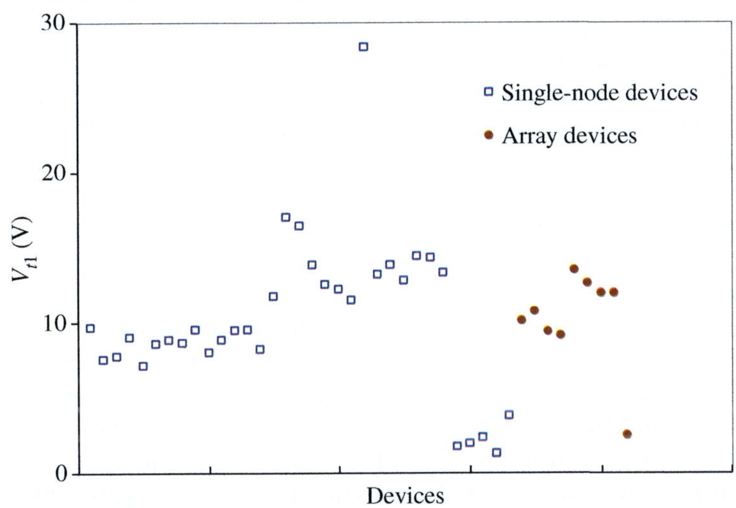

Figure 17.11 Measurement shows that the critical ESD triggering voltage can be adjusted by device designs, e.g., $V_{t1} = 1–28$ V.

17.3 Graphene ESD Protection Switch

What are the desired properties for an ideal ESD protection device? Higher carrier mobility, higher thermal conductivity, mechanically tougher are some of the properties, to name a few, that may translate into higher I_{t2}, smaller size (J_{t2}), less overheating, etc. Isn't graphene attractive in these terms? Graphene, as a magic 2D material, features ultrahigh electron mobility (~5000 cm^2/V-s), extremely high thermal conductivity (κ ~ 5.30 × 10^3 W/mK), super high Young's

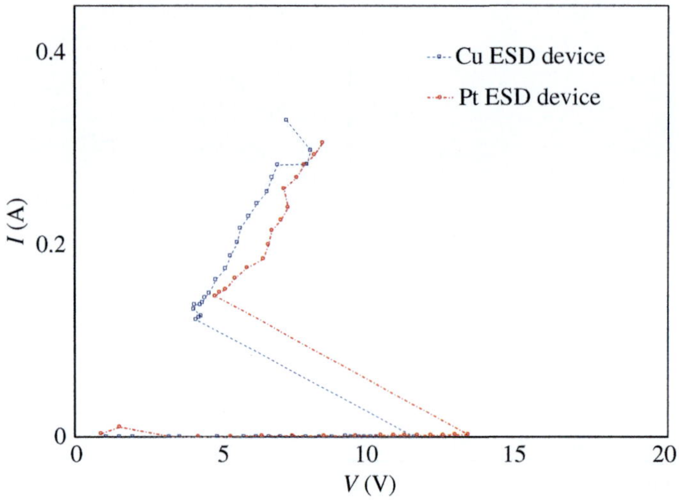

Figure 17.12 TLP testing shows different V_{t1} and V_h for prototype single-node nano-crossbar ESD switch devices (20 μm × 20 μm in node size) using Cu and Pt as electrodes, respectively.

modulus (~1 T Pa), and very light mass density, etc., making it a possibly favorable candidate for making excellent ESD protection devices [4–7]. Figure 17.14 depicts a unique graphene nano-electromechanical system (gNEMS) switch ESD protection structure, which is a two-terminal mechanical switch containing a vacuum cavity between a suspended graphene membrane as the top electrode (A) and a bottom Si/metal electrode (K) [8–11]. The gNEMS ESD switches are fabricated through CMOS-compatible processes and reside in the BEOL deck in CMOS ICs. As an ESD protection device, the gNEMS ESD switch is connected to a pad on an IC chip. In normal IC operations, the gNEMS device stays OFF, not interfering with IC functions. When an ESD pulse appears at the pad, the strong transient electrostatic force will pull down the suspended graphene film toward the bottom electrode until it touches the K terminal, hence turning ON the gNEMS switch to discharge the ESD pulse and protecting the IC. After the ESD transient flies over, the strong electrostatic force will pull up the bended graphene membrane back to the original location, hence, turning OFF the gNEMS device. There are several obvious advantages associated with a gNEMS ESD switch: ideally "zero" C_{ESD}, I_{leak}, self-generated noise, and noise coupling effect; above-IC in BEOL deck consuming no extra Si asset, and hence layout-friendly. Prototype gNEMS ESD switch devices were fabricated using polycrystalline graphene ribbons grown by CVD method and characterized to validate the new ESD protection concept [8]. Figure 17.15 presents the DC-sweeping test of prototype gNEMS devices with a fixed cavity depth ($d = 350$ nm) and varying graphene ribbon width (W = 7, 10, 15 μm) and length (L = 10, 15, 20 μm), which clearly shows the expected I–V switching effect. Figure 17.16 depicts the measured transient ESD discharging I–V characteristics by TLP zapping for a sample gNEMS ESD switch (d = 350 nm, W = 5 μm, L = 7 μm), which demonstrates the desired symmetric ESD switching and discharging functions. The slight asymmetry observed is attributed to the physical asymmetry of the prototype gNEMS devices fabricated as depicted in Figure 17.14. Figure 17.17 shows the measured ESD triggering V_{t1}, ranging from 8.5 to 17.5 V, for sample gNEMS switches with design variations. The measured leakage current is very low, I_{leak} ~ 3–13 pA. Uniquely, gNEMS is a *mechanical* switch in nature, which is entirely different from any traditional ESD protection structures that are electron devices. Yet, due to the high Young's modulus of graphene, the gNEMS can be

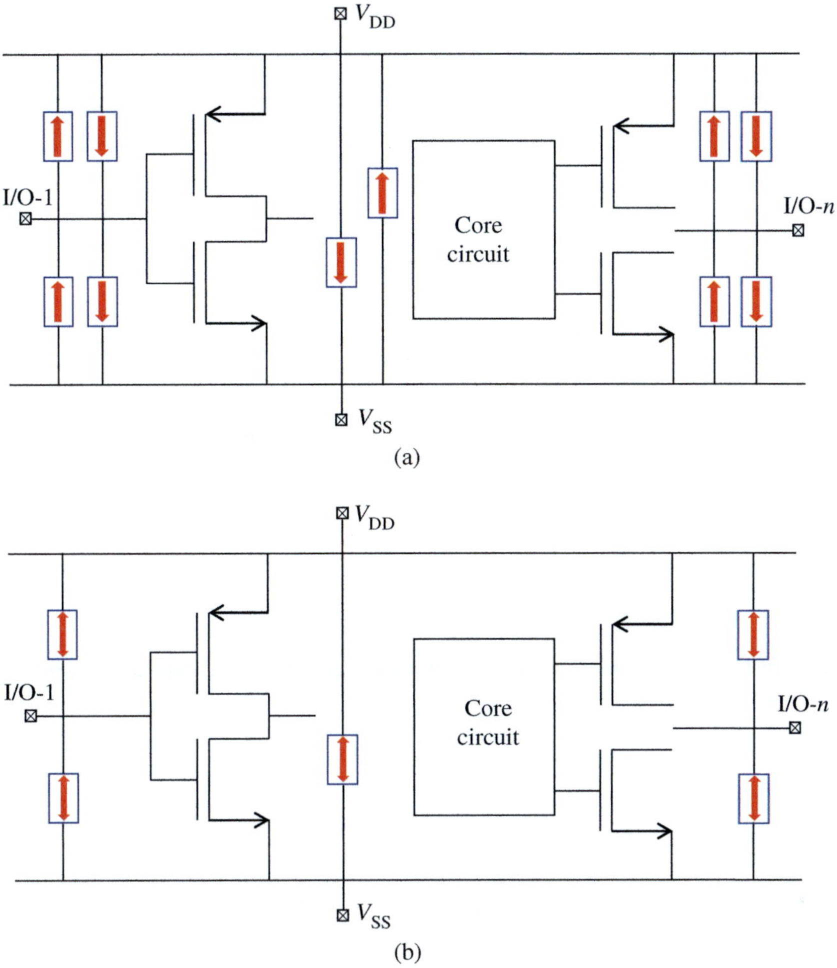

Figure 17.13 Illustration for using the symmetric dual-polarity nano-crossbar ESD switches to simplify full-chip ESD protection designs: (a) large numbers of ESD protection devices needed if using one-directional ESD protection structures, and (b) using dual-directional nano-crossbar ESD switch to reduce the number of ESD protection devices needed on a chip.

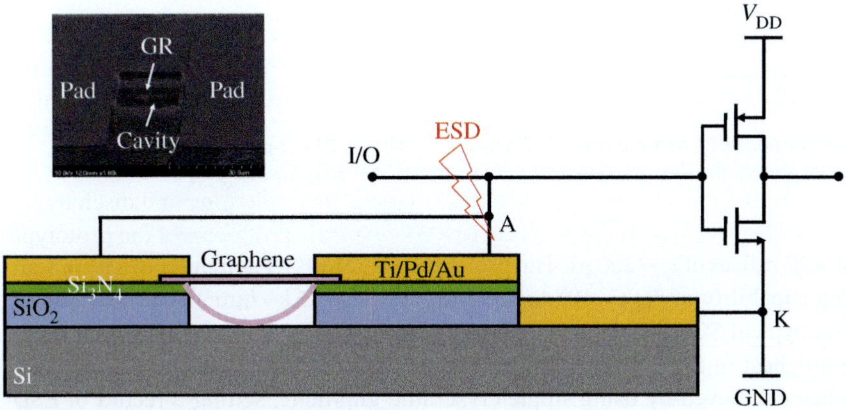

Figure 17.14 The concept of above-IC graphene gNEMS ESD switch structure, its on-chip ESD protection scheme, and an image of a fabricated gNEMS device.

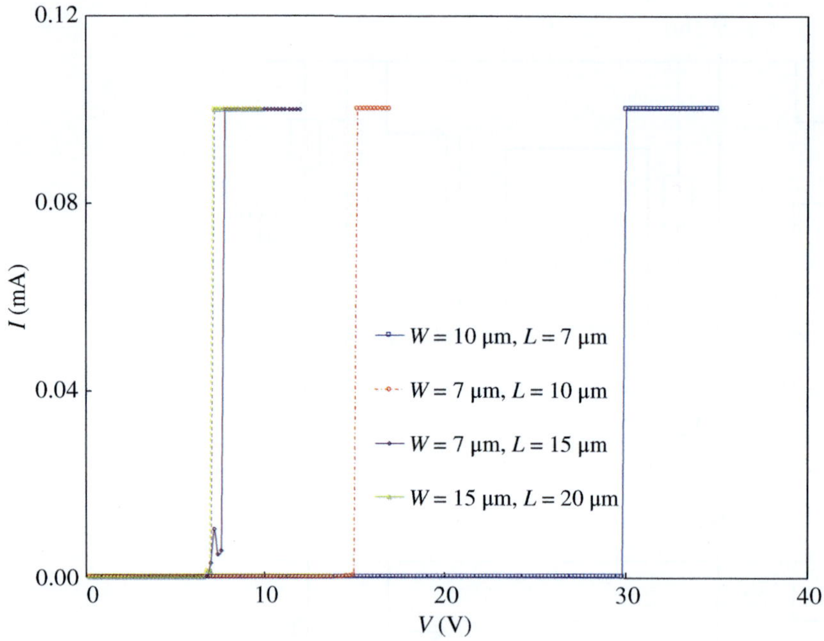

Figure 17.15 DC-sweeping test shows switching effect of the prototype gNEMS devices fabricated using poly-crystalline graphene ribbons (d = 350 nm).

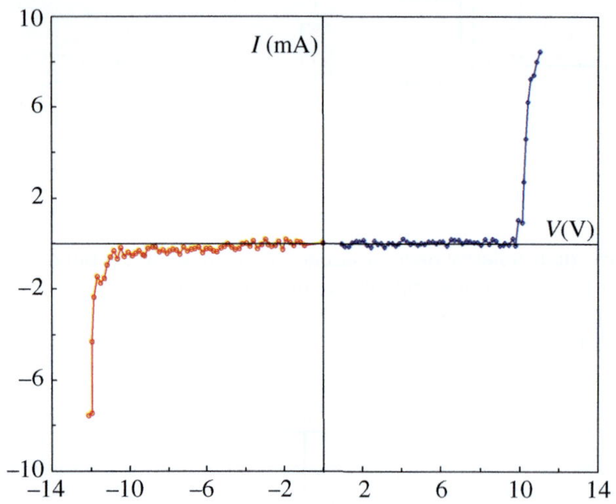

Figure 17.16 TLP measurement of a sample gNEMS device (d = 350 nm, W = 5 μm, L = 7 μm; poly-crystalline graphene) shows the desired symmetric dual-directional ESD discharging I–V characteristics.

triggered by ultrafast ESD pulses of $t_r \sim 200$ ps. The prototype gNEMS ESD demonstrates ultrahigh ESD current-handling capability of $J_{max} \sim 10^8$ A/cm^2 (i.e., ESDV > 1.5 kV/μm^2) that is orders of magnitude higher than typical SCR ESD protection structures (e.g., ~7.5 V/μm^2), which are often considered to be the toughest in-Si PN-based ESD protection device. Continuous improvement in gNEMS devices has been achieved by using single-crystalline graphene, setting a record of ESD

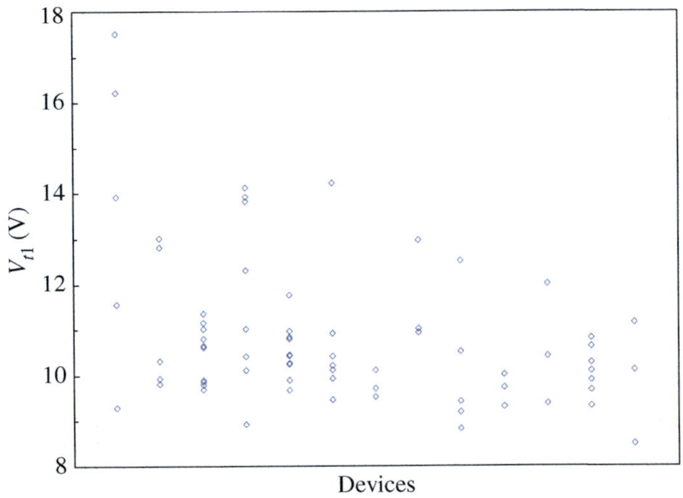

Figure 17.17 TLP measurement shows adjustable ESD V_{t1} values for prototype gNEMS switch devices, fabricated using poly-crystalline graphene ribbons, which are determined by design variations.

current handling capability of $J_{t2} \sim 1.19 \times 10^{10}$ A/cm^2 (ESDV ~ 178 kV/μm^2) under TLP stressing and $J_{t2} \sim 6.09 \times 10^9$ A/cm^2 by VFTLP zapping [12, 13].

17.4 Graphene ESD Protection Interconnects

As known, ESD metal interconnects have been a design challenge because wide metals are needed to carry out large ESD currents, which produce significant amount of ESD-metal-induced capacitance that is becoming increasingly unacceptable to advanced ICs. ESD metal interconnects design has been overlooked for long, though very important. Often, on-chip ESD protection devices are designed to be very robust; however, ESD failures frequently occur in the ESD metal interconnects. The conservative way of using excessive ESD metal coverage will seriously affect IC performance. On the other hand, reducing ESD metal width without careful and accurate experimental validation will readily lead to ESD failures in ESD metal interconnects. Graphene nanoribbons can be a good candidate for ESD interconnects, replacing Cu or Al metals [14]. Obviously, this is due to the excellent properties of graphene: exceptionally high thermal conductivity (~13 times of that of Cu) and ultrahigh electron mobility result in much higher current handling capability of graphene ribbons (GR), about 10 times better than that of Cu. Extraordinary mechanical strength also makes GR very tough to resist ESD-induced damages. Graphene ribbons are recently studied comprehensively for their ESD protection potentials by TLP and VFTLP ESD stressing, using a large set of GR wire samples (~6000) grown by CVD method, featuring various GR width and length dimensions suitable for typical IC interconnects [14]. Figure 17.18 depicts a scheme of using GRs as ESD interconnects on an IC chip. Figures 17.19 and 17.20 present the measured transient ESD discharging I–V characteristics for sample monolayer GR wire ($L = 12$ μm and $W = 5$ μm) stressed by TLP and VFTLP ESD pulses, respectively, revealing the ESD failure threshold current (I_C) and voltage (V_C). The measured ESD current handling capability for the GR wires, achieving very high $J_C \sim 10^8$ A/cm^2 under both TLP and VFTLP stressing, is orders of magnitude higher than that for Cu metal interconnects. Figure 17.21 depicts TLP-measured ESD failure threshold voltage (V_C) and

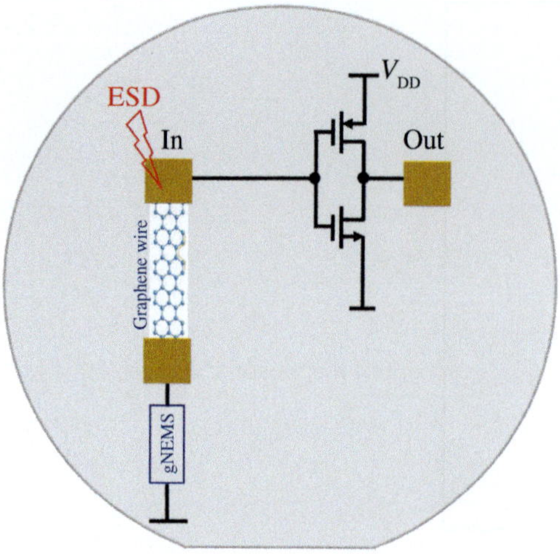

Figure 17.18 Illustration of using graphene wires as ESD interconnects on a chip.

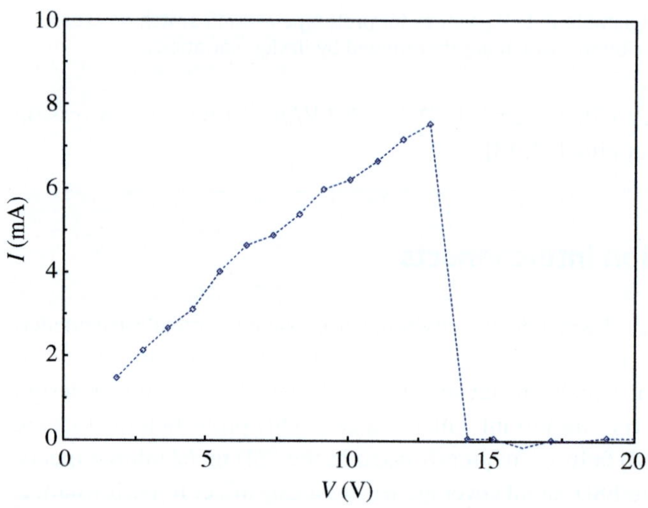

Figure 17.19 Measured ESD discharging I–V curve for a sample GR wire ($L = 12\,\mu\text{m}$, $W = 5\,\mu\text{m}$) by TLP testing ($t_d = 100\,\text{ns}$, $t_r = 10\,\text{ns}$) shows the ESD failure threshold current and voltage.

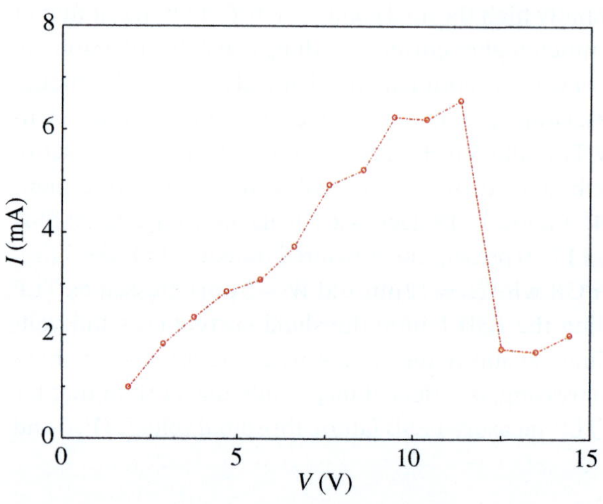

Figure 17.20 Measured ESD discharging I–V curve for a sample GR wire ($L = 12\,\mu\text{m}$, $W = 5\,\mu\text{m}$) by VFTLP testing ($t_d = 5\,\text{ns}$, $t_r = 200\,\text{ps}$) shows the ESD failure threshold current and voltage.

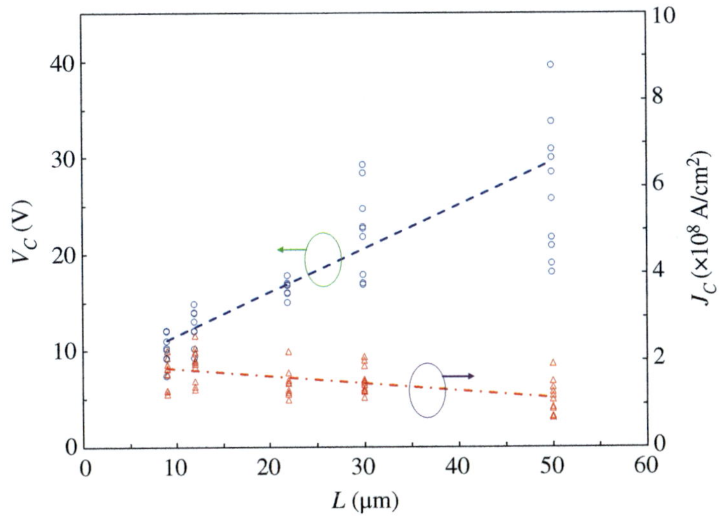

Figure 17.21 Measured ESD failure threshold voltage and current density for GR wire samples of varying L ($W = 5\,\mu m$) by TLP testing ($t_d = 100$ ns, $t_r = 10$ ns) shows that V_C increases significantly as L increases due to increased R, while J_C changes slightly against L.

current density (J_C) for monolayer GRs of varying ribbon length (at $W = 5\,\mu m$). It is readily observed that V_C increases monotonically as L increases due to higher resistance for a longer GR wire. The measured J_C seems to decrease slightly as L increases, possibly attributed to more defects in longer GR wires associated with the to-be-improved CVD method used to grow large-size graphene films. Figure 17.22 presents the TLP-measured ESD failure threshold current and current density for GR wires with varying ribbon width (at fixed $L = 12\,\mu m$). It is found that I_C increases almost linearly with the GR width due to continuous reduction in resistance as W increases. The measured J_C seems to be insensitive to the GR width. Figure 17.23 is the TLP-measured ESD failure threshold current density for GR sample ($L = 12\,\mu m$, $W = 5\,\mu m$) under TLP pulses with varying pulse duration ($t_d = 75$–150 ns at $t_r = 10$ ns). It is readily observed that J_C continuously decreases as the TLP pulse becomes longer, apparently due to the ESD energy accumulation effect. Figure 17.24 compares the TLP-measured ESD failure threshold current for GR wires ($W = 5\,\mu m$ with varying L) made of monolayer and bilayer graphene films. The measurement statistics clearly shows that bilayer GR wires can handle much higher ESD current than their monolayer counterparts. While more research is still needed to develop graphene ribbon wires into robust ESD interconnects, the potential is well recognized in experiments for future on-chip ESD protection solutions.

17.5 Future ESD Protection Outlook

What will future on-chip ESD protection solutions look like? The answer may not be straightforward at this moment. Nevertheless, it is clear that on-chip ESD protection design is becoming more and more challenging for future chips, not just for high-performance ICs, but also being complicated by heterogeneous chips of diversified functionalities. One thing is clear that reducing ESD protection robustness for future chips is not the right way to go.

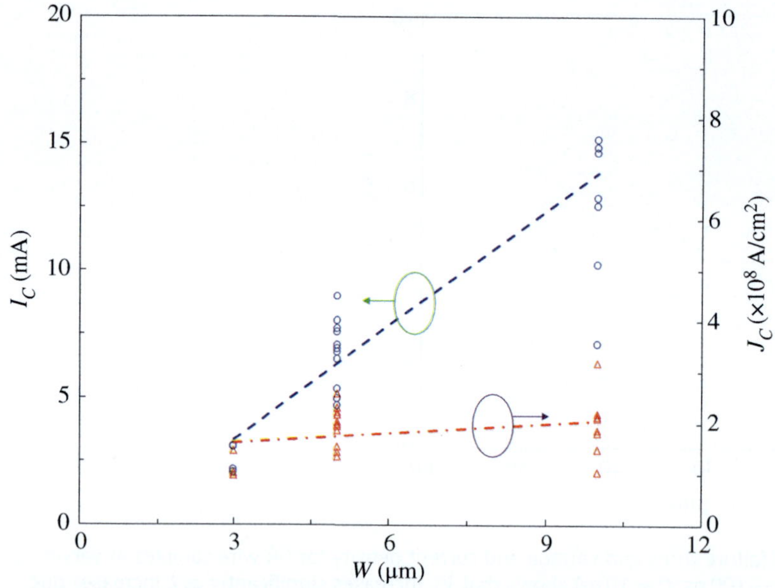

Figure 17.22 Measured ESD failure threshold current for GR wire samples of varying W ($L = 12\ \mu m$) by TLP testing ($t_d = 100$ ns, $t_r = 10$ ns) shows that I_C increases significantly as W increases due to reduced R, while J_C is relatively insensitive to W.

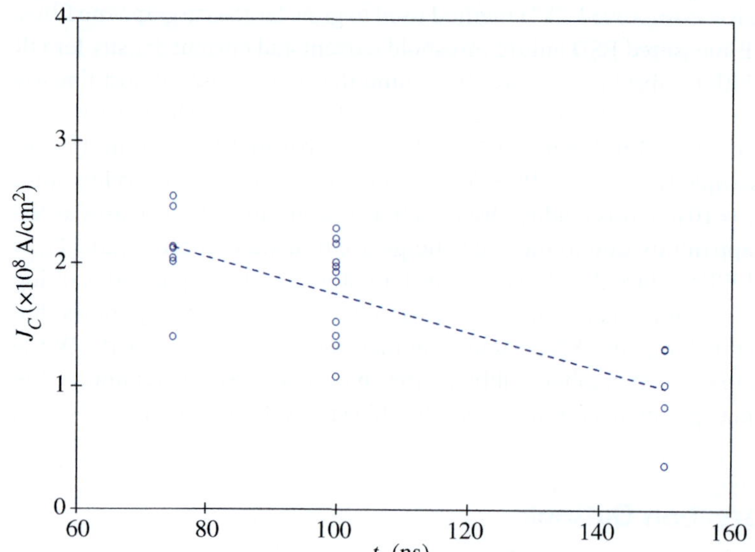

Figure 17.23 Measured ESD failure threshold current density for GR wire samples ($L = 12\ \mu m$, $W = 5\ \mu m$) by TLP of varying pulse duration ($t_d = 75$, 100, and 150 ns at $t_r = 10$ ns) shows that J_C increases significantly for longer TLP pulses due to energy accumulation effect.

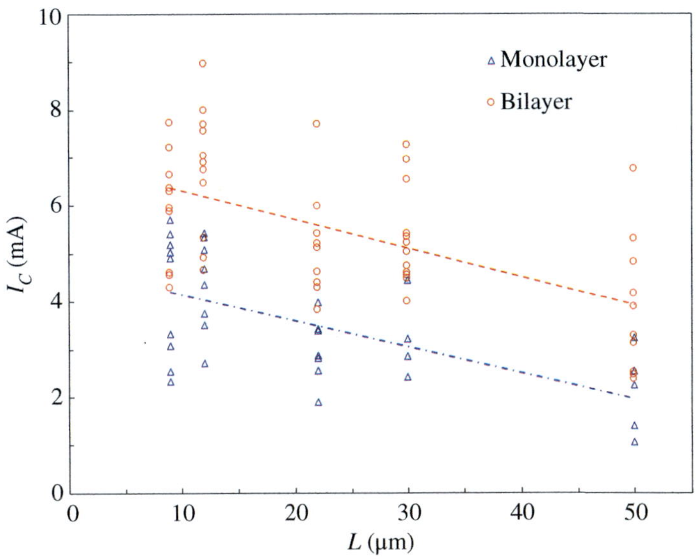

Figure 17.24 Measured ESD failure threshold current for GR wire samples of varying L ($W = 5\ \mu m$) made of both monolayer and bilayer graphene by TLP testing ($t_d = 100$ ns, $t_r = 10$ ns) shows higher J_C for bilayer GR wires due to lower R.

Understanding that ESD protection is complicated by the multiple-coupling effects, i.e., electro-thermal-transient-materials-process-device-circuit-layout-system coupling effects, and recognizing that the ESD-induced design overhead is directly associated with the traditional in-Si PN-based ESD protection structures, it is imperative to explore revolutionary and disruptive ESD protection solutions, from ESD discharging mechanisms to ESD protection device concepts to ESD-circuit integration. The concepts of above-IC nano-crossbar array ESD switches and graphene NEMS ESD switches, as well as graphene ESD interconnects discussed earlier are all encouraging trials in the right direction for future ESD protection. The author envisions that one pathway to future ESD protection for complex heterogeneous chips will be using non-traditional (i.e., non-PN-based electronic devices) ESD switching devices built in the back-end-of-the-line in CMOS ICs. A mechanical switch, if allowing ~100 ps switching speed, such as the graphene gNEMS ESD switch discussed earlier, can be a potential ESD protection solution for future chips. Using *in-BEOL above-IC non-PN-based ESD protection structures* can not only dramatically eliminate the ESD-induced parasitic effects and reduce ESD consumption of precious Si asset, but also entirely change the future ESD protection design practices. Figure 17.25 depicts an outlook for future ESD protection designs: ideal ESD switches, such as phase-changing nano-crossbar array ESD switches and graphene gNEMS ESD switches, in a combination of graphene ribbon ESD interconnects, can be designed and "dropped" in the BEOL deck after the fabrication of core ICs. Therefore, IC designers in fabless design houses can focus on designing the core ICs for the best performance, while foundries can provide the *drop-in ESD protection* solutions by integrating the fully validated robust ESD protection structures into the BEOL deck of fabricated IC chips. It would be a paradigm change in future on-chip ESD protection designs. Really, the sky will be the only limit for future ESD protection.

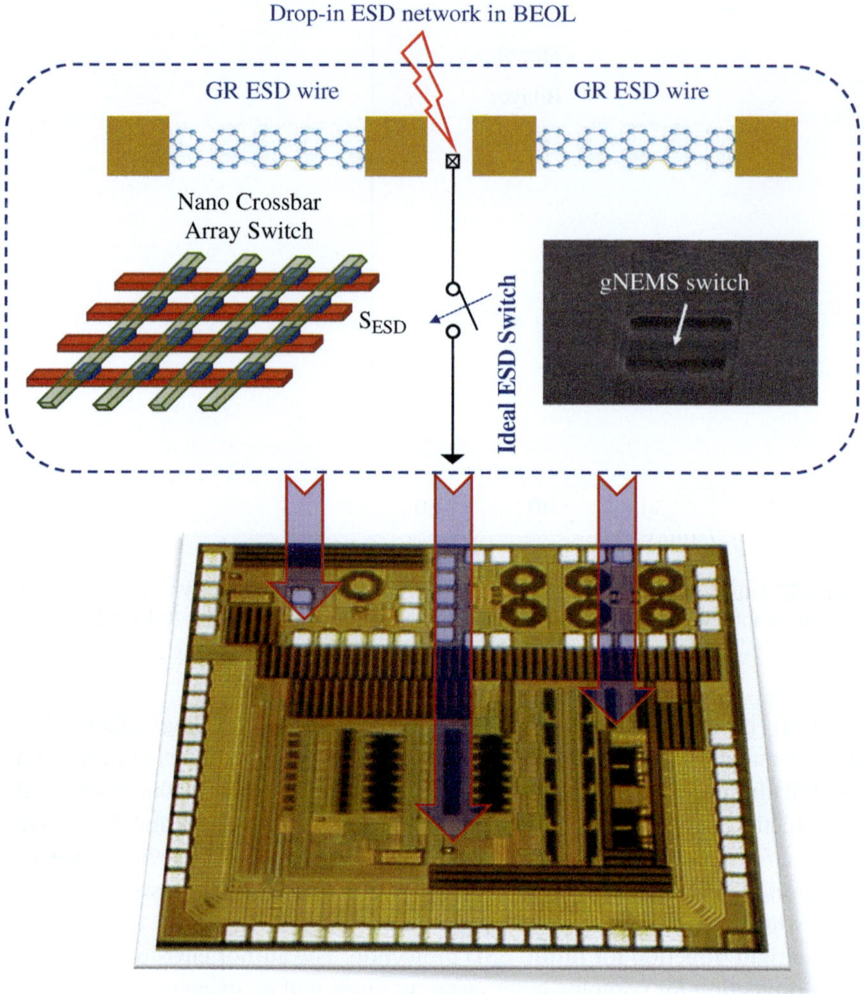

Figure 17.25 An outlook for future on-chip ESD protection: ideal ESD switches can be dropped into the BEOL deck of IC cores fabricated by a foundry.

17.6 Summary

This chapter discusses some disruptive ESD protection concepts to potentially overcome the ESD-induced design overhead problem inherent to traditional in-Si PN-junction-based ESD protection solutions that have been widely used for decades. Exploratory above-IC ESD protection designs, such as nano-crossbar phase-changing ESD switch array, graphene gNEMS ESD switch, and graphene ESD interconnects, show very encouraging results. It is envisioned that future on-chip ESD protection can be realized by designing various "ideal" ESD switch devices that can be "dropped" into the BEOL deck of IC cores designed for high performance. It is the time to rethink about future ESD protection solutions, which will be entirely different from the existing ESD protection structures and design practices that have been enjoyed in the past few decades ever since ICs were invented.

References

1 Zhang, L., Huang, R., Gao, D. et al. (2009). Unipolar resistive switch based on silicon monoxide realized by CMOS technology. *IEEE Electron Device Lett.* 30 (8): 870–872. https://doi.org/10.1109/LED.2009.2024650.

2 Lin, L., Zhang, L., Wang, X. et al. (2011). Novel nanophase-switching ESD protection. *IEEE Electron Device Lett.* 32 (3): 378–380. https://doi.org/10.1109/LED.2010.2099100.

3 Wang, L., Wang, X., Shi, Z. et al. (2013). Dual-directional nano crossbar array ESD protection structures. *IEEE Electron Device Lett.* 34 (1): 111–113. https://doi.org/10.1109/LED.2012.2222337.

4 Geim, A.K. and Novoselov, K.S. (2007). The rise of graphene. *Nat. Mater.* 6 (3): 183–191. https://doi.org/10.1038/nmat1849.

5 Neto, A.C., Guinea, F., Peres, N.M.R. et al. (2009). The electronic properties of graphene. *Rev. Mod. Phys.* 81 (1): 109. https://doi.org/10.1103/RevModPhys.81.109.

6 Bunch, J.S., van Der Zande, A.M., Verbridge, S.S. et al. (2007). Electromechanical resonators from graphene sheets. *Science* 315 (5811): 490–493. https://doi.org/10.1126/science.1136836.

7 Milaninia, K.M., Baldo, M.A., Reina, A., and Kong, J. (2009). All graphene electromechanical switch fabricated by chemical vapor deposition. *Appl. Phys. Lett.* 95 (18): 183105. https://doi.org/10.1063/1.3259415.

8 Ma, R., Chen, Q., Zhang, W. et al. (2016). A dual-polarity graphene NEMS switch ESD protection structure. *IEEE Electron Device Lett.* 37 (5): 674–676. https://doi.org/10.1109/LED.2016.2544343.

9 Zhang, W., Ma, R., Chen, Q. et al. (2016). The electro-mechanical responses of suspended graphene ribbons for electrostatic discharge applications. *Appl. Phys. Lett.* 108: 153103. https://doi.org/10.1063/1.4946007.

10 Ng, J., Chen, Q., Xie, Y. et al. (2017). Comparative study between the fracture stress of poly- and single- crystalline graphene using a novel NEMS structure. *Micro Nano Lett.* 12 (11): 907–912. https://doi.org/10.1049/mnl.2017.0422.

11 Chen, Q., Ng, J., Li, C. et al. (2017). Systematic transient characterization of graphene NEMS switch for ESD protection. *Micro Nano Lett.* 12 (11): 875–880. https://doi.org/10.1049/mnl.2017.0420.

12 Li, C., Di, M., Pan, Z., and Wang, A. (2021). A study of materials impacts on graphene electrostatic discharge switches. *Proceedings of IEEE Electron Devices Technology and Manufacturing Conference (EDTM).* https://doi.org/10.1109/EDTM50988.2021.9420816.

13 Li, C., Chen, Q., Ng, J. et al. (2021). Design, fabrication and characterization of single-crystalline graphene gNEMS ESD switches for future ICs. *IEEE Transactions on Device and Materials Reliability (TDMR).* https://doi.org/10.1109/TDMR.2021.3090311.

14 Chen, Q., Ma, R., Zhang, W. et al. (2016). Systematic characterization of graphene ESD interconnects for on-chip ESD protection. *IEEE Trans. Electron Devices* 63 (8): 3205–3212. https://doi.org/10.1109/TED.2016.2582140.

Index

Practical ESD Protection Design, First Edition. Albert Wang.

Printed and bound by CPI Group (UK) Ltd, Croydon, CR0 4YY
04/05/2026
14870569-0001